D1273535

Fundamentals of Geomorphology

Second Edition

Fundamentals of Geomorphology

Second Edition

R J Rice

Longman Scientific & Technical

Copublished in the United States with
John Wiley & Sons, Inc., New York

Longman Scientific & Technical
Longman Group UK Limited
Longman House, Burnt Mill, Harlow
Essex CM20 2JE, England
and associated companies throughout the world.

Copublished in the United States with,
John Wiley & Sons, Inc., 605 Third Avenue, New York, NY 10158

© R. J. Rice 1977, 1988

First published 1977
Second edition 1988

British Library Cataloguing in Publication Data
Rice, R. J.
    Fundamentals of geomorphology. – 2nd ed.
    1. Landforms
    I. Title
    551.4   GB401.5

ISBN 0-582-30151-3

Library of Congress Cataloging-in-Publication Data

Rice, R. J. (Roger John), 1931–
    Fundamentals of geomorphology.
    Bibliography: p.
    Includes index.
    1. Geomorphology.   I. Title.
GB401.5.R52 1988   551.4   85-23962
ISBN 0-470-20690-X (Wiley, USA only).

Set in 9/11pt Linotron 202 Rockwell Light
Produced by Longman Singapore Publishers (Pte) Ltd.
Printed in Singapore.

Contents

Acknowledgements

We are grateful to the following for permission to reproduce copyright material:

Academic Press for fig. 7.5 from fig. 29, p. 161 (M. I. Budyko, 1974); American Association for the Advancement of Science for fig. 2.2 (C. Emiliani & N. J. Shackleton, 1974) and AAAS and the authors for fig. 18.11 from fig. 1 (T. C. McIntyre *et al.* CLIMAP, 1976), copyright 1974 & 1976 American Association for the Advancement of Science; The American Association of Petroleum Geologists for fig. 5.3 (Harrison & Mathur, 1964) and fig. 5.4 (K. O. Emery *et al.* 1970) from *Bulletin of American Association of Petroleum Geologists* Nos 3 & 54; American Geophysical Union and the authors for fig. 3.10 (B. J. Isacks & Opdyke, 1968); American Journal of Science and the authors for fig. 8.4 (S. A. Schumm & R. J. Chorley, 1964) and fig. 19.7 (S. A. Schumm & R. W. Lichty, 1965); Edward Arnold Publishers Ltd. for fig. 15.7 from *Beaches and Coasts* by Cuchlaine King, 1972; The author for fig. 4.11 (E. H. Brown, 1960); Cambridge University Press for fig. 8.8 (data) (Carson & Kirkby); Chapman & Hall for figs. 10.1, 10.3, 10.4, 10.5 from figs. 15a & b, 22, 25 & 28, pp. 48–88 (R. A. Bagnold, 1941); The authors for fig. 19.5 from fig. 6.1 (R. J. Chorley & B. A. Kennedy, 1971); Constable Publishers for figs. 19.1 & 19.2A (data) from *Geographical Essays* by W. M. Davis 1954; Elsevier Publishing Co. for fig. 17.4 (data) (L. W. Wright, 1970); Geological Society of America Inc. and the authors for fig. 8.12 (Prior & Stephens, 1972) and fig. 9.14 (S. A. Schumm & H. R. Kahn, 1972); Institute of British Geographers for fig. 6.3 (A. Goudie *et al.* 1970), fig. 15.8 (G. de Boer, 1964), fig. 16.8 (J. T. Andrews) and IBG and the author for fig. 19.8 combining figs. 5, p. 77 & 7, p. 80 (Schumm, 1975) and fig. 6 (Schumm, 1979); International Glaciological Society and the authors for fig. 14.7 from figs. 3 & 4 (R. J. Ray *et al.* 1983); The author for fig. 6.2 (André Journaux); The author for fig. 19.3 (Lester King, 1962); Macmillan Journals Ltd. and the authors for fig. 18.2 from figs. 1, 2, 4 & 5 (A. Street & A. Grove) © 1976 Macmillan Journals Ltd.; Macmillan, London & Basingstoke for figs. 19.2 (B & C) from *Morphological Analysis of Landforms* by W. Penck; National Science Foundation for fig. 3.12 (A. S. Laughton, 1972); Nature Magazine and the authors for fig. 3.5 (F. J. Vine & D. H. Matthews, 1963); Pergamon Press Ltd. for fig. 11.7 (B) (W. S. B. Paterson, 1981); Princeton University Press for fig. 18.8 (S. A. Schumm, 1965) © 1965 Princeton University Press; Quaternary Research for fig. 2.2 from fig. 8 (M. B. Cita *et al.* 1977); Editor of Radiocarbon for fig. 2.6 (curve) from fig. 2, p. 110 (Klein *et al.* 1982) in *Radiocarbon* 24; The Royal Society and the authors for fig. 1.2 from fig. 3, p. 320 (D. G. King-Hele, 1980) and fig. 11.4 (J. W. Glen, 1955); Scottish Geographical Society and the author for fig. 12.9 (Moisley and Linton, 1960); Seismological Society of America for fig. 3.8 (Barazangi and Dorman, 1969); The author for fig. 9.6 (A. Sundborg, 1956); University of California Press for fig. 12.8 (Francois E. Matthes, 1966), copyright © 1950 by The Regents of the University of California; University of Chicago Press and the authors for fig. 18.1 from fig. 4 (J. D. Milliman and R. H. Meade, 1983); John Wiley & Sons Inc. for fig. 13.10 (J. Lundqvist, 1965); The authors for fig. 4.12 (S. W. Wooldridge and D. L. Linton, 1955).

Whilst every effort has been made to trace the owners of copyright, in a few cases this has proved impossible and we take this opportunity of offer our apologies to any authors whose rights may have been unwittingly infringed.

Preface to the Second Edition

The second edition of this book follows the lines of the first to the extent that many of the remarks in the original preface are still apposite. The work stems from a course of lectures given to first-year undergraduates in the Department of Geography at the University of Leicester. The aim of that course – and therefore of the book – continues to be to build on the foundations normally provided by sixth-form studies at school. In order to facilitate the transition from school to university work a conventional arrangement of the material has been retained, but the opportunity has been taken to introduce a number of significant changes. The most obvious has been to include a totally new chapter on aeolian activity in Part 2 of the book, a change that reflects a widespread resurgence of interest in wind as a geomorphic agent. This has demanded some reordering of the chapters in Part 2 which now covers a wider range of subaerial processes than was formerly the case.

A second significant modification has been to expand the material that was formerly incorporated into the last chapter on subaerial denudation and fashion it into a completely new concluding Part 5 entitled 'Concepts of landform evolution'. Of the two chapters in Part 5, one is devoted to the rate of geomorphic change, and the other to some of the conceptual frameworks within which landform studies are normally pursued. Both these topics are ones that have attracted growing interest from geomorphologists during the last decade, so that it is entirely appropriate that they should receive additional attention in this second edition. The opportunity has also been taken to expand on current ideas regarding the frequency and amplitude of Pleistocene climatic changes, and to say a little about the astronomical theory of climatic change that now commands such widespread support.

When the original preface was written in the mid-1970s the previous decade had witnessed a revolution in the earth sciences with formulation of the concept of plate tectonics. The intervening years have been a period of consolidation rather than dramatic change, and little alteration has been made to Part 1 of the book other than to update where it seemed desirable. Parts 3 and 4 on glacial and marine geomorphology have similarly been retained without major structural change, although the opportunity has been taken in a number of instances to indicate the trends of recent research findings.

As in the first edition, a list of references has been provided at the end of each chapter, together with some suggested further reading. In a volume ranging over such a broad range of subject matter it is impossible to attempt a full coverage of the literature, and the objective in each case has been to provide a few key starting points that should enable the reader to trace back through the literature specific topics that particularly interest him. With this consideration in mind, few actual references to the 'classic' i.e. pre-1950, literature have been incorporated in the bibliographies.

Finally, I am delighted to be able to acknowledge once more a deep personal indebtedness to my family. I am grateful to my mother, Mrs A. E. Rice, for converting my faltering attempts at typing into an excellent final copy. To my wife and children I again extend my thanks for their continued support and tolerance during the period of writing.

R. J. Rice
Leicester
October 1986

Preface to the First Edition

This book stems from a course of lectures given to first-year undergraduates in the Department of Geography at the University of Leicester. The aim of the course – and therefore of the book – has been to build on the foundations normally provided by sixth-form studies at school. At the same time a conscious effort has been made to shift the student's thinking from some of the more sterile aspects of school work in geomorphology towards paths that seem, in the author's view potentially much more rewarding. The term 'fundamentals' in the title is intended to imply an introductory survey from which the reader can progress smoothly to more advanced studies in future years. Limitations of space have precluded full coverage of the multifarious aspects of modern geomorphology; instead the aim has been to provide a conspectus of selected trends within the subject. Similar considerations of space have prevented the full development of a number of more difficult topics that have nevertheless been introduced in outline; the justification resides in the view that exposure to the underlying ideas will indicate to the reader the direction in which future work is likely to lead and, with luck, whet his appetite for these later studies.

By contrast with several recent texts on geomorphology, a relatively full treatment is accorded to relevant aspects of physical geology. Many reasons might be adduced in support of this return to an older tradition, but the foremost is the general philosophy that the surface of the globe is best viewed as the interface between two energy systems, one fuelled by the sun and the other by internal processes within the earth. The overriding importance of the former cannot be denied, but equally it would seem foolish to ignore the role of 'endogenetic' forces at the very time when geophysicists are emphasizing the general mobility of the crust.

Throughout the book the approach may be described as quantitative but non-mathematical. Experience has taught that even able undergraduates may be deterred by a too rigorously mathematical introductory course; on the other hand it is vital that such students should come to appreciate, often for the first time, the necessity for quantitative measurement and analysis. A primary objective of the book has therefore been to instil such an appreciation without ever assuming more than a most rudimentary knowledge of mathematics.

Authorship of a book that spans so many aspects of geomorphology means that I have inevitably had to rely upon persons and writings too numerous to mention individually. At the end of each chapter I have included a list of articles and books which proved particularly useful in the initial drafting; the lists are by no means exhaustive and to any individual who finds his original ideas incorporated without due acknowledgement I tender sincere apologies. I should particularly like to thank my mother, Mrs A. E. Rice, for her speedy and painstaking typing on my behalf. Last but by no means least, I should like to express my gratitude to my immediate family for their forbearance during the period of the writing.

R. J. Rice
Leicester
January 1976

Part 1
THE GEOLOGICAL FRAMEWORK

Chapter 1

Gross morphology and structure of the earth

The aim of this first chapter is to review the way in which knowledge of the shape and constitution of the earth has been gathered since the initial attempts to measure the size of the sphere over 2 000 years ago. It is concerned with the study of the earth as it exists today; consideration of the way internal processes may have caused it to evolve through time forms the essential theme of Chapters 3–5. Even the seemingly limited objective of describing the shape and relief of the solid globe faces formidable problems. Water obscures the exterior relief over more than two-thirds of the earth so that, for many centuries, scientific investigation was confined to the continents. Bathymetric sounding of the oceans did not begin in a systematic fashion until the second half of the nineteenth century, and only since 1950 has it proceeded at a rapid rate. Even more recent is the innovation of using submersibles to view the nature of the ocean floors directly; as yet relatively few areas have been studied by this method.

Visual examination of the materials composing the earth is extraordinarily restricted. The depth to which mines and boreholes have penetrated is a minute fraction of the total radius of the sphere. Even at the present day the deepest mines go down little more than 3 km and the longest boreholes penetrate less than one five-hundredth part of the distance from the surface to the centre of the earth. Moreover, the distribution of boreholes is extremely uneven, with study of the materials of the ocean floors lagging far behind that of the continents owing to the immense problems of boring in deep water. Although some nineteenth-century oceanographic expeditions had managed to sample the top 30 cm of sediment, as late as 1960 very few cores exceeding 20 m in length had been recovered from deep-water sites. The normal instrument at that time was the piston corer which uses hydrostatic pressure to force the sampling tube into the sediments. It was with the launch of the American research ship *Glomar Challenger* in 1968 that rotary drilling was first adopted. As part of the DSDP (Deep-Sea Drilling Project) and IPOD (International Program of Ocean Drilling) this vessel, prior to decommissioning in 1983, bored well over 500 holes in all oceans of the world with the exception of the high arctic; the deepest penetration was about 1 750 m, with the ship at times operating in over 7 300 m of water. The successor vessel, *JOIDES Resolution*, has the design capability to exceed that performance in the course of the more recently established ODP (Ocean Drilling Program).

In the following account attention will first be directed to the basic shape and gross relief features of the earth. Thereafter consideration will be given to the two major lines of evidence bearing upon the internal constitution of the earth; namely, variations in the strength of the earth's gravity field and the way in which vibratory motions are transmitted through the globe.

Shape and dimensions of the earth

The Greeks had already conjectured the spherical form of the earth by the sixth century BC, and the first important attempt to measure its size is commonly credited to Eratosthenes in the third century BC. He employed the simple observation that, during the summer solstice, the sun is directly overhead at Aswan whereas it is 7° 12′ from the zenith at Alexandria (Fig. 1.1). Assessing the distance between Alexandria and Aswan as 5 000 stadia, he calculated the overall circumference of the earth as 250 000 stadia. The measurements were necessarily crude but if, as is believed, the stadium was 185 m the computed dimension is only about 15 per cent too large. Later Greek and Arab astronomers made many similar assessments using the same basic principles, but interest in the whole subject was dramatically revived in the seventeenth century when Sir Isaac Newton predicted that the earth would prove to be a spheroid flattened at the poles. Early support for this contention came from an accurate pendulum clock that, regulated to keep time in Paris, lost almost 2½ minutes a day when transferred to equatorial latitudes. This suggested that gravitational attraction might vary with latitude and

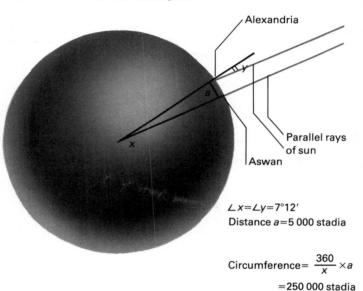

Fig. 1.1. The simple geometrical principles by which Eratosthenes calculated the overall size of the globe.

$\angle x = \angle y = 7°12′$

Distance $a = 5\,000$ stadia

$$\text{Circumference} = \frac{360}{x} \times a$$

$$= 250\,000 \text{ stadia}$$

thus imply a changing distance from the centre of the earth. On the other hand, survey work in France failed to reveal the expected poleward increase in the length of a degree of latitude, and in order to resolve the problem the French Academy of Sciences in the 1730s dispatched expeditions to what are now Ecuador and Finland to measure the length of a meridian degree at contrasting latitudes. The results clearly confirmed Newton's original prediction.

The investigation sponsored by the Academy of Sciences heralded the birth of the modern science of geodesy. At first this depended almost entirely on painstaking measurement of huge arcs across the surface of the globe, but since the 1950s the trajectory of artificial satellites has provided the primary data. Two concepts are basic to geodetic investigations. The first is the spheroid, an imaginary surface which may be visualized as the shape the earth would take if, while retaining its present mass and motion, its materials were redistributed to give uniform concentric shells, the outermost of which would be a continuous ocean about 2 400 m deep. The combined gravitational and centrifugal forces would produce an ellipsoid with a semi-major axis length of 6 378.14 km, a semi-minor axis length of 6 356.75 km and a flattening of 1/298.3. The second concept is the geoid which may be visualized as the surface described by mean sea-level over the oceans and by the level to which the sea would rise in hypothetical interconnected tunnels under the continents. The geoid lacks the geometrical simplicity of the spheroid since it reflects the gravitational attraction of the irregularly distributed features of the earth's outer shell. Yet, for reasons that will become apparent below, comparison of the forms of the spheroid and geoid discloses much less discrepancy than might be anticipated from such obvious topographic features as ocean basins, continental masses and mountain ranges. On the other hand, very precise measurements on the orbits of artificial satellites imply undulations in the geoid not related in any obvious way to surface features (Fig. 1.2). The cause of these undulations remains uncertain, but they presumably reflect deep-seated variations within the earth and may possibly have great significance in explaining some of the problems that have been encountered in attempts to reconstruct past sea-levels (see Ch. 16).

Fig. 1.2. The form of the geoid as revealed by modern satellite measurements (after King-Hele, 1980). The lines portray deviations in metres from a regular spheroid with a flattening of 1 : 298.255; the areas of depression are shaded. The implication is that a swimmer proceeding from south of India to north of New Guinea would find himself 175 m further from the centre of the earth without ever having to swim uphill.

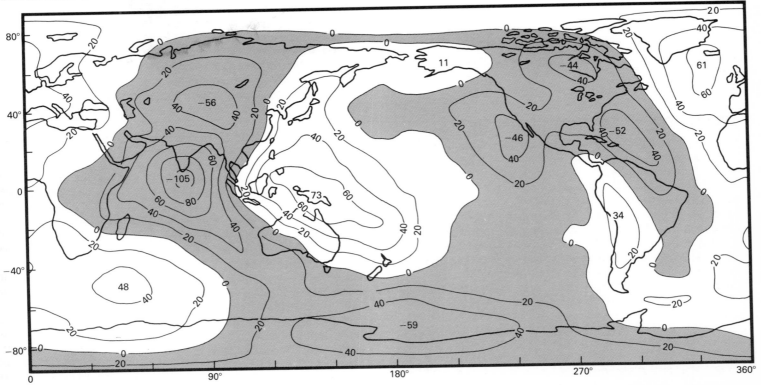

Distribution of relief on the earth's surface

Compared with only a few decades ago, the form of the solid outer surface of the globe is now very thoroughly surveyed. Over the continental areas the pace of mapping has been much accelerated by the use of stereoscopic aerial photography, while over the oceans the continuous recording echo-sounder, capable of producing a complete profile of the submarine topography beneath a moving vessel, has been widely employed. For accurate oceanographic work a source of sound pulses, commonly a 'sparker' generating the pulses by means of an electrical discharge, is towed behind a research vessel just below the surface, together with a receiving

Fig. 1.3. The principle of the continuous recording echo-sounder. Reflections are regularly received not only from the sea floor itself but also from suitable surfaces within the submarine sediments. Consequently, much valuable information about submarine structure is normally obtained in the course of echo-sounding.

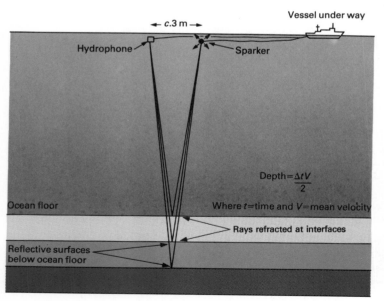

hydrophone (Fig. 1.3). Although the speed of sound in water varies with pressure, temperature and salinity, a precision of ± 2 m may be attained in water that is 5 000 m deep. The result is that, with very few exceptions, the basic form of the ocean floors is now well known. A further development of this technique is to use 'side-looking' sonar systems in which narrow beams of acoustic pulses are projected laterally from a transmitter towed a few hundred metres above the sea floor. This has proved particularly useful for the plotting of topographic detail on selected sections of the ocean floor.

One of the most formidable problems facing any surveyor interested in the bedrock relief is a perennial cover of ice. This applies both to continental areas such as Antarctica and Greenland with their massive ice-sheets, and to the Arctic Ocean with its persistent cover of sea ice. However, for each of these unfavourable environments methods have now been devised that permit accurate and speedy charting of the rock surface. In the arctic this has involved the use of submarines, while over the major ice-sheets planes equipped with special radar have added rapidly to our knowledge of the bedrock topography (see p. 212). As a consequence, further exploration is unlikely to reveal basic lineaments not already discovered, but will be concerned mainly with elaborating topographic detail.

The most striking aspect of the form of the solid surface is the dominance of two distinct levels. From a hypsographic curve (Fig. 1.4) it can be seen that approximately 30 per cent of the earth's surface lies between +2 000 and −200 m, and a further 50 per cent between −3 000 and −6 000 m. The former comprises the continental level, the latter the oceanic level. Of very much smaller extent are the intermediate slopes, the high mountain chains and the oceanic deeps. Complicating this simple division of global relief is the fact that not only the oceanic level but also part of the continental level is currently submerged. As will be shown below, in terms of deep-seated structure the most fundamental distinction is that between ocean basins and continents, the latter to include the drowned fringes of the continental level. On the other hand, erosional and depositional processes beneath the sea are so different from those on land that it is natural to seek a distinction between subaerial and submarine environments. Accommodation of these two principles requires that a threefold division of global relief is recognized, namely, the continents, the submerged continental margins and the ocean basins.

Relief of the continents

The importance and variety of relief patterns on the continents is acknowledged in the everyday use of such expressions as mountain range, hill country or coastal plain. They indicate an intuitive recognition of an association of features repeated at intervals over the globe. At first sight it might seem that if greater precision could be accorded these descriptive terms it would be a simple task to map their distribution over the land surfaces of the world. However, such is the complexity of relief forms that their scientific classification has proved a most formidable problem. Moreover, geomorphologists have often tended to adopt genetic classifications in which as much attention is paid to geological structure and presumed evolutionary history as to pure morphology. Thus Fenneman in his classic division of the United States into physiographic provinces wrote:

> Unity or similarity of physiographic history is a formula which *almost* designates the basis here in mind for the delineation of provinces. It implies that the topography throughout the province is all to be explained and described in the same story. It does not imply that it is superficially the same throughout the area. . . . Hence in basing provinces on topography it is understood that the features throughout a province are *essentially related* rather than superficially alike.

Most workers seeking a purely morphological classification acknowledge a minimum of four variables to be considered: relief amplitude, slope angle frequency, drainage texture and characteristic profile form. The need for the first two is obvious. The third, although closely related to relief and slope, is necessary to distinguish between a landscape of ravines and one with more widely spaced valleys. The last is essential to distinguish, for example, between dissected tablelands and plains with upstanding hills, since both could conceivably have identical values for relief and slope. Having identified four critical parameters, there remains the problem of specifying the method by which actual measurement and analysis should be made. No general agreement on this point has ever been reached, and rigorous application of various quantitative measures of form has so far achieved limited success.

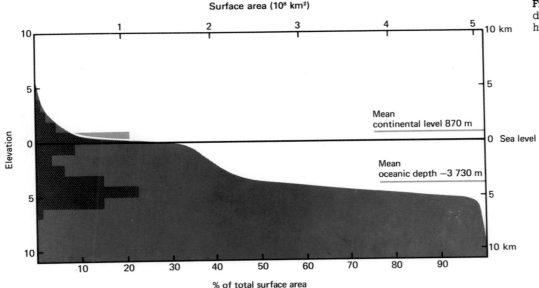

Fig. 1.4. The distribution of global relief, depicted by means of both a hypsographic curve and a bar graph.

Surface area (10⁸ km²)

Elevation

% of total surface area

Mean continental level 870 m

Sea level

Mean oceanic depth −3 730 m

Fig. 1.5. The relief of the continents classified into seven terrain types defined purely on the basis of their morphology. The key shows diagrammatically the relationship between the different terrain types and the criteria employed in their definition (simplified from an original map in Trewartha, Robinson and Hammond, *Fundamentals of Physical Geography*).

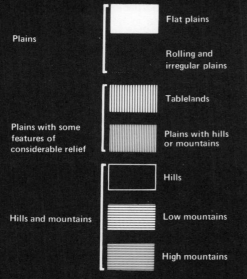

Plains
- Flat plains
- Rolling and irregular plains

Plains with some features of considerable relief
- Tablelands
- Plains with hills or mountains

Hills and mountains
- Hills
- Low mountains
- High mountains

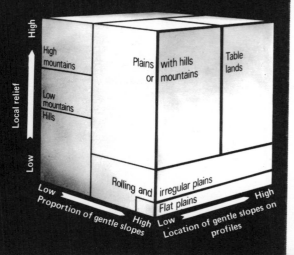

High

Local relief

Low

High mountains

Low mountains

Hills

Plains with hills or mountains

Table lands

Rolling and irregular plains

Flat plains

Low — Proportion of gentle slopes — High

Low — Location of gentle slopes on profiles — High

ICE SHEETS

The American geographer Hammond devised a seven-category classification based upon three major characteristics: relative amount of gently sloping land, local relief and generalized slope profiles. The way in which the principal terrain types are related to each other, together with their world distribution, is shown in Fig. 1.5. The most striking aspect of the world map is the spotty distribution. Unlike the ocean basins to be described below, there is no gross patterning common to all the continents and very few systematic arrangements can be discerned. The one terrain type well represented on all land masses is the plain, over one-third of the total continental area falling into that category. High mountains, on the other hand, are totally absent from the Australian mainland and occupy a very small proportion of Africa. Many disparities of this sort can be related to recent geological history. For instance, the comparatively large proportion of Eurasia and the Americas occupied by high mountains appears to be related to the participation by those continents in recent orogenic activity. Conversely the great extent of rolling and irregular plains in Africa and Australia can be associated with a relative lack of such activity. However, comments of this nature are prompted by a knowledge of tectonic history; in terms of pure geometrical pattern, the most abiding impression is of a haphazard arrangement of terrain types.

Relief of the continental margins

By convention the continental margins are divided into two separate units, the continental shelf and the continental slope, well exemplified in the submarine topography off the coast of the north-eastern United States (Fig. 1.6).

(a) The continental shelf

With a mean seaward gradient of less than a quarter of one degree, the shelf comprises the underwater extension of the continental coastal plain. Its width averages about 60 km, but ranges from zero around parts of the Pacific to 900 km around parts of the Arctic Ocean. Its seaward edge is marked by a rapid increase in gradient known as the shelf break. This normally lies at a depth of about 200 m but considerable variations are found. It is often particularly deep in high latitudes, occurring at 350–400 m around the Arctic Ocean and parts of the North Atlantic, and at 300 m along the fringes of the Antarctic continent. Were sea-level to fall so as to expose the

shelf, the land area of the globe would increase from 29 to almost 33 per cent of the total surface.

In areas of potential economic exploitation the shelf has often been surveyed in great detail and even examined visually by means of submersibles. It has been possible to recognize topographic forms directly comparable to those found on the adjacent land mass. This is true, for instance, in the apparent seaward extension of river valleys around many coasts of the world, and also in submerged glacial landforms traceable off the coasts of both Europe and North America. Yet such comparisons should not be allowed to obscure the greater overall smoothness of the submarine topography, especially along the outer parts of the shelf, which results primarily from a cover of unconsolidated sediments.

(b) The continental slope

Lying immediately below the shelf break, this is one of the most distinctive relief features of the whole globe. It is almost always present to mark the true edge of the continent and its total area is actually greater than that of the shelf. In detail its form is quite variable. Local gradients exceeding 25° have been recorded, but a more representative value would be 2–5°. To those familiar with continental relief these angles may seem quite low, but it must be remembered that they can be maintained over distances of 50 km or more and so represent changes in elevation of several thousand metres. Although the slope sometimes passes directly into deep oceanic trenches, it is normally fringed seawards by the continental rise which, with gradients of under a quarter of a degree, merges gently into the true ocean floor. The slope is conventionally regarded as terminating at the 2 000 m isobath but its actual foot varies considerably in elevation, ranging between 1 500 and 4 000 m around the Atlantic. It is not unusual to find step-like features along the face of the slope and these are occasionally so wide as to constitute extensive underwater plateaux; the Blake Plateau off the Atlantic coast of Florida, for instance, is nearly 300 km wide at depths between 700 and 1 000 m and is bounded seawards by an exceptionally steep lower segment of the continental slope. Further common diversifying features are submarine canyons. These vary in size and form, from deep sinuous gorges to broad flat-floored troughs. Some are confined to the face of the slope, whereas others extend back across the shelf to link with valleys on the continental surface. The Hudson Canyon

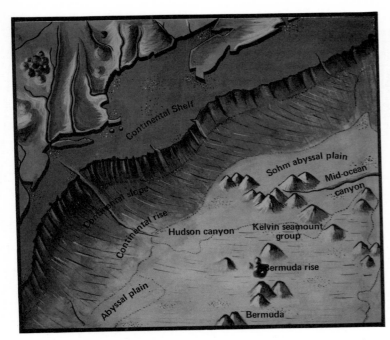

Fig. 1.6. A physiographic diagram of the submarine topography off part of the Atlantic seaboard of North America.

(Fig. 1.6) is one of the best explored and consists of an Upper Gorge incised into the shelf and slope, and a Lower Gorge incised into the rise. The largest submarine canyons attain a size comparable to that of the Grand Canyon in Arizona.

Relief of the ocean basins

Unlike the continents, the ocean basins all display a common relief pattern. This consists of a broad central ridge flanked on both sides by low, extremely flat plainlands. In addition, certain basins exhibit deep marginal trenches. Recognition of this gross patterning permits description of the ocean basins under three separate headings.

(a) Mid-ocean ridges

Although the existence of shoaling in the mid-Atlantic was recorded over 100 years ago it took many further decades of bathymetric survey to establish the continuity of the relative shallows. When this was followed, in the mid-twentieth century, by recognition of a virtually continuous submarine ridge-system traceable across the globe for a distance of about 60 000 km (Fig. 1.7), it heralded a revolution in the earth sciences (p. 35 et seq.). The term mid-ocean ridge is not entirely appropriate as the feature is often positioned asymmetrically within the ocean basin, and its enormous dimensions are scarcely conveyed by the word 'ridge'. A better idea of its size comes from an appreciation of the fact that it occupies about one-third of the total width of the Atlantic and that its crest at −2 000 m is commonly only half as deep as the fringing plainlands.

The most distinctive topography is to be found near the ridge crest. Here the relief is at its most rugged with abrupt falls of 1 000 m or more from the culminating peaks to the highest of a series of flanking plateaux. The culminating peaks themselves are often cleft by a deep central fossa. In the 1970s a 50 km section of the mid-Atlantic ridge at 37°N was the subject of intensive study in the FAMOUS project (French-American Mid-Ocean Undersea Study). Exceptionally detailed bathymetric charting, augmented by visual examination from submersibles, revealed a central trench 30 km wide and up to 1 500 m below the adjacent edges of the ridge. The greatest depths were encountered in an inner rift only 1–3 km wide, and the whole area appeared to be traversed by innumerable fissures trending parallel to the axis of the ridge. A trench of comparable size is usually well developed elsewhere in the Atlantic and also in the Indian Ocean, but is much less obvious in the Pacific where ridge relief is more subdued. As early as the 1950s it was noted from bathy-metric charts that the mid-Atlantic ridge between West Africa and Brazil is abruptly offset, with individual displacements of the central fossa locally exceeding 100 km. Subsequent work has shown that such lateral shifts are characteristic of virtually all mid-oceanic ridges.

(b) Ocean-basin floors

The floors of all the major ocean basins lie at approximately the same depth, generally between 4 500 and 5 500 m. Their most character-istic feature is the abyssal plain where gradients do not normally exceed 1 in 1 000. Abyssal plains are found in each of the major ocean basins, although they are commonly restricted to a series of

Fig. 1.7. The world distribution of the mid-oceanic ridge system and the major ocean trenches.

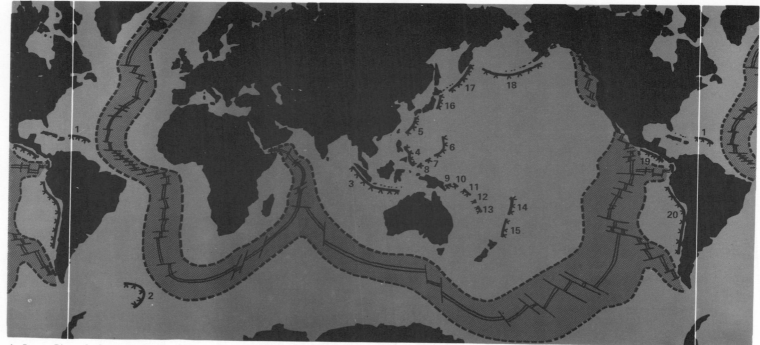

1. Puerto Rico 2. South Sandwich 3. Java 4. Philippine 5. Ryukyu 6. Marianas 7. Yap 8. Palau 9. New Britain 10. N. Solomons 11. S. Solomons
12. New Hebrides (N) 13. New Hebrides (S) 14. Tonga 15. Kermadec 16 Idzu-Bonin 17. Kuril 18 Aleutian 19. Middle America 20. Peru-Chile

individual depressions rather than occupying a single very large unit (Fig. 1.6). The sills between the depressions are often rougher in surface texture but may not rise greatly above the level of the plains; where narrow they may be traversed by channel systems leading from the higher to the lower plain through 'abyssal gaps'. On a much grander scale are long steep-sided troughs rather misleadingly called mid-ocean canyons since they do not normally occur near the centre of the oceans. The best known has been traced over 2 000 km along the foot of the continental rise off eastern Canada where it is typically 5–8 km wide and 20–200 m deep. Comparable but smaller canyons have been recorded from several of the other ocean basins.

A number of distinctive forms diversify the abyssal plains. These include abyssal hills which range up to a few hundred metres in height and often occur in large groups, particularly on the floor of the Pacific. More spectacular are the so-called seamounts. By definition these are submerged conical peaks rising at least 1000 m from the adjacent sea floor. It is estimated that there are between 10 000 and 20 000 scattered through all the oceans of the world. They are not confined to the ocean-basin floors but also occur on the continental slopes and mid-ocean ridges. Although many appear to be scattered in almost random fashion, others fall into obvious linear patterns that may be traceable for 1 000 km or more. A small proportion of seamounts are distinguished by having prominent flat tops and are then known as guyots. Despite being found in each of the ocean basins the total number of guyots does not exceed a few hundred, and although their summits mostly lie at depths of 1 000–2 000 m they exhibit little consistency in elevation. It is on the summits of seamounts and guyots that many of the coral reefs in the Pacific are to be found.

A final noteworthy feature of the ocean-basin floors is the presence of linear zones of relatively rugged relief. These were first detected off the Pacific coast of North America before the Second World War, but their full extent was only appreciated decades later when it was shown that at least five separate zones of greater surface roughness can be traced some 2 000 km through the abyssal plains and hills of the Pacific floor. Each zone consists of narrow but immensely long troughs and ridges separated by scarps several hundred metres high and sloping at angles of 5–10°.

(c) Oceanic trenches

It was realized over 100 years ago that the deepest parts of the ocean are not centrally placed but lie relatively close to the continental margins where they form narrow elongated basins. The *Challenger* voyage of the 1870s recorded its greatest depth of 8 000 m in what is now called the Marianas trench. Later soundings by other research vessels penetrated even deeper in the Kuril and Kermadec trenches, but in 1950 a new HMS *Challenger* returned to the Marianas trench and recorded a depth of 10 863 m. This stood as a record for several years, but soundings in the same area later exceeded 10 900 m. Some uncertainty attended a number of these very deep soundings since false echoes from the steep trench walls can pose problems of interpretation. However, in 1960 the bathyscaphe *Trieste* was piloted by Piccard and Walsh into the deepest part of the Marianas trench and recorded a depth of 10 910 m on a pressure gauge, thus providing powerful confirmation of the previous echo-sounding results.

Twenty separate trenches are now recognized, the majority being grouped around the edges of the Pacific basin (Fig. 1.7). They display a number of features in common. Characteristically they descend on one side far below the level of the adjacent ocean-basin floor, and on the other are flanked by a steep continental slope culminating in either a major mountain chain or a line of oceanic islands. Normally about 100 km wide and 1 000–4 000 km long, they are all more or less arcuate in plan. The slopes tend to be steeper on the landward side where they may average 10–15°

compared with 5–10° on the seaward side. Several detailed surveys have disclosed crenulate trench walls with the gradients, if anything, seeming to increase with depth before abruptly giving way to narrow flat floors that are typically 1–3 km wide.

The earth's gravitational attraction

In outlining the shape and gross relief of the earth emphasis has inevitably fallen on contrasts between the continents and ocean basins. The question that naturally follows is the extent to which these contrasts reflect deep-seated structural variations lying beyond the limits of direct observation. One field of study shedding light on this problem is investigation of the earth's gravitational attraction. Basic to the science of gravimetry is Newton's law that two bodies attract one another with a force proportional to the product of their masses and inversely proportional to the square of the distance between them. Mathematically this may be written as $F = G\ MM'/d^2$, where F is the force between two bodies of mass M and M', d is the distance between them and G is the value known as the gravitational constant. By manipulation of this apparently simple law, major discoveries about the nature of the earth's interior have been made. One of the most important relates to the mean density of the materials composing the globe; a second relates to the internal distribution of different materials.

The mean density of the earth

When Newton published his gravitational law in 1687 he recognized that it offered the basis for measuring directly the total mass and thereby the mean density of the earth. However, realization of this objective was delayed for over a century by the extremely delicate experimental work that was required. The most celebrated early attempt to 'weigh the earth' was made by Cavendish in his home at Clapham Common in 1798. He endeavoured to measure the deflective force when large lead weights were swung close to two lead balls suspended from a very sensitive torsion balance (Fig. 1.8). In this remarkably elegant and carefully controlled experiment, he calculated the ratio of the pull exerted by one of the weights to the pull on the same ball exerted by the earth. In this way he was able to compute the total mass of the earth and thereby estimate its mean density as '5.48 times greater than that of water'.

This classic investigation was repeated with gradually increasing

Fig. 1.8. A schematic illustration of the apparatus used by Cavendish to measure the gravitational constant (G). In the Newtonian expression $F = G(MM_1/d^2)$ the gravitational constant is the only unknown quantity: a fine fibre of known torsion (yielding a value for F) is used to suspend a bar from which hang two lead balls (of mass M); two larger lead balls (of mass M_1) can be rotated so as to lie alternately on either side of the small ones (at distance d). Once G is known the same formula can be applied to the attraction of the earth for one of the lead balls, the only unknown in that case being the mass of the earth.

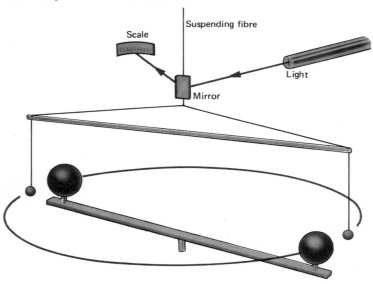

refinements during the next one and a half centuries, but with the launching of artificial satellites in the 1950s a totally new method of computing the mean density of the earth became available. From precisely measured satellite trajectories, it is now estimated that the total mass of the earth is 5.974×10^{27} g, and its mean density 5 512 kg m^{-3}*. Since the corresponding figure for rocks exposed at the surface ranges between 2 600 and 3 300 kg m^{-3}, there must clearly be a concentration of much denser material towards the centre of the globe; several lines of argument conduce to the view that the density here approaches a value of 11 000 kg m^{-3}.

* $=5.512$ g cm^{-3} (see Appendix).

The earth's gravity field

As indicated on p. 2 eighteenth-century workers had already deduced that gravity varies from one region to another, and since that time much effort has been devoted to the accurate assessment of these variations. For that purpose it was first essential to establish a suitable unit of measurement. Consideration of Newton's law shows that the gravitational force varies with the mass not only of the earth but also of the other body involved. This is clearly unsatisfactory as a measure of the earth's gravitational attraction, and in consequence the latter is normally expressed in terms of an acceleration, conventionally designated g. Being the quotient of force and mass, g is independent of the second body involved. The acceleration of gravity describes the rate at which any body falling in a vacuum at the earth's surface increases its velocity per unit of time. The unit of acceleration is the gal (named after Galileo) and is equal to 1 cm s^{-1} s^{-1}. Numerous observations have shown that the acceleration of gravity is about 980 cm s^{-1} s^{-1}. Regional deviations from this figure are so small that in practice it is essential work with a unit equal to only one-thousandth part of the gal, known as the milligal (mGal).

A second necessity before the regional variations could be accurately measured was development of suitable instruments. Specialist modern laboratories can determine the rate of acceleration of a falling body to an accuracy of 1 part in 10 million, and such stations form the basis of a global network for absolute gravity measurements. This network can then be employed for calibrating simpler portable instruments that allow rapid local surveys to be conducted. The most widely used type of gravimeter depends upon the extension of a delicate spring; some idea of the remarkable sensitivity of modern instruments comes from the realization that a 100 mm spring will extend by only 10^{-5} mm for each milligal increase. Despite problems arising from the motion of the transporting vehicle, an instrument of this design has been adapted for use in both ships and aircraft, so that a relatively comprehensive global cover is now available.

Gravimetric surveys have revealed significant regional variations. Some of this variation is readily explicable. On an idealized globe that is non-rotating, spherical, smooth and composed of homogeneous concentric shells there would be no regional deviations. If each constraint is relaxed in turn so that the idealized globe conforms more and more closely to the actual earth, the cause of some of the regional variation is immediately apparent. If it is first assumed that the globe rotates, a 'centrifugal force' is introduced which diminishes from a maximum at the equator to zero at the poles; since the centrifugal effect acts in the opposite direction to gravitational acceleration this means, for instance, that an object will weigh less at the equator than at the poles. Lines of equal value of g, known as isogals, would run parallel to the lines of latitude on such a rotating sphere. As seen on p. 3, the angular velocity of the earth is sufficient to distort the globe into an oblate spheroid so that perfect sphericity is the second constraint that must be relaxed. The values of gravity on a rotating ellipsoid can be calculated, and again the isogals should run along the parallels of latitude. Although the situation now being envisaged is still idealized, it does begin to approximate conditions over large areas of the ocean basins.

For much of the earth's surface, however, the third constraint of smoothness must obviously be relaxed. Since ascent of a mountain involves movement away from the centre of the earth, g is going to decrease, and if the material above sea-level is assumed to have zero density this will occur at a rate of 0.3086 mGal m^{-1}. For purposes of standardization observed gravity values are normally referred to the sea-level datum by adding an amount for elevation known as the free-air correction. However, the actual gravity at sea-level would be affected by the overlying rock whose mass can be estimated by employing an appropriate density; the acceleration due to this mass is known as the Bouguer correction factor and has to be subtracted from the observed gravity value. In practice small additional adjustments are made for such factors as the gravitational pull of the sun and moon, and it might then be anticipated that the resultant value would correspond to that for the appropriate latitude. In fact, however, there are often residual discrepancies which are known as Bouguer anomalies.

If the foregoing arguments are sound, Bouguer anomalies must result from the one constraint not yet relaxed, namely, the arrangement of materials of different density within the earth. As early as the eighteenth century one of the expeditions dispatched by the French Academy of Sciences to measure the length of a meridian degree noted that the Andes exerted less influence on a plumb-bob than would be predicted by a simple assumption of concentric shells. The unexpectedly slight attraction of mountain masses was confirmed during later surveys along the Himalayan foothills where the pull due to topography was found to be only one-third of that anticipated. Two famous hypotheses were offered in explanation. In 1854 Pratt surmised that the density of elevated mountains might be in inverse relation to their height, while a year later Airy attributed the weak gravitational attraction to a root of light crustal material extending down into a denser substratum. Both hypotheses predicate a relatively shallow level within the earth's interior above which the mass per unit area is approximately equalized; this notion of hydrostatic equilibrium, termed 'isostasy' by Dutton in 1889, is of crucial importance since it implies that any substantial addition or subtraction of material at the surface of the globe will cause local depression or uplift of the crust.

It should be emphasized that only major relief at the earth's surface is isostatically compensated. Small features still have a direct effect on the local gravitational field, and this property is being exploited over the ocean basins to produce what is called geotectonic imagery. A satellite termed Seasat, launched in 1978, was equipped with a radar altimeter capable of measuring the height of the ocean surface to within less than 10 cm. The attraction of sea floor features lacking isostatic compensation will deform the ocean surface, and in a remarkable technical achievement that includes removal of the effects of wind-generated waves, these minor undulations are being converted into images of the subjacent topography; however, unlike conventional sea-floor maps, these geotectonic images portray only the small-scale forms that lack individual isostatic compensation, and much larger forms such as the mid-oceanic ridges scarcely feature at all.

The above review of a highly complex subject serves to show how detailed measurement has demonstrated a clear relationship between large-scale surface topography and deep-seated structure. However, several possible models of density distribution within the earth fit observed gravity values so that gravimetry alone does not define the structures. Highly significant supplementary information comes from a second line of investigation, the study of earthquake shock-waves.

The transmission of seismic waves through the earth

Study of vibratory motions of the earth is the province of science known as seismology. Vibratory motions may be initiated in a vast number of ways, ranging from the pounding of surf on the beach and of traffic on the roads, to underground nuclear explosions, earthquakes and even the gravitational pull of the moon. Most information about the earth's interior has come from the study of earthquake shock waves originating in displacements along faults. Waves generated by the grinding together of the two sides of a fault radiate from the focus in all directions and may then be recorded on instruments known as seismographs located in observatories throughout the world. The primary function of such equipment is to time and analyse earthquake shock waves so as to provide evidence regarding the structure and composition of the earth's interior.

The nature of earthquake waves

Two major types of wave may be distinguished, surface waves and body waves. The former travel across the surface of the earth, spreading out from the epicentre rather like ripples on a pond. Such is the sensitivity of modern instruments that, after a major earthquake, surface waves can be detected after making several complete circuits of the globe, reappearing in more subdued form at intervals of $2\frac{1}{2}$–$2\frac{3}{4}$ hours. More significant to present purposes, however, are the body waves which travel through the earth's interior. Two types of body wave may be distinguished; P-waves are compressional with the transmitting medium vibrating backwards and forwards parallel to the direction of propagation, while S-waves are shear waves in which the transmitting medium oscillates in a direction perpendicular to that of propagation. In the seismograph record of a single earthquake the P-waves arrive first, the S-waves shortly thereafter and the surface waves last, the sequence reflecting contrasts in the velocities of the different waves.

By comparing the records from three widely spaced seismographs, the time and focus of an earthquake can be fixed and the rate of travel of the various seismic waves thereby computed. Even in the early days of seismology it was realized that the time taken for a body wave to reach the recording instrument is not simply proportional to the distance travelled, but that those waves which penetrate moderately deeply into the interior move rather faster. One consequence is that the waves do not follow straight-line paths but paths that are slightly convex towards the centre of the earth. Waves encountering abrupt changes in physical properties may be either reflected or refracted depending upon their angle of incidence.

Evidence relating to the internal structure of the earth

The single most important conclusion deriving from the study of body waves is the existence of a number of abrupt discontinuities within the earth's interior. In 1913 Gutenberg demonstrated the presence of a clearly defined core with markedly different properties from those of the rest of the interior. Its effect is seen most obviously in the 'shadow zone' that extends from 103° to 142° from the epicentre and in which no direct body waves are recorded (Fig. 1.9). Beyond 142°

Fig. 1.9. The internal structure of the earth as revealed by the propagation of seismic waves. The actual pattern of travel from a single large earthquake may be extraordinarily complex; in addition to the small sample of direct waves depicted below there are numerous reflected waves of which just one example is shown by the dashed line.

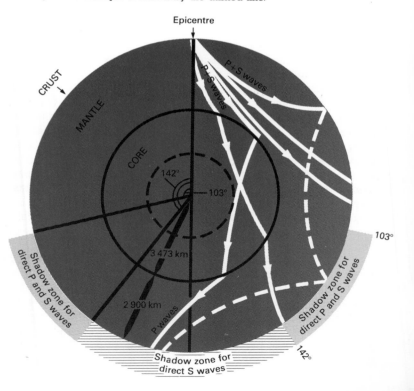

no normal S-waves are recorded but P-waves reappear; their transmission times, however, are much retarded in comparison with those received at 103°. In its failure to propagate S-waves and slow transmission of P-waves the core is exhibiting typical properties of a liquid; its diameter is about 6 945 km. In 1936 Lehmann showed that hitherto puzzling wave refractions, and accelerated movements through the core, could be explained by the existence of an inner solid core with a diameter of about 2 800 km. The composition of the core remains uncertain although it is thought to consist predominantly of iron with minor additional elements that would give density values consistent with those calculated from gravity observations.

A second major discontinuity was discovered in 1909 by the eminent seismologist, Mohorovicic. While examining records of the Kulpa valley earthquake in Yugoslavia he realized that, at distances between 200 and 720 km from the epicentre, both the P- and S-waves displayed two distinct bursts of movement. He inferred that the separate bursts must have followed different routes and concluded that the only satisfactory explanation lay in the presence of a shallow discontinuity within the earth. One route was assumed to have been confined to the overlying rock layer, the other to have descended deeper and to have been bent by refraction; analysis of arrival times indicated faster transmission rates beneath the discontinuity. Subsequent investigators have honoured Mohorovicic by naming the discontinuity after him (unfortunately often contracted to Moho), and have shown that it is a virtually world-wide feature lying at a depth of 10 km beneath the surface of the oceans and at a depth varying between 20 and 75 km beneath the continents. The greatest figures

occur beneath the continental mountain chains so that the form of the discontinuity resembles a greatly exaggerated mirror image of the earth's surface. As one of the major features in the overall structure of the globe, it is conventionally regarded as separating the crust from the underlying mantle. In 1923 an Austrian seismologist, Conrad, detected, at a depth of 10–25 km, a less distinct discontinuity that is often regarded as dividing the continental cust into upper and lower parts.

The Conrad and Mohorovicic discontinuities were detected because the transmission speeds of seismic waves change abruptly at certain depths. In the upper continental crust the velocity of P-waves is about 6.1 km s^{-1}, increasing to 6.9 km s^{-1} in the lower crust. These velocities are consistent with a granitic composition above the Conrad discontinuity, and a mixed gabbroic and granitic composition below it. By contrast, the crust beneath the oceans is now believed to consist of a predominantly basaltic stratum overlying a thicker gabbroic stratum; these are often referred to as Layers 2 and 3 respectively, with Layer 1 comprising the overlying sediment cover. Beneath the Mohorovicic discontinuity the speed of P-wave transmission leaps abruptly to over 8 km s^{-1}, and the physical properties of this upper part of the mantle suggest a composition resembling that of the olivine-rich rock known as peridotite.

Although the seismic studies of Mohorovicic and Conrad were of signal importance in demonstrating basic structural contrasts between continents and ocean basins, more recent attention has focused on what may be an equally fundamental phenomenon known as the 'low-velocity layer'. Here, in a thick but variable layer of the mantle

Fig. 1.10. Schematic sections to illustrate the characteristic structure of the outer layers of the globe.

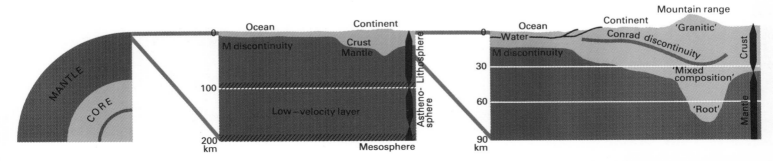

ranging between about 70 and 250 km below the surface, the general tendency for seismic wave velocity to increase with depth is temporarily reversed. Gutenberg first detected this feature in the 1920s when he noticed that waves from earthquake foci at depths of much over 50 km travel less quickly than might be expected on theoretical grounds. More recent investigations have demonstrated P-wave velocities of about 7.8 km s^{-1}, together with a significant attenuation of S-waves, as characteristic of a substantial layer that extends rather deeper beneath the continents than beneath the oceans. The probable cause of this low-velocity zone is partial melting of the solid rock at the temperatures and pressures ruling at that depth. In recognition of the potentially crucial role that might be played in tectonic movements by a mechanically weak low-velocity layer, a nomenclature to supplement the traditional core–mantle–crust subdivision became desirable; the terms commonly employed are asthenosphere for the low-velocity layer itself, and lithosphere for the overlying mantle and crust. Major structural features of the earth's outer shells are illustrated in Fig. 1.10. Much uncertainty still surrounds the precise values for the viscosity of the lower mantle and the asthenosphere, but a difference of at least an order of magnitude may be involved. Estimates for the lower mantle generally lie around 10^{22} P, and for the asthenosphere between 10^{20} and 10^{21} P.

References

Ballard, R. D. and **van Andel Tj**. (1977) 'Project FAMOUS: morphology and tectonics of the inner rift valley at 36° 50′N on the Mid-Atlantic Ridge', *Geol. Soc. Am. Bull.* **88**, 507–30.

Fenneman, N. M. (1916) 'Physiographic divisions of the United States', *Ann. Ass. Am. Geogr.* **6**, 19–98.

Francheteau K. (1983) 'The oceanic crust'. *Sci. Am.* **249(9)**, 68–84.

Hammond, E. H. (1954) 'Small-scale continental landform maps', *Ann. Ass. Am. Geogr.* **44**, 33–42.

King-Hele, D. G. (1980) 'The gravity field of the Earth', *Phil. Trans. R. Soc.* **A294**, 317–28.

Soller, D. R., Ray, R. D.. and **Brown, R. D.** (1982) 'A new global crustal thickness map', *Tectonics* **1**, 125–49.

Selected bibliography

A valuable modern text on the oceans of the world is provided by J, Kennett, *Marine Geology*, Prentice-Hall, 1982. Details about individual ocean basins are to be found in the seven-volume series *The Ocean Basins and Margins* (ed. AEM Nairn and FG Stehli), Plenum. An up-to-date account of the overall structure of the globe is offered by MHP Bott, *The Interior of the Earth* (2nd end), Arnold, 1982; an excellent brief summary is also to be found in the September 1983 issue of the *Scientific American*.

Chapter 2
Concepts of time

In chapter 1 the earth was viewed as an unchanging planet on which measurements could be made to determine its form and structure. Much of the rest of this book is concerned with modifications the global surface undergoes, whether as a result of deep-seated tectonic movements or the activity of erosional agents. This clearly introduces the dimension of time which must be examined in some detail before proceeding to the dynamics of surface change.

Much of classical geology was concerned with establishing the relative ages of the rocks exposed at the surface. A few fundamental principles sufficed in the execution of this work. Outstanding was that of superposition, by which the upper beds in an uninverted succession were dated as younger than the lower ones. Of almost equal importance was palaeontological dating by which distinctive fossil assemblages were recognized as characteristic of certain time periods throughout the world; this method was employed even before Darwin provided the philosophical basis through his concept of evolution. These principles were applied with such skill that nineteenth-century workers were able to establish a stratigraphic column which is still used world-wide today with only minor modifications. The standard nomenclature is shown in Table 2.1.

The geological time-scale

The nineteenth-century investigations could only provide the basis for relative dating, and their reliance on palaeontological evidence meant that correlations between rocks older than the origin of life were wellnigh impossible. Varied arguments were advanced in estimates of the length of geological time, but the crude assumptions underlying all such calculations made the results very suspect. As Sollas wrote in 1900: 'How immeasurable would be the advance of our science could we but bring the chief events which it records into some relation with a standard of time.' In practice the basis for such an advance had been established several years earlier when Henri Becquerel and Marie Curie had identified the radioactivity of the elements uranium and thorium. By 1899 Rutherford had demonstrated the association of radioactivity with atomic disintegration and had postulated the emission of two different types of radiation which he named alpha and beta. In 1903 he showed that alpha radiation consists of a stream of positively charged particles similar to the nuclei of helium atoms (2 protons + 2 neutrons), and later that beta radiation consists of negatively charged electrons shot from the nucleus when a neutron is converted into a proton. He maintained that emission of radiation leads to a change in the radioactive element, and in a classic paper with Soddy argued that 'the proportional amount of radioactive matter that changes in unit time is a constant . . . for each type of active matter a fixed and characteristic value'. The tools were thus at hand for the great advance Sollas had desired, and within a short time several workers were attempting to date rocks experimentally. By 1910 the results were already beginnng to suggest a very much longer span for geological time than the 30 or 40 m.y. that had previously been supposed.

The principle employed was simple. Radioactive disintegration is a random process dependent solely on the number of atoms of the unstable element present at any particular instant of time. A set proportion of these atoms, represented by the value known as the decay constant, will disintegrate in the next unit of time; expressed in an alternative way, half the atoms will disintegrate in a period known as the half-life of the element concerned. Daughter elements will be produced and if the quantity of parent nuclide and daughter element can be accurately measured the period during which radioactive decay has been progressing can be calculated. The original determinations were based upon the transformation in igneous rocks of either uranium or thorium into lead (Fig. 2.1). One limitation of the early U/Pb and Th/Pb techniques was the shortage of suitable minerals containing these nuclides, and the later development of methods based upon other unstable isotopes was important in both enlarging the potentialities for dating and affording opportunities of

Table 2.1 Geological time-scale (after Harland *et al.*, 1982)

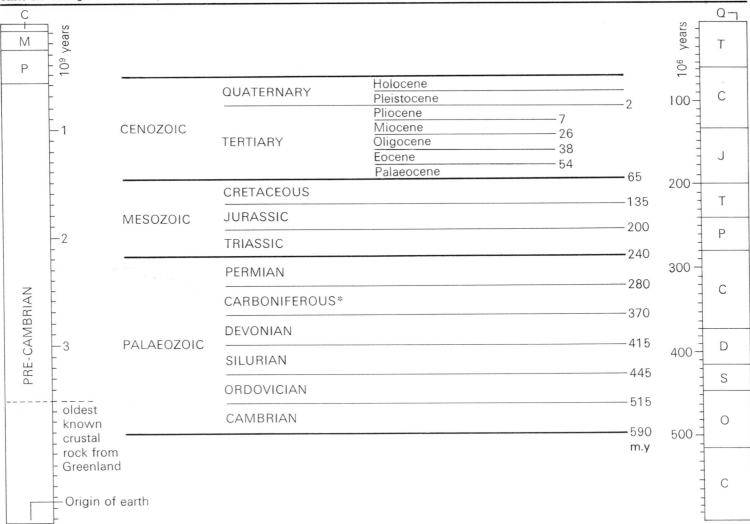

*Known as Pennsylvanian and Mississippian in N. America.
The divided bars to the left and right show the relative duration of the different divisions of geological time.

Fig. 2.1. Uranium and thorium dating. The decay series for U^{238}, U^{235} and Th^{232} are shown, each horizontal transformation representing emission of an alpha particle, each diagonal transformation representing emission of a beta particle. The half-lives of nuclides in common use for dating purposes are also indicated.

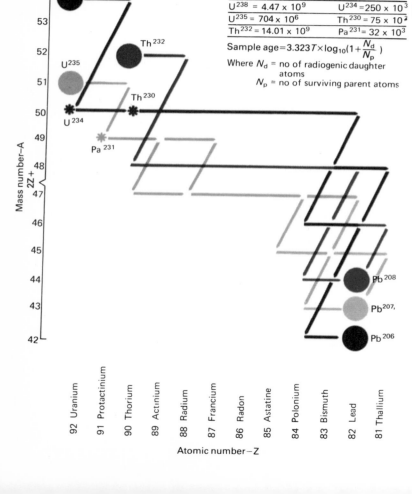

Half-life (T) in years

$U^{238} = 4.47 \times 10^9$	$U^{234} = 250 \times 10^3$
$U^{235} = 704 \times 10^6$	$Th^{230} = 75 \times 10^3$
$Th^{232} = 14.01 \times 10^9$	$Pa^{231} = 32 \times 10^3$

$$\text{Sample age} = 3.323T \times \log_{10}(1 + \frac{N_d}{N_p})$$

Where N_d = no of radiogenic daughter atoms

N_p = no of surviving parent atoms

testing one method against another. A vital technical innovation was the development of mass spectroscopy permitting detailed isotopic analyses of a mineral. Nowadays much dating is based either upon the decay of the Rb^{87} isotope of rubidium to the Sr^{87} isotope of strontium by the emission of beta particles, or upon the transformation of the K^{40} isotope of potassium into the Ar^{40} isotope of argon by the capture of an electron. Like the earlier procedures both the Rb/Sr and K/Ar methods are subject to a number of potential errors, but these can be minimized by modern technical refinements.

Through use of the above nuclides with their immense half-lives, ages have been assigned to the previously established divisions of the geological column (Table 2.1, p. 19). Furthermore, by a clever manipulation of data from those nuclides producing lead derivatives, it has been possible to estimate that the continental crustal material was first formed about 4.5×10^9 years ago. Since stony meteorites when subjected to K/Ar and Rb/Sr dating have yielded a similar age, and lunar rock samples recovered during manned space flights range from 3 to 4.5×10^9 years old, there is a growing belief that the earth may be one of a whole group of solar planets that originated a little over 4 500 m.y. ago.

Problems of late Cenozoic chronology

The radiometric techniques just discussed have provided a dating framework for the whole of the stratigraphic column. However, for the geomorphologist it is the chronology and time-scale of the late Cenozoic era that is particularly important. Some workers have claimed that the vast majority of landforms are Quaternary in age and the rest no older than late Tertiary. This has been contested by others and there is an obvious problem of defining what is meant by the age of a landform that undergoes slight but continuous modification. Nevertheless, it can hardly be denied that the last few million years are of critical importance to the geomorphologist and require much closer examination than the preceding aeons of geological time.

General considerations

There are at least four major problems confronting any attempt to formulate a satisfactory chronology for the late Cenozoic era. Most of the earlier part of the stratigraphic column was constructed on the basis of marine sediments and faunas raised above modern sea-level.

Continuous marine successions of late Cenozoic age are rare on the continents at the present day. The contemporaneous terrestrial beds tend to be very fragmentary and to span only a short interval at any one locality. In consequence it is necessary to devise methods of correlating one section with another and building up a composite picture from many diverse sources. A second problem arises from the relatively brief duration of the late Cenozoic era. This means that biological evolution during that period was so restricted that one of the basic principles on which the rest of the stratigraphic column is founded is of severely limited application. A third difficulty results from the rapidity of environmental change during the late Cenozoic era. This is attested by an immense amount of both biological and physical evidence pointing unequivocally to rapid fluctuations of climate. The exact number of major oscillations remains uncertain and, with relatively little to distinguish between them, it has proved extremely difficult to build up long chronological sequences from the highly fragmented terrestrial evidence that is normally available. A final problem is the absence of any single method of radiometric dating that can satisfactorily span the whole of the period under discussion. The usefulness of methods based upon radiogenic lead declines rapidly for rocks younger than 100 m.y. while the Rb/Sr technique is difficult to apply to materials younger than 20 m.y. Although K/Ar dating has been used on rocks less than 30 000 years old, the amount of radiogenic argon in such cases is so slight that the accuracy of the results is often open to doubt. Alternative methods are available for application to younger materials (see below, p. 27), but most suffer from restrictions in the age range for which they are appropriate.

The Pliocene–Pleistocene boundary

Despite much discussion there is still no universal agreement on the criteria to be adopted in fixing the boundary between Pliocene and Pleistocene times, and without such agreement there is no way of determining the age of the boundary. Some workers have wanted to employ the evidence of fossils so as to maintain consistency with earlier geological eras, but as already pointed out, evolutionary changes in the late Cenozoic tend to be overshadowed by the effects of rapid climatic fluctuations. Other workers have argued that the most distinctive attribute of the Pleistocene is the growth of large continental ice-sheets over Europe and North America, and that the onset of the Pleistocene should therefore be equated with the first dramatic fall in temperature. This idea was implicitly endorsed by the International Geological Congress in 1948 when it recommended that the boundary should be fixed by reference to sections in southern Italy which show a sudden influx of 'cold species' at the end of a long succession of marine sediments known to be of Pliocene age. Yet if the evidence of glaciation were to be adopted as the basic criterion, it would push the Pliocene–Pleistocene boundary back to several million years ago at least. In both California and Alaska glacial deposits interstratified with dated igneous rocks must be more than 2.7 m.y. old. If the existence of an ice-sheet in Antarctica were to be admitted as proof of glaciation, a wholly illogical situation would arise since it now seems probable that the southernmost continent was covered with ice at least 10 m.y. ago during Miocene times. This clearly illustrates the potential pitfalls of an unwise change in the criteria for subdividing geological time.

Several investigators have argued that if the characteristic feature of Pleistocene times is large-scale temperature fluctuations, the onset of these should be held to mark the Plio–Pleistocene boundary. This raises problems of definition since there is no reason to believe that climate was unchanging during earlier geological periods. Yet there does seem to be some justification for the view that the amplitude and rapidity of fluctuations was especially great during the Pleistocene epoch. Workers in several different regions have concluded that during early and middle Cenozoic times mean annual temperatures showed an irregular decline on which were superimposed minor oscillations of a degree or two. Yet about 2.4 m.y. ago the climatic fluctuations abruptly began to increase in amplitude with many areas experiencing changes of 5°C or more. The precision and global applicability of the estimates may still be open to some question, but the approach does afford one of the more widely accepted bases for defining the beginning of the Pleistocene epoch. Moreover, the date of 2.4 m.y. has the great advantage of coinciding with a major reversal in the alignment of the earth's magnetic field (see p. 38), an event that can be detected in sedimentary successions in many parts of the world and thus provide a valuable means of correlation.

The Pleistocene–Holocene boundary.

So far as we know, there is nothing to distinguish the amelioration of climate at the end of the most recent Pleistocene cold period from the many others that preceded it. This has led several eminent workers to argue that it would be more logical to abandon the term Holocene and to recognize that we live during one of the climatic fluctuations of the Pleistocene; if such a view is adopted the recommended term for the present-day warm interlude is the Flandrian stage. On the other hand, in middle latitudes there are obvious advantages in distinguishing between glacial and post-glacial events, and the term Holocene was originally intended as a designation for the post-glacial period. Yet, because the melting of an ice-sheet is transgressive of time, what is literally post-glacial in one locality is contemporaneous with what is glacial in another. In such circumstances it is probably best to fix an arbitrary boundary even if it appears to make little physical sense in some parts of the globe. The most widely acknowledged boundary is that originally proposed by Scandinavian pollen analysts who identified a period of rapid warming that led to accelerated withdrawal of the ice margin across the Baltic region. The start of this warming phase has subsequently been fixed at about 10 000 years ago, a date now adopted by convention as the beginning of the Holocene. Throughout this book the term Holocene will be retained, even though the logic of the argument that we live within a climatic fluctuation of the Pleistocene epoch is overwhelming.

Nature of the evidence bearing on late Cenozoic chronology

The evidence of deep-sea cores

It has already been stressed that, on the continents, late Cenozoic deposits are sporadic in occurrence and rarely span a long time interval at any single locality. With the growing capacity to recover cores from the ocean floor it became possible to sample marine successions that appear to have accumulated as a relatively continuous rain of very fine sediment. The materials themselves are divisible into lutite or terrigenous debris and the tests and hard parts of such minute marine organisms as Foraminifera and diatoms. These microfossils occur in vast numbers and, because the organisms have distinct environmental requirements, are capable of yielding important evidence regarding the changing temperature of the oceanic water. Most information has been derived from pelagic temperature-sensitive Foraminifera such as *Globorotalia truncatulinoides* which is characteristic of warm tropical conditions, and *Globigerina pachyderma* found today even at the North Pole. The relative frequency of these and other species varies at different levels within individual cores and is thought to indicate numerous fluctuations in water temperature and therefore in climate (Fig. 2.2).

Careful analysis of different horizons within deep-sea cores has revealed other systematic changes. Most striking of these is the oxygen isotope variation in the calcareous shells of Foraminifera. The ratio of the isotopes O^{16} and O^{18} theoretically depends upon two major factors, the isotopic composition and temperature of the sea-water in which the shells were secreted. If one assumes for the moment a constant composition, the O^{18}/O^{16} ratio may simply be regarded as a function of temperature, with shells formed in cold water relatively richer in O^{18} than those formed in warm water. The rate of enrichment is extremely small, but when a mass spectrometer is used for analysing the tests from oceanic cores significant fluctuations in the O^{18}/O^{16} ratio are found. The results are normally expressed as parts per mil (‰) deviation from what is known as the PDB standard, that is, the isotopic composition of a Cretaceous belemnite from the Peedee formation in South Carolina (Fig. 2.3). At first sight the method appears ideal for determining the changes in sea-water temperature throughout the late Cenozoic, but in practice interpretation of the O^{18}/O^{16} values has raised many controversial issues. These have centred around the other major factor controlling the constitution of pelagic shells, namely the isotopic composition of the water.

Water molecules with O^{16} atoms are preferentially evaporated with the result that atmospheric processes temporarily deplete the oceans of that isotope. If the water is returned immediately in the form of local precipitation the effect is minimal. However, areas of evaporation and precipitation do not coincide and the mass transfer of moisture in its gaseous state leads to variations in the composition of modern sea-water. It has therefore been necessary to define a standard mean ocean water (SMOW) to which other measurements may be referred. The most severe problem, however, arises from possible longer-term changes in isotopic composition. The outstanding control is acknowledged to be the growth of ice-sheets which will

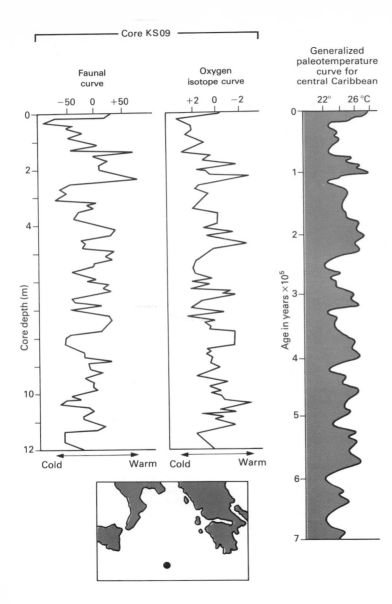

Core KS09

Faunal curve

−50   0   +50

Oxygen isotope curve

+2   0   −2

Generalized paleotemperature curve for central Caribbean

22°   26 °C

Core depth (m)

0

2

4

10

12

Cold   Warm   Cold   Warm

Age in years ×10⁵

0

1

2

3

4

5

6

7

store water of different make-up from that in the sea and thus affect the residue left in the ocean basins. Yet the exact magnitude of this effect has proved difficult to measure. Two approaches have been adopted.

The first involves calculation of both the volume and likely isotopic composition of the major continental ice-sheets that periodically developed over the Northern Hemisphere during Pleistocene times. As will be discussed in Chapter 11, progress has undoubtedly been made in reconstructing the three-dimensional forms of these ice-sheets. Moreover, increasing knowledge of the constitution of the present Greenland and Antarctic ice-sheets provides a reasonable guide to the isotopic values to be employed in any computations. Yet it has to be remembered that, from basic physical principles, the O^{18}/O^{16} ratios in the ice will have varied with the temperatures prevailing at the time of precipitation. For anything beyond a first approximation this means that the researcher must be able to model the shape, surface temperature and pattern of movement of ice-sheets that no longer exist. It should therefore cause little surprise that assessments of the average constitution of the European and North American ice-sheets have varied widely, although current estimates, expressed as a deviation from SMOW, congregate in the −30 to −35‰ range with the implication that much of the measured fluctuation in the deep-sea cores is attributable to this source.

The second approach has been to concentrate attention on the shells of benthic organisms recovered in the deep-sea cores. Since the oceanic bottom water is so cold, normally 1.5–2°, its temperature cannot have declined substantially during a glaciation, whereas its isotopic make-up will have changed just like the rest of the oceanic

Fig. 2.2. Examples of the evidence now available from deep-sea cores regarding Pleistocene environmental changes. On the left are the results of analyses carried out on a core (KS09) from the eastern Mediterranean. The faunal curve is based upon changing abundances of cold-water indicators (*Globerigina bulloides*, *G. incompta*, *G. quinqueloba*, *Globorotalia scitula* and *Globigerinita glutinata*), and warm-water indicators (*Globigerinoides ruber*, *Hastigerina siphonifera*, *Orbulina universa* and *Globorotalia truncatulinoides*); the oxygen isotope curve is based on samples of *Globigerinoides ruber*. The record probably covers the last 630 000 years. On the right is a portrayal of the likely water-temperature changes in the central Caribbean during the last 700 000 years, again based on oxygen isotope analysis (Mediterranean data from Cita *et al.*, 1977; Caribbean data from Emiliani and Shackleton, 1974).

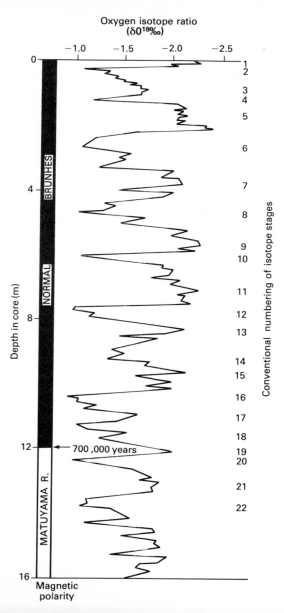

water; this means any isotopic variations in benthic shells are presumably attributable to fluctuations in composition rather than temperature. Since the tests of the bottom-dwelling organisms do exhibit significant variations, this has tended to confirm the volume of continental ice-sheets as the dominant control of O^{18}/O^{16} ratios, with only a lesser influence being exercised by concurrent temperature changes.

Whatever the exact balance between composition and temperature effects, the observed O^{18}/O^{16} fluctuations undoubtedly reflect environmental changes of great global significance. Potentially, for instance, they not only afford a means for reconstructing the volume of ice stored on the continental surfaces but thereby also provide a guide to late Cenozoic sea-level changes. To facilitate correlation of the records from different cores, it is accepted practice to number the major fluctuations from the top downwards, with the even numbers signifying relatively cold/glacial periods, and odd numbers warm/interglacial periods. As can be seen from Fig. 2.3, the deep-sea cores indicate that there have been at least nine cyclical fluctuations in the last 700 000 years, a period covering less than half the full duration of the Pleistocene.

The evidence of shallow marine and estuarine sediments.

On a number of coasts around the world there are important sequences of shallow-water sediments spanning much of the Pleistocene epoch. Correlation between sites poses many problems, but significant contributions to Pleistocene chronology have come from careful faunal and floral analysis of the sediments. A good example is afforded by the sequences around the southern margin of the North

Fig. 2.3. The record of oxygen isotope fluctuations in core V28–238 raised from the equatorial Pacific Ocean near the Solomon Islands in 1971. This core has frequently been adopted as a standard to which finds from other localities are referred. Warm conditions with limited ice cover are indicated by peaks to the right, cold conditions with extensive ice-sheets by peaks to the left. The fluctuations are numbered from the present backward through time, so that odd numbers represent 'interglacial' conditions and even numbers 'glacial' conditions. A critical dating horizon of 700 000 years BP is afforded by the transition from normal to reversed polarity at a depth of 12 m (see Ch. 3 for further details of the palaeomagnetic record).

Sea in both England and the Netherlands. In East Anglia the so-called crag deposits consist of shelly marine sands and estuarine silts and clays. They are particularly interesting since they underlie what are believed to be among the earliest true glacial deposits anywhere in Britain; moreover, as the lowest crag members are classified on faunal grounds as Pliocene, the succession almost certainly spans some part of the local pre-glacial Pleistocene. The sequence has been studied by means of cliff sections and a specially commissioned borehole at Ludham in Norfolk (Table 2.2). Both Foraminifera and pollen bear witness to major climatic oscillations with four cold phases

Table 2.2 The succession of Pleistocene glacial and interglacial episodes currently recognized in Britain. Note that only the three most recent glaciations are represented by tills, the earlier ones being inferred from biological evidence in East Anglia (after West, 1980)

Glacial	Interglacial	
Devensian		
	Ipswichian	Tills locally interbedded with organic deposits indicative of relatively warm conditions
Wolstonian		
	Hoxnian	
Anglian		
	Cromerian	
Beestonian		
	Pastonian	Estuarine sands, silts and peats, exposed in cliff and artificial sections in East Anglia
Pre-Pastonian		
	Bramertonian	
Baventian		
	Antian	Crag deposits, mainly marine sands and silts, sampled in the Ludham borehole, Norfolk
Thurnian		
	Ludhamian	

separated by intervals that were probably no cooler than today. Since three further cold phases are identified from the stratigraphy of later ice-deposited sediments, a minimum a seven glacial periods are commonly recognized in the British succession. This is still much less than the number of climatic cycles recorded in the deep-sea cores, although it would be wrong to seek exact one-to-one equivalence in the two lines of evidence. It is generally accepted, for instance, that the last glaciation, called the Devensian in Britain, was of a complex nature and is represented in the deep-sea cores by stages 2 and 4. The disparity in the number of recognized cold phases may also reflect breaks in the continuity of the crag succession since Dutch workers maintain that a rather fuller sequence of climatic oscillations can be identified in the equivalent beds in the Netherlands. Yet it remains true that only limited success has so far attended attempts to correlate the evidence from deep- and shallow-water marine successions.

The evidence of continental sequences

It would be quite impossible in a few pages to survey the multifarious evidence on which the Pleistocene chronology of the continental land surfaces has been built up during the last few decades. Interpretation of the evidence is often highly controversial and remarkably few generalizations can be made that would meet with universal assent. Initially a distinction must be drawn between areas known to have been covered by the Pleistocene ice-sheets and those which were not. In glaciated areas attention was long concentrated on till sheets laid down one on top of the other and believed to bear witness to multiple glacial advances. With increasing use of pollen analysis equal attention has been devoted to biogenic materials interleaved with the tills. Accompanying this emphasis on pollen-bearing sediments has been more thorough study of such fossil remains as non-marine molluscs and beetles. Most fossiliferous sediments are of such restricted extent that correlation from one site to another must depend either on the use of more continuous marker horizons such as till sheets and spreads of glacio-fluvial gravel, or on some intrinsic quality of the fossil record itself. For example, *Abies* appears to be abundant in the penultimate interglacial of north-western Europe but rare in the last interglacial, while *Corylus* pollen becomes frequent at a much earlier stage in the last than in the penultimate interglacial. These and similar diagnostic features have permitted direct correlation over relatively long distances although the possibility of closely similar sequences being repeated at different times must always be borne in mind, especially when the precise number of climatic fluctuations remains one of the unknowns; it is worth stressing, however, that the record of terrestrial sediments in most glaciated areas is now acknowledged as very short and therefore confined to the later part of the Pleistocene.

For each major glaciated region there is a recognized nomenclature covering the period during which large continental ice-sheets developed. The names in common usage for the Alps, north European plain, British Isles and North America east of the Rockies are shown in Table 2.3. There are problems of correlation between these four regions, but the relationships shown are those commanding widest support at the present time. More uncertainty attends the earlier than the later glacial and interglacial periods, and it would be foolish to deny that further research may well lead to substantial reorganization of the lower part of the table.

Before leaving this review of glaciated regions, it is worth noting that in Antarctica and Greenland there exist long stratigraphic columns of ice that, with careful decipherment, can yield important information about conditions at time of accumulation. Several lengthy cores have now been obtained from each ice-sheet, and it seems certain that we still have available for study ice that originated during the last glaciation. Dating of the oldest ice poses more of a problem.

Table 2.3 The terminology of glacial and interglacial periods in four areas of the Northern hemisphere

North-western Europe	British Isles	Alps	North America
Weichselian	Devensian	Würm	Wisconsin
Eemian	Ipswichian	Riss/Würm	Sangamon
Saalian	Wolstonian	Riss	Illinoian
Holsteinian	Hoxnian	Mindel/Riss	Yarmouth
Elsterian	Anglian	Mindel	Kansan
Cromerian	Cromerian	Günz/Mindel	Aftonian
		Günz	Nebraskan

Although very thin but countable annual layers believed to be about 14 000 years old have been detected, most of the techniques suitable for analysing and dating the upper layers are inapplicable to the extremely attenuated and deformed basal layers where age has to be estimated largely by reference to assumed annual accumulation rates and models of flow in an ice-sheet.

This is the procedure adopted by Dansgaard and associates in an analysis of a 1 390 m core from Greenland (see Fig. 2.5). They divided the core into 7 500 layers which were analysed for their O^{18} content; as noted earlier the O^{18}/O^{16} ratio is a sensitive indicator of the snow temperature at the time of precipitation. They found that a value of $-28‰$ deviation from SMOW, characteristic of modern snow at the point where the core was taken, persists with only minor variations to a depth of over 1 100 m. Below that there is a sudden fall to values consistently below $-35‰$ and occasionally below $-40‰$. Figures of this order extend to a depth of 1 330 m, beyond which there is a rise to values at least as high as those found in snow falling today. The core between 1 100 and 1 330 m is believed to consist of ice accumulating during the last glacial period, while the basal layers may conceivably date from the last interglacial. However, given the uncertainties over dating, attention is probably best concentrated on the upper part of the core. The potentiality of identifying small but significant temperature fluctuations superimposed on the larger-scale changes is exemplified by work on the period for which comparison can be made with historical records. Thus it is possible to detect relatively warm periods around 1930, 1750 and 1550, together with colder intervals around 1820 and 1690. The latter coincide with well-authenticated glacier advances and periods of unusually extensive pack ice in the North Atlantic. An illustration of an earlier fluctuation recorded in the Greenland core is a brief but exceptionally warm interlude near the end of the last glaciation occurring just over 11 000 years ago; this appears to coincide with vegetational changes already well established by the pollen record of north-western Europe.

In unglaciated regions the nature of the evidence relating to late Cenozoic climatic fluctuations varies widely. Pollen analysis has again made a major contribution, although a shortage of sites conducive to sustained accumulation of biogenic materials is a limiting factor. However, several long polleniferous cores have been recovered since 1960 from tectonically active basins in southern Europe; some of these apparently extend back to Pliocene times and imply numerous climatic oscillations comparable to those inferred from work in north-western Europe. Another major source of information in extra-glacial regions is the sedimentary sequence in long-lived lake basins. For instance, studies of the deposits in Lakes Biwa and Baikal in Asia have yielded faunal and floral records going back several million years for parts of Japan and the USSR respectively.

However, some of the most interesting results of palaeolimnology have come from modern desert areas where temporary lakes existed during wetter climatic phases but dried up during more arid periods. An excellent illustration is afforded by the Great Salt Lake basin in Utah (Fig. 2.4). Here abandoned strandlines denoting a water-level some 330 m higher than at present (see Fig. 4.7) are believed to date from a period broadly contemporaneous with the last glaciation in Europe and North America. A 307 m core of clastic and evaporite sediments from the floor of the old lake bears witness to numerous earlier alternations between deep water and desiccated salt flats. The uppermost 110 m of the core has been analysed in detail and attests to twenty-five phases of desiccation in a period believed to cover the last 700 000 years (Fig. 2.5). A similarly complex sequence of environmental changes has been inferred from study of the loess deposits in central Europe, and with some justification it has been claimed that these lines of evidence provide a potentially better means of correlation with the marine record than do the traditional glacial sequences. Least information is currently available for humid tropical regions, but even here there are clear indications of significant fluctuations during the later part of the Pleistocene. For example, in part of the Amazon basin polleniferous fluvial sediments indicate local replacement of the rain forest by savanna, probably as a result of a decrease in precipitation.

Radiometric and other dating techniques appropriate to the Pleistocene and Holocene epochs

Radiocarbon dating

Of several techniques introduced during the last few decades to provide reliable dates for the Pleistocene and Holocene epochs, by far the most important has been that based upon the C^{14} isotope of carbon. The method depends on the fact that the atmosphere and hydrosphere represent reservoirs of radioactive carbon which are tapped by living organisms, both animal and plant, to build their various stuctures and tissues. The source of the radioactive carbon lies in cosmic ray bombardment which converts a minute proportion of atmospheric nitrogen into C^{14}. This in turn is rapidly oxidized to CO_2 and widely dispersed, not only through the atmosphere but also through all surface waters. There are three isotopes of carbon, C^{12}, C^{13}, and C^{14}, the first being by far the most common with only one atom in a million million being of the C^{14} variety. Of the three, C^{14} is the only unstable isotope and has a half-life of 5 730 years. Prior to recent interference by Man, an approximate long-term equilibrium existed between the quantity of new carbon arising from cosmic radiation and that lost through disintegration. Since the radioactive isotope is evenly distributed throughout the lower atmosphere, and since living organisms absorb CO_2 with little or no discrimination between the various isotopes, each organism during its lifetime incorporates a constant proportion of unstable carbon. After death, replenishment with C^{14} ceases and there follows a continuous decline in the C^{14} content as radioactive decay proceeds. The ratio of radioactive to stable carbon is therefore a measure of the age of such diverse organic materials as bones, tusks, shells, hides, peat and wood.

Analysis has traditionally been made by counting the number of radioactive disintegrations per minute per gram of carbon. Since only one-millionth of the C^{14} atoms present in a sample will decay in any 3-day measuring period, and since the number of disintegrations will halve for every 5 730-year increase in the age of the sample, the radioactivity becomes extremely difficult to separate from background radiation even when the laboratory is deeply buried and complex shielding of the equipment is employed. This problem normally limits radio-carbon dating to materials under 50 000 years old, although alternative measuring procedures, most notably by mass spectrometer, are now offering the prospect of a more extended dating ability. Residual uncertainty in the laboratory determination of age is reflected in the best estimate usually being accompanied by a ± figure. This corresponds to one standard deviation in the counting values and implies, by normal statistical arguments, that there is a 68 per cent probability of the true age lying within one standard deviation and a 95 per cent probability of its lying within two standard deviations.

Fig. 2.4. The prominent shoreline that marks the former level of Lake Bonneville near Salt Lake City, Utah. The modern Great Salt Lake lies some 300 m lower in the centre of the basin to the left of the photograph.

Fig. 2.5. Pleistocene environmental changes attested by relatively full continental sequences. In no instance is the dating certain, but the evidence from each points unequivocally to numerous rapid fluctuations.

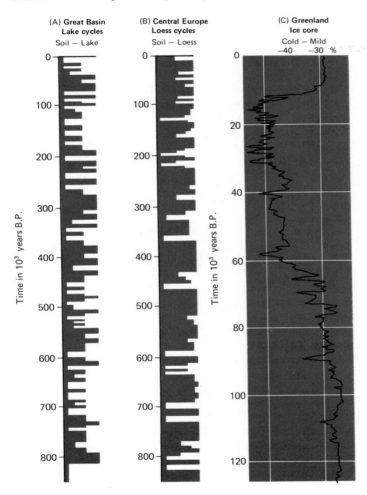

Precautions need to be taken when collecting samples for dating. Contamination by either younger or older material may seriously affect the results. For example, penetration by modern rootlets will make an ancient peat appear much younger than its true age. Conversely, carbonates dissolved from ancient rocks and absorbed in photosynthesis by submerged aquatic plants will make them appear older than their true age. Although hazards of this type can sometimes be detected and allowed for in modern laboratory techniques, they need to be constantly borne in mind. It is worth noting that modern carbon presents the more serious form of contamination owing to its greater potential effect on the calculated age (Fig. 2.6).

The principle of radio-carbon dating rests upon one fundamental assumption that deserves rather closer examination, namely that the concentration of atmospheric C^{14} has remained constant through time. The proposition is not difficult to test since the known age of selected sample materials can be compared with their 'radio-carbon age'. Such research has revealed systematic deviations most easily explained as due to temporal changes in atmospheric C^{14}. For instance, between AD 1500 and 1700 there seems to have been an abnormally high concentration of C^{14}, so that organic materials formed in 1700 appear on radio-carbon dating to be little more than 100 years old. By analysing annual rings from the bristlecone pine (*Pinus longaeva*) in California, the world's oldest living tree, it has been possible to assess C^{14} variations during the last 7 000 years (Fig. 2.6). It is now believed that the excess of C^{14} between 4000 and 3000 BC was so great that radio-carbon dates from this period are about 700 years too young. Such discrepancies are of prime importance to the archaeologist and have generated important new hypotheses about the evolution and spread of cultures; but they must also be remembered by the geomorphologist as he endeavours to measure with increasing precision the rates of erosional and depositional processes.

The causes of the changes in abundance of C^{14} remain uncertain, although the most likely explanations involve variations in the intensity of the earth's magnetic field and thereby in cosmic ray activity in the atmosphere. It is obviously important to test for discrepancies between 'true' and 'radio-carbon' ages prior to 7 000 years ago, but unfortunately the older the material the more difficult it becomes to apply independent checks, and attempts to compare radio-carbon ages with those derived from counting annual layers of varves in lake-

30

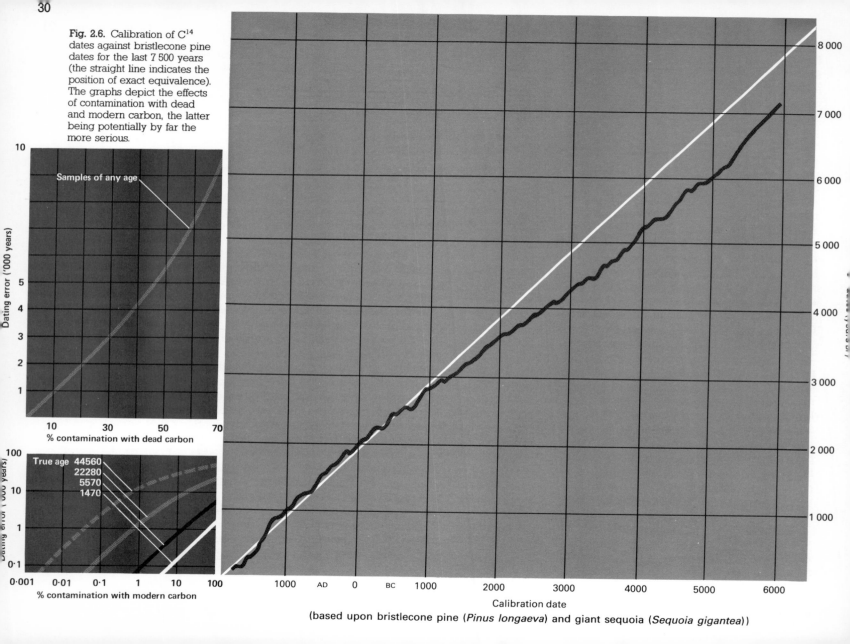

Fig. 2.6. Calibration of C^{14} dates against bristlecone pine dates for the last 7 500 years (the straight line indicates the position of exact equivalence). The graphs depict the effects of contamination with dead and modern carbon, the latter being potentially by far the more serious.

Samples of any age

Dating error ('000 years)

10
5
3
2
1

10   30   50   70
% contamination with dead carbon

True age 44560
22280
5570
1470

100
10
1
0·1

Dating error ('000 years)

0·001 0·01 0·1 1 10 100
% contamination with modern carbon

8 000
7 000
6 000
5 000
4 000
3 000
2 000
1 000

1000 AD 0 BC 1000 2000 3000 4000 5000 6000
Calibration date
(based upon bristlecone pine (*Pinus longaeva*) and giant sequoia (*Sequoia gigantea*))

floor sediments have yielded conflicting results. Nevertheless, it is still reasonable to conclude that the basic dating technique provides a good guide to age over the time range for which it can be employed, although the apparent precision of the figures must not lead to their being treated uncritically.

The great contribution of radio-carbon dating has been to provide a time-scale for the whole of the Holocene and the final stages of the Pleistocene. The subdivision of Holocene time was originally based upon a sequence of floral changes detected by palaeobotanists in Scandinavia; the distinctive pollen zones that were recognized have since been accorded radio-carbon dates and are still in widespread use (Table 2.4). For the Pleistocene vital correlations have been established between widely separated regions and vastly contrasting environments. Deep-sea cores have been calibrated, glacial advances and retreats dated and the histories of different continents, and even of the two hemispheres, compared. A signal advance made possible by the proliferation of radiocarbon dates has been the ability to map the environments of the whole globe as they existed at some specified time in the past. In the early 1970s the North American CLIMAP project (Climate Long-range Investigation Mapping And Prediction) initiated a major reconstruction of the globe as it was 18 000 years ago. The choice of the 18 000-year period had a twofold significance: firstly it lay well within the limit of C^{14}-dating which provided the primary basis for correlation, and secondly it represented the time of maximum ice extent at the culmination of the last glaciation. A second investigation, comparable in concept to the CLIMAP project but undertaken by other workers, was directed at reconstructing climatic changes in the world's major deserts. The basis for this study was C^{14}-dating of organic materials indicative of former lake levels, and attempts were made to trace changing patterns of relative lake elevations at 1 000-year intervals from 30 000 years ago to the present day. These two studies are more fully discussed and illustrated on pp. 380–4.

Uranium-series dating

Although the creation of lead from uranium or thorium is extraordinarily slow, the transformation actually takes place through a series of intermediate short-lived daughter products (Fig. 2.1). Most of these are too transient for dating purposes, but U^{234}, Th^{230} and Pa^{231}, with half-lives of 250 000, 75 000 and 32 000 years respectively, have

Table 2.4 Subdivision of the Holocene, together with an indication of the associated vegetational changes in England and Wales (after West, *Pleistocene Geology and Biology*, 1977).

Time based on C14 dating	Blytt and Sernander periods	Pollen zone	Zone characteristics for England and Wales	
		VIII modern	Afforestation	
1000	Sub-Atlantic	VIII	Alnus–Quercus Betula (–Fagus–Carpinus)	
AD				
BC 1000				Deforestation
2000	Sub-Boreal	VIIb	Alnus–Quercus–Tilia	
3000			Ulmus decline	
4000	Atlantic	VIIa	Alnus–Quercus–Ulmus–Tilia	
5000	Boreal	VI	Pinus–Corylus (c) Quercus–Ulmus–Tilia (b) Quercus–Ulmus	
6000			(a) Ulmus Corylus	
7000		V	Corylus–Betula–Pinus	
8000	Pre-Boreal	IV	Betula–Pinus	

all attracted attention as potential sources of dating. To understand the methods that have been evolved it is essential to realize that the rate of formation of any intermediate nuclide is a function not only of the abundance and decay rate of its parent but also of its own decay rate. Eventually an equilibrium must be established when the so-called 'activity ratio' of parent and daughter will be unity. However, natural processes operating at the surface of the globe often segregate an isotopic mix in which the activity ratio is significantly different from unity. If such a mix thereafter constitutes part of an isolated system, the ratio will proceed towards equilibrium at a predictable rate. An example will illustrate the principles involved. In sea-water, and also in modern coral reefs which incorporate a few parts per million of uranium in their structure, there is found to be an activity ratio between U^{238} and U^{234} of 1.14; in other words there is an initial daughter excess. Once a reef has ceased to grow, and provided that it forms a closed chemical system with no later contamination, the ratio for the next 1.5 m.y. will in theory steadily decline towards unity. The same reef may well provide an example of daughter deficiency dating. This arises because, although the coral contains U^{234}, it does not, at the start, normally contain any Th^{230} which is effectively insoluble in sea-water. However, with the passage of time this deficiency in Th^{230} will be reduced by the decay of the U^{234} until, after about 500 000 years, an equilibrium will be established. In this way both U^{234}/U^{238} and Th^{230}/U^{234} activity ratios are indicators of age, and both have been used to derive apparently reliable dates for raised coral reefs in Barbados and New Guinea that are well over 200 000 years old.

Cave deposits constitute a second type of material to which uranium-series disequilibrium dating has been applied. Thorium is insoluble in normal fresh water as in sea-water, so that any of that nuclide found, for instance, in stalactites and stalagmites is assumed to originate from radioactive decay. Age is then determined from the extent to which the shorter-lived thorium nuclide has grown into equilibrium with its uranium parent. By comparison with the coral reefs discussed above, much greater variability in the initial U^{234}/U^{238} ratio has to be accommodated in the calculations, but by manipulation of the observed values of U^{238}, U^{234} and Th^{230} consistent dating of speleothems has been achieved. Old stalagmites from caves in both Britain and Canada have been assigned ages of over 300 000 years in this way, while in Arkansas dates in excess of 700 000 years have

been claimed, although these are based solely on U^{234}/U^{238} ratios and depend upon major assumptions about the initial ratio at the time the system became closed.

Sea-floor sediments are a third type of material where intermediate products in the uranium decay series have been used for dating. The method here involves analysis of Pa^{231} and Th^{230} in deep-sea cores, and depends on the assumption that there is a constant ratio in sea-water of the parent nuclides U^{238} and U^{235}. The daughter products become attached to clay minerals settling to the ocean bottom and, once isolated from further replenishment, continue to disintegrate at their respective decay rates. In theory the Pa^{231}/Th^{230} ratio should be 10.6 : 1 at the time of deposition and gradually increase with age. With dwindling amounts of both isotopes the method should be capable of dating sediments up to about 200 000 years old, but considerable doubt has seen cast upon the reliability of some of the results obtained since not all the conditions upon which the method is based are necessarily met; in particular there appears to be fractionation of the two nuclides in their settlement to the sea floor so that there is not a uniform initial ratio of 10.6 : 1.

The foregoing review illustrates the potentialities of uranium-series dating and significant results are increasingly being obtained for that important period lying just beyond the range of radio-carbon techniques. Yet it is worth emphasizing that each case mentioned involves the fundamental assumption that, after some clearly defined starting-point, the transformation of unstable nuclides took place within a closed system. Any breach of that assumption vitiates the results obtained, so that the most stringent possible testing should always be applied before the dating is accepted as reliable.

Non-radiometric dating

A wide range of techniques has been devised for use in special circumstances. Reference has already been made to the counting of annual accretionary layers in such diverse materials as tree rings, ice-cores and the sediments deposited in glacial lakes. Ages of old moraines have been inferred from the thickness of the weathering rind on the constituent boulders, while very young moraines have been dated from the sizes of the lichens growing on them. Examples of this type could be multiplied many times, but the conditions under which most can be applied are severely circumscribed and many produce only relative ages that need calibration by some radiometric scale if actual ages are to be obtained. However, two techniques offering the promise of more general contributions to Pleistocene chronology are amino-acid and thermoluminescent dating.

Amino-acid dating is based upon the principle that virtually all the protein in the hard parts of living organisms, such as bones and shells, consists of long chains of amino acids in what is known as the L-configuration. After death the proteins slowly break down with the release of free amino acids, a proportion of which convert from L-isoleucine to D-alloisoleucine in a process called racemization. The rate of racemization varies from one species to another, and also with the prevailing temperature. In warm climates the process may take about 1 m.y. to complete, but in high latitudes over 3 m.y. may be required. This control by temperature precludes the D/L ratio in fossil materials being directly convertible into age, and for the moment the method is best confined to relative dating within a small geographical area likely to have experienced a common climatic history. So far most work has been undertaken on marine shells collected from old shorelines, and these have allowed the identification of local 'amino-zones' specified by the range of observed D/L ratios for a particular organism. In the longer term it should prove feasible to calibrate the amino-zones by radiometric dating and thus obtain a reliable guide to their actual ages.

The principles of thermoluminescence, or TL, have been employed mostly by archaeologists concerned with the dating of pottery. However, the underlying idea can also be applied to sediments that have lain buried for some considerable time. When a sample of such material is heated to a temperature of, say, 300 °C in an oxygen-free gas, there is a weak but measurable emission of light comprising what is known as the 'glow curve'. If, after cooling, the process is repeated no comparable light is emitted. This is because the original TL is dependent upon energy stored in the lattice structure of such minerals as quartz and feldspar. This energy derives from nuclear radiation that detaches electrons from their parent atoms and displaces them to defects or 'traps' in the crystal structure. On heating, these electrons escape from their traps and are responsible for the emission of photons of visible light. The energy stored in the lattice is a function of the natural radiation at the point where the burial occurred, and of the duration of that burial. The former can be measured in several different ways, and the response of the sample to a given dosage can be determined experimentally by artificial irradiation. From these two values and the glow curve the period of burial can theoretically be calculated.

The word 'theoretically' should be emphasized since there remain many problems associated with the implementation of TL dating. In the case of pottery the zero point when all the electron traps in the mineral lattice were emptied is obviously the time of firing; with sediments the nature of the corresponding zero point is less clear, although it is assumed to be the last occasion on which the material was subjected to 'optical bleaching' by prolonged exposure to sunlight. It is for this reason that TL dating applies to the period since sedimentation occurred. Further uncertainty arises through the possible occurrence of TL attributable to other agencies than nuclear radiation, such as pressure, friction and chemical reactions. Finally, although the environmental radiation may reasonably be regarded as constant, its effect on granular materials will be much influenced by the amount of interstitial water, and this is a variable for which it is difficult to make any precise allowance. No exact age range for TL dating can be given since it will vary according to local circumstances, but it most commonly lies in the range 10^3–10^6 years. Most work on TL dating in the 1960s and 1970s was carried out in the USSR, and only since then has its potential been fully appreciated in western Europe and North America. On the basis of Soviet experience, calculations of absolute ages by TL are likely to underestimate true ages by a substantial margin, although relative ages appear reasonably reliable. Even the latter would represent a signal advance in the dating of non-organic Pleistocene materials, and there seems good reason to devote more attention to TL dating than has been the practice in the past.

Discussion of a third important dating method, that based upon periodic reversals of the earth's magnetic field, is deferred to the general review of palaeomagnetism contained in Chapter 3.

References

Belova, V. A. (1975) *A History of Vegetation in the Basin of the Baikal Rift Zone*, Nauka Press, Moscow.

Chappell, J. (1974) 'Geology of coral terraces: Huon Peninsula', *Geol. Soc. Am. Bull.* **85**, 533–70.

Cita, M. B. *et al* (1977) 'Palaeoclimatic record of a long deep-sea core from the eastern Mediterranean', *Quat. Res.* **8**, 205–35.

CLIMAP Project Members (1976) 'The surface of the ice-age earth', *Science N. Y.* **191**, 1131–44.

Dansgaard, W. (1981) 'Ice core studies: dating the past to find the future'. *Nature, London* **290**, 360–1.

Dansgaard, W. et al. (1971) 'Climatic record as revealed by the Camp Century ice core', in *Late-Cenozoic Glacial Ages* (ed. K K Turekian) Yale Univ. Press.

Davies, K. H. (1983) 'Amino-acid analysis of Pleistocene marine molluscs from the Gower peninsula', *Nature, London* **302**, 137–9.

Dreimanis, A. et al. (1978) 'Dating methods of Pleistocene deposits and their problems. I – Thermoluminescence dating. II – Uranium-series disequilibrium dating', *Geoscience Canada* **5** 55–60 and 184–8.

Eardley, A. J. et al. (1973) 'Lake cycles in the Bonneville basin, Utah', *Geol. Soc. Am. Bull.* **84**, 211–16.

Emiliani C. and Shackleton N. (1974) 'The Brunhes Epoch: Isotopic paleotemperatures and geochronology', *Science* **183**, 511–14.

Flenley, J. R. (1984) 'Andean guide to Pliocene–Quaternary climate', *Nature, London* **311**, 702–3.

Hammen T. van der (1974) 'Pleistocene changes of vegetation and climate in tropical South America', *J. Biogeogr.* **1**, 3–26.

Harland, W. B. *et al* (1982) *A Geologic Time Scale*, CUP

Horie, S. (ed.) (1976) *Palaeolimnology of Lake Biwa and the Japanese Pleistocene*, Kyoto Univ.

Kukla, G. J. (1977) 'Pleistocene land–sea correlations I. Europe', *Earth Sci. Rev.* **13**, 307–74.

Lorius, C. et al. (1979) 'A 30 000 yr isotope climatic record from Antarctic ice', *Nature, London* **280**, 644–8.

Mesolella, K. J. et al. (1969) 'The astronomical theory of climatic change: Barbados data', *J. Geol.* **77**, 250–74.

Osmond, J. K. (1979) 'Accumulation models of ^{230}Th and ^{231}Pa in deep sea sediments', *Earth Sci. Rev.* **15**, 95–150.

Schroeder, R. A. and **Bada, J. L.** (1976) 'A review of the geochemical applications of the amino-acid racemization reaction', *Earth Sci. Rev.* **12**, 347–91.

Shackleton, N. and **Opdyke, N.** (1973) 'Oxygen isotope and palaeomagnetic stratigraphy of equatorial Pacific core V28–238: oxygen isotope temperatures and ice volumes on a 10^5- and 10^6-year scale', *Quat. Res.* **3**, 39–55.

Street, F. A. and **Grove, A. T.** (1979) 'Global maps of lake-level fluctuations since 30 000 B.P.', *Quat. Res.* **12**, 83–118.

Thompson, G. M. et al. (1975) 'Uranium-series dating of stalagmites from Blanchard Springs Caverns, USA', *Geochim. Cosmochim. Acta* **39**, 1211–18.

West, R. G. (1980) 'Pleistocene forest history in East Anglia', *New Phytol.* **85**, 571–622.

Wintle, A. G. and **Huntley, D. J.** (1982) 'Thermoluminescence dating of sediments', *Quat. Sci. Rev.* **1**, 31–54.

Selected bibliography

A modern review of geological dating is provided by W. B. Harland *et al.*, *A Geologic Time Scale*, CUP, 1982. Techniques especially appropriate to the Pleistocene epoch are discussed in R. Berger and H. E. Suess, *Radiocarbon Dating*, Univ. Calif. Press, 1979; M. Ivanovich and R. S. Harmon *Uranium-Series Disequilibrium: applications to environmental problems*, OUP, 1982; and M. J. Aitken, *Physics and Archaeology* (2nd edn), OUP, 1974.

The journal *Radiocarbon*, in addition to publishing lists of dates from individual laboratories, also includes many of the 'state-of-the-art' papers presented to the International Radiocarbon Conferences that are currently held every 2 or 3 years.

Chapter 3
Global tectonics

With an established time-scale, it is now feasible to examine the evolution of the gross relief forms described in Chapter 1. Before about 1960 conventional wisdom held that the pattern of continents and ocean basins is unchanging, although for many years a number of individuals had voiced the contrary opinion that large-scale translocations of the continents had occurred. The most passionate and influential advocate of this latter idea was Alfred Wegener. Publication of his famous book *Die Entstehung der Kontinente und Ozeane* in 1915 was followed by decades of bitter controversy since so much of the evidence adduced by Wegener and his disciples was capable of alternative interpretations. Although a meteorologist by profession, Wegener was first drawn to the idea of continental drift by the congruence of the coastlines on either side of the Atlantic Ocean. Initially rejecting the hypothesis as improbable, he was led to re-examine it by reading literature describing palaeontological and geological similarities on the bordering continents. It was both the strength and weakness of his position that it impinged upon so many scientific fields; on the one hand an enormous wealth of corroborative evidence could be accumulated, but on the other numerous workers in very diverse specialisms had to be convinced by his arguments. So widespread were the repercussions of the whole concept that by 1928 Wegener admitted he could no longer keep abreast of the rapidly growing literature.

Rather than attempt to summarize all the conflicting arguments of the early period, it is easier to review the research that provoked such a profound change of attitude on the part of the scientific community. Without doubt the most influential was that concerned with rock magnetism and the reconstruction of the ancient geomagnetic field; but once the likelihood of continental movement had been established, a wide range of further investigations added confirmatory evidence. Two of the most important, heat flow and seismic studies, will be briefly examined before attention is finally turned to the unifying concept of plate tectonics.

Palaeomagnetism

The geomagnetic field

As early as 1600 Sir William Gilbert, physician to Queen Elizabeth I, showed that the earth's magnetic field resembles that which would be produced by a giant bar magnet located near the centre of the earth and aligned approximately along the axis of rotation. Extensions of the long axis of the magnet intersect the surface at points known as the North and South magnetic poles. This dipole field may be described in terms of the magnitude and direction of the local magnetic force. The magnitude reaches its maximum value close to the magnetic poles and its minimum around the magnetic equator. The direction is usually specified in terms of inclination and declination (Fig. 3.1). The inclination is the angle made by a freely suspended magnetized needle and the horizontal. With a perfectly regular dipole field such a needle would stand vertically at the magnetic poles and horizontally at the magnetic equator; between these extremes the angle would vary as a function of latitudinal distance from the magnetic pole. The declination is the angular difference between the geographic meridian and the horizontal component of the earth's magnetic field. Its value reaches the extremes of zero and 180° along the great circle passing through both the geographic and magnetic poles.

The observed geomagnetic field differs in a number of important respects from the simple dipole field so far assumed. Intensity, for example, exhibits substantial longitudinal variations, while inclination and declination display considerable regional divergences from the anticipated dipolar pattern. It should be emphasized that these latter divergences are not due to shallow crustal peculiarities such as local iron-ore bodies but are part of an integral pattern that seems to have its origin much deeper within the earth. In addition there are significant temporal changes. Continuous recording stations widely distri-

Fig. 3.1. The geomagnetic field: (A) inclination; (B) declination.

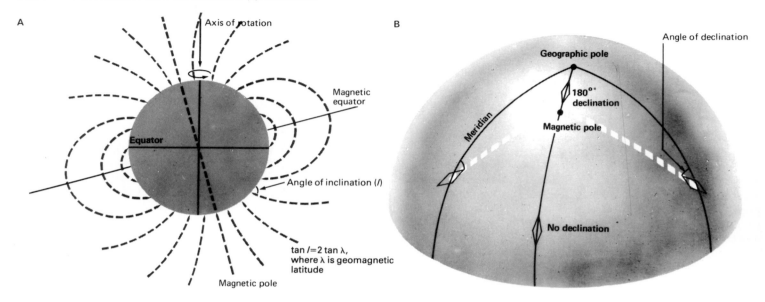

buted over the globe reveal small daily fluctuations in intensity, declination and inclination. These same properties also vary on a longer, secular time-scale. Ever since the seventeenth century it has been known that the positions of the magnetic poles slowly move. When first recorded the declination at London was about 10 °E; by 1800 it had reached over 20 °W but is now decreasing and in 1980 was about 6.5 °W. In the same period the inclination has varied between 75 and 66° and the intensity has gradually weakened. Viewed globally many of the regional divergences from the simple dipole field migrate westwards at an annual rate of 0.18° of longitude, that is, complete one revolution in about 2 000 years. On an even longer time-scale the most striking change is undoubtedly reversal of the earth's polarity so that what is now the North Pole becomes the South Pole and vice versa.

The most acceptable hypotheses of the origin of the geomagnetic field envisage that the outer core of the earth, being liquid and an electrical conductor, acts as a 'self-exciting dynamo'; the precise mechanism is still far from clear, but it seems that some form of energy in the earth's interior is ultimately converted into electric currents which encircle the core and produce the dipole magnetic field. Most significantly, all the widely supported hypotheses demand a permanently dipolar field with an alignment approximately parallel to the axis of rotation.

Remanent magnetization

When an igneous rock containing iron-rich minerals cools from the molten state it becomes magnetized in accordance with the prevailing geomagnetic field. Laboratory investigations show that most of the magnetization is acquired at a critical temperature, commonly around 550°, known as the Curie point of the rock. So long as the rock is not reheated to near its Curie point, the acquired magnetization, although very weak, is remarkably stable and may be retained for hundreds of millions of years. In other words, when an iron-rich lava cools through the Curie temperature a permanent record is frozen into the material. The record is contained in the thermal remanent magnetization (TRM) of the rock, its strength varying according to the mineralogical composition of the rock and the intensity of the geomagnetic field at the time of cooling. Modern remarkably sensitive magnetometers make it possible to measure not only the TRM of igneous rocks but also the detrital remanent magnetization (DRM) of many sedimentary rocks. As detrital particles settle through water, any magnetic fragments tend to align themselves in conformity with the ambient magnetic field. In this way iron-bearing grains assume a preferred orientation that confers a weak DRM on the sediment and ultimately on the resulting rock. This is sometimes destroyed by later chemical processes so that a sedimentary rock with a weaker and less stable magnetization presents greater problems of interpretation than an igneous rock; nevertheless, with adequate precautions in sampling and measurement it may still yield much valuable information.

The basis of palaeomagnetic reconstruction

When a rock sample is selected for palaeomagnetic study, its orientation is first carefully measured and any necessary compensation for tectonic disturbance calculated. Subsequent analysis in a magnetometer can theoretically specify the magnitude, declination and inclination of the local force at the time the remanent magnetization was induced. On the assumption that the total geomagnetic field was dipolar in form it is then possible to reconstruct the contemporaneous positions of the magnetic poles. Of course there are several sources of potential error. Any allowance for tectonic disturbance is often difficult to assess to an accuracy better than ±5°. Secular fluctuations distorting the pure dipolar field place an additional constraint on the precision of the reconstruction. In practice the remanence of several samples of the same age may be measured, calculations made for individual samples and then statistical procedures employed to define confidence limits for the positions of the poles.

Most early palaeomagnetic investigations were concerned with Cenozoic lavas in such countries as Japan, Italy and France. Then, in the 1950s, studies by a distinguished physicist, Blackett, on the Triassic sandstones of the British Isles revealed polar positions some 200 m.y. ago that were dramatically different from those of today. One obvious explanation, that the magnetic poles had slowly migrated over the earth's surface, led to attempts at drawing so-called polar

wandering curves. These involve plotting the apparent polar positions at different geological periods and joining the points by a smooth curve. When this was done for a single continent the consistent pattern that emerged seemed to support the polar wandering hypothesis. However, when a corresponding curve was calculated for a second continent it was found to differ from the first. It was this vital discovery that led to rejection of the polar wandering hypothesis and revival of continental drift as the only satisfactory explanation for the palaeomagnetic findings. If the continents have shifted their relative positions, all contemporaneous rocks from an individual continent should yield a unique polar position, whereas those from another continent should yield a different polar position. In essence this is what palaeomagnetic investigations found.

The assumed identity between magnetic and geographic poles, justified on theoretical grounds, does not in fact permit detailed reconstructions of continental movement. Whereas inclination defines palaeolatitude, declination does not define palaeolongitude; former positions with respect to some arbitrary datum such as the Greenwich meridian remain indeterminate. Nevertheless, relative displacements can be evaluated by drawing polar wandering curves. In theory, as long as two continental blocks are fused together, or not moving with respect to each other, they should yield an identical curve. Once they start drifting apart their individual curves should begin to separate. By this principle the predictions of any particular model of continental drift can be tested against palaeomagnetic observations. For instance, Wegener suggested that the major continental blocks once formed a supercontinent which he termed Pangaea (Fig. 3.2). He argued that this existed until at least Permian times since evidence for Permo-Carboniferous glaciation in South America, Africa, India and Australia required contiguity of these continents if the global ice-cover were to be kept to acceptable proportions. The final break-up of Pangaea was believed to have begun in Jurassic times, with the Atlantic Ocean gradually opening up as the Americas drifted westwards and the remnant of the supercontinent fragmenting into Europe, Africa, Asia, Australia and Antarctica. If Wegener's reconstruction of Pangaea was correct, all continents when fitted into their respective positions should indicate a single palaeomagnetic pole during late Palaeozoic times. In broad outline this proves to be the case, and although recent work has led to some modification in detail, Wegener's original thesis has stood the test of time remarkably

Fig. 3.2. The 'supercontinent' Pangaea as reconstructed by Wegener for Upper Carboniferous times.

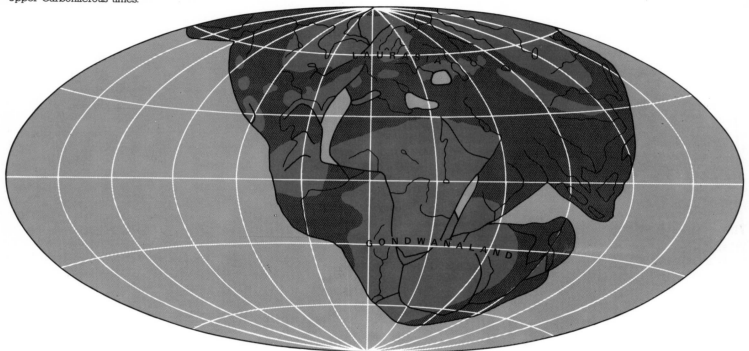

well. The polar wandering curves for individual continents begin to split in late Triassic times. This is most clearly seen in a comparison of curves for Africa, South America and North America. Australia and Antarctica appear to have moved away as a separate unit at about the same period, but did not split from each other until much later, possibly as recently as early Cenozoic time. This was also the date of the final and very rapid northward movement of India.

Reversals of polarity

When, in 1906, the French worker, Brunhes, discovered that a certain Cenozoic lava was reversely magnetized the initial reaction was to interpret the result as a freak and to invoke a 'self-reversal' mecha-nism by which a rock might become magnetized in the opposite sense to the ambient field. Laboratory experiments showed that such self-reversal requires quite exceptional circumstances, and when later workers found reversed magnetization to be almost as common as normal it was obvious that the geomagnetic field must have changed. This was confirmed when radiometric dating proved the known field reversals to be synchronous all over the globe. Detailed analysis of the transition period suggests that, prior to a reversal, a rapid decline in geomagnetic intensity is accompanied by growing instability of the poles until, quite abruptly, they move to approximately antipodal positions and the field progressively regains its normal strength. The whole process appears to take about 10 000

Fig. 3.3. The chronology of polarity reversals: (A) reversals during the last 165 m.y. together with an indication of distinctive anomalies used for dating the ocean floors (Fig. 3.6); (B) details of the reversals during the last 5 m.y.; (C) an illustrative core showing the nature of the evidence that may be derived from this source.

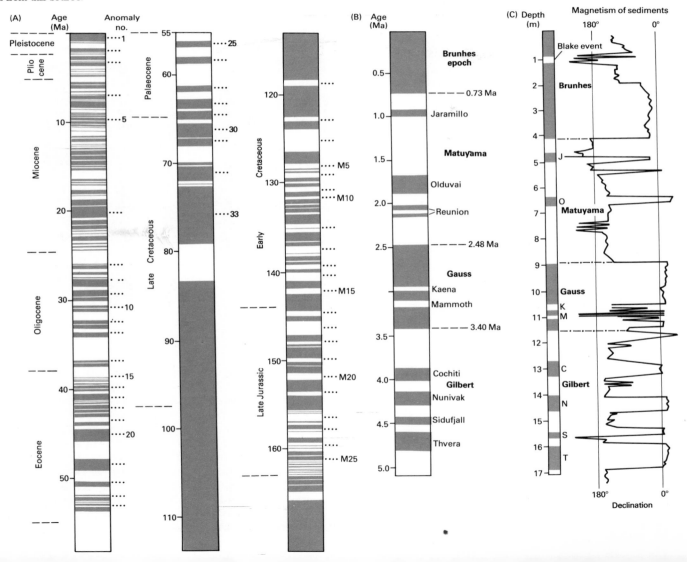

years, with the actual movement of the poles being completed in less than 5 000 years.

Research has now disclosed well over 100 polarity reversals during the Mesozoic and Cenozoic eras. A numbering system has been devised to cover the whole of this period (Fig. 3.3), but for the upper Cenozoic a more elaborate classification based upon names is in common use. The last 5 m.y. are divided into 'epochs', each dominated by either reversed or normal polarity, lasting about 10^6 years, and accorded the name of a famous research worker in the field of geomagnetism. Within each epoch are shorter 'events' with an approximate duration of 10^4–10^5 years, while even briefer episodes during which the poles migrate for 10^2–10^3 years as far as the opposite hemisphere are designated 'excursions'.

The magnetic polarity record, calibrated by radiometric dating, has proved of immense value. Owing to the number of reversals an isolated rock sample can rarely be dated on the basis of its remanent magnetization alone. However, where long sequences are available, and especially if the sequence continues up to the present day, correlation with the established time-scale is frequently possible. The following section on the oceans affords one such example, while Fig. 3.3c illustrates another example. A deep-sea core provides a continuous sedimentary sequence that is often most easily dated by reference to changes in its remanent magnetization. Correlation is possible not only between marine cores from widely different areas, but also with continental sequences that themselves retain a similar geomagnetic record. Identification of the Brunhes–Matuyama boundary has proved particularly significant in many different contexts (see, e.g. Fig. 2.3), and there seems little doubt that magnetostratigraphy will continue to play a major role in late Cenozoic dating.

Variations in magnetic intensity over the oceans.

In the late 1950s a routine magnetic survey of the north-eastern Pacific by a research team from the Scripps Institute at La Jolla, California, disclosed a remarkable pattern of linear zones, each trending approximately north–south and averaging 30–50 km in width, in which the intensity of the geomagnetic field is alternately above and below average (Fig. 3.4). The interest that was aroused stimulated a similar survey in the western Atlantic where an analogous magnetic

striping was discovered even though no such orderly arrangement is present over the intervening continent. At first the origin of the linear patterning defied explanation, but in a famous paper in 1963

Fig. 3.4. The linear pattern of magnetic anomalies as first detected over the north-eastern Pacific by Raff and Mason (Geol. Soc. Am. Bull. 72, 1961).

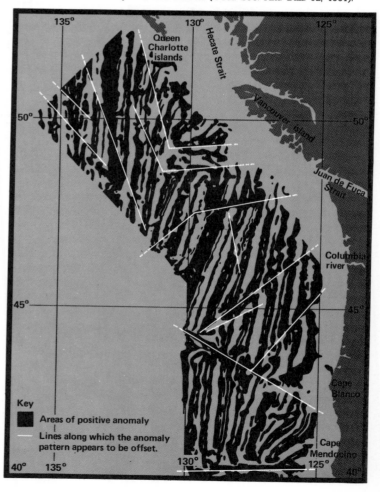

Key

■ Areas of positive anomaly

— Lines along which the anomaly pattern appears to be offset.

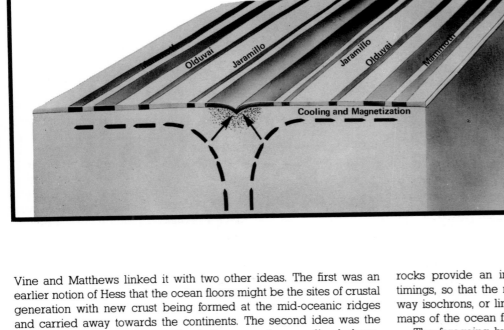

Fig. 3.5. Schematic diagram of the explanation offered by Vine and Matthews (1963) for the linear magnetic anomalies over the oceans.

Vine and Matthews linked it with two other ideas. The first was an earlier notion of Hess that the ocean floors might be the sites of crustal generation with new crust being formed at the mid-oceanic ridges and carried away towards the continents. The second idea was the then relatively new concept of polarity reversals, as outlined above. Vine and Matthews suggested that, if the hypothesis of sea-floor spreading were correct, hot new crust might be imprinted with the contemporaneous magnetic field as it cooled through the Curie point at the mid-oceanic ridge (Fig. 3.5). The linear striping would then be attributable to periodic changes in polarity since crust that had cooled during an epoch of normal polarity and subsequently moved away from the ridge would yield a zone of positive anomalies, while reversely magnetized crust would yield a zone of negative anomalies; the anomaly patterns on either side of a mid-oceanic ridge should form a mirror image of each other, as they indeed do. The whole process has been likened to a giant tape-recorder in which the tapes move away from the central ridge and, as they do so, pick up messages about the nature of the geomagnetic field. Continental rocks provide an independent record of the messages and their timings, so that the movement of the 'tapes' can be deduced. In this way isochrons, or lines joining points of equal age, can be drawn on maps of the ocean floors (Fig. 3.6).

The foregoing hypothesis suggests two important inferences that may be drawn from the pattern of anomalies. Firstly, the age at which local sea-floor spreading began can be assessed. As an illustration, the oldest anomaly in the North Atlantic between Norway and Greenland is number 24 which indicates that section of ocean began to open up under 60 m.y. ago. Secondly, rates of sea-floor spreading can be computed. This was first attempted after an aeromagnetic survey in 1963 of the Reykjanes ridge southwest of Iceland where each flank was estimated to have moved at about 10 mm yr^{-1} for the last 4 m.y., that is, at a total spreading rate of 20 mm yr^{-1}. Since then all the mid-oceanic ridges have been surveyed and comparable spreading rates calculated. The greatest rate, 160 mm yr^{-1} for the last 9 m.y., has been found along the East Pacific ridge between the equator and 30°S. Here, as elsewhere, the rates of opposing

Fig. 3.6. The age of the Atlantic floor as deduced from the pattern of magnetic anomalies.

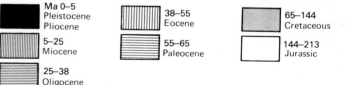

Ma 0–5 Pleistocene Pliocene	38–55 Eocene	65–144 Cretaceous
5–25 Miocene	55–65 Paleocene	144–213 Jurassic
25–38 Oligocene		

flanks calculated over periods of millions of years appear to be the same, but for briefer intervals there may be detectable asymmetry in the spreading.

Further tests of sea-floor spreading

A corollary of sea-floor spreading is that the oceanic crust becomes progressively older as distance from the mid-oceanic ridge increases. Two simple tests of this proposition have been devised (Fig. 3.7). Firstly, if oceanic islands, mainly composed of basaltic outpourings, are formed close to a spreading centre and then slowly transported away by the moving crust, the youngest should lie closest to the central ridge and the oldest furthest from it. Using the date of the oldest known rock on each island in the Atlantic as an indication of the age of the island itself, Wilson in 1963 showed that this relationship generally holds true. Secondly, the age of the earliest sediments on the sea floor should increase with distance from the spreading centre. In general, the DSDP proved this relationship to apply. Over by far the greatest proportion of the ocean floor the basal sediments are dated as either Cretaceous or Cenozoic, with Jurassic material confined to a very few marginal localities. The sedimentary cover also displays an expectable increase in thickness away from the mid-oceanic ridges, although this could in part be ascribed to the greater availability of eroded material near the continental margins. However, even in the western Pacific where the abyssal floor is separated from Asia by deep trenches the sediment thickness still appears to increase with distance from the mid-oceanic ridge.

Heat-flow measurements

Early observations in mines revealed an average increase in rock temperature of about 30 °C for every kilometre of descent. As it was believed that radioactive minerals in the continental crust provided the main heat source, it was argued that the heat flow beneath the oceans was likely to prove much smaller. Since the early 1950s, however, techniques have been available for the direct measurement of heat flow through the ocean floor, and thousands of such measurements have now been made by forcing a probe equipped with thermistors into the oceanic sediments; in addition, temperatures have also been measured in the holes drilled by the *Glomar Challenger*. The results have totally confounded the earlier predictions. On

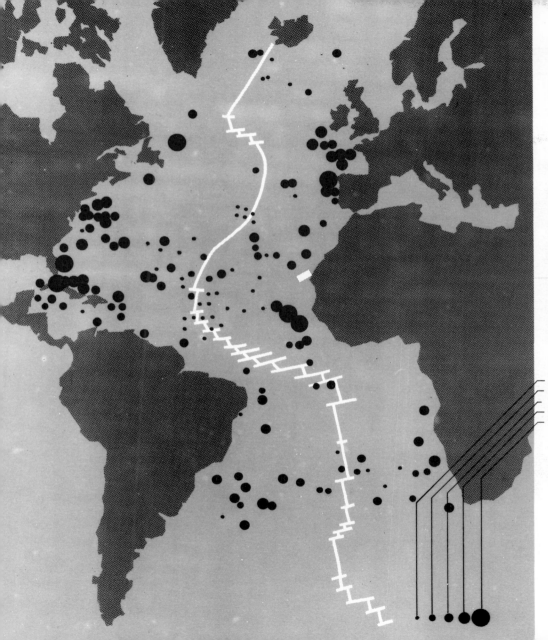

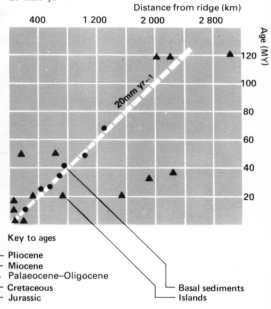

Fig. 3.7. The age of basal sediments recovered from deep-sea cores. The graph shows the age of the Atlantic islands expressed as a function of distance from the mid-oceanic ridge. Also shown is the age of the basal sediments recovered on an early leg of the DSDP. The dashed line corresponds to a spreading rate for each flank of 20 mm yr⁻¹.

Distance from ridge (km)

400   1 200   2 000   2 800

Age (MY)

20mm yr⁻¹

Key to ages

— Pliocene
— Miocene
— Palaeocene–Oligocene
— Cretaceous
— Jurassic

— Basal sediments
— Islands

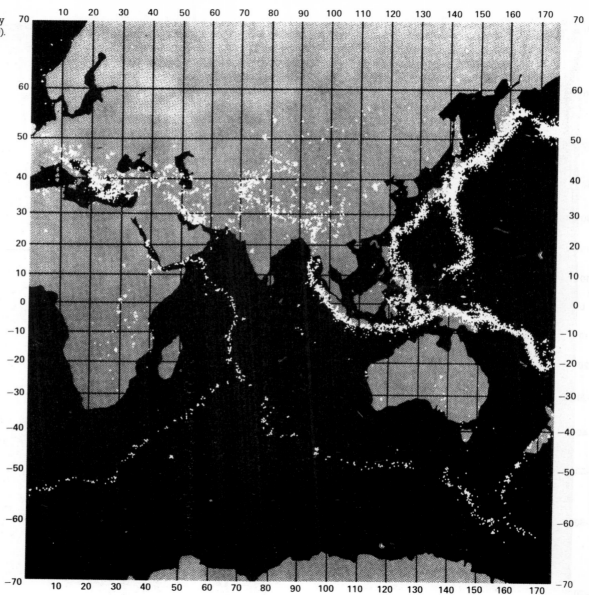

Fig. 3.8. Areas of current seismicity (after Barazangi and Dorman, 1969).

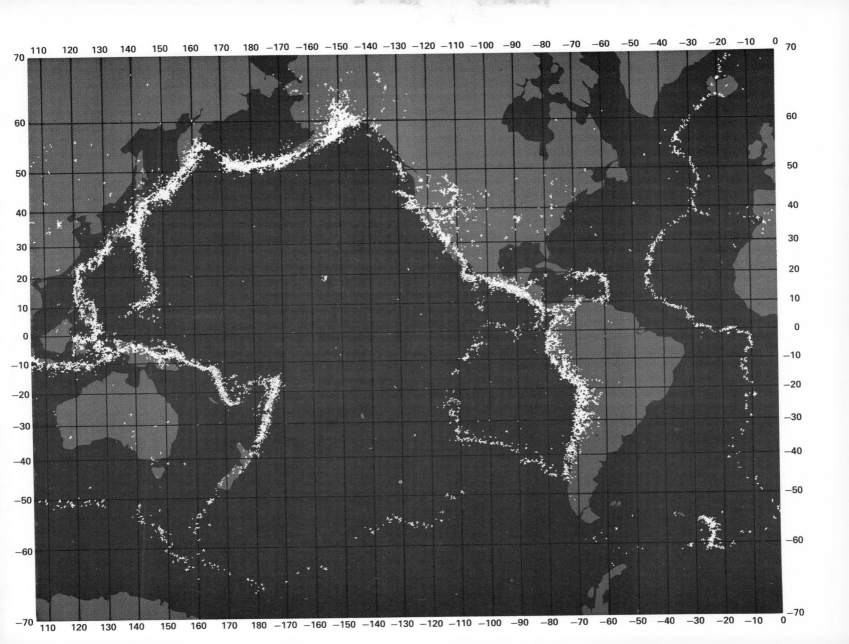

average the geothermal flux over the oceans has proved to be about 70 per cent higher than that over the continents, the figures being 95 and 55 mW m^{-2} respectively. In both types of area substantial regional differences exist. Over ancient shields such as characterize Brazil, Arctic Canada and European Russia a typical value is about 45 mW m^{-2}, while in zones of more recent volcanicity or mountain building the corresponding figure is around 70 mW m^{-2}; it is now estimated that, of the heat observed at the surface, some 70 per cent is derived from the mantle rather than shallow radiogenic sources with the implication that the temperature differences at the Moho may amount to several hundred degrees. Within the oceanic areas even greater contrasts exist. As might be expected from the hypothesis of sea-floor spreading, maximum values are normally observed over the mid-oceanic ridges where figures in excess of 100 mW m^{-2} are not uncommon. High values are also encountered on the continental side of major island arcs, whereas values in the order of 40 mW m^{-2} are typical of abyssal plains and deep-sea trenches far removed from any spreading centre. One estimate based upon these figures is that within the upper mantle at a depth of 100 km there may be a temperature difference of 400 °C between areas covered by continents and those covered by new oceanic floor.

Many authorities have seen in these results confirmation of a very old idea, that of thermal convection currents within the mantle. They envisage hot currents rising beneath the mid-oceanic ridges, with the trenches sited along zones of convergence where material in the convective cells begins to turn down; friction along the upper side of a descending limb is suggested as the heat source for the island arcs. Uncertainty still surrounds the precise nature of the energy source for the convective cells, and the depth within the mantle at which the cells operate, but widespread acceptance of the basic idea has focused attention on details of heat flow at mid-ocean ridges. The results have been striking. At −2 000 m on the floor of the Red Sea, believed to be a relatively new spreading centre, metalliferous brines with a salinity of up to 257‰, and a temperature of up to 62 °C have been located in a series of enclosed basins. Sea-water percolating through recently emplaced basalts is thought to be not only heated but also made chemically very active by magmatic gases so that, as it rises seawards, it leaches such metals as zinc, iron, manganese and lead from the fresh igneous rock together with salts from any earlier sediments. Near the East Pacific Rise ocean-floor materials show an analogous metallic enrichment, again attributable to precipitation from sea-water that had earlier percolated along contraction cracks in cooling basaltic magma. Yet the most spectacular discovery of such hydrothermal activity came in the late 1970s when the submersible *Alvin* was used to investigate parts of the East Pacific Rise. Near Baja California, for instance, water at over 350 °C was found jetting from a line of vents surrounded by mineralized chimneys up to 5 m high. Near by, cooler water at about 20 °C welled from springs that supported a unique community of clams, mussels, limpets, crabs and giant tube worms. This is a quite extraordinary environment when it is remembered that only a few metres away the temperature of the basal oceanic water is only 1.8 °C.

Seismic studies

The analysis of seismic waves has made a major contribution in the study not only of the internal constitution of the earth (p. 14) but also of the fracture zones within the crust and upper mantle. By plotting all earthquake epicentres over a period of several years, it is possible to identify the main areas of current fault activity (Fig. 3.8). Seismicity is a highly localized phenomenon. Most epicentres lie in a circum-Pacific belt, with a secondary concentration extending from Indonesia through south-central Asia to the Mediterranean regions of Europe. Within the oceanic areas there is a striking localization along the mid-oceanic ridges. In addition to indicating the distribution of active faults, careful analysis of modern seismograph traces allows the sense of motion along individual faults to be determined. The method depends on the first displacements for a single earthquake being at some stations compressional, and at others dilatational. These two types of first-motion are found to be arranged in quadrants separated by two mutually perpendicular planes (Fig. 3.9). It can be shown that one of the latter must correspond to the fault plane, and it is usually easy to choose the more likely on geological grounds. The sense of relative displacement can then be inferred: it should be parallel to the fault plane, normal to the intersection of the nodal planes and towards the compressional quadrant. First-motion studies of this type have been particularly significant in elucidating displacements associated with two groups of features, namely transform faults and Benioff zones.

Transform faults

The presence of major linear offsets affecting the trend of mid-

Fig. 3.9. Schematic representation of first-motion studies. On the left the two-dimensional diagram depicts the compressional and dilatational quadrants, together with loops indicating the intensity of the P-waves, i.e. the maximum intensity is recorded 45° from the fault trace. On the right the distribution of the four quadrants is shown on the surface of the globe. In both cases there are two possible planes along which movement may have occurred; the choice is normally made on the basis of known geology.

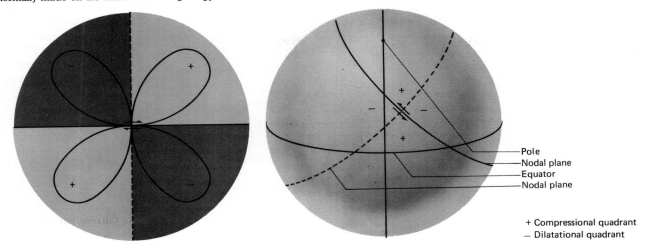

Pole
Nodal plane
Equator
Nodal plane

\+ Compressional quadrant
− Dilatational quadrant

oceanic ridges, first noted during bathymetric charting, became particularly obvious in the course of detailed magnetic surveys (Figs 3.4 and 3.6). Because the zones of magnetic anomaly vary in width and strength, the pattern on either side of the offsets can be matched and the amount of misalignment estimated; apparent displacements frequently exceed 10 km and not uncommonly surpass 100 km. Since they affect magnetic patterns formed in the immediate geological past, the dislocations must be currently active. Termed transform faults by Wilson in 1965, they should possess properties quite different from those of the ordinary transcurrent fault illustrated in Fig. 4.1. In the first place, there is no reason to believe that the separate segments of a ridge were ever in continuous alignment; the close matching of anomaly patterns is simply due to almost identical sequences being generated from each ridge segment. It follows that shearing should normally be confined to that part of the dislocation lying between the ridge axes, with the pace of relative displacement equal to the overall spreading rate. Moreover, movement should be in the opposite sense to that which would be predicted by analysis as a conventional transcurrent fault. Seismic studies have confirmed these properties. Movement is restricted to the inter-ridge section of the offsetting faults, is predominantly of a strike-slip nature, and the first-motion evidence is consistent with transform but not transcurrent faulting. Many other earthquakes associated with the mid-oceanic ridges occur along the axis of individual ridge segments and are predominantly of a dip–slip nature.

Benioff zones

On a sphere of constant size, creation of new crust along the mid-oceanic ridges must obviously be matched by destruction elsewhere (Fig. 3.10). There are many reasons for believing that the oceanic trenches mark such zones of crustal destruction, but probably the most convincing comes from plots of earthquake foci in the vicinity of the trenches. The foci define steeply sloping zones of earthquake activity which plunge at angles between 30° and 90° from the floors of trenches beneath the adjacent island arcs or continental margins. Named Benioff zones to honour the eminent seismologist who

Fig. 3.10 A schematic sketch showing the generation of new lithosphere at so-called spreading centres and its destruction at steeply dipping Benioff zones (after Isacks *et al.*, 1968).

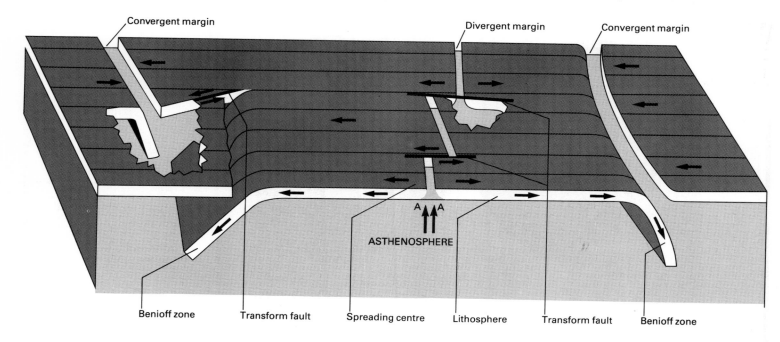

provided the first detailed description, these seismic zones are distinguished by being the location of the deepest recorded earthquakes, many having been detected at depths of over 500 km. First-motion studies indicate compression parallel to the dip of the Benioff zone, apart from a very shallow section near the trench where there appears to be tension. These results suggest that a rigid crustal slab moving away from the mid-oceanic ridge is first buckled downwards with consequent tension in its upper layers, and then plunges much deeper into the mantle with resulting compression. Lighter crustal material, possibly accompanied by deformed ocean-floor sediments, is carried down to considerable depths and helps explain the negative gravity anomalies recorded over most island-arc systems; erasure of the magnetic anomaly pattern at the trenches may be similarly explained by the heating of the descending slab to above its Curie point. The rate of underthrusting computed for various arc-trench systems has been shown to vary from about 20 mm yr^{-1} for the South Sandwich trench to over 100 mm yr^{-1} for the Tonga and Kermadec trenches.

Plate tectonics

Basic concepts

Earlier in this chapter reference was made to two far-reaching concepts that have had a profound effect upon the development of the earth sciences. The first is the idea of continental drift, and the

Fig. 3.11. The major lithospheric plates into which the surface of the globe is divided. Some of the boundaries are still tentative, especially in Asia, and there are numerous small 'platelets' not yet adequately defined. Compare with Fig. 3.8.

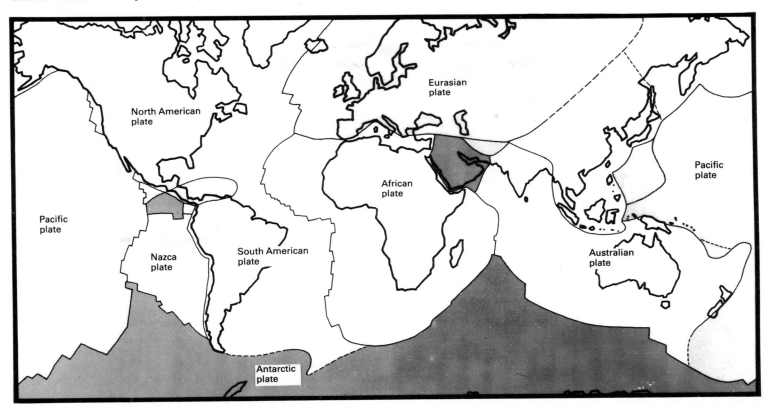

second that of sea-floor spreading. One arose from prolonged comparative study of the continents, the other from later investigations of the ocean floors. They are related but not identical ideas, and it is largely a fusion of the basic tenets of each that gave birth to the unifying concept of plate tectonics.

It is a fundamental postulate of much recent geophysical theory that the lithosphere is composed of a series of internally rigid plates (Fig. 3.11). Originally six major plates were identified, but it is now recognized that there must be at least a further twenty smaller plates

that may move independently of their larger counterparts. Each plate may or may not carry continental crust and there is no simple relationship between the pattern of the plates and the distribution of the continents. The Pacific plate, for instance, is almost entirely 'oceanic' and the Eurasian plate very largely 'continental'. Investigation of the nature and movement of the plates is the field of study known as plate tectonics. Much research is devoted to studying the margins of the plates for it is here that most seismic, volcanic and tectonic activity seems to be concentrated. Three contrasting types of plate margin are recognized, divergent, convergent and conservative. At a divergent or constructive margin new material is generated and, as already seen, this takes place along the mid-oceanic ridges. At a convergent or destructive margin plate material is consumed by subduction into the underlying mantle, normally along a Benioff zone. At a conservative margin two plates slip past each other with neither experiencing a change in surface area; a transform fault constitutes the obvious example of a conservative margin.

If the entire surface of the globe is covered with plates, most of which are accreting along one margin and being consumed along another, the pattern relative to any fixed frame of reference must undergo slow but almost constant modification. For example, at most convergent margins only one plate is actively subducted which means that the margin itself must migrate to accommodate the advance of the second plate. Equally the positions of many of the divergent margins must move relative to each other. This may be illustrated by the African plate which is flanked to both east and west by active oceanic spreading centres. The absence of any known convergent margin beneath Africa can be explained if the two mid-oceanic ridges are themselves moving apart. In other words, not only do the plates move but the pattern they form also varies with time. Evaluation of changes in the local pattern of plate boundaries, and of the relative velocities of the plates themselves, often involves analysis of a 'triple junction'. These are points at which three plates meet and are a necessary corollary of mobile accreting plates on a spherical surface. Given that there are three different types of boundary, it has been calculated that there are no less than sixteen possible combinations of ridge, trench and transform fault at a triple junction, of which at least six are of common occurrence. Still further complexity is introduced by the occasional spreading centre which, once established, nevertheless ceases to function and is replaced by a new centre near by (see, e.g. Fig. 3.12). It is obvious that, despite the seeming simplicity of the basic concept of plate tectonics, it is ultimately endeavouring to describe an extremely complicated global system.

Plate geometry

There are certain geometrical controls which must govern the movement of plates. For instance, the passage of a rigid plate from one position to another must always be describable in terms of a simple rotation about an axis passing through the centre of the sphere. Each point on the plate then follows a small circle path about the pole position of that rotation axis. This means that in the case of a conservative margin, where material is neither created nor destroyed, the motion of the plates must be along small circle routes around the pole of rotation. A further consequence of the geometrical properties of the sphere is that the rate of relative movement at divergent and convergent margins must vary as the sine of the angular distance from the rotation pole; this is exactly analogous to the variations in velocity at different geographical latitudes on the rotating earth. An early test of the idea of rigid plates moving over the earth's surface was made by comparing these geometrical requirements with observational data from the mid-Atlantic ridge. The pole of rotation was first defined by drawing great circles normal to the known transform faults. Observed spreading rates were then analysed to see whether they exhibit the expected variation with angular distance from the rotation pole. Given the uncertainties involved, the observational data accorded remarkably well with the geometrical predictions. Since then poles of rotation have been determined for all the main lithospheric plates as they appear to be moving at the present time.

It is relatively easy to calculate the rotations that would be necessary to move the American, African and Eurasian continental blocks from their respective pre-drift locations in Pangaea to their modern positions. This clearly does not define the actual routes taken since the continents could, for instance, have followed zigzag paths with many successive rotations over short angular distances. With increasing information about the age of the ocean floor it became possible to determine in much greater detail the likely movements of the plates (Fig. 3.12). It has been claimed that a minimum of six poles of rotation is needed to simulate the movements of the North American and Eurasian continental blocks since late Mesozoic times. Differing angular velocities about these poles produced variations in the rate of sea-floor spreading and continental separation. Whether the rate also varied during the operation of a single pole of rotation is less certain since modern work has tended to cast doubt on earlier claims that both the Atlantic and Pacific floors exhibit major changes in spreading rates during the last 10 m.y.

One reason for the widespread acceptance of plate tectonics has been its predictive capacity, a point well illustrated by the history of

Fig. 3.12. The opening of the North Atlantic based upon interpretation of magnetic anomalies (after Laughton, 1972). Spreading centres operative at each period are shown and it is notable how these have periodically changed their location.

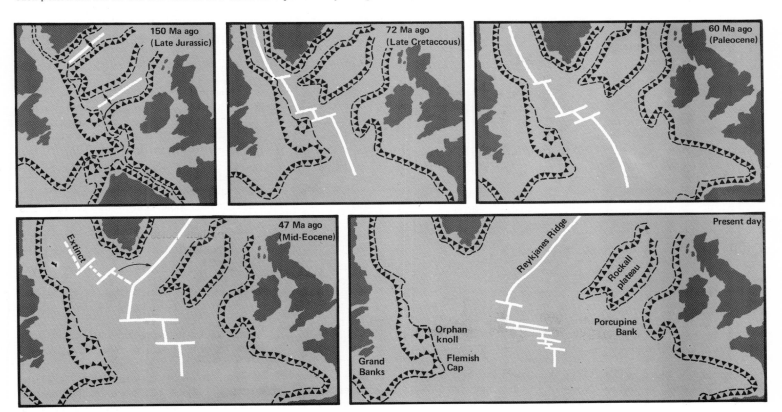

research into the opening of the Atlantic Ocean. In the mid-1960s a computer was used to ascertain the mathematically best solution to the original fit of the continents bordering the Atlantic. It was then discovered that the fit could be greatly improved if it were assumed that the Iberian peninsula had undergone a counter-clockwise rotation after the splitting from the Americas had begun. Apart from some early palaeomagnetic data there was at that time no direct evidence that such a rotation had occurred, but in the 1970s investigations in the Bay of Biscay disclosed a pattern of magnetic anomalies indicative of an old spreading centre that had functioned for about 10 m.y. in late Mesozoic times. In other words, at this latitude a temporary triple junction had developed on the mid-Atlantic ridge and had forced Iberia to rotate in a counter-clockwise direction.

Matters of dissent and controversy

The concept of lithospheric plates, incorporating many aspects of the earlier continental drift and sea-floor spreading hypotheses, appears to provide a convincing if partial explanation of the gross relief features of the globe. As the preceding section illustrates, it offers the potentiality for reconstructing the evolution of the continents and ocean basins in very considerable detail. Yet it would be unwise to omit all reference to dissentient views, even though the number of 'fixists' who deny the reality of continental drift now seems extremely small. Such critics can point to a long list of assumptions upon which many interpretations in plate tectonics have to be based since no direct observations of deep-seated processes are possible. Moreover, these same critics claim to discern an element of data selection in certain palaeomagnetic investigations. For example, many rocks have undergone a secondary magnetization which needs to be removed in the laboratory if the most stable magnetization is to be revealed. This most stable direction is then compared with other observations and, if consistent, is normally deemed primary in origin. Such a procedure can, of course, lead to circularity of argument since only those results conforming with the initial hypothesis are regarded as satisfactory. There can also be ambiguities in the interpretation of DSDP oceanic cores. The basal sediments, for instance, are occasionally baked and therefore older than the underlying rock. In such cases it is possible that the drilling has terminated in a sill rather than in true basement rock which consequently has not been dated. Yet these and other points raised by opponents of plate tectonics certainly do not justify rejection of the whole concept.

One weakness of Wegener's original formulation of continental drift was a failure to identify any adequate cause for the movement.

It is now generally accepted that, if the plate model is correct, displacement within the mantle is required. A few workers have argued that the inferred properties of the mantle preclude any form of convection cell, but most have seen in the lower viscosity of the asthenosphere an indication of suitable conditions. Yet it remains a fact that so far no model of convective cells commands general assent. Some authorities envisage the cells affecting the whole thickness of the mantle, while others believe that cells are more likely to be restricted to the upper 700 km within which earthquakes are known to occur. Nor should it be assumed that drag by the cell is entirely responsible for displacement of the crust; calculations suggest that a 'gravitational push' owing to elevation and wedging apart at the mid-oceanic ridge, combined with a 'density drag' owing to cooling of the plate on its way towards the subduction zone, are probably more important in maintaining the crustal movement.

Finally, another source of controversy is the likelihood of long-term changes in the overall dimensions of the globe. Hypotheses of both contraction and expansion have found limited support at different times in the past, but their advocates have never mustered sufficient evidence to convince the many sceptics. A change in the size of the earth is not necessarily in conflict with the main thesis of plate tectonics, and in any case as far as the geomorphologist is concerned it is the internal energy system as it has been working over the last few million years that is of primary importance, not hypothetical longer-term changes such as are envisaged by proponents of an expanding or contracting earth.

References

Barazangi, M. and **Dorman, J.** (1969) 'World seismicity maps compiled from ESSA, Coast and Geodetic Survey, Epicenter data, 1961–1967', *Seism. Soc. Am. Bull.* **59**, 369–80.

Chase, C. G. (1978) 'Plate kinematics: the Americas, East Africa and the rest of the world', *Earth Plan. Sci. Lett.* **37**, 355–68.

Corliss, J. B. *et al.* (1979) 'Submarine thermal springs on the Galapagos rift', *Science N.Y.* **203**, 1073–83.

Davies, G. F. (1980) 'Review of oceanic and global heat flow estimates', *Rev. Geophys. Space Phys.* **18**, 718–22.

Edmond, J. M. and **von Damm, K.** (1983) 'Hot springs on the ocean floor', *Scient. Am.* **248**(4), 70–85.

Erickson, A. J. *et al.* (1975) 'Geothermal measurements in deep-sea drill holes', *J. Geophys. Res.* **80**, 2515–28.

Hekinian, R. *et al.* (1983) 'Intense hydrothermal activity at the axis of the East Pacific Rise near 13 °N: submersible witnesses the growth of a sulfide chimney', *Mar. Geophys. Res.* **6**, 1–14.

Isacks, B. J. *et al.* (1968) 'Seismology and the new global tectonics', *J. Geophys. Res.* **73**, 5855–99.

Laughton, A. S. (1972) 'The southern Labrador Sea – a key to the Mesozoic and early Tertiary evolution of the North Atlantic', in *Initial Reports of the Deep Sea Drilling Project*, vol. 12, Nat. Sci. Found.

Mason, R. G. and **Raff, A. D.** (1961) 'Magnetic survey off the west coast of North America, 32 °N latitude to 42 °N latitude', *Geol. Soc. Am. Bull.* **72**, 1259–66.

McElhinny, M. W. and **Embleton, B. J. J.** (1974) 'Australian palaeomagnetism and the Phanerozoic plate tectonics of eastern Gondwanaland', *Tectonophysics* **22**, 1–29.

Monin, A. S. *et al.* (1981) 'Visual observations of Red Sea hot brines', *Nature, London* **291**, 222–5.

Morgan, W. J. (1968) 'Rises, trenches, great faults and crustal blocks', *J. Geophys. Res.* **73**, 1959–82.

Norton, I. O. and **Sclater, J. G.** (1979) 'A model for the evolution of the Indian Ocean and the break-up of Gondwanaland', *J. Geophys. Res.* **84**, 6803–30.

Phillips, J. D. and **Forsyth, D.** (1972) 'Plate tectonics, palaeomagnetism and the opening of the Atlantic', *Geol. Soc. Am. Bull.* **83**, 1579–1600.

Pollack, H. N. and **Chapman, D. S.** (1977) 'Mantle heat flow' *Earth Plan. Sci. Lett.* **34**, 174–84.

Ries, A. C. (1978) 'The opening of the Bay of Biscay – a review', *Earth Sci. Rev.* **14**, 35–63.

Sclater, J. G., Jaupart, C. and **Galson, D.** (1980) 'The heat flow through oceanic and continental crust and the heat loss of the earth', *Rev. Geophys. Space Phys.* **18**, 269–311.

Spiess, F. N. *et al.* (1980) 'East Pacific Rise: hot springs and geophysical experiments', *Science N.Y.* **207**, 1421–32.

Sykes, L. R. (1967) 'Mechanism of earthquakes and nature of faulting on the mid-ocean ridges', *J. Geophys. Res.* **72**, 2131–53.

Vine, F. J. and **Matthews, D. H.** (1963) 'Magnetic anomalies over oceanic ridges', *Nature, London* **199**, 947–9.

Wegener, A. (1967) *The Origin of Continents and Oceans* (trans. J. Biram from 4th edn.) Methuen.

Wesson, P. (1972) 'Objections to continental drift and plate tectonics', *J. Geol.* **80**, 185–97.

Williams, C. A. (1975) 'Sea-floor spreading on the Bay of Biscay and its relationship to the North Atlantic', *Earth Plan. Sci. Lett.* **24**, 440–56.

Wilson, J. T. (1963) 'Evidence from islands on the spreading of ocean floors', *Nature London* **197**, 536–8.

Wilson, J. T. (1965) 'A new class of faults and their bearing on continental drift', *Nature, London* **207**, 343–7.

Selected bibliography

The recent literature on plate tectonics is vast. A particularly readable and well-illustrated account is provided by F. Press and R. Siever, *Earth* (3rd edn) Freeman, 1982. Less comprehensive but useful reviews are to be found in K. C. Condie, *Plate Tectonics and Crustal Evolution* (2nd end) Pergamon, 1982, and M. W. McElhinny *The Earth: its origin, structure and evolution*, Academic Press, 1979. By contrast, a collection of earlier dissentient views that still makes interesting reading is *Plate Tectonics – assessments and reassessments*, Am. Ass. Petrol. Geol. Mem. 23, 1974. A modern view of the displacement of the continents is portrayed cartographically in A. G. Smith and J. C. Briden, *Mesozoic and Cenozoic Paleocontinental Maps*, CUP, 1977.

Chapter 4
Localized movements of the continental crust

Simultaneously with the large-scale movements described in Chapter 3, the continental blocks are subject to important deforming stresses, many but not all of them directly attributable to a positioning at the margin of a lithospheric plate. It is to these more localized movements of the continental crust that the present chapter is devoted. Two fundamental causes of the movements may be distinguished. The first is the loading and unloading that results from such events as the growth and decay of an ice-sheet, or the impounding and drainage of a deep lake. During loading the crust is downwarped in the form of a large basin, while unloading is followed by progressive recovery. Deformation of this type is clearly predicted by the concept of isostasy. The second and much more common cause of crustal movement is tectonic in origin. It is due to deep-seated changes taking place either near the base of the crust or in the mantle. The nature of the changes cannot be observed directly but must be inferred from the effects seen at the surface. The most obvious are rock structures such as folds and faults which have long been the subject of study and classification by geologists (Fig. 4.1).

Classification of tectonic activity has generally stressed the distinction between the radial and tangential components of earth movement. Gilbert, for instance, emphasized the contrast between what he termed epeirogenic and orogenic activity. In the former, displacement was held to be primarily radial, allowing extensive areas of the crust to be uplifted or depressed with little internal deformation. In the latter the tangential component was held to be dominant, resulting in folding and dislocation of strata and ultimately, as the name implied, in the building of major mountain chains.

Beloussov, a Russian geophysicist, has argued that this simple twofold classification is unsatisfactory in a number of respects. In particular he underlined the importance of what he called oscillatory movement. This undulatory motion involves alternating elevation and depression without any permanent record of the displacements being imprinted on the local rocks. Working in Africa, on the other hand, King has argued that yet another type of movement needs to be distinguished, namely cymatogeny. This consists of broad arching of the continental crust, accompanied by the simultaneous generation of gneissose rock at depth. Uplift at the crest may amount to 1 000 m or more, while the overall width of the structure extends to hundreds of kilometres.

The ideas of Gilbert, Beloussov and King illustrate the considerable diversity of views that exist regarding the classification of earth movements. They also exemplify the various lines of evidence that may be employed in elucidating the nature of the movements. Gilbert laid stress on the evidence of geological structure: epeirogeny is attested by horizontal strata elevated high above their original level of accumulation, while orogeny is indicated by crumpled rocks that have clearly been subject to powerful deforming stresses. Beloussov, on the other hand, emphasized the need to investigate contemporary crustal mobility by modern survey techniques, while King stressed the importance of correctly interpreting surface morphology which he believes can provide one of the fullest guides to recent tectonic history. Although geological structure undoubtedly provides the best indication of the broad pattern of tectonic activity, it is essential for the geomorphologist to recognize that it relates specifically to the rocks and not to the surface of the globe. It is extremely difficult to translate the evidence of folded strata that were once deeply buried into, say, simple vectors of surface displacement. Both modern survey techniques and geomorphic studies have the advantage of dealing directly with surface form, but are severely limited in other respects. Although current instrumental surveys can attain remarkable standards of accuracy, the results often have to be compared with much older records of doubtful reliability. Moreover, the period that can be covered by this approach is so short that the question inevitably arises as to how representative it is of the vastly longer span of geological time. Geomorphic evidence falls into an intermediate category, lacking the precision of modern survey methods but covering a much longer time interval.

Fig. 4.1. Conventional classification of structures resulting from folding and faulting.

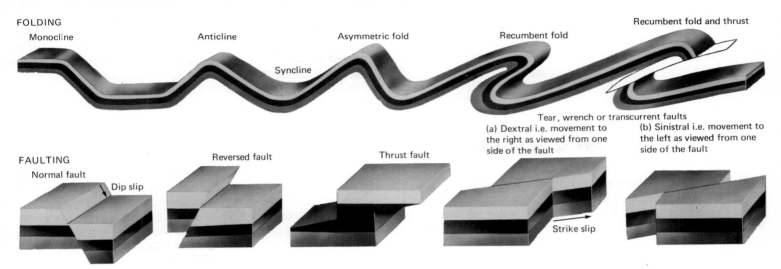

FOLDING

Monocline   Anticline   Asymmetric fold   Recumbent fold   Recumbent fold and thrust

Syncline

Tear, wrench or transcurrent faults
(a) Dextral i.e. movement to   (b) Sinistral i.e. movement to
the right as viewed from one   the left as viewed from one
side of the fault   side of the fault

FAULTING

Normal fault   Reversed fault   Thrust fault

Dip slip   Strike slip

Each of the major lines of evidence outlined in the preceding paragraph will now be examined in more detail. In practice this will involve considering successively longer periods of time; for contemporary movement, also known as neotectonics, a representative figure is up to 10^3 years, for geomorphic evidence it is 10^3–10^6 years, and for geological evidence it is over 10^6 years.

Evidence for contemporary crustal movement

Tidal gauge records

Detection of short-term changes in the relative level of land and sea requires accurate instrumentation, normally provided nowadays by continuous-recording tidal gauges. Some ports have very long histories of measuring tidal heights, but there are often difficulties in comparing old and modern records owing to changes in the type of instrument used. Even where acceptable modern records are available, interpretation demands considerable care. For example, collation of tidal data from many sites around the globe indicates that, during the present century, there has been a rise in sea-level averaging about 1 mm yr^{-1} so that this correction factor needs to be applied if tidal gauge data are to be used for assessing local land movements. It must also be recognized that sea-level is affected by a variety of factors that are neither obvious nor easy to evaluate. It is known, for instance, that sea-level fluctuates seasonally on many coasts owing to changes in wind and current direction. Given this sensitivity it is obviously hazardous to compute a global rate of sea-level change and unthinkingly attribute any local residual entirely to crustal deformation.

Despite these limitations, careful analysis of tidal data has revealed that many coastal areas are currently suffering displacement. The highest rates tend to be concentrated where the crust is still rebounding after removal of a massive ice-sheet. Thus the northern shores of the Gulf of Bothnia are rising at 9 mm yr^{-1}, parts of Alaska at over 10 mm yr^{-1} and areas of northern Britain at about 3 mm yr^{-1}. Such results are consistent with long-term trends that have been recognized for many years. More surprising is the rate of movement implied by gauges sited beyond the limits reached by the Pleistocene ice-sheets. The most detailed information comes from the United States. On the Atlantic seaboard south of New York much of the coastline appears to be sinking at a rate of at least 1 mm yr^{-1} with some segments exceeding a rate of 2 mm yr^{-1}. On the Pacific seaboard the tidal gauge at San Francisco is sinking while that at Crescent City in northern California is rising, the records from the two sites implying a differential movement since 1940 of about 3 mm yr^{-1}. Viewed in the context of a human lifetime all these rates may seem of relatively small consequence, but if sustained for a prolonged geological period each would produce major topographic changes.

Archaeological evidence

Coastal settlements often provide clues to sea-level change over a rather longer interval than that so far considered. A harbour built at Torne on the Gulf of Bothnia in 1620 had to be abandoned within a century because of the shoaling of the approaches. Along the coasts of the Aegean several settlements have been totally submerged by subsidence of the land. However, the best-known example of a town affording evidence for rapid changes in level is Puteoli on the Bay of Naples. Much of the original building is dated to the first century AD when the site obviously stood above sea-level. Markings left on marble columns by rock-boring marine molluscs indicate a submergence prior to 1500 of over 6 m. Yet it is known that by the mid-sixteenth century the locality was undergoing uplift and it is possible that total emergence followed. Beginning in the early nineteenth century subsidence again became the dominant movement; in 1828 the base of the columns was about 0.3 m below sea-level, but by 1950 this figures had increased to over 2 m. In the 1970s the pattern of movement dramatically reversed. In three years a rise in ground-level of 1.7 m was recorded, and after a pause, renewed elevation in the early 1980s totalled 1.1 m in 2 years. This uplift was accompanied by almost continual earthquakes, and the history of Puteoli affords a noteworthy indication of the rapidity with which uplift and subsidence can follow each other, at least in a region of active vulcanicity.

A subsidence of 4 m or more during the last two millennia has been claimed for south-eastern England on the basis of Roman remains found well below high-water mark in the London area. However, such evidence needs to be treated with caution since factors other than true crustal subsidence might explain many of the observed relationships. In 1972 Akeroyd plotted data from some twenty-eight sites in eastern England between the north Kent coast and the Fenlands (Fig. 4.2) and concluded that, during the last few millennia, the region has probably subsided but only at an average rate of 1 mm yr^{-1}. This secular downwarping is one of the factors contributing to the need for the Thames barrier, constructed in the 1980s, to protect London from flooding.

Geodetic or high-precision levelling

There is no single definition of what constitutes a high-precision standard of levelling, and as instruments have improved more stringent requirements have been laid down. In the 1950s some surveys specified that closing errors on traverses should not exceed $4.0\sqrt{K}$ mm, where K is the length of the traverse in kilometres. This would require the closure on a 25 km traverse to be 20 mm or less to avoid the need for a resurvey. Such a standard is now almost routine and often an accuracy of better than $1.0\sqrt{K}$ mm is attained. By surveys of this precision repeated at intervals of several years quite small changes in surface elevation can be detected, and numerous countries have now established special units charged with the task of measuring 'recent crustal movement'. Additionally, in many critical localities special instruments have been installed to provide a continuous monitoring of surface deformation. The most common instrument is the tiltmeter, of which there are several different designs; when properly positioned in a deep tunnel, mine or borehole this can detect a tilt of only 1 microradian, that is, less than 0.0001°.

There remains the problem of interpreting any measured displacement since by no means all changes in surface elevation represent deep-seated crustal movements. Tiltmeters often record short-term cyclical fluctuations due to such natural phenomena as tidal pull, variations in groundwater level and transient meteorological phenomena. Longer-term changes may have an anthropogenic origin. Subsidence due to mining can easily exceed 1 m in only a few decades, and an analogous surface depression often follows extraction of oil, gas or water. In California pumping of irrigation water lowered parts of the Central Valley by as much as 7 m between 1943 and 1964, while further south abstraction of oil and gas

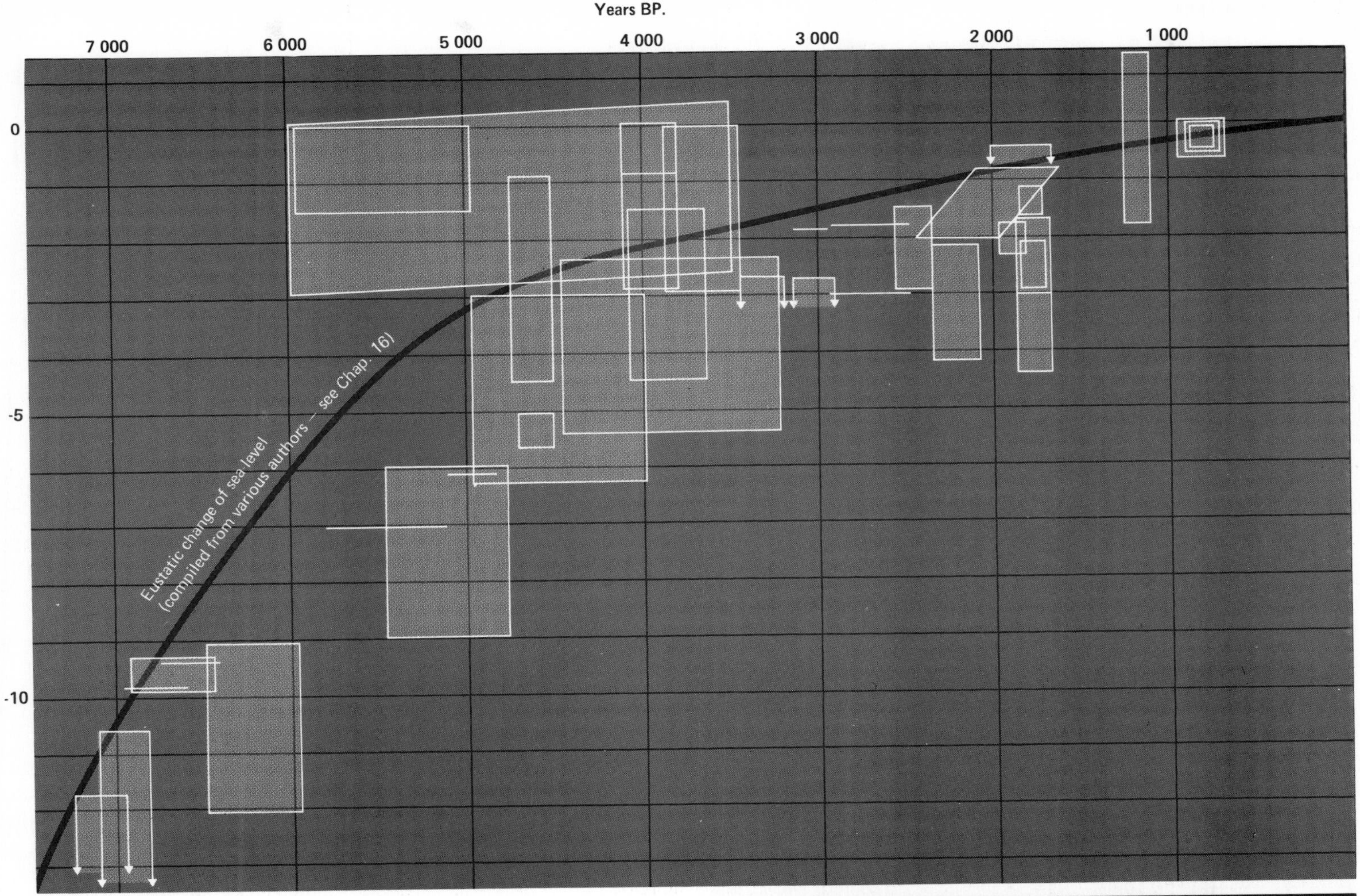

Fig. 4.2. A plot of archaeological and geomorphological sites in south-eastern England yielding estimates of sea-level during the last 7 000 years (after Akeroyd, 1972). Also shown is a curve depicting world-wide sea-level changes in the same period. The concentration of symbols below the eustatic curve suggests that the whole region has probably been subject to slight subsidence.

Key Each box or horizontal line represents a site where an estimate of former mean sea-level has been made. The dimensions of the symbol indicate the uncertainty attaching to the estimate:

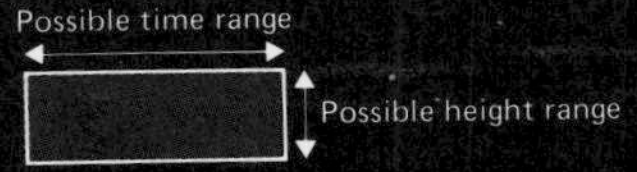

Possible time range

Possible height range

Symbols with arrows are used for sites where the maximum, but not the minimum, possible height of mean sea-level may be estimated. Where the age is based upon a C14 date, the potential time range is indicated as one standard deviation.

around Los Angeles harbour caused a basin-shaped depression about 7 km across and 8 m deep at its centre. The previously rapid subsidence at Los Angeles was halted by pumping several million litres of sea-water per day back into the porous rock to replace the oil and gas being removed. In Europe the potentially disastrous sinking of Venice is attributable to a combination of groundwater pumping, which is now strictly controlled, compaction of geologically young sediments, and true tectonic downwarping around the northern margin of the Adriatic Sea. The weight of impounded water behind large dams, and even the weight of closely spaced high-rise city buildings, has been shown to cause local depression of the earth's surface.

Even after due allowance for these anthropogenic factors, there remains abundant evidence for deep-seated crustal movement. The idea of using accurate levelling to monitor natural changes in surface elevation is an old one, but was first employed on a large scale during the late nineteenth century. Following a disastrous earthquake in 1891 the Japanese instituted a programme of systematic levelling aimed at establishing the nature and rate of crustal movement. In 1923 Tsuboi reported on the evidence that had accumulated in central Japan. He found that uplift was taking place here at an average rate of 5 mm yr^{-1}, with values ranging locally from 80 mm to 0.75 mm. More recent investigations have not only confirmed the general magnitude of these figures, but also demonstrated a highly complex mosaic of small blocks all rising at different and irregular rates. Two illustrative examples may be briefly quoted. For over 50 years parts of the Niigata region in western Honshu rose at about 2 mm yr^{-1}, but in 1955 the uplift gradually accelerated to 10 mm yr^{-1}. After 1959 the upheaval slowed down and even turned into a slight fall; following a major earthquake in 1964 a subsidence of about 100 mm was recorded. The second example concerns a remarkable series of tremors at Matsushiro in central Honshu during 1965–66. In the course of many thousands of tremors, several different types of physical change were noted. One of the most prominent was erratic uplift and tilting of the land surface. By careful analysis of the changes taking place it proved possible for several months to issue public forecasts of likely earthquake intensity. Particularly violent phases in April and August 1966 were successfully foretold in this way. The importance attached to the evidence of levelling in Japan is reflected in plans to resurvey traverses totalling 20 000 km at regular 5-year intervals.

The pioneer work in Japan stimulated surveys in many other parts of the world, most frequently in regions of active seismicity where such research has obvious practical value. In the USSR, for instance, repeated surveys of part of the Caucasus have disclosed marked differential displacement, with certain blocks being elevated at about 4 mm yr^{-1} and a number of marginal basins subsiding at about the same rate. The pattern is very reminiscent of that recorded in Japan, but the rates of movement are on the whole rather lower. Some of the most striking results of geodetic levelling come from California where over 80 000 km of repeated traverses have defined quite clearly regions of both elevation and depression. Repeated surveys in the Los Angeles area have disclosed considerable variations in the rates at which local hill ranges are being uplifted; among the fastest is the Tehachapi (for location see Fig. 4.5) which is rising at 15 mm yr^{-1}. Even more dramatic was the discovery of the 'Palmdale Bulge' on the San Andreas fault which was at first thought to have been uplifted by 200 mm in little more than a year, possibly as precursor to a major earthquake. However, no earthquake followed and some doubt has been cast on the accuracy of this particular measurement. Yet such is the precision of modern instruments that the interval between surveys can be reduced to less than a year and reliable results still obtained. Fifteen separate traverses across southern California have detected annual vertical displacements ranging from zero to 26 mm. The exceptionally high figures at one or two sites suggest movement is probably pulsatory and that decadal averages may conceal a very episodic motion.

As might be expected, severe earthquakes are often accompanied by measurable surface displacements close to the fault trace. Following the San Fernando earthquake in California in 1971 parts of the San Gabriel Mountains were found to have been abruptly elevated by as much as 2.3 m. Near Valdivia in Chile it was a fortunate coincidence that a section of the Inter-American Highway had been surveyed shortly before the whole area was shaken by an exceptionally large earthquake in 1960. A new survey showed part of the route had been depressed by 2.3 m, while immediately to the west displaced shoreline features such as the upper limit of mussel and seaweed growth registered an uplift of 5.7 m. In Algeria instrumental levelling after the El Asnam earthquake of 1980, one of the largest ever recorded in north-western Africa, showed elevation along a thrust fault amounting to 5.0 m. One problem in these instances is to know the frequency with which similar displacements can be expected to recur. In Chile historical records since the early sixteenth century suggest destructive earthquakes are experienced at intervals of less than 100 years, and differential movements near Valdivia of at least 10 mm yr^{-1} are by no means improbable.

Examples so far have all been drawn from areas of known Cenozoic mountain-building. It should not be assumed, however, that crustal movement of tectonic origin is confined to such regions. In the United States (Fig. 4.3) repeated surveys have disclosed a broad area of subsidence depressing New York city by about 4 mm yr^{-1} and parts of Delaware and coastal Virginia by 2 mm yr^{-1}. By contrast, parts of Georgia and Alabama are being elevated at over 5 mm yr^{-1}. In the European USSR a similarly intricate pattern of uplift and subsidence has been demonstrated, with recorded movements commonly between +3 and −3 mm yr^{-1}. In addition there are local centres of very rapid uplift, with one dome-shaped area in the Ukraine reported to be rising at 10 mm yr^{-1}. Where levelling along the same traverse has been repeated several times during the last few decades, the rate of movement has often proved far from constant. In several instances an actual reversal in the direction of displacement has led Soviet workers to propound the idea that the crust may experience small-scale undulatory movements that make total stability the exception rather than the rule. In Great Britain the Ordnance Survey has adopted a conservative attitude, preferring to attribute differences between successive surveys to human or instrumental failings rather than crustal movement. Nevertheless, much of the levelling evidence is consistent with movements inferred from other sources, even if not by itself constituting proof. For example, Ordnance Survey data may be interpreted as indicating uplift in

Fig. 4.3. Provisional maps of contemporary crustal movement in the United States and European Russia as revealed by geodetic surveys.

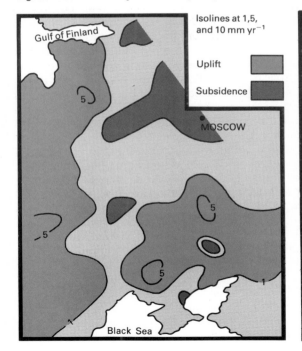
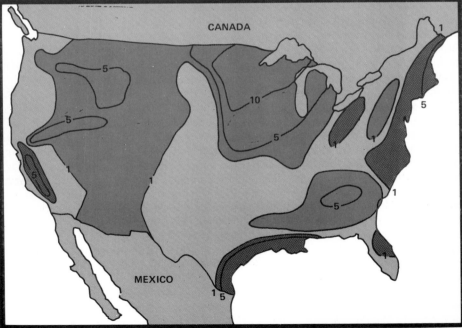

south-eastern Scotland of about 0.5 mm yr^{-1}, conforming with tidal gauge analyses that suggest a tilting of Great Britain from north-west to south-east (Fig. 4.4).

Inadequate care over such matters as the stability of benchmarks reoccupied during successive surveys has led to a justifiable questioning of the reliability of some traverses. Yet it can scarcely be doubted that vertical movements currently taking place in many areas are of sufficient rapidity to be measured by modern survey techniques. Much interest will attach to the extension of these techniques to all parts of the world since at the moment the global coverage is very uneven.

Measurement of horizontal displacements

Although relatively few measurements of contemporary horizontal motion have yet been made, advances in the design of electromagnetic distance measuring (EDM) devices could soon revolutionize the position. These instruments normally depend on determination of the phase difference of a modulated carrier signal returning from a reflector; the source of the signal nowadays is commonly a laser, and a precision of about 10^{-7} or 0.1 mm in 1 km can be attained. Coupled with a retro-reflector on a near-earth satellite, an EDM instrument may even permit direct measurement of oceanic widening by plate tectonics; already a comparable system has been employed in measuring a Californian baseline, over 896 km long, between Quincy and San Diego on opposite sides of the San Andreas fault (Fig. 4.5), which appears to be shortening at an average rate of 9 ± 3 cm yr^{-1}. It has been known for over a century that this dislocation is characterized by lateral rather than vertical motion (Fig. 4.6). In 1857 the Fort Tejon earthquake caused visible surface ruptures over a distance of 300 km with horizontal movement in places amounting to 9 m. The 1906 San Francisco earthquake and the 1940 Imperial Valley earthquake were accompanied by lateral displacements of 6.5 and 5 m respectively, in all cases the motion being dextral as the coastal margin of California moved northwards relatively to the rest of the state. In the circumstances it is scarcely surprising that most of the world's attempts at measuring horizontal movement have been concentrated on the San Andreas fault.

It was after the 1906 San Francisco earthquake that Reid proposed the 'elastic rebound theory' of earthquake activity that now forms a corner-stone of seismology. It envisages progressive accumulation of

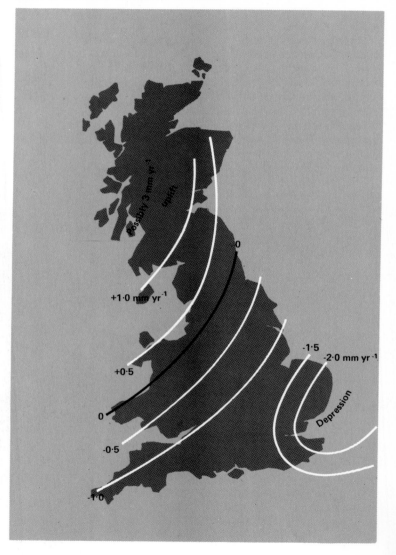

Fig. 4.4. A tentative estimate of the rate of vertical surface displacement in Great Britain during the last few centuries.

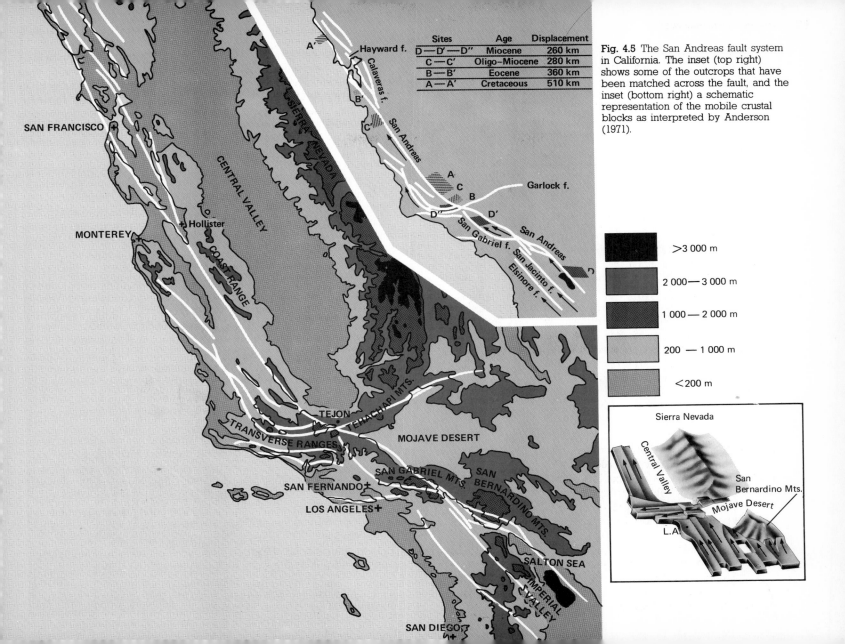

Sites	Age	Displacement
D — D' — D''	Miocene	260 km
C — C'	Oligo–Miocene	280 km
B — B'	Eocene	360 km
A — A'	Cretaceous	510 km

Fig. 4.5 The San Andreas fault system in California. The inset (top right) shows some of the outcrops that have been matched across the fault, and the inset (bottom right) a schematic representation of the mobile crustal blocks as interpreted by Anderson (1971).

>3 000 m

2 000 — 3 000 m

1 000 — 2 000 m

200 — 1 000 m

<200 m

SAN FRANCISCO

MONTEREY

Hollister

SIERRA NEVADA

CENTRAL VALLEY

COAST RANGE

Hayward f.

Calaveras f.

San Andreas

Garlock f.

San Gabriel f. San Andreas

San Jacinto f.

Elsinore f.

TEJON

TEHACHAPI MTS.

TRANSVERSE RANGES

SAN FERNANDO

LOS ANGELES

MOJAVE DESERT

SAN GABRIEL MTS.

SAN BERNARDINO MTS.

SALTON SEA

IMPERIAL VALLEY

SAN DIEGO

Sierra Nevada

Central Valley

San Bernardino Mts.

Mojave Desert

L.A.

elastic strain energy in a zone on either side of a fault; it may be likened to the accumulation of such energy in a bow that is flexed before the shooting of an arrow. Eventually the stress along the fault will exceed the frictional resistance and slip is then initiated. This will occur first at a single point, but release there will throw increased stress on adjacent segments so that movement is propagated along the fracture at an estimated velocity of 3–4 km s⁻¹; on major dislocations the length of the rupture during a single earthquake may reach 1 000 km. Movement ceases at the point where the stress no longer exceeds the frictional resistance and the fault remains locked in position. The longer a fault segment remains locked, the greater the potential accumulated strain energy and the more devastating the shock may be when it finally comes. This is one reason for concern about the interval that has now elapsed since the last major movement along the San Andreas fault near San Francisco.

One implication of the elastic rebound theory is that, with increased knowledge of fault dynamics, it may be possible to foretell when a major earthquake is imminent. The two variables that would need to be measured are the amount of accumulated strain energy and the frictional resistance of the fault. Such considerations have elicited much support for attempts to monitor the gradual deformation of the crust and so get an approximate idea of the build-up of strain energy. By the use of repeated triangulation surveys, modern EDM devices and specially installed strain gauges and creepmeters it has been established beyond doubt that surface strain near the San Andreas fault can be measured; yet the pattern which emerges is so complex that many uncertainties remain.

Along some branches of the fault system south of San Francisco continuous creep has been observed; the most famous instance is at Hollister where the creep is gradually splitting apart the buildings of a winery at a rate of about 12 mm yr⁻¹. Only very small tremors accompany this creep, apparently indicating that lateral slip can occur without intermittent locking. On the other hand, the creep is slower than the average movement along certain other sections of the fault system, and EDM devices suggest that strain may be accumulating in the crust for 20 km on both sides of Hollister. If that is the case, the stored energy may well be released by infrequent earth-

Fig. 4.6 The scar of the San Andreas fault as viewed from the air over southern California. Note the dextral offsetting of a number of the drainage courses, the most obvious example being near the centre of the photograph.

quakes of greater magnitude. Further south, where the fault trace for a short distance trends nearly east–west, the crustal strain indicates compression almost normal to the fracture. This area has not experienced a major earthquake during the last 100 years, and it seems possible that the east–west segment of the fracture is subject to exceptionally powerful locking with rare but very intense earthquakes. Still further south in the Imperial Valley, systematic triangulation with increasingly sophisticated EDM instruments has established what appears to be the simplest sequence of strain accumulation and release on the San Andreas system. Surveys in 1935, 1941, 1954 and 1967 had already disclosed deformation apparently affecting a zone up to 60 km wide on both sides of the Imperial fault. In the 1970s most work was concentrated close to the fault trace itself and numerous surface changes were monitored. Aseismic creep averaging about 6 mm yr^{-1} was detected at a number of sites and, following a swarm of minor tremors in 1975, new surface ruptures were mapped. This revealed how complex the fracture pattern actually is and emphasized the difficulty of evaluating accumulated strain. That not all the energy was being released by these minor movements became apparent when, in late 1979, a major earthquake reactivated the Imperial fault with a maximum measured displacement of 0.7 m.

Geomorphic evidence for crustal movement

Deformed strandlines

Of all types of geomorphic evidence, that presented by deformed shorelines is the most convincing. This is because the initial horizontal form is beyond dispute, whereas the original shape of a river terrace or erosion surface may be a matter of controversy. Although instances of disturbance from deep-seated tectonic causes are known, most attention has focused on strandline deformation arising from crustal unloading. Two circumstances in particular can lead to isostatic uplift and warping of strandlines. The simpler, though less frequent, is the desiccation of a lake that was sufficiently large to have caused local depression of the crust. A good illustration is provided by Lake Bonneville in Utah (Fig. 4.7). This forerunner of the Great Salt Lake has left striking testimony to its old levels in prominent strandlines, not only around the margins of the basin but also on a number of central islands. It formed during a cooler, wetter climatic phase and at its maximum covered 50 000 km^2 and reached a depth of 330 m. The total mass of the water was 9×10^{12} tonnes, a large proportion of which was abruptly evacuated when a natural dam at Red Rock Pass was breached in a catastrophic event, probably some 16 000 years ago. Mapping of the highest shoreline shows that it has undergone broad domed uplift, with the greatest displacement of 64 m sited where the water was deepest. Over a large central area the water averaged 290 m deep, and if the density of the displaced subcrustal material is assumed to be 3 300 kg m^{-3} a total recovery of over 85 m might be expected. Such calculations suggest uplift may still be incomplete, and this accords with levelling in 1911 and 1958 that indicated doming to be continuing at about 1 mm yr^{-1}. Because the forces acting to restore the basin to isostatic equilibrium decrease as uplift occurs, the rate of movement itself declines. As with radiometric dating, the passage of a certain unit of time sees the rate halved, and it can be shown that for Lake Bonneville the appropriate unit is about 2 800 years. This implies that, had there been instantaneous removal of the whole weight of Lake Bonneville, some 42 m of uplift would have taken place in the first 2 800 years, that is, at an average rate of about 15 mm yr^{-1}. In practice the unloading was not instantaneous and the uplift was correspondingly slower; nevertheless, the picture that emerges is of a mobile subsurface layer of much lower viscosity than is appropriate to the relatively rigid lithosphere. This underlines the important role now believed to be played by the asthenosphere in surface deformation.

The more common cause of deformed strandlines is the melting of an ice-sheet. Large lakes may develop around the decaying ice-mass, or the sea may invade areas vacated by the retreating ice-front. As melting continues the load of the crust is reduced and the land differentially uplifted. Any beaches that have developed are elevated and warped. From the height and distribution of the raised beaches the history of isostatic recovery can be deciphered. However, two intrinsic problems, absent in the case of a desiccated lake, are estimating the original ice thickness and calculating the amount of uplift that may have preceded local deglaciation. Moreover, the bearing of raised marine shorelines on the subject of isostatic rebound can only be fully appreciated in the context of world-wide changes in sea-level. For this reason the main discussion of post-glacial recoil is deferred until Chapter 16 in which Quaternary

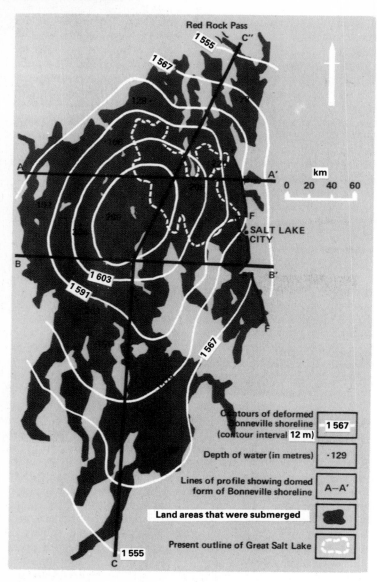

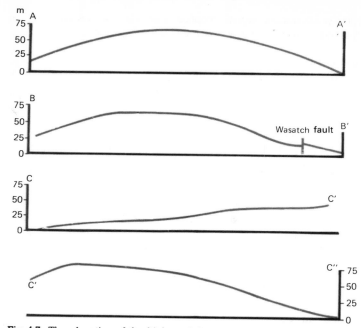

Fig. 4.7. The elevation of the highest deformed strandline around Lake Bonneville in Utah, USA (data from Crittenden, 1963).

sea-level oscillations are discussed in detail. Here it will suffice to note some of the main conclusions arising from the study of elevated strandlines in Scandinavia and North America. In the former area, the oldest shorelines are held to indicate a recoil of some 800 m in the last 13 000 years; the negative gravity anomalies that still characterize the Gulf of Bothnia attest to 100–150 m of isostatic recovery yet to come. In North America the highest recorded marine shorelines lie at an elevation of about 300 m in an area of Hudson Bay that only became ice-free 8 000 years ago; estimates of uplift still to come vary between 100 and 250 m. These figures from recently deglaciated regions amply confirm the general mobility of the crust when subject to vertical stress, with rates of uplift capable of averaging 50 mm yr^{-1} for extended periods. It is estimated that a volume of 7×10^6 km^3 of rock must have migrated inwards beneath Scandinavia during a

13 000-year period, again emphasizing the low viscosity of the displaced material at the level of the asthenosphere.

Aggradational features

Aggradational features display the results of recent crustal movement in a variety of ways. The most direct is by a simple faulted dislocation of the completed surface form (Fig. 4.8). River terraces, glacial moraines and alluvial fans are among the many features known to have been modified in this way. For example, alluvial fans accumulating against an active fault scarp may bear testimony to several discrete phases of tectonic movement. Along the western face of the Sierra Nevada in Southern Spain, spasmodic uplift has clearly initiated several distinct cycles of fan development. Some of the earliest accumulations have been disturbed during the more recent faulting; in a number of instances the apex of an individual fan has been elevated as part of the Sierra Nevada massif so that it now lies separated from its original lower slopes by a prominent fault bluff.

If accumulation and tectonic disturbance occur simultaneously, patterns of sedimentation are likely to be affected. The clearest evidence of the earth movements may then lie in the stratigraphy of the deposits, and it is in this way that many large deltas bear witness to continual downwarping. Immense thicknesses of shallow-water debris imply that local subsidence has occurred *pari passu* with the influx of new sediment. In part this may be ascribed to isostasy which demands that each new increment of load be compensated by further depression of the crust. However, the density of deltaic material is so small compared with that of the mantle that this cannot be the full explanation; there must be additional internal processes involved. In the upper layers of the Mississippi delta, the most thoroughly studied in the world (Fig. 4.9), a soil horizon has been identified that is believed to date from mid- to late-Wisconsinan times. Near the extremity of the modern delta this marker level today lies at a depth of almost 300 m. Of this figure approximately 100 m is attributable to the fall in sea-level that occurred during the last glaciation. The

Fig. 4.8. A 2 m high fault scarp cutting across the floodplain of a small stream in Montana, USA. The scarp was initiated at the time of the Hebgen earthquake in 1959; it produced a knickpoint in the stream long profile that first migrated up the valley but later degenerated into a series of minor rapids.

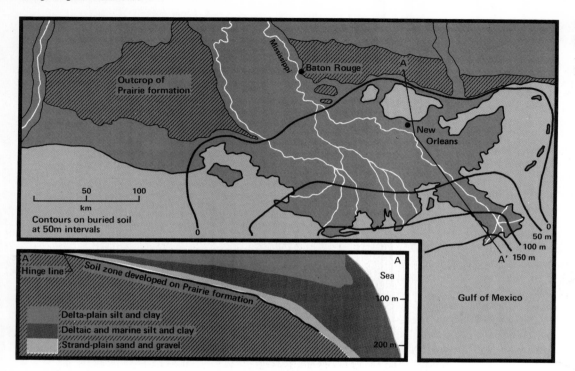

Fig. 4.9. Contours on the downwarped soil zone beneath the Mississippi delta. The inset shows a section, with a vertical exaggeration of ×225, along the line A–A′ through New Orleans.

rest is presumably due to downwarping which has thus averaged about 10 mm yr^{-1} for the last 20 000 years.

Although horizontal fault displacements are most frequently attested by valleys offset into characteristic zigzag patterns, as is well seen along parts of the San Andreas fault (Fig. 4.6), the fullest record of movement may be preserved in the morphology of river terraces. Where a fault is active during the fashioning of a suite of terraces the earlier members of the succession will be displaced more than the later since the effects are cumulative. If the terraces can be dated, the history of tectonic activity may be decipherable in considerable detail. After painstaking survey of the Branch River terraces in New Zealand (Fig. 4.10) Lensen concluded that, during the last 20 000 years, lateral movement along the Wairau fault has averaged at least 3.4 mm yr−1.

Uplifted and warped erosion surfaces

Many authorities have seen, in the present morphology of the continents, the imprint of intermittent tectonic uplift. The basic concept requires a prolonged period of stability, during which the landscape evolves towards a surface of low relief related to the contemporaneous sea-level, followed by uplift when the subdued surface is first elevated and then progressively consumed as a new landscape develops at a lower level. If repeated several times this sequence results in a staircase of gently sloping platforms separated by steeper bluffs. Within this basic conceptual framework several models of landscape evolution have been proposed; these are further discussed in Chapter 19, but for the moment such details are unimportant.

There can be few upland areas in the world where the idea of alternating planation and uplift has not been applied at some stage

Fig. 4.10. The faulted terraces of the Branch River in South Island, New Zealand. The large block diagram represents the current position, the smaller insets the sequence of erosion and displacement phases by which it has evolved (based on data in Lensen, 1968).

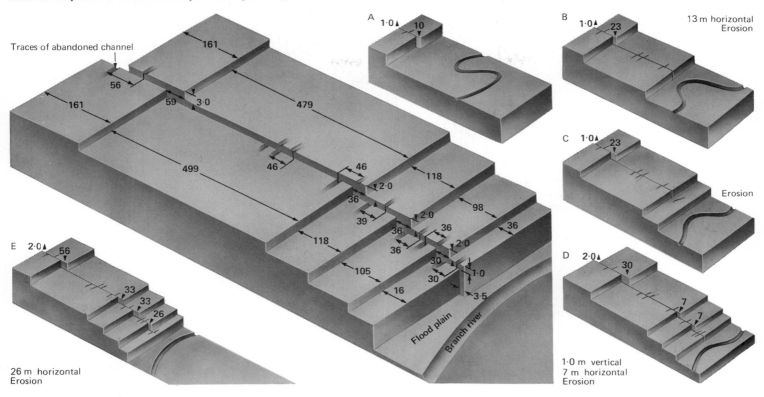

in the history of geomorphological investigation. Contrasting examples are afforded by the Welsh uplands and the Rocky Mountains. In a meticulous morphological survey of Wales, Brown in 1960 identified three major plateau surfaces at levels of approximately 525–600, 375–425 and 250–300 m (Fig. 4.11). Each was interpreted as the product of Cenozoic subaerial erosion during an early phase in the history of the present drainage. He found little evidence of warping so that the form of the three features implies either intermittent uplift of the Welsh massif as a structural unit or spasmodic falls in sea-level. Both explanations pose problems. If the latter were true,

similar features might be expected in other upland areas on the oceanic margins, and although analogous surfaces have been identified, there is little consistency in altitude. Moreover, world-wide sea-level changes of the requisite amplitude would have immense implications for the volume of sea-water or the capacity of the ocean basins. It therefore seems more likely that the Welsh massif has been uplifted as a block. Yet such a block must presumably be limited in size, although so far no flexed or faulted margins to the erosion surfaces have yet been satisfactorily defined, in part reflecting the crudity of the analytical tools available to the geomorphologist.

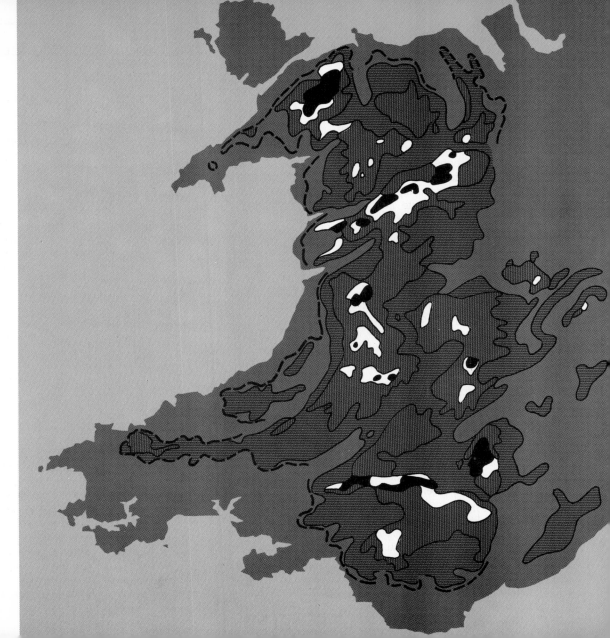

Fig. 4.11. Simplified diagram of the erosion surfaces of Wales as mapped by Brown (1960).

Key

Upstanding residuals

High Plateau (525–600 m)

Middle Peneplain (375–425 m)

Low Peneplain (250–300 m)

Approximate run of a former coastline, now at 200 m (? of Calabrian age)

Many investigators have noted that the Rocky Mountain crestline in parts of Colorado consists not of jagged peaks but of undulating areas that are interpreted as the residuum of an old erosion surface undergoing dissection. The surface has been traced southwards to a point where it passes beneath a cover of lavas radiometrically dated at 30–35 m.y. old. As the surface bevels mid-Eocene and older rocks it is generally regarded as of late Eocene age, and has been claimed to be identifiable over an area of at least 10 000 km². Where still preserved beneath volcanic materials it can be shown to have suffered differential uplift of between 1 500 and 3 000m and locally to have been displaced along faults by as much as 300 m.

These two examples illustrate both the value and limitations of erosion surfaces as guides to crustal uplift. Their greatest potential benefit lies in coverage of regions and periods for which alternative evidence is often almost entirely lacking. Unfortunately, however, morphology by itself is often capable of several different interpretations. The gross form of the Welsh upland, for instance, was ascribed by Jones in 1961 to the stripping away of a former cover of Triassic sediments and exhumation of a much older landscape, so that few inferences could be drawn about Cenozoic history. More recently Battiau-Queney has offered another contrasting interpretation which not only involves an extremely long pre-Cenozoic denudational history but also attributes several aspects of the present morphology to localized tectonic deformation rather than subaerial erosion. Furthermore, it is worth recording at this point that an influential group of workers has expressed scepticism about the traditional way of interpreting accordant summit levels, contending that in many cases they are not remnants of earlier landscapes and so have little historical significance (see Ch. 19). While this viewpoint has had a salutary effect in compelling a more critical examination of the concepts and evidence on which denudation chronologies have been formulated, it has not entirely undermined the belief that the relief of many upland areas holds one of the keys to local tectonic history. Nevertheless, the fact of so many conflicting hypotheses emphasizes the equivocal nature of much of the morphological evidence on which the geomorphologist has to operate.

Geological evidence for crustal movement

General principles

The writings of classical Greece clearly envisage relative movements of land and sea to explain the presence of marine fossils in rocks now far above sea-level. After William Smith, the eighteenth-century founder of stratigraphy, had demonstrated that a single sheet of rock was of the same age at points many kilometres apart and at levels differing by hundreds of metres, it was obvious that differential movement of the land must be involved. With the further development of simple stratigraphic principles, it became possible to unravel the structures and to assess the nature and age of the earth movements that produced them. The present disposition of the thick stratum of Chalk in south-eastern England provides an illustrative example.

Originally laid down as a continuous horizontal sheet of marine sediments, the Chalk now remains as a dissected remnant underlying the London and Hampshire basins, but removed by erosion from the crest of the major upfolds in the Weald and Isle of Wight. The deformation suffered by the Chalk can be depicted by means of structure contours drawn on a suitable datum plane such as that between the Chalk and overlying Tertiary beds (Fig. 4.12). Where denudation has destroyed the datum plane its elevation can be estimated by assuming appropriate thicknesses for the intervening strata and projecting upwards from a lower datum surface. In this way the structural relief of south-eastern England can be shown to exceed 1 500 m. This is a measure of the differential movement to which the region has been subject during Cenozoic time. It is, however, possible to identify much more precisely the stages through which the present structural relief evolved by considering the evidence of the strata that overlie the Chalk. In both the London and Hampshire basins an angular unconformity between the Chalk and the oldest Cenozoic beds indicates a major tectonic phase, possibly accounting for some 400 m of relative uplift, beginning in late Cretaceous times and extending into Palaeocene times. Further unconformable relationships within the Eocene and Oligocene sediments testify to continuing movement, and although the amplitude of each folding episode was small the cumulative effect was considerable. The absence of Miocene sediments makes it difficult to determine how much tectonic deformation occurred in mid-Tertiary times, but modern estimates

Fig. 4.12. Contours of the sub-Eocene surface in south-eastern England (after Wooldridge and Linton, 1955). The map and the insets illustrate alternative ways of depicting the minimum amount of Cenozoic tectonic deformation. In both cases, however, it needs to be remembered that the 'structural relief' evolved slowly over a long period and may have borne little direct relationship to surface topography (see, for instance, Jones, 1980).

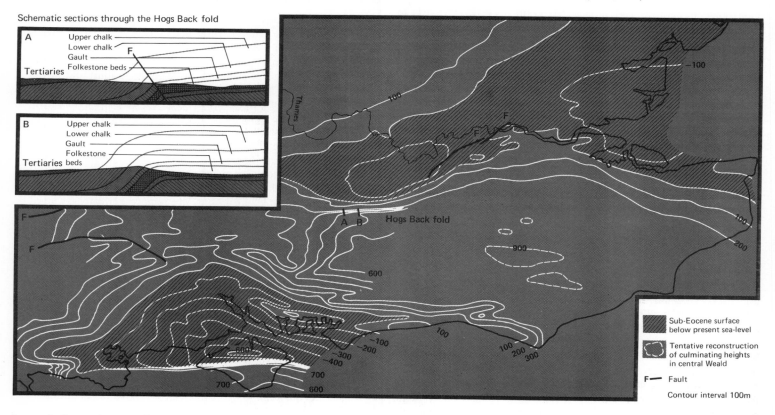

Schematic sections through the Hogs Back fold

have challenged an earlier view that much of the structural relief dates from this interval. Fossiliferous sediments attesting to a marine transgression in late Pliocene times are sporadically preserved in East Anglia and on the dip-slope of the North Downs in Kent, the contrast in elevation between the various sites implying a differential movement of at least 180 m. Early Pleistocene marine deposits preserved in these same areas seem to be similarly tilted, thus confirming the considerable amplitude of warping in south-eastern England during Quaternary times.

It cannot be emphasized too strongly that the contours and geological cross-sections of Fig. 4.12 depict structures without any direct reference to former surface topography. Tectonic displacements are cumulative and have been going on spasmodically since late Mesozoic times; throughout this period the crustal movements were undoubtedly accompanied by erosion, and it must not be assumed, for instance, that the Wealden area ever reached an elevation of 900 m. Indeed the Chalk cover over the Weald was almost certainly breached by mid-Eocene times, and it appears likely, although difficult to prove, that for prolonged periods in the Cenozoic era there was little disparity in the mean rates of crustal uplift and sub-aerial denudation.

Concepts of orogenesis

In a classic nineteenth-century study of the Appalachians, Hall and Dana drew attention to the fact that the very thick sequences of shallow-water rocks forming those mountains must have accumulated in a continually subsiding zone to which they applied the term geosyncline. This idea was adopted and elaborated by workers on both sides of the Atlantic with general agreement that a complete orogenic or mountain-building cycle consists of three distinct phases: firstly a prolonged period of geosynclinal sedimentation, secondly a briefer episode of tectogenesis in which the strata are intensely crumpled and intruded by plutonic rock, and finally a rapid uplift of the deformed rock to produce the actual mountain chain. Subdivision of these phases often proved possible. For example, the tectogenic phase commonly involves an early period in which material known as flysch is eroded from growing structural highs and deposited in nearby basins; the later folding affects not only the true geosynclinal sediments but also the flysch. The close of the tectogenic phase and early part of the uplift is often accompanied by deposition in marginal depressions of coarse debris known as molasse. By dating both the flysch and the relatively undisturbed molasse, particularly in areas of recent mountain-building, it was shown that the tectogenic and uplift phases could be of only short duration.

In the 1960s these traditional views of orogenesis had to be assimilated into the new concept of plate tectonics. No radical revision was necessary, and it can be argued that the idea of mobile lithospheric plates provided the fundamental mechanism for mountain-building that had previously been lacking. It was immediately apparent that orogenesis was a feature of convergent plate margins, and that, at a simple level, the tectogenic phase could be equated with crustal subduction and the ensuing uplift with buoyancy of the subducted material. Geosynclinal sedimentation as formerly envisaged needed some modification, and the thick sedimentary sequences such as were orginally discerned in the Appalachians were ascribed to accumulation along a passive but subsiding continental margin. Yet the greatest change came in the acknowledgement of the diversity of orogenic activity with no single model satisfying all circumstances; attempts at classification are more fully discussed in Chapter 5.

The implications alluded to in the previous paragraph do not nullify the basic proposition that orogenic activity in any one continental area appears to be spasmodic. It was recognition of this fact that enabled nineteenth-century geologists in Europe to identify three Phanerozoic mountain-building periods:

1. The Caledonian, during which folding was predominantly along north-east–south-west axes in an area extending from northern Norway through Scotland to central Ireland. Prolonged sedimentation appears to have characterized this region from late Pre-Cambrian times until the culminating tectogenic activity at the close of Silurian times; however, radiometric dating has permitted identification of further tectogenic phases spanning a total period of at least 100 m.y.

2. The Hercynian, during which fold axes ranged from east-west in Ireland to north–south in parts of the Massif Central and Spanish Meseta. The main phase of movement was concentrated in late Carboniferous times, but a secondary phase was experienced in some areas in early Permian times.

3. The Alpine, during which the structures in the mountain chains that today characterize much of central and southern Europe were formed. There is still uncertainty about the exact sequence of events during the Alpine orogeny, but the phase of maximum tectonic activity is traditionally placed in early to middle Miocene times.

This dating of orogenies was originally worked out in Europe, and as geologists from that continent moved into other regions they tended to employ the same framework for interpreting local tectonic histories, even to the extent of applying the terms Caledonian, Hercynian and Alpine on a virtually global scale. Too often significant time differences were ignored so that a world-wide cyclical pattern of cordilleran growth and destruction began to be envisaged that was largely an artefact of the method of investigation used; a close parallel is found in the way Scottish raised beaches were formerly assigned to the '100-foot', '50-foot' and '25-foot' levels, even though the field observations indicated significant deviations from these values (see p. 332). The development of plate-tectonic theory led to a critical reappraisal of the evidence since there was no intrinsic reason for orogenesis at convergent plate margins to display a strongly cyclical pattern. This new assessment suggested that during almost the whole of Phanerozoic time mountain-building was in progress in one area or another (Table 4.1); on a global scale the extent of cordilleran growth

Table 4.1 A schematic portrayal, continent by continent, of the age of the world's major orogenies since the end of Pre-Cambrian times

Period	W. Europe (traditional)	European Russia	North America	Japan	Australia	South America	Africa
T	Alpine	Alpine	Laramide	Oyashima			
C							
J		Cimmeridian	Nevadan	Sakawa			
T				Akiyoshi			
P			Appalachian				
C	Hercynian	Hercynian	? ?				
D			Acadian				
S	Caledonian	Caledonian					
O			Taconic				
C							

may have fluctuated through time, but this could be explained by random variations in the number of continental blocks sited at the margins of convergent lithospheric plates. It also became clear that many major mountain chains have experienced much longer and more complicated tectonic histories than was formerly suspected. The Alps, for instance, have evolved as a complex orogenic belt where the early folding and thrusting is now dated as Mesozoic in age, and important pulses of further deformation followed in both the late Eocene and late Oligocene before the traditional culmination in Miocene times.

If it is true there is always orogenic activity taking place, it should be possible to find regions where it is occurring at the present day. As early as 1835 Charles Darwin during the voyage of the *Beagle* seems to have had no doubt he was witnessing mountain-building in progress when he experienced several severe earthquakes in the Andean region. Since evolving geological structures cannot normally be observed directly, contemporary surface uplift as outlined earlier in this chapter constitutes the main source of evidence. Of course, the time-scale over which such measurements can be made is almost infinitesimal compared with the time-scale of an orogeny and there is justifiable hesitation in applying a multiplication factor of 10^5 or 10^6. Nevertheless it is worth remembering that uplift at 1 mm yr^{-1}, if sustained for only 1 m.y., will elevate the surface by 1 000 m and thereby engender a significant mountain range. Although such a calculation ignores any concurrent erosion, it still demonstrates that there is no inherent difficulty in the view that orogenic uplift is taking place at the present day.

Large-scale horizontal displacements

Just as, for many years, the majority of geologists rejected continental drift, so they also tended to minimize potential lateral offsets along faults. Often observed relationships across a fault can be explained by either dip-slip or strike-slip, and in such cases the tendency was to assume dip-slip and discount the possibility of significant strike-slip. However, a number of individual workers adopted a different stance and argued in favour of large-scale horizontal translocations; it was their views that generally prevailed in the 1960s with the development of plate-tectonic theory. An excellent example of changing interpretation is afforded by the Great Glen fault in Scotland. Prior to the Second World War this was usually regarded as a normal fault with a vertical throw of hundreds or even thousands of metres. Then in 1946 Kennedy argued that the dislocation is a trans-current fault with a horizontal displacement of approximately 100 km, basing this interpretation on masses of granite on opposite sides of the fault that are so matched in structural detail that they are best explained as one original stock split into two; on geological grounds he dated the main movement as Devonian in age. In the 1960s the Great Glen fault was incorporated into plate-tectonic theory as part of an ancient transform system whose continuation is now found in the Canadian Maritime Provinces, but which was split into two by the opening of the North Atlantic in Mesozoic times.

In his description of the Great Glen fault Kennedy likened the feature, particularly in its straightness and abnormally wide shatter belt, to the San Andreas fault. This was a bold pronouncement since at the time it was far from universally agreed that there had been large-scale horizontal displacements in California. As early as 1925 a few writers were arguing that strike-slip on the San Andreas system might amount to between 10 and 40 km, but it was not until 1953 that Hill and Dibblee assembled data which, they claimed, demonstrated offsets totalling at least 500 km since Cretaceous times. Their basic idea was that markers could be identified on both sides of the fault showing diminishing displacement with diminishing age (Fig. 4.5, p. 61). According to their reconstruction, Cretaceous beds are offset 510 km, Eocene beds 360 km, Miocene beds 280 km and Pleistocene beds 15 km. In a later study Crowell argued for a displacement of 260 km since early Miocene times but was doubtful about the earlier history of the fault; he believed it possible, however, that there had been a displacement of as much as 1 000 km in the basement rocks since Eocene times. Increasing uncertainty with age is to be expected, since the older the materials that are involved the wider their separation and the greater their liability to independent metamorphism after dismemberment along the fault. If a figure of 260 km of movement since early Miocene times is adopted, the rate of displacement has averaged just over 10 mm yr^{-1}. This is somewhat lower than many estimates of current movement and suggests we may be living through a period of unusually intense activity. On the other hand, it should be stressed that the San Andreas fault system is extremely complex and much work is still required to establish the true rate of displacement on individual branches. In plate-tectonic theory the system as a whole is again interpreted as a large transform fault (see below p. 79).

References

Akeroyd, A. (1972) 'Archaeological and historical evidence for subsidence in southern Britain', *Phil. Trans. R. Soc.* **A272**, 151–69.

Anderson, D. L. (1971) 'The San Andreas fault', *Sci. Am.* **225**, 53–68.

Battiau-Queney, Y. (1984) 'The pre-glacial evolution of Wales', *Earth Surf. Proc. Landf.* **9**, 229–52.

Beloussov, V. V. (1962) *Basic Problems in Geotectonics*, McGraw-Hill.

Bird, P. and Rosenstock, R. W. (1984) 'Kinematics of present crust and mantle flow in southern California', *Geol. Soc. Am. Bull.* **95**, 946–57.

Bollinger, G. A. (1973) 'Seismicity and crustal uplift in the southeastern United States', *Am. J. Sci.* **273A**, 396–408.

Brown, E. H. (1960) *The Relief and Drainage of Wales*, Univ. Wales Press.

Brown, E. H. (1979) 'The shape of Britain', *Trans. Inst. Brit. Geogr.* **4**, 449–62.

Brown, L. D and Oliver, J. E. (1976) 'Vertical crustal movements from leveling data and their relation to geologic structure in the eastern United States', *Rev. Geophys. Space Phys.* **14**, 13–35.

Chi, S. C. et al. (1980) 'Leveling circuits and crustal movements', *J. Geophys. Res.* **85**, 1469–74..

Crittenden, M. D. (1963) 'New data on the isostatic deformation of Lake Bonneville', *U.S. Geol. Surv. Prof. Pap.* 454-E.

Crowell, J. C. (1979) 'The San Andreas fault system through time', *J. Geol. Soc. Lond.* **136**, 293–302.

Curtis, B. F. (1975) *Cenozoic History of the Southern Rocky Mountains*, Geol. Soc. Am. Mem. 144.

Harrison, J. C. (1976) 'Tilt observations in the Poorman mine near Boulder, Colorado', *J. Geophys. Res.* **81**, 329–36.

Hicks, S. D. (1978) 'An average equipotential sealevel series for the United States', *J. Geophys. Res.* **83**, 1377–9.

Hill, M. L. and Dibblee, T. W. (1953) 'San Andreas, Garlock and Big Pine faults, California', *Geol. Soc. Am. Bull.* **64**, 443–58.

Johnson, C. E. et al. (1982) 'The Imperial Valley, California, earthquake of 15 October 1979', *U.S. Geol. Surv. Prof. Pap.* 1254.

Jones D. K. C. (1980) 'The tertiary evolution of south-east England with special reference to the Weald', *I. B. G. Spec. Pub.* **11**, 13–47

Jones, O. T. (1961) 'The relief and drainage of Wales: an essay review', *Geol. Mag.* **98**, 436–8.

Kennedy, W. Q. (1946) 'The Great Glen fault', *Q. J. Geol. Soc. Lond.* **102**, 41–75 (for recent review see also J. C. Briden, *et al.* (1984) 'British palaeomagnetism, Iapetus Ocean and the Great Glen fault', *Geology* **12**, 428–31).

King, L. C. (1962) *Morphology of the Earth*, Oliver and Boyd.

Lensen, G. J. (1968) 'Analysis of progressive fault displacement during down-cutting at the Branch River, South Island, New Zealand', *Geol. Soc. Am. Bull.* **79**, 545–55.

Mason, R. G. et al. (1979) 'Mekometer measurements in the Imperial Valley, California', *Tectonophysics* **52**, 497–503.

Nilsen, T. H. (1984) 'Offset along the San Andreas fault of Eocene strata from the San Juan Bautista area and western San Emigdio Mountains, California', *Geol. Soc. Am. Bull.* **95**, 599–609.

Philip, H. and Meghraoui, M. (1983) 'Structural analysis and interpretation of the surface deformation of the El Asnam earthquake of October 10, 1980', *Tectonics* **2**, 17–49.

Rikitake, T. (1968) 'Earthquake prediction', *Earth Sci. Rev.* **4**, 245–82.

Rona, P. A. (1974) 'Subsidence of Atlantic continental margins', *Tectonophysics* **22**, 283–99.

Ruegg, J. C. et al. (1982) 'Deformations associated with the El Asnam earthquake of 10 October 1980: geodetic determination of vertical and horizontal movements', *Seism. Soc. Am. Bull.* **72**, 2227–44.

Savage, J. C. and Burford, R. O. (1970) 'Accumulation of strain in California', *Seism. Soc. Am. Bull.* **60**, 1877–96.

Savage, J. C. et al. (1979) 'Geodolite measurements of deformation near Hollister, California, 1971–78', *J. Geophys. Res.* **84**, 7599–615.

Sieh, K. E. and Jahns, R. H. (1984) 'Holocene activity of the San Andreas fault at Wallace Creek, California', *Geol. Soc. Am. Bull.* **95**, 883–96.

Slater, L. E. and Burford, R. O. (1979) 'A comparison of long baseline strain data and fault creep records obtained near Hollister, California', *Tectonophysics* **52**, 481–96.

Smith, D. E. et al. (1979) 'The measurement of fault motion by satellite laser ranging', *Tectonophysics* **52**, 59–67.

Smith, P. J. (1980) 'Palmdale bulge: fact or fiction?', *Nature London* **283**, 247.

Wooldridge, S. W. and Linton, D. L. (1955) *Structure, Surface and Drainage of Southeastern England*, Philip.

Selected bibliography

There has been a rapid increase in the literature on current crustal movement, with many countries establishing their own national surveys to monitor neotectonic movement. The Proceedings of international symposia on this subject are to be found in vols 52 (1979) and 71 (1981) of the journal *Tectonophysics*.

Chapter 5
Patterns of global relief and crustal mobility

As the two previous chapters demonstrate, the trend of recent investigations has undoubtedly been towards recognizing both the rapidity and pervasive nature of crustal movement. In defining areas of relative vertical stability and zones of intense crustal deformation, the model of mobile lithospheric plates as currently formulated has obvious implications for the geomorphologist. In the former areas one might anticipate that the continental surfaces would be dominated by low relief produced as a result of sustained denudation uninterrupted by uplift; similarly the older parts of the ocean floors away from the spreading centres should be dominated by smooth surfaces of accumulated sediment. The plate margins, on the other hand, should be the location of topography reflecting active tectonism, and comparison of Figs. 1.5 and 3.11 confirms, with a few noteworthy exceptions, this general accordance between plate edges and mountainous terrain. In the following account the pattern of topographic relief at divergent and convergent plate margins will first be examined. The possibility of vertical movements affecting the central regions of the plates will then be discussed, before attention is finally turned to two additional topics appropriately treated within the framework of plate tectonics, namely volcanic landforms and eustatic changes of sea-level.

Divergent plate margins

It is axiomatic that nearly all divergent margins should occupy oceanic positions. The characteristic topography of the mid-oceanic ridges has already been described and will not be discussed further.

However, great interest attaches to locations where oceanic spreading centres abut against continental blocks since here it may be possible to discern what happens during the early phases of continental separation. Two such locations attracting particular attention have been the Gulf of Aden and the Gulf of California, the former being especially instructive because it has been claimed to represent the initial stages of a split between Africa and Arabia.

The Gulf of Aden and Red Sea region

The Gulf of Aden and Red Sea (Fig. 5.1) together constitute a marine inlet over 3 000 km long but nowhere wider than 350 km; from the head of the Red Sea to the Mediterranean is a mere 120 km. The inlet transects a complex dome-shaped structure of Pre-Cambrian rock which reaches elevations of 2 000 m on the African side and 3 000 m on the Arabian side. Within 150 km of these crests the structure slopes outwards to 1 000 m in Africa and 1 500 m in Arabia. Bathymetric surveys have charted depths exceeding 2 200 m in both the Red Sea and the Gulf of Aden, but the two areas possess contrasting forms. A transverse profile of the central Red Sea (Fig. 5.1, section B–B') shows marginal shelves less than 200 m deep, steep slopes leading down to the main trough at 600–1 000 m, and then further steep slopes descending to a so-called inner or axial trough at 2 000 m; this axial trough is restricted to the central segment of the Red Sea. By contrast the Gulf of Aden, except at its western end, has a much broader central deep flanked by steep slopes that begin within a few kilometres of the coast (Fig. 5.1 section D–D'). Along the middle of the Gulf the floor is diversified by a topographically rough zone, known as the West Sheba ridge, with prominent north-east – south-west lineaments of which the most striking is the Alula-Fartak trench descending to over 5 350 m.

The major relief features are associated with distinctive geophysical properties. The dome-shaped shield is generally characterized by low gravity values, but by contrast the whole of the Red Sea south of the Gulf of Aqaba exhibits high values, with particularly large Bouguer anomalies over the axial trough. Across the Gulf of Aden gravimetric surveys have revealed a relatively uniform distribution of positive Bouguer anomalies with a small super-imposed negative anomaly over the central rough zone. Most models of likely structures based upon these findings involve almost the whole of the Gulf of Aden and the Red Sea being underlain by oceanic rather than

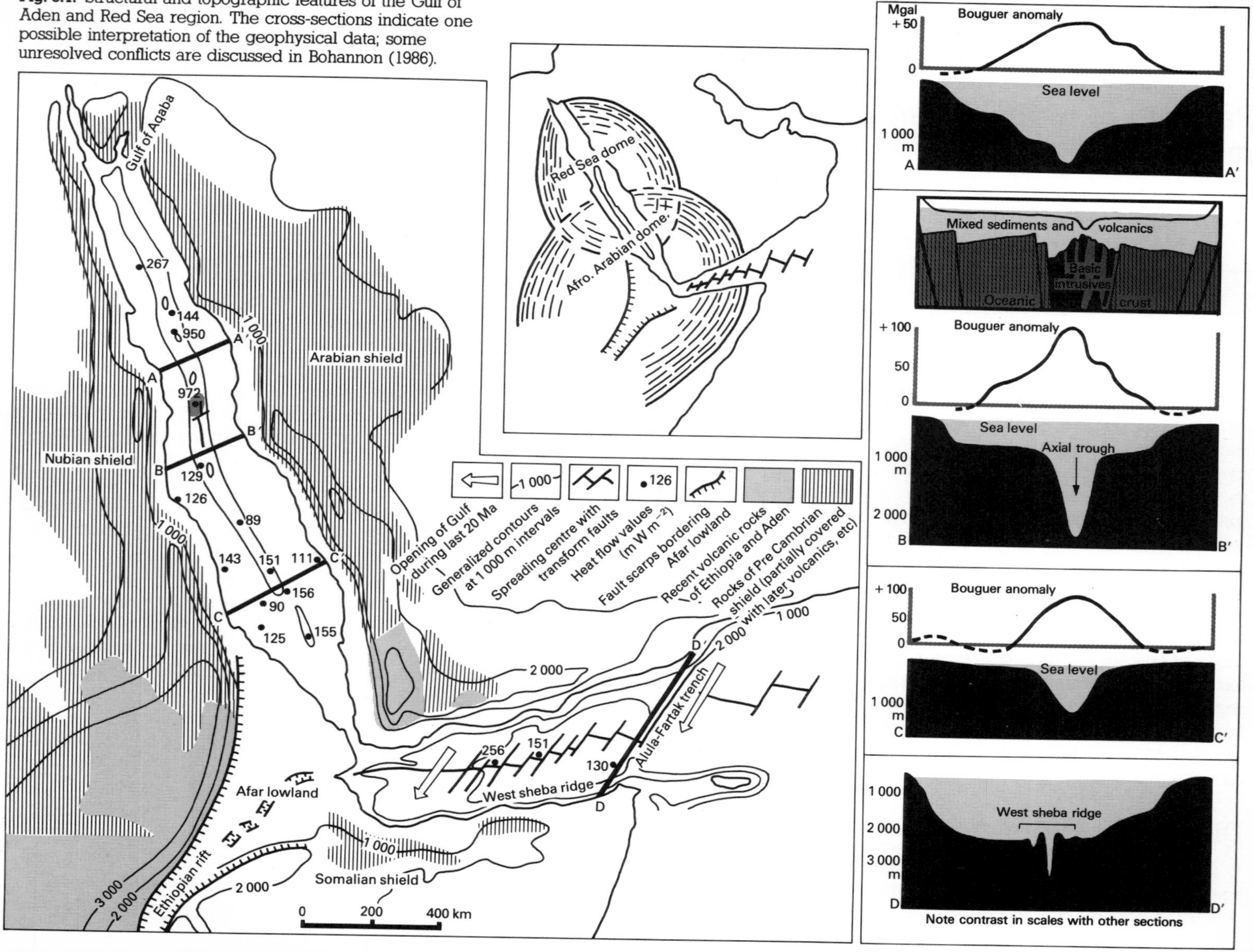

Fig. 5.1. Structural and topographic features of the Gulf of Aden and Red Sea region. The cross-sections indicate one possible interpretation of the geophysical data; some unresolved conflicts are discussed in Bohannon (1986).

continental crust. Nearly all submarine heat-flow measurements are well above the world mean. The highest values tend to be concentrated along the West Sheba ridge and the axial trough (see p. 46), but the entire Red Sea floor is associated with flows about twice the global average.

Magnetic surveys in the Gulf of Aden have revealed a prominent pattern of linear anomalies approximately parallel to the West Sheba ridge. The pattern is typical of that over mid-oceanic ridges and displays the effects of a series of transform faults. Dating of the lineations has proved difficult, with the earliest recognizable anomaly originally dated as 10 m.y., but later assigned an age of 40 m.y. by one group of workers. The latter envisaged a very spasmodic separation of Arabia and Somalia, with periods of active sea-floor spreading at about 10 mm yr^{-1} interspersed with stable phases when thick sequences of sediment accumulated. A subdued pattern of linear anomalies has also been detected across almost the full width of the Red Sea, but clearly defined anomalies are restricted to the axial trough. The sequence within the trough is thought to represent the results of active sea-floor spreading for the last 3–4 m.y., but outside that feature the pattern has proved difficult to interpret. The weakness of the signal is probably due to a cover of sediments which boreholes have shown locally to exceed 3 000 m in thickness. This sedimentary infill has again led some workers to argue for an episodic opening, the earliest phase of which may date back over 20 m.y., with the most recent phase represented by the current axial trough.

Seismic activity along the trough, and also along the West Sheba ridge, implies that both the Red Sea and the Gulf of Aden are presently active centres of crustal spreading and transform faulting. This view has been confirmed by studies of the Afar triangle where the two marine inlets converge. In the 1970s French workers established a geodetic network in this dry and inhospitable lowland, and after a resurvey following prolific basaltic fissure eruptions in 1978 were able to show a very localized crustal extension of up to 2.4 m. It appears that the easternmost segment of the Gulf of Aden spreading centre is here actually exposed above sea-level, and much attention is therefore being devoted to understanding the tectonic structure of this highly complex area.

Detailed reconstruction of events in the whole Gulf of Aden–Red Sea region has proved controversial. In addition to the spreading centres already discussed, a third has been identified in the Ethiopian rift system so that the region probably constitutes a triple junction. The main plates involved are the Nubian, Somalian and Arabian, but there may also be small rigid 'platelets' in the Afar lowland. The principal evolutionary phases seem to involve an initial complex doming followed by the development of linear collapse structures that then became the sites of oceanic spreading. The early stages of collapse may have been accomplished by both flexuring and faulting, but a former belief that very substantial thinning of the continental crust was involved has had doubt cast upon it by the almost universal occurrence of oceanic crust beneath both marine inlets. The original doming is thought to have elevated a relatively featureless erosion surface which received its final shaping in early Cenozoic times. The present radial drainage pattern reflects this domed uplift, with only short streams flowing into the rifted depressions. The asymmetry of the resultant divides has led to many instances of stream piracy, and it may even be possible to reconstruct drainage lines directly dismembered by the opening of the Gulf of Aden although the evidence on this latter point remains inconclusive. The initiation of neither the Red Sea nor the Gulf of Aden is yet firmly dated, but a mid-Cenozoic age appears most likely. Vast quantities of sediment partially filled the narrow, steep-walled sutures that first transected the dome, but sea-floor spreading later exposed genuine oceanic crust such as is seen today along the axial trough and the West Sheba ridge.

The Ethiopian rift mentioned in the previous paragraph was for many years treated as part of a unified graben system traceable from the Dead Sea in the north, via the Red Sea and East Africa, to the Zambezi valley in the south (Fig. 5.2). The northern and central sections of this system have been shown above to constitute an area of recent continental separation, and it is tempting to regard the East African rifts as incipient centres of the same process. However, among other things, the age of the structures argues against such a simple interpretation. The lower Zambezi graben appears to date back at least to early Mesozoic times, and further north the initial tectonism is certainly no younger than that around the Red Sea and Gulf of Aden. Nevertheless, there are notable similarities between the two regions. The rifting is concentrated close to the crest of a series of strongly arched structures, with warped erosion surfaces defining the flanks of individual domes. The faulting has induced

Fig. 5.2. The East African rift system. The areas of domed uplift are outlined by shading, with the amplitude indicated by contours drawn on tilted erosion surfaces of mid-Cenozoic age. Both dating and details of form have been the subject of much controversy, but the broad pattern of movement seems clear. The inset shows schematically the nature of the uplift and its effect on local drainage development approximately along the line of the equator. Again details of dating are uncertain, but (A) probably represents conditions in early Cenozoic times, and (D) conditions since mid-Pleistocene times.

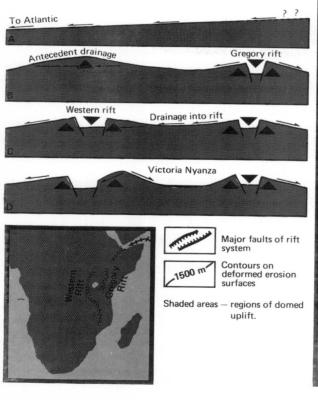

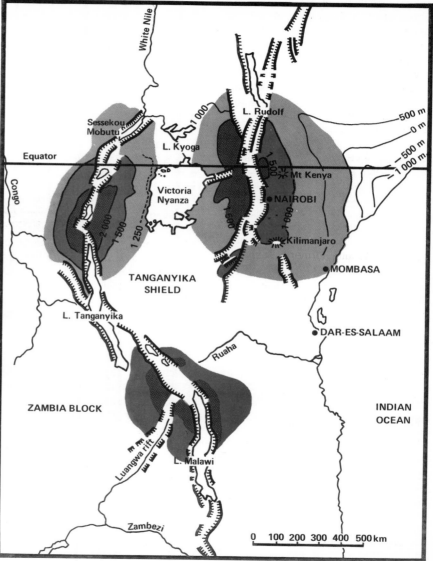

striking instances of stream piracy and drainage reversal, most clearly seen in the region of Victoria Nyanza. This area originally drained westwards to the Congo, and during a period of crustal stability extending from Mesozoic into early Cenozoic times was reduced to an erosion surface of very subdued relief. Mid-Cenozoic arching along the line of the Western Rift required vigorous erosion by the west-flowing rivers to maintain their antecedent courses. This upwarping was later accompanied by a central rift that deepened so rapidly in Miocene times that the connection with the Congo was finally severed. The streams of southern Uganda continued to flow into the rift until accelerated Pleistocene movements completed the disruption of the drainage; part was reversed to flow into the newly formed Victoria Nyanza, and part drained down over-steepened gradients to the rift floor. From the elevation of erosion surfaces along the rift margins differential warping of about 850 m since late Oligocene times has been postulated for south-western Uganda. The depression of the graben floor must have been even greater because over 2 600 m of infill has been proved in a borehole near Lake Sessekou Mobutu (formerly Lake Albert). Comparable movements seem to have characterized the Gregory rift in Kenya. Domed Cenozoic uplift is here estimated at more than 1 500 m, with clear evidence of a linear downwarp along the crest preceding the actual rifting.

As already mentioned, there is no reason for supposing that the East African rift system represents a line of currently active continental separation. Yet it does seem to have evolved in a manner very reminiscent of the early development of the Red Sea, and it may even represent a site of aborted lithospheric spreading that failed to develop to the point of full continental dismemberment.

The Gulf of California

Oceanographic research within the Gulf of California (Fig. 5.3) has revealed many indications of an offset spreading centre that appears to be a northward continuation of the East Pacific Rise. The more significant discoveries include a distinctive pattern of deep rhomboid basins, active fault scarps, above-average heat-flow values and Bouguer gravity anomalies suggesting an axial intrusion of high-density material. Although magnetic surveys have failed to disclose decipherable anomaly patterns, there is probably a single spreading centre offset by numerous transform faults. The latter are aligned

approximately parallel to the San Andreas fault which may simply be the largest of their kind. Support for this interpretation comes from further spreading centres, accompanied by hydrothermal vents and sulphide deposits, at the northern end of the San Andreas fault off the coast of Oregon, Washington and British Columbia. There are indications of tectonic movements around the Gulf reminiscent of those already described from Africa. In north-western Mexico the Sierra Madre Occidental has been elevated along its western edge by as much as 2 000 m, resulting in excavation of spectacular gorges that rival in scale the Grand Canyon of Arizona. Meanwhile, the northern segment of the Baja California peninsula is estimated to have suffered differential uplift and tilting of about 1 500 m. The tectonic instability of the region since mid-Cenozoic times is thus not in doubt, and it is even possible to discern some evidence for a collapsed dome. By Miocene times an elongated trough had already developed along the line of the Gulf, but the creation of new oceanic crust and the north-westerly drift of Baja California was delayed until about 6 m.y. ago. This dating has been established by a collaborative French–American–Mexican project known as RITA (named from the Rivera and Tamayo fracture zones) which has made an intensive study of the entrance to the Gulf. Submersibles have been used to identify the main spreading centre which is marked by, among other features, the presence of massive sulphide deposits and a benthic fauna of giant size. Both magnetic anomaly patterns and submarine topography point to a complex history for the Tamayo fracture which is confirmed as a transform fault. The final extension of the spreading centre into the interior of the Gulf probably occurred only 3.5 m.y. ago.

Despite the apparent analogies with events in the Afro-Arabian region, it should be noted that recent plate evolution off the Mexican coast has been particularly complex. At present a convergent margin represented by the Middle America trench ends immediately south of the entrance to the Gulf but is believed to have extended much further north in mid-Cenozoic times. Within the relatively recent geological past, therefore, this section of the continental edge was probably experiencing subduction rather than lateral extension.

Characteristic features of continental separation

The foregoing review of divergent margins in Africa and North America suggests that continental separation is likely to be preceded

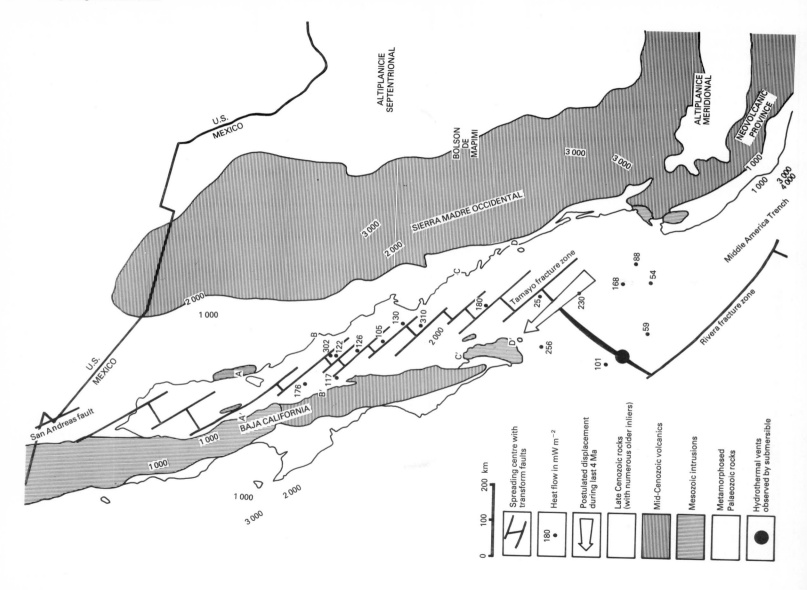

ALTIPLANICIE SEPTENTRIONAL

ALTIPLANICIE MERIDIONAL

NEOVOLCANIC PROVINCE

U.S.
MEXICO

BOLSON DE MAPIMI

3 000

3 000

3 000

Middle America Trench

1 000
1 000
3 000
4 000

SIERRA MADRE OCCIDENTAL

3 000
2 000

2 000

1 000

U.S.
MEXICO

•88
•54

168•

Tamayo fracture zone

•25
230•

2 000

•59

•256

101•

Rivera fracture zone

B
302•
122•
126•
105•
130•
•310
180•

176•

A'

B' 117•

C'

D'

San Andreas fault

A'

BAJA CALIFORNIA

1 000

1 000

2 000

1 000

2 000

3 000

200 km

100

0

Spreading centre with transform faults

180 • Heat flow in mW m⁻²

Postulated displacement during last 4 Ma

Late Cenozoic rocks (with numerous older inliers)

Mid-Cenozoic volcanics

Mesozoic intrusions

Metamorphosed Palaeozoic rocks

Hydrothermal vents observed by submersible

by massive upwarping and central collapse before the actual drifting starts. In many ways this is a predictable sequence since oceanic spreading zones are invariably marked by broad ridges which, if they developed beneath a continental block, would lead to linear arching of the surface. Various reasons have been advanced for the topographic highs along the centres of spreading. Seismic and gravity studies show abnormal structural conditions beneath the mid-oceanic ridges. The Mohorovicic discontinuity is either absent or modified,

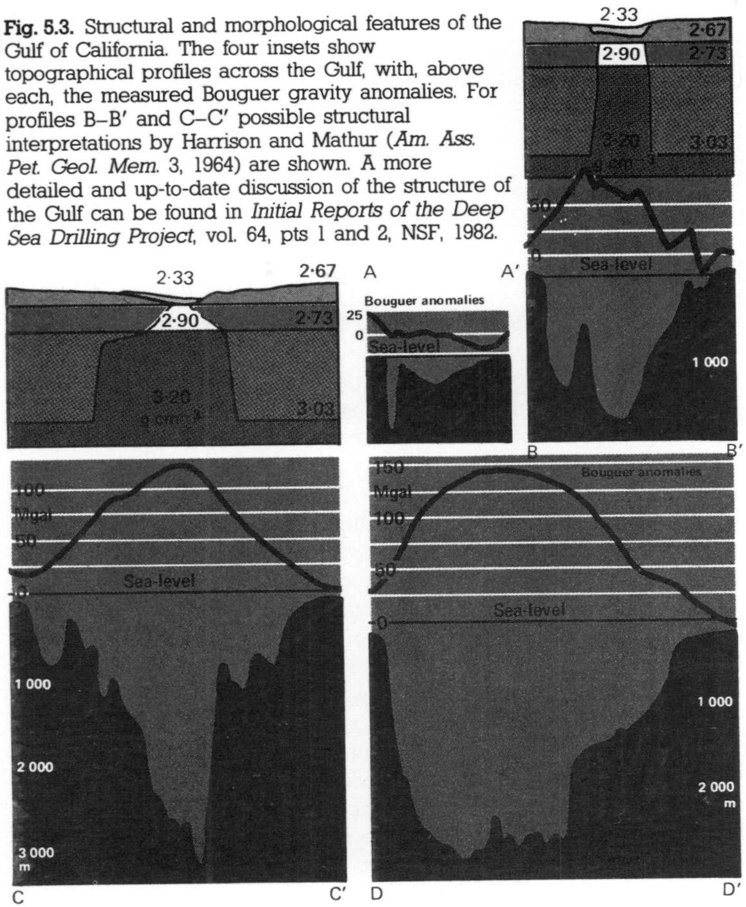

Fig. 5.3. Structural and morphological features of the Gulf of California. The four insets show topographical profiles across the Gulf, with, above each, the measured Bouguer gravity anomalies. For profiles B–B′ and C–C′ possible structural interpretations by Harrison and Mathur (*Am. Ass. Pet. Geol. Mem.* 3, 1964) are shown. A more detailed and up-to-date discussion of the structure of the Gulf can be found in *Initial Reports of the Deep Sea Drilling Project*, vol. 64, pts 1 and 2, NSF, 1982.

and the transmission of seismic waves is unexpectedly slow. Bouguer gravity anomalies can be interpreted as indicating mantle material of unusually low density. All these properties could be attributed to thermal expansion of the hot upper mantle, a condition also implied by the high heat-flow measurements. A possible supplementary factor in elevation of the crust is the occurrence of phase changes in the mantle. Phase in this context refers not only to the liquid and solid states but also to the structural changes which solid materials may undergo as temperature and pressure vary; volumetric changes arising from rearrangement in the packing of atoms can amount to 10 per cent or more. Probably the simplest overall model envisages a divergent margin lying above the upward-moving limb of a pair of convective cells. Thermal expansion in the mantle elevates the topmost brittle crust by 1 000–3 000 m, and then splits it symmetrically to cause separation of any continental mass that may be involved; associated adjustments in the lower crust and the rest of the lithosphere take place by means of viscous creep.

Older examples of continental separation

So far attention has been confined to continental areas where separation by sea-floor spreading has been active at the most since mid-Cenozoic times. There are many continental margins which, on the hypothesis of rigid lithospheric plates, must have originated rather earlier by separation along oceanic spreading centres. These have been termed passive or Atlantic-type margins, but the extent to which they display common topographic forms directly attributable to their early tectonic history is still uncertain. In most instances continental separation took place at such a distant period that morphological expression of domed uplift might be expected to have been destroyed by erosion. More likely to be preserved is the partially attenuated margin of the continent, subject to subsidence as it moved away from the spreading centre.

Both the North American and European borders of the Atlantic appear to fit this evolutionary model reasonably well. The continental shelves bear thick sedimentary accumulations of Cretaceous and Cenozoic age resting directly on a Palaeozoic basement (Fig. 5.4). The beds generally thicken seawards and at the continental slope may attain an aggregate thickness of many kilometres. In the United States it has been estimated that the total volume of Mesozoic and later sediments between the Appalachians and a line 1 000 km

Fig. 5.4. Section through the continental margin off Cape Hatteras, North Carolina, USA. Note that a ridge (? reef complex) appears to have served as a dam until the end of Cretaceous times (after Emery, *et al.*, 1970).

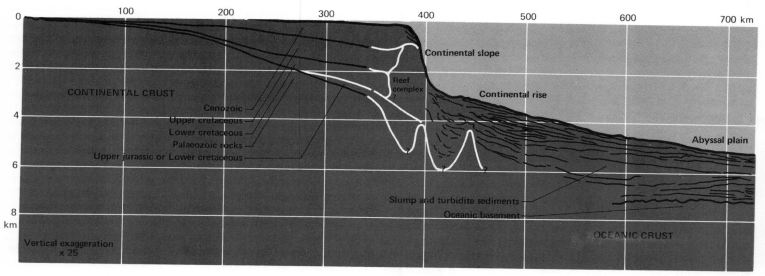

seaward of the coast amounts to 8.7 million km³; this is equivalent to a mean thickness over the whole area of about 2 500 m. The Cenozoic contribution to this total is 2.8 million km³, or a mean thickness of 800 m. Two implications of these vast quantities deserve mention. The first is the substantial denudation required to supply all the detritus. Although in the south a high proportion of the material consists of biogenic carbonates, further north it is dominantly terrigenous in origin. Even at a conservative estimate at least 2 500 m of rock seems to have been eroded from the adjacent Appalachian region during Cretaceous and Cenozoic times. The second implication is that the Palaeozoic basement must have sunk *pari passu* with sedimentation so as to accommodate such a thickness of shallow-water marine strata. It has been calculated that the mean rate of subsidence must have been about 0.01 mm yr⁻¹ throughout Cenozoic times. In part this is ascribable to isostatic depression beneath the growing weight of new sediment, but there must also be an element of tectonic movement. One way of assessing the subsidence that may be directly attributed to sea-floor spreading is to consider

the relationship between age and depth of the Atlantic Ocean. The modern mid-oceanic ridge rises on average to within 2 500 m of sea-level whereas the oldest ocean floor lies at over 5 500 m. In other words, as oceanic crust moves away from the ridge its level declines by some 3 000 m in 150 m.y. This yields a mean rate of 0.02 mm yr⁻¹, rather larger than the subsidence calculated for the shelf. However, plots of depth against age for the ocean floor suggest that new crust moving away from a spreading centre descends relatively rapidly for 80 m.y. (in conformity with the equation $d = 2\ 500 + 350t^{1/2}$, where t is the time in millions of years) but thereafter much more slowly. This could well explain the low figure on the Atlantic shelf during Cenozoic times, and there is certainly no gross discrepancy between the two independent methods of calculation.

Of the gross morphological features along a passive margin, it is the continental shelf that seems most readily ascribed to sea-floor spreading. However, a few workers have also sought evidence for the domed uplift that is believed to precede actual rupturing. It has been suggested, for instance, that the basic form of the Atlantic seaboard in Canada may be attributable to events that occurred when North America split from Europe and Africa. Broad arching in Permo-Triassic times was immediately succeeded by deep rifting to produce the numerous graben that are still a prominent feature of Atlantic coast tectonics. As the land masses separated, the trailing edge of the continent subsided to form the continental shelf. Marine transgression ensued, but the more elevated remnants of the original arch were never submerged so that they have been the site of continuous subaerial erosion since early Mesozoic times. It seems clear that the modern relief cannot incorporate remnants of the original Mesozoic landscape because the amount of erosion has been too great. Rather, the present landforms are the product of prolonged erosion acting on a region of resistant rocks last subject to major tectonic uplift at the time of continental separation; this would not preclude the spasmodic elevation that would periodically be required to maintain isostatic equilibrium. It is worth recalling that, despite much argument about the evolution of the river pattern in the northern Appalachians, many workers have contended that the original drainage flowed westwards from an axis of uplift near the continental edge and was progressively reversed by piratical streams draining to the Atlantic coast.

If a comparison is made with the Atlantic margin of Europe, certain similarities are immediately discernible. Deep graben filled with Triassic sediments separate many of the upland areas of Britian, and analogous structures have now been located on parts of the continental shelf to the west. The main cover of the shelf is of Cretaceous and Cenozoic age, and although less thick than on the other side of the Atlantic still attests to long-term subsidence. In upland Britain attempts to trace the evolution of the present drainage system have commonly envisaged an original system of rivers flowing eastwards away from the continental edge. Yet there are equally striking contrasts. Exploration of oil and gas resources in the North Sea has shown that this intra-continental basin has suffered a complex tectonic history, including an early phase of rifting that has even been interpreted by some workers as an aborted line of continental separation. In north-western Europe the Cretaceous transgression was extremely widespread and resulted in a bed of chalk that provides a valuable datum for deciphering subsequent earth movements. In addition to being buried beneath 3 000 m of later sediments in the central North Sea basin, this bed has undergone flexuring that, by virtue of its

contemporaneity with prolonged orogenic activity in the Alps, has been referred to as the 'outer ripples of the Alpine storm'. These tectonic disturbances, subsequent to those associated with the rupture from North America, have greatly complicated the structural evolution of the Atlantic fringe of Europe. This is also evident in the contortion of strata on the continental shelf, contrasting with the relatively undisturbed disposition of the beds off the North American coast.

Attention has been concentrated on the continental margins around the North Atlantic because these are among the most thoroughly studied in the world. They have the disadvantage that continental separation is believed to have begun in the south 200 m.y. ago so that there has been an immensely long period of subsequent erosion. In the north the opening between Greenland and Scandinavia is more recent, having been dated to about 60 m.y. ago. The gross relief of the Scandinavian peninsula appears to correspond reasonably well to the model of continental rupture by sea-floor spreading. An elongated marginal block, faulted along its western edge, is tilted eastwards so as to direct most of the drainage to the Baltic rather than the Atlantic. However, at least two complicating factors need mention. As noted above, north-western Europe suffered major tectonic dislocation during the Alpine orogeny, and many workers have ascribed uplift of the Norwegian mountains to this cause. Secondly, part of the basin-shaped form of Scandinavia is due to depression beneath the Pleistocene ice-sheets and the potential effects of glaciation need to be borne in mind when evaluating the present relief pattern.

Convergent plate margins

The lack of identity between the pattern of plates and continents results in three different types of convergent margin – ocean-to-ocean, ocean-to-continent and continent-to-continent. The associated structural and relief features are so contrasted as to require separate consideration.

Ocean-to-ocean type

The convergence zone of two oceanic plates is characterized by trenches and island arcs. The essential topographic features of the trenches, already described on p. 11, need no further elaboration. For purposes of a simple classification, the associated strings of islands are sometimes divided into single-arc systems in which the trench is bordered at some distance by a long line of volcanic islands, and double-arc systems in which a parallel festoon of mainly sedimentary islands is interposed next to the trench. The single-arc system is well exemplified by the Kuril Islands, the double-arc by Indonesia, and the progression from one to the other by the Aleutians. In relating surface form to inferred structure, it is presumed that the descending lithospheric plate reaches a sufficient temperature to generate a regular line of volcanoes at a distance of about 200 km from the trench. Later erosion of the volcanic peaks, together with incorporation of ocean-floor sediments carried by the subducted plate, constructs near the trench a thick accretionary wedge of sediment that is folded, faulted and metamorphosed before being uplifted into a second line of islands. Arcs were at one time interpreted as mountain chains *in statu nascendi*, but several examples are now known to be older than the analogous mountain ranges into which they were supposed to develop. It appears therefore that arc–trench systems, despite being very active parts of the earth's crust, can endure for long periods without major modification, although when one of the plates carries a continental passenger so that the margin is converted into the continent-to-ocean type potentially dramatic results ensue.

Continent-to-ocean type

The mountains along the western border of North and South America provide the best example of forms and structures believed to result from an oceanic plate thrusting beneath the edge of a continental block. In North America the relationships are complicated by the sea-floor spreading in the Gulf of California, but the Andean area seems to provide a relatively simple illustration of the continent-to-ocean type of plate margin. The Peru–Chile trench extends as a structural feature some 4 500 km from the equator to 44° S, although near its northern and southern ends it becomes so filled with debris that it shoals to less than half its maximum depth. The middle reach of the trench descends to about 8 000 m while many Andean ranges rise to over 5 000 m. The maximum relief amplitude of about 15 000 m is the greatest found anywhere on earth within a horizontal distance of 400 km. Seismic activity defines a prominent Benioff zone dipping rather irregularly eastwards beneath the continent. The

Mohorovicic discontinuity declines from 11 km beneath the Pacific to over 70 km below the Andes and then rises steeply again beneath the eastern edge of the mountains. The central section of the cordillera is divisible into three physiographic and structural units. On the east a mountain system rising to well over 5 000 m consists mainly of metamorphosed Palaeozoic sediments. On the west a second mountain system of comparable height is composed of abundant Mesozoic and Cenozoic volcanics, interlayered with shallow-water marine deposits; in addition, numerous batholithic intrusions ranging in age from Jurassic to late Cenozoic tend to become younger in an easterly direction. Between these two elevated units is the region known as the altiplano which has been a basin of accumulation for continental clastic debris since at least Cretaceous times; the combined thickness of sediments and volcanic materials may here exceed 10 000 m.

It appears that the Andean chain first began to form in early Mesozoic times as the Nazca plate being generated at the East Pacific Rise underthrust a continental margin laden with a thick sequence of Palaeozoic marine sediments. A volcanic arc developed above the subducting plate and the accumulated Palaeozoic rocks suffered their first major deformation. This earliest episode may have preceded the opening of the South Atlantic, but later developments were almost certainly linked with the westward drift of the South American continent. Activity quickened in late Mesozoic and early Cenozoic time, one of the traditional periods of Andean orogeny. The volcanic and associated marine sediments in the west were sharply folded and intruded by granite plutons, while the earlier rocks of the east suffered additional compression and uplift. This pattern continued into later Cenozoic times with plutonic activity slowly migrating eastwards. The migration of plutonism away from the trench is puzzling, but has been attributed to changing thermal conditions along the upper surface of the descending plate; as a plate descends it cools the adjacent mantle so that progressively deeper penetration is required to maintain magma production. An economically important hypothesis, although not universally agreed, is that subducted metalliferous sea-floor sediments make a major contribution to the hydrothermal ore deposits that characterize the whole cordillera.

Although phases of slightly more intense activity may be distinguished, the central Andes appears to have suffered almost continuous tectonic deformation since early Mesozoic times. The volume of clastic material testifies to concomitant erosion on a prodigious scale. It seems unlikely that in the environment of such an actively growing mountain range topographic forms of any great antiquity can be preserved. Nevertheless, many workers have discerned in the relief of the Andes indications of uplifted erosion surfaces, explicable only if elevation has been spasmodic but extremely rapid.

The concept of plate subduction along a Benioff zone seems to place a natural limit of a few hundred kilometres on the width of a mountain chain. Where a cordilleran system exceeds that width, a more complicated evolutionary history must be envisaged. This applies to the western cordillera of the United States extending 1 500 km from Colorado to the Californian coast. It is presumed that much of the orogenesis is due to westward drift of the continent, but the width of the mountain belt must result from more complex tectonics than those of simple subduction. The exact pattern of plate movement has yet to be agreed, but some of the results are clear. Whereas undisturbed Cretaceous and Cenozoic sediments of the Atlantic continental shelf have an aggregate volume of 8.7 million km^3, for the Pacific coast the corresponding figure is no more than 2 million km^3 of which at least half is pyroclastic in origin. However, an immense amount of Mesozoic and Cenozoic material, possibly totalling over 5 million km^3, has been returned to the continent in the form of the thermally metamorphosed, contorted and uplifted materials known as mélange that comprise the various coastal mountain ranges. As North America has drifted westwards, the detritus from erosion of the continent has apparently been subject to continual orogenic recycling.

Continent-to-continent type

Continental crust is not readily subducted, and the most complex orogenic conditions prevail when both convergent lithospheric plates carry continental 'passengers'. Such a situation is illustrated by the Alps and Himalayas lying between Eurasia on the north and Africa and India on the south. The precise sequence of events has varied greatly along different segments of the mountain system. Large volumes of sediment in the Alpine system were laid down in the Tethyan Sea between Europe and Africa. In early Mesozoic times the sea was truly oceanic in the east but narrowed westwards where it was dominated by extensive shelves on which thick sedimentary

sequences accumulated. Frequent shifts in the location of spreading centres and subduction zones led to fragmentation of the continental margins with the production of numerous 'micro-plates' which, by their relative rigidity, much influenced the pattern of later deformation. Closure of the Tethyan Sea that began in later Mesozoic times squeezed the 'micro-plates' between the jaws of the Africa–Europe vice and eventually, in early to mid-Cenozoic times, led to the development of huge nappes and overthrust sheets. The uplifted areas were subject to intense erosion with deposition of thick molasse sequences in intermontane and marginal basins.

The Himalayas were initiated when the Indian and Asian continental blocks came into collision about 45 m.y. ago. Prior to that event a northward-dipping subduction zone existed beneath Tibet, and it was plate consumption along this zone that ultimately led to India being 'sutured' to Eurasia. The precise processes responsible for the major mountain chain that then developed along the line of contact remain obscure, partly because uncertainty still surrounds the deep-seated structures beneath the Himalayas. There is no evidence for conventional subduction, and yet interpretation of the magnetic anomalies in the Indian Ocean implies crustal shortening of at least 1 700 km since the two continental blocks first collided. Much of this has probably been absorbed by indentation of the Asian plate; the Tibetan crust has been thickened to some 70 km, and it has been argued that huge slices of continental rock have been squeezed out eastwards along a series of transcurrent faults. The remaining shortening has apparently been absorbed by the frontal edge of the Indian plate fragmenting along a group of northward-dipping low-angle shears that have resulted in successive slices of continental crust being thrust one below the other. Concurrent uplift has characterized not only the main Himalayan chain but also many areas as far north as the Tien Shan on the Sino-Soviet border.

Where two continental blocks collide structural conditions are so complex and varied that it is difficult to generalize about cordilleran development. Appropriate models of orogenesis will undoubtedly be developed in the future, but for the moment it is worth reiterating that, on the modern view of a mobile earth, mountain-building is a persistent activity that is certainly taking place at the present time. Uplifted erosion surfaces that were formerly interpreted as characteristic of a late stage of orogenesis must presumably now be seen as an integral part of long and involved sequences of events that may be repeated many times in the evolution of a cordilleran belt.

Intra-plate movements

Emphasis in the concept of lithospheric plates is on horizontal displacements, with major vertical movements being envisaged only at plate margins. Yet, as indicated in Chapter 4, there is abundant evidence for epeirogenic movement in areas far from accepted plate edges, and some geodetic levelling appears to indicate that a continental surface can undergo almost constant vertical adjustment. It is clearly possible to exaggerate the rigidity of lithospheric plates and the very shape of the geoid requires that, as they move across the surface, they experience some deformation. It should also be recalled that many areas nowhere near a plate margin suffer the occasional severe earthquake. For example, although California is renowned for its seismicity, two of the largest earthquakes in the United States have been centred near Charleston on the Atlantic seaboard and near New Madrid in the Mississippi valley. Attempts are now being made to evaluate current stress distributions within the major continental blocks. Combined data from observations on recent tectonism, from first-motion studies of earthquakes and from direct measurement of *in situ* stress reveal the existence of relatively uniform 'stress provinces' that may each cover an area of 10^6 km^2 or more. For example, maximum compressive stresses for much of North America trend ENE – WSW and for western Europe NW – SE. Abrupt transitions between compressive and extensional provinces in the western United States seem to signify a crustal or uppermost mantle source for the stresses.

It is clear that plate surfaces must not be assumed to be totally inert regions. The evidence from north-western Europe suggests that orogenesis produced by continental collision can cause appreciable deformation 1 000 km or more from the actual plate margin. In addition, it was noted earlier that many geomorphologists have invoked domed uplift to explain specific aspects of continental relief. In some instances this uplift could be due to initiation of a new spreading centre, but in others it seems unrelated to current plate margins. Africa probably provides the clearest examples of such epeirogenic uplift and depression, but most continental blocks have suffered some degree of deformation in this way. On the oceanic parts of the plates, DSDP investigations by the *Glomar Challenger*

showed that broad topographic swells, up to 500 000 km^2 in area and 2 000 m in relative relief, result from outpourings of exceptional volumes of volcanic rock. These so-called mid-plate rises imply unusual mantle conditions of localized extent and there is no intrinsic reason why they should be confined to oceanic areas.

The hypothesis of lithospheric plates undoubtedly provides the most satisfactory explanation yet offered for global relief patterns. Yet many features still defy full explanation, and it is essential to realize that, despite its apparent simplicity, the plate concept will require, and is capable of, much future refinement.

Volcanic landforms

Just as plate margins are the site of the most intense tectonic deformation, so they are also the location of much of the most spectacular volcanic activity. A distinction may be drawn between divergent margins where, once continental separation is complete, nearly all extruded material is basaltic in character, and convergent margins where the range of chemical composition is much wider. Since there is a close relationship between magma composition and the resulting surface morphology, a contrast may be drawn between the volcanic landforms characteristic of mid-ocean ridges and those found in association with subduction zones. In addition, volcanic features situated in a number of other locations require brief examination.

Mid-ocean ridges

Vulcanicity along the mid-ocean ridges mainly comprises quiet fissure eruptions which extrude vast quantities of lava over the sea floor. The material is almost entirely a form of silica-rich and potassium-poor basalt known as tholeiite. Eruptions from more centralized vents, which are capable of building huge cones or shields rising several thousand metres from the sea floor, tend to be of alkali basalt. Examples of oceanic islands formed in this way near a mid-ocean ridge include Tristan da Cunha and Gough Island in the Atlantic, Prince Edward and St Paul Islands in the Indian Ocean and Easter Island in the Pacific.

The only large island built up by sustained vulcanicity astride a mid-oceanic ridge is Iceland (Fig. 5.5). Some 100 000km^2 in area and underlain by unusually thick oceanic crust, this landmass is divisible into three parts, a median trough-faulted zone generally

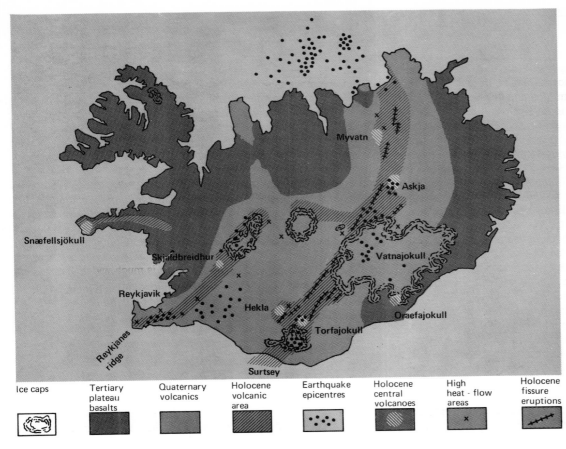

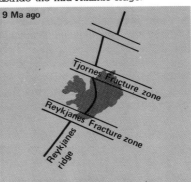

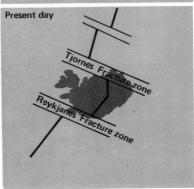

Fig. 5.5. The major structural and morphological features of Iceland. The inset diagrams show one possible interpretation of the growth of Iceland astride the mid-Atlantic ridge.

| Ice caps | Tertiary plateau basalts | Quaternary volcanics | Holocene volcanic area | Earthquake epicentres | Holocene central volcanoes | High heat-flow areas | Holocene fissure eruptions |

between 100 and 200 km wide, and two flanking plateaux mainly constructed of innumerable basaltic flows. The range of volcanic forms is very wide. Large areas have been covered with tholeiitic fissure eruptions, the lava sometimes being of such low viscosity that it flowed more than 100 km before solidifying. Gently sloping shield volcanoes have been constructed by protracted emission of lava with only small quantities of pyroclastic debris: Skjalbreidur, for example, is 600 m high, 10 km in diameter and has slopes that average about

7°. Complex strato-volcanoes with abundant pyroclastic material are less well developed in Iceland than in many continental areas, but Hekla and Oraefajokull are more appropriately described as strato-volcanoes than as simple shield volcanoes. In one or two localities a viscous rhyolitic magma has been extruded to form steep-sided domes. The occurrence of such acidic rock in this mid-oceanic situation has generally been attributed to differentiation within a magma chamber by the preferential settling of early-formed crystals like olivine. However, modern experimental work suggests a similar effect could result from different degrees of partial melting in a single mantle rock such as peridotite.

Lying astride an active spreading centre, Iceland might be expected to suffer gradual stretching with the emplacement of new volcanic material along the median zone. There is much evidence that such processes are operative. The oldest rocks on the island are about 20 m.y. old and are generally found in the extreme east and west; if they originated near the median zone they imply a mean spreading rate of about 20 mm yr^{-1}. A striking characteristic of the median zone is the profusion of fissures which, on selected traverses, have been shown to occupy some 6 per cent of the total distance and to represent a lateral extension of between 10 and 20 mm yr^{-1}. Repeated EDM surveys have also indicated widening of the central graben. In the 1960s a specially instrumented line 25 km long was found to have lengthened 65 mm in 3 years. Even more dramatic results were obtained near the north coast in the late 1970s. Within 3 years local extension of some 3 m was measured, and over a period of several months the ground surface was found to be rising at 7–10 mm day^{-1} for a week or more and then abruptly subsiding again.

Many seamounts and guyots may have originated by volcanic activity close to a spreading centre, and been carried to their present oceanic positions by plate movement. It should be possible to determine the proportions initiated in this way when their ages are better known and it can be seen whether they correspond to the date of the nearby sea floor. Many if not all guyots are believed to owe their flat tops to wave action; the later subsidence that this implies may be ascribed to the lowering of the plate surface as it moved away from the mid-oceanic ridge. The great thickness of coral on many Pacific islands, built on a volcanic cone but extending to depths far below those at which coral will grow today, might be explained by the same mechanism (see p. 357).

Subduction zones

Subduction along a Benioff zone appears to engender large-scale production of magma which rises to the surface along conduits through the overlying crust. This is most clearly seen in the development of volcanic island arcs, but is also responsible for extrusive activity in many orogenic regions. The characteristic lava of island arcs is andesite rather than basalt. Magmatic differentiation and partial melting have both been invoked to explain this composition, but an additional factor may well be the veneer of saturated sea-floor sediment on the downgoing slab that promotes generation of an andesitic magma when reheating occurs. Where the Benioff zone extends beneath a continent, assimilation of other rock types may be involved, and such circumstances seem particularly favourable for the extrusion of lavas of widely different composition.

It is this variety of erupted materials that contributes to the wide diversity of volcanic landforms. In general the greater the acidity of the magma, the higher the proportion of pyroclastic debris and the more explosive the average eruption (Table 5.1). Using the percentage of fragmentary materials as a measure of explosive violence, Rittman in 1962 estimated that areas in the central Pacific would have an explosion index of 3, whereas the Indonesian island arc would have an index of 99. Characteristic volcanic forms along Benioff zones include many features built either wholly or in large measure of pyroclastic fragments. Cinder or scoria cones can be constructed very rapidly around the pipe from which the material is being ejected. Of much greater extent are plains built by ash flows

Table 5.1 Regional contrasts in the nature of volcanic activity (after Rittman)

Location	Explosion Index
Island arcs	95
Cordilleras (Andes)	97
Grecian Islands	83
Southern Italy	41
African rift system (continental)	40
Iceland	39
Atlantic and Indian Oceans	16
Pacific Ocean	3

Explosion index – fragmentary material expressed as percentage of all material emitted.

in which fine solid particles are suspended in volcanic gases. When first erupted these flows move rapidly with many of the properties of a fluid, but after settling and cooling they form the rock type known as ignimbrite or welded tuff. Total thicknesses of over 100 m have been recorded, and in Sumatra one area of ignimbrite covers more than 25 000 km^2.

It is, however, the composite strato-volcano that must be regarded as the most characteristic form associated with an active Benioff zone. Eruptions that build volcanoes composed of both lava and detrital fragments are commonly classified as either Strombolian, Vulcanian or Pelean depending on their nature and degree of violence. Strombolian eruptions are either continuous or have a cycle of activity that is repeated after only a short period of dormancy. The magma is often basaltic in composition and the proportion of pyroclastic material low. Vulcanian eruptions tend to be more explosive and to be associated with increasingly viscous lava that solidifies rapidly. The interval between eruptions is greater and activity is often renewed by ejection of vast amounts of ash as any blockage in the vent is destroyed. In a Pelean eruption the solid summit of the volcano is first elevated and then disintegrated as one or more 'glowing avalanches' or clouds of ash descends the slopes of the cone. The Pelean eruption is usually associated with the most acidic magmas. Occasionally where a magma is extremely viscous and pressure from beneath is limited, a volcanic dome may be built over the vent in preference to the more common strato-volcano.

Composite cones possess a simple outward form but an intricate structure. The external slopes are typically either straight or gently concave with angles between 10 and 35°. Although the main conduit rises vertically through the volcano, subsidiary dikes often feed small lateral cones; it is not uncommon for the top of the main cone to be composed primarily of pyroclastic material while the base contains proportionately more lava extruded from fissures on the lower slopes. Nearly all composite volcanoes have a central crater, but its size and form may vary considerably. Following an eruption and the emptying of the top of the conduit by a fall in magma level, the unsupported walls of the crater tend to collapse; if the material is mainly pyroclastic, loose screes develop, whereas if it is mainly lava, steep rock faces can be preserved until the next eruption. After a long and complex history, many composite cones finally collapse to form large depressions known as calderas. These may be 10 km or more in diameter. Famous examples include Mt Mazama, the crest of which disappeared some 6 000 years ago to form Crater Lake in Oregon, Monte Somma which collapsed in AD 79 overwhelming the Roman towns of Pompeii and Herculaneum with ash and mudflows, and Krakatoa in the Sunda Strait which was destroyed with exceptional explosive violence in 1883.

The potential of strato-volcanoes for devastating eruptions was dramatically illustrated in 1980 by Mt St Helens in the north-western United States. After 123 years of dormancy renewed activity was first signalled by earth tremors that were correctly interpreted as indicating the movement of magma into the base of the mountain. An ominous bulge was then noticed growing on the northern flank of the cone, and a few weeks later the main eruption began with an earthquake that triggered a huge landslide on the bulge. A devastating blast of ash, gas and steam exploded from the breach and laid waste a zone extending 20 km from the volcano on a front almost 30 km wide. Even at a distance of 20 km fully grown trees were blown over like matchsticks. Mudflows of saturated ash and debris coursed down local valleys, filling their floors to depths of 60 m in places. Ash that was shot into the sky covered large areas of the State of Washington to a depth of 10 cm or more. Less than 2 years after the Mt St Helens eruption, the volcano of El Chichon in Mexico experienced an even more violent although less well publicized phase of activity. Both these volcanoes lie above subduction zones dipping eastwards beneath the margin of the North American continental block.

Vulcanicity in other regions

Although most recent vulcanicity is concentrated along the mid-oceanic ridges and subduction zones, several other localities have experienced major eruptions during Mesozoic and Cenozoic times. One important group of eruptions is that responsible for the extrusion of huge quantities of basalt on to the continental surfaces. Examples covering well over 100 000 km^2 each are afforded by the Columbia plateau basalts in the United States, the Deccan traps in India and the Parana basalts in South America. Smaller areas include the Drakensberg region of South Africa and the Antrim plateau in Northern Ireland. The volumes of lava are enormous, with the Parana basin alone estimated to contain some 500 000 km^3. Several of the outpourings might be related to extrusive activity near a spreading centre during a phase of continental separation. For instance, the

Parana basalts are Jurassic in age and may be linked to the initial rupture between South America and Africa; similarly the Karroo lavas in South Africa may be related to an early phase of Gondwanaland fragmentation. A third possible example is afforded by the Antrim plateau basalts and the associated extrusives of western Scotland. Most evidence points to these rocks being no younger than Oligocene, and at least one radiometric assay has yielded an age of 74 m.y. This is entirely consonant with vulcanicity at the time northwestern Europe was actively splitting from Greenland and North America. In all these three cases the basalt might be construed as an outpouring near a divergent plate margin that has been carried to its present position by plate movement. The Columbia plateau basalts and Deccan traps, although less readily explained in this way, are generally acknowledged as a manifestation of local crustal tension; however, the underlying cause of that tension remains to be determined.

A second group of eruptive centres occurs in association with continental rift systems. The outstanding example is that of East Africa. Here some of the volcanic rocks are very alkaline in composition, and in an extreme form consist almost entirely of carbonates with little or no silica. The ash from such volcanoes may be so rich in sodium carbonate that after heavy rain it renders local waterholes highly poisonous for both cattle and wildlife. There has been much discussion regarding the origin of 'carbonatites', and it may be significant that other major rift systems, such as the Rhine valley in Germany and Lake Baikal in Siberia, have experienced highly alkaline vulcanicity; both structures and igneous activity seem to connote rather unusual conditions within the mantle and lower crust.

A third group of volcanoes not directly related to plate margins includes those that lie in distinctive linear patterns across the ocean basins. They are particularly characteristic of the Pacific and the best example is provided by the Hawaiian archipelago and its northward continuation in the Emperor seamounts, together forming a single chain that can be traced for nearly 6 000 km. Radiometric dating of the lavas has shown a systematic increase in age from the active centres of Mauna Loa in the south-west, to almost 30 m.y. at the northward extremity of the archipelago, and to 64 m.y. near the northern end of the seamounts. These dates all contrast with the local oceanic crust which is thought to be more than 70 m.y. old. It is believed that the whole volcanic chain results from the gradual move-

ment of the Pacific plate over an almost static hot-spot in the mantle, an idea largely confirmed in the late 1970s by one of the DSDP cruises of the *Glomar Challenger* specifically designed to test the hypothesis. Movement of the plate relative to the hot-spot was calculated as averaging 80 mm yr^{-1}, and an abrupt alteration in the alignment of the volcanoes was attributed to a sudden change in the pattern of sea-floor spreading 43 m.y. ago. The idea of hot-spots has been tentatively extended to other lines of volcanic islands on the Pacific floor, and even to extrusive activity in other regions, but the supporting evidence is not always entirely convincing.

Eustatic changes of sea-level

Mid-oceanic ridges are of suffecient size to provide a mechanism by which world-wide changes in sea-level could be induced. The idea of such changes is a recurrent theme in geomorphological studies concerned with the pre-glacial evolution of uplands in both Europe and North America. However, until the advent of plate tectonics, it was difficult to envisage on a globe of constant size any process that could significantly change world sea-level other than the growth and decay of ice-sheets. This glacio-eustatic effect is considered in Chapter 16, and attention will here be confined to the range of sea-level changes that might theoretically result from variations in the dimensions of the ocean basins.

At present the total area of the oceans is 362×10^6 km^2 and the volume of water $1\ 350 \times 10^6$ km^3, giving a mean depth of 3 730 m. On a smooth sphere with all the water concentrated in a single uniform ocean, the water would be about 2 440 m deep and its surface measured from the centre of the earth would lie some 240 m above present sea-level. This constitutes a theoretical maximum height for the ocean surface. If progressively more of the crust is then visualized as rising above the ocean-level, that level must in turn sink. The present hypsographic curve represents one particular state of this relationship, and the fundamental distinction between continents and ocean basins suggests it is unlikely to alter radically with time. However, sea-level would fall well below its present altitude if the mid-oceanic ridges did not occupy substantial parts of the ocean basins. The total volume of the ridges has been estimated at 119×10^6 km^3, which is equivalent to a water layer over the oceans about 330 m thick. Given constancy in the size of the globe and the volume of water, it is reasonable to conclude that sea-level has never been more than 240 m above, nor more than 330 m below, its present height. It is most unlikely that either of these extremes has even been approached: the first demands total disappearance of the continents, for which there is no evidence, and the second implies an unexplained destruction of the matter composing the present mid-oceanic ridges. More realistically the latter represent a partial conversion of mantle material from high to low density; the likely volumetric expansion has been calculated at no more than 27×10^6 km^3, which is equivalent to a variation in sea-level of about 74 m. This value would be reduced still further on the very reasonable assumption that the absence of spreading centres would also involve the disappearance of ocean trenches.

The above figures are purely hypothetical and merely demonstrate the capacity of mid-oceanic ridges to affect global sea-level. Whether they have actually done so is still a matter of conjecture. The problem is intimately linked with the periodicity of major tectonic processes in general. It is possible to envisage a steady-state situation in which plates are constantly moving but there is very little change in sea-level. On the other hand, if all spreading centres were periodically to become inactive and the ocean floors to subside, long-term eustatic changes might be expected. The pattern of magnetic anomalies over the ocean floors should theoretically provide a guide to changes in spreading rates, but the global synchroneity of any observed variations is still very uncertain. Another possible method of assessing constancy in plate movement is to examine changes in the rate of orogenic activity at convergent margins. Several writers have reviewed the evidence from the cordilleran chain along the Pacific coast of the Americas, but have come to different conclusions. In 1971 Damon argued in favour of periodicity, whereas a year later Gilluly collated dates from deep-seated plutons that seem to imply virtually continuous orogenic activity since early Mesozoic times. There has even been disagreement whether orogenic movements are associated with fast or slow motion. At first sight rapid motion might seem likely to promote orogenesis, but some workers have maintained that the onset of mountain-building at a convergent margin could itself reduce the previous spreading rate. Whatever the final outcome of these deliberations, it is worth emphasizing that within the relatively short time-scale relevant to geomorphological as distinct from geological investigations, plate-tectonic theory seems unlikely to yield mechanisms for major global sea-level changes. Of greater significance may be the palaeo-geoidal arguments of Mörner that are discussed further in Chapter 16.

References

Abdallah, A. et al. (1979) 'Relevance of Afar seismicity and volcanism to the mechanics of accreting plate boundaries', *Nature, London* **282**, 17–23.

Bishop, W. W. and Trendall, A. F. (1966–67) 'Erosion surfaces, tectonics and volcanic activity in Uganda', *Q. J. Geol. Soc. Lond.* **122**, 385–420.

Bjornsson, A. et al. (1979) 'Rifting of the plate boundary in North Iceland, 1975–1978', *J. Geophys. Res.* **84**, 3029–38.

Bohannon, R. G. (1986) 'How much divergence has occurred between Africa and Arabia as a result of the opening of the Red Sea?' *Geology* **14**, 510–13.

Brown, C. and Girdler, R. W. (1982) 'Structure of the Red Sea at 20 °N from gravity data and its implication for continental margins', *Nature, London* **298**, 51–53.

Cronin, T. M. (1981) 'Rates and possible causes of neotectonic vertical crustal movements of the emerged southeastern United States coastal plain', *Geol. Soc. Am. Bull.* **92**, 812–33.

CYAMEX Scientific Team (1979) 'Deep tow studies of the Tamayo transform fault', *Mar. Geophys. Res.* **2**, 37–70.

CYAMEX Scientific Team (1981) 'First manned submersible dives on the East Pacific Rise at 21 °N (project RITA): general results', *Mar. Geophys. Res.* **4**, 345–79.

Damon, P. (1971) 'The relationship between late Cenozoic volcanism and tectonism and orogenic-epeirogenic periodicity', in *Late-Cenozoic Glacial Ages* (ed. K. K. Turekian), Yale Univ. Press.

Decker, R. W. et al. (1971) 'Rifting in Iceland – new geodetic data', *Science N. Y.* **173**, 530–2.

Doornkamp, J. C. (1968) 'The nature, correlation and ages of planation surfaces in southern Uganda', *Geogr. Annlr.* **50-A**, 151–61.

Emery, K. O. et al. (1970) 'Continental rise off eastern North America', *Am. Ass. Petrol. Geol. Bull.* **54**, 44–108.

Flemming, N. C. and Roberts, D. G. (1973) 'Tectono-eustatic changes in sea level and seafloor spreading', *Nature, London* **243**, 19–22.

Gansser, A. (1973) 'Facts and theories on the Andes', *J. Geol. Soc. Lond.* **129**, 93–131.

Gilluly, J. (1973) 'Steady plate motion and episodic orogeny and magmatism' *Geol. Soc. Am. Bull.* **84**, 499–514.

Girdler, R. W. and Evans, T. R. (1977) 'Red Sea heat flow', *Geophys. J. R. Astr. Soc.* **51**, 245–51.

Girdler R. W. et al. (1980) 'A geophysical survey of the westernmost Gulf of Aden', *Phil. Trans. R. Soc.* **A298**, 1–43.

Jackson, E. D. and Koisumi, I. (1980) *Initial Reports of the Deep Sea Drilling Project*, Vol. 55, Washington (US Govt Printing Office).

Jordan, T. E. et al. (1983) 'Andean tectonics related to geometry of subducted Nazca plate', *Geol. Soc. Am. Bull.* **94**, 341–61.

Koski, R. A., Clague, D. A. and Oudin, E. (1984) 'Mineralogy and chemistry of massive sulfide deposits from the Juan de Fuca Ridge', *Geol. Soc. Am. Bull.* **95**, 930–45.

Kulm, L. D. et al. (1981) *Nazca Plate: crustal formation and Andean convergence*, Geol. Soc. Am. Mem. 154.

Lipman P. W. and Mullineaux, D. R. (1981) 'The 1980 eruptions of Mount St Helens, Washington', *U. S. Geol. Surv. Prof. Pap.* 1250.

Menard, H. W. (1969) 'Elevation and subsidence of the oceanic crust', *Earth Plan. Sci. Lett.* **6**, 275–84.

Molnar, P. and Tapponnier, P. (1977) 'The collision between India and Eurasia', *Sci. Am.* **236** (4), 30–41.

Officer, C. B. and Drake, C. L. (1982) 'Epeirogenic plate movements', *J. Geol.* **90**, 139–53.

Officer, C. B. and Drake, C. L. (1985) 'Epeirogeny on a short geologic time scale', *Tectonics* **4**, 603–612.

Parsons, B. and Sclater, J. G. (1977) 'An analysis of the variation of ocean floor bathymetry and heat flow with age', *J. Geophys. Res.* **82**, 803–27.

Patriat, P. and Achache, J. (1984) 'India–Eurasia collision chronology has implications for crustal shortening and driving mechanism of plates', *Nature, London* **311**, 615–21.

Richardson R. M., Solomon, S. C. and Sleep, N. H. (1979) 'Tectonic stress in the plates', *Rev. Geophys. Space Phys.* **17**, 981–1019.

Rittman, A. (1962) *Volcanoes and their Activity*, Wiley.

Tarantola, A., Ruegg, J. C. and Lepine, J. P. (1980) 'Geodetic evidence for rifting in Afar, 2: Vertical displacements', *Earth Plan. Sci. Lett.* **48**, 363–70.

Veevers, J. J. (1981) 'Morphotectonics of rifted continental margins in embryo (East Africa), youth (Africa-Arabia) and maturity (Australia)', *J. Geol.* **89**, 57–82.

Watts, A. B. (1982) 'Tectonic subsidence, flexure and global changes of sea level', *Nature, London* **297**, 469–74.

Selected bibliography

A stimulating discussion of the apparent relationships between surface form and deep-seated structural movements is provided by C. D. Ollier, *Tectonics and Landforms*, Longman, 1981. The crucial importance of contrasting relationships at different points on a lithospheric plate is being increasingly recognized in the compilation of research papers related to one specific environment: J. K. Leggett (ed.) *Trench–forearc Geology: sedimentation and tectonics on modern and ancient active plate margins*, Blackwell, 1982: A. W. Bally, P. L. Bender, T. R. McGetchin and R. I. Walcott (eds) *Dynamics of Plate Interiors*, Am. Geophys. Union and Geol. Soc. Am. 1980: 'The Evolution of passive continental margins in the light of recent deep drilling results', *Phil. Trans. R. Soc.* **A294**, 1980. Likewise, modern ideas on orogeny are reviewed in A. Miyashiro, K. Aki and A. M. C. Sengör, *Orogeny*, Wiley, 1979 and K. J. Hsü (ed.) *Mountain-building Processes*, Academic Press, 1982.

A useful source on volcanoes and volcanic landforms is H. Williams and A. R. McBirney, *Volcanology*, Freeman Cooper, 1979.

Part 2
SUBAERIAL DENUDATION

Chapter 6
Weathering and the regolith

As the essential precursor to transport of rock material by subaerial agencies, weathering forms a fitting link between the internal processes reviewed in Part 1, and the external processes that form the subject of this Part 2. Weathering may be defined as the response of rock materials that were once in equilibrium with conditions in the lithosphere to the changed conditions at the earth's surface. It is in essence a form of metamorphism since the rocks are adjusting to a change in temperature and pressure in the presence of circulating liquids and gases. Ultimately a new equilibrium may be established with little physical or chemical alteration taking place in the weathering products. There are, however, many complex weathering cycles through which material may pass before such a state is achieved, and in the meantime the incompletely altered material has commonly migrated downslope under the influence of gravity. This mobility of the surface mantle or regolith also has the effect of exposing fresh rock to the weathering processes.

Two basic type of weathering may be distinguished, mechanical and chemical. The former refers to those processes involving disintegration or comminution of the original rock, the latter to those involving decomposition or chemical alteration. It must be emphasized that rock material is rarely affected by one process operating alone, but is normally subject to several interactive processes tending to induce concurrent disintegration and decomposition. Although stress is often laid on mechanical weathering in certain environments and chemical weathering in others, this merely indicates their relative importance without implying that either is totally absent.

Mechanical weathering

Four major sources of stress potentially capable of fragmenting bedrock are dilatation, thermal expansion, crystal growth and organic activity.

Dilatation

A deeply buried block of bedrock is subject to extreme confining pressures by the column of overlying material that may have a density of about 2 700 kg m^{-3}. As denudation removes the top of the column these pressures diminish and the block may adjust to this unloading by upward expansion. The clearest manifestation is the development of a closely spaced joint system, and on many quarry faces the rock within a few metres of the ground surface is seen to be divided into innumerable small cuboidal blocks, whereas the same material at depth remains massive with only a few widely spaced joints. Pressure release is a phenomenon well known to mine and quarry operators since it occasionally leads to explosive shattering of freshly opened faces, and it has also been encountered during deep-sea drilling when basalt pebbles brought to the surface have 'popped' or fractured spontaneously. Direct measurement of dilatation has been attempted by comparing the dimensions of a block of rock before and after its removal from the foot of a deep quarry wall; in this way linear expansion of about 1‰ has been recorded for granites in Georgia, USA.

The most striking natural dilatation features occur in massive igneous rocks where jointing parallel to the surface sometimes produces remarkably regular sheeting. This is particularly spectacular in the Sierra Nevada of California (Fig. 6.1) where pseudo-bedding layers at the surface may be no more than 0.1 m thick and yet the rock at depth is sufficiently massive to sustain sheer walls along the glaciated Yosemite trough many hundreds of metres high. Pressure-release structures are not limited to igneous rock, nor are they necessarily due to vertical unloading. Distinctive scars along canyon walls in the arid south-western United States have been ascribed to rockfalls controlled by joints arising from lateral expansion in massive sedimentary strata. It seems reasonable to infer that where layered structures develop parallel to the surface they will tend to perpetuate existing topographic forms; however, the main significance of the joint systems is probably the opportunity they

Fig. 6.1. Dilatational sheeting reminiscent of normal bedding planes developed on the surface of igneous rocks in the Sierra Nevada, California, USA. It is instructive to compare this superficial layering with the deep unjointed rock of similar composition that is capable of sustaining the huge vertical faces illustrated from the nearby Yosemite valley in Figs 12.1 and 12.2.

afford for deep penetration by other weathering agencies.

Thermal expansion

The idea of rock disintegration by thermally induced expansion and contraction has a long history. Many early desert explorers commented on the numerous shattered pebbles they noticed lying on the ground, and some even claimed the actual fracturing was accompanied by an audible report like a rifle-shot. The abundance of broken pebbles was commonly attributed to the large diurnal temperature ranges experienced in deserts, possibly accentuated by the lack of vegetation and soil cover to provide protection from the direct rays of the sun. Allusion was also made to the poor thermal conductivity of most rock materials resulting in a thin outer rind being heated and cooled over a much wider temperature range than the interior of a pebble. Finally, it was held that the varied albedos and coefficients of thermal expansion possessed by different rock minerals might be capable of inducing disintegrative stresses.

As early as 1915 Tarr was casting doubt on the more extravagant claims. Experimenting with granite blocks, he calculated the stresses due to solar heating as appreciably smaller than the elastic strength of the rock. A variety of later investigators have almost without exception failed to demonstrate fragmentation due solely to thermal changes. The best-known research on thermoclastis is that of Griggs who heated and cooled a granite cube 89 400 times through a temperature range of 30–140 °C. If each temperature cycle is equated with a single day, this is equivalent to 245 years of thermal weathering. Yet at the end Griggs was unable to detect any significant deterioration in rock structure. The general validity of many similar laboratory tests can hardly be doubted, but they do not dispose of the field observation that desert surfaces are often strewn with broken pebbles. Either an alternative process is responsible, or the laboratory tests do not reproduce in some vital particular the natural conditions of the desert floor. Ollier has pointed out that specimens heated in an oven are completely unconfined, whereas

the partial burial of many pebbles observed in the field might possibly generate extra disintegrative stresses; along the same lines, it may also be better to employ a radiant heat source than the ambient temperature of an oven. Other workers have stressed the possible importance of exposed rock faces much larger than those normally studied in the laboratory. Still others, citing additional experiments by Griggs that demonstrated the relative effectiveness of cooling granite blocks in moist rather than dry air, have maintained that weakening by chemical alteration must precede any final thermal rupturing; yet some fractured desert pebbles are composed of quartzitic materials notably resistant to chemical change.

For the present the importance of thermoclastis as a weathering mechanism remains unproven, but further research should shed more light on the problem. In the laboratory cabinets are being designed that can be programmed to reproduce much more faithfully the varied microclimatic circumstances in which natural weathering occurs. In the field modern thermometry is being employed to monitor actual changes in rock temperature. It has been shown, for instance, that rock surfaces in deserts may reach 80 °C with a diurnal range of at least 40 °C, and that temperature gradients with depth can surpass 0.15 °C mm^{-1}. Probably the most intractable problem is assessing the potential effect of long-term fatigue failure which several authors have suggested may play a crucial role in thermoclastis. Two other heat sources that might contribute to thermal rupturing are fires and lightning, but again any quantitative estimate of their significance is difficult to obtain.

Crystal growth

The growth of foreign crystalline solids within a rock is known to generate large disruptive stresses. Both the freezing of water and the precipitation of solids from solution have been invoked as powerful disintegrative agencies. Water has the unusual property of reaching its maximum density at 3.98 °C. Above and below this temperature its density falls from the standard value of 1 000 kg m^{-3}, reaching 999.8 kg m^{-3} at 0 °C and declining to 917.0 kg m^{-3} on the transformation into ice. This phase change from water to ice thus involves a volumetric expansion of just over 9 per cent, but the temperature at which it occurs varies according to the confining pressure; for every increase of 10 MN m^{-2}, the temperature at which ice forms declines by about 1 °C. The best-known illustration of the force

generated during freezing is the fracture of domestic water pipes. As the air cools below 0 °C part of the water in the pipes begins to freeze, expanding in volume as it does so. This increases the pressure on the residual water which remains liquid until the temperature has fallen sufficiently to cause freezing at this higher pressure. In practice domestic copper piping succumbs to the stresses that result from sustained temperatures only slightly below 0 °C. Theoretically with a temperature of −22 °C, a pressure of 216 MN m^{-2} could be exerted on the confining walls of a rock joint; were the pressure to exceed that value the molecular structure of the ice would change, with denser packing leading to higher densities and diminishing volume. Such extreme pressures are not normal, if for no other reason than that any confining ice-seal would first be extruded from the joint. Estimates of the maximum pressures likely to be generated in rock fissures vary considerably, but claimed values of around 15 MN m^{-2} are of the same order as the measured tensile strength of many common rocks.

In practice the situation is undoubtedly more complex than the foregoing analysis indicates. Much of the water in a fine-grained porous rock is held not in wide fissures directly open to the atmosphere but as thin, tightly bonded films that will not change to ice until the temperature falls well below 0 °C; even at subzero temperatures water may still migrate through the finer pore spaces (see also p. 280). These and other factors in frost-riving or cryoclastis are capable of systematic study in the laboratory, and there are now many examples of specially designed freezing chambers having been used to examine the effects on rock samples of different temperature changes. Maritime and continental environments have been simulated by employing so-called Icelandic and Siberian cycles with minimum temperatures of around −8 and −28 °C respectively; each Icelandic cycle has normally been programmed to last 1 day, each Siberian cycle 2 or more days. In a conventional experiment either the weight of the spalled debris or the changed weight of the sample itself is measured and the result expressed as a percentage of the original weight. Several workers began by testing both wet and dry rock samples, but abandoned the latter when it became obvious that they were virtually unaffected by the temperature changes.

Among the most extensive investigations are those undertaken by the laboratories of the Centre de Géomorphologie at Caen in Normandy. Here 1 000 samples can be treated at a time in each of two chambers, and even though several hundred cycles are often run, more than 1 000 samples can be tested in a single year. Experiments have been made on representative rocks from north-western France as well as on those from other parts of Europe (Fig. 6.2). Many limestones have been shown to be extremely susceptible to frost action. Chalk in particular was completely fragmented by 100 cycles of either Icelandic or Siberian severity, whereas certain basalts and other fine-grained igneous rocks remained unaltered after 300 such cycles. In almost no instances were Siberian cycles more effective than Icelandic, an observation supported by more detailed studies which imply that most cryoclastic disruption occurs as the temperature drops from 0 to −5 °C. The frequency with which the temperature crosses the freezing-point, therefore, seems to be more significant than the absolute range, although for the greatest impact the temperature probably needs to drop sharply and to remain below 0 °C for several hours in each cycle.

All these laboratory tests amply confirm the potential effectiveness of freeze–thaw activity, although the susceptibility of different lithologies varies widely. Porous rocks are often very liable to fragmentation, and they tend to break into smaller pieces; impervious rocks, on the other hand, are more resistant and split readily only along joints whose spacing is the primary determinant of the size of any debris. Further field investigations are now becoming desirable so that a clear comparison can be made between natural conditions and those being created in the laboratory; several workers have queried whether field circumstances are ever as favourable as those being

Fig. 6.2. Laboratory experiments simulating freeze–thaw action: (A) susceptibility of different rock types to 'Siberian' cycles; (B) relationship between susceptibility and porosity for a range of igneous and metamorphic rocks (data from Centre de Géomorphologie, Caen, France).

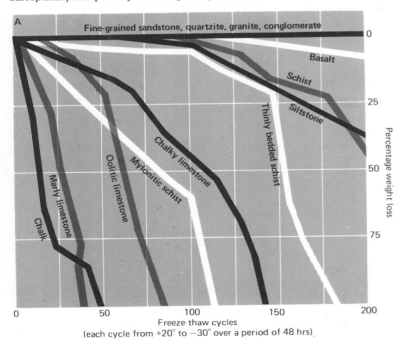

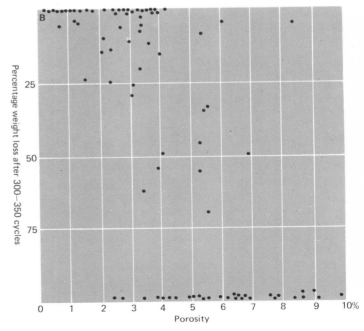

employed in experimentation.

A wide variety of soluble salts can be precipitated in the joints and pore spaces of a rock. The potentially deleterious effect of salt crystallization has long been recognized by engineers and architects concerned with the durability of building stones, and the deteriorating façades of many historic monuments are now attributed to this form of weathering. At least three sources of disintegrating stresses have been distinguished: the initial force of crystallization, volumetric changes due to hydration, and thermal expansion cycles. Growth of a crystal can continue even though it is already pressing against confining walls, one of the controlling factors being the degree of supersaturation of the surrounding solution; for instance, with a supersaturation ratio of two, gypsum can theoretically crystallize against a pressure of about 30 MN m^{-2}. Hydration, which involves the salts absorbing water into their lattices, leads to pronounced swelling; in certain circumstances the hydration of thenardite (Na_2SO_4) to mirabilite ($Na_2SO_4 . 10H_2O$) is estimated to exert a pressure of almost 50 MN m^{-2}. The third noteworthy property of some precipitated salts is a coefficient of thermal expansion that exceeds that of most host rocks and might therefore engender disruption.

Many laboratory experiments have emphatically confirmed the disruptive capabilities of precipitated salts. As early as 1828 Brard submerged various rocks in a solution of Na_2SO_4 and found that after only a few cycles of wetting and drying most lithologies showed signs of disintegration. More recently Goudie and Cooke have experimented with a wide range of solutes and lithologies. The normal procedure has been to immerse a rock sample in a saturated solution for 1 hour, then oven-dry it for a few hours at 60–80 °C and later at 30 °C, the complete cycle lasting 1 day. Certain salts are clearly much more effective than others, just as certain lithologies are much more susceptible than others (Fig. 6.3). Attempts to differentiate between the three potential stress sources mentioned above have cast doubt on the importance of thermal expansion and most of the effects seem to stem from initial crystallization and subsequent hydration cycles.

The weathering potential of precipitated salts is indisputable, but the environmental conditions under which that potential is most fully realized are less clear. The necessary aqueous solutions are present in a wide variety of circumstances, but perhaps the most favourable are those where saline groundwater can be drawn to the surface by capillary rise. The high supersaturation needed for the full development of crystallization stresses implies rapid evaporation and possibly a lack of protective soil and vegetation cover; blocks of hard sandstone 18 × 38 × 44 mm, taken from Britain and placed on the floor of a saline depression in southern Tunisia were found to have almost totally disintegrated within 6 years. Another crucial influence will be the microclimate since both crystallization and hydration processes are temperature- and humidity-dependent. Taken together all these factors suggest that salt weathering is likely to be particularly important in desert areas, and possibly also along coasts. However, its contribution in other environments must not be overlooked. As an illustration, a major factor in the weathering of urban building stones is now recognized to be the high concentration of SO_2 in the city atmosphere. This acidulates the rainwater and renders limestone buildings particularly prone to corrosion. Following a complex sequence of chemical reactions, gypsum crystallizes just below the surface and flakes of rock constantly peel away. Measurements on the balustrade that was added to St Paul's Cathedral in London in 1718 showed that, by 1980, degradation had totalled over 20 mm, implying a mean annual rate of about 0.08 mm. Of even greater concern, on drip sites at the base of the balustrade lowering of 0.139 mm was recorded within a 12-month period between 1980 and 1981.

Organic activity

Although it is widely acknowledged that growing plant roots, which in some circumstances can extend to depths of tens of metres, have the capacity to wedge open bedrock joints, the actual forces exerted are difficult to ascertain since roots usually follow openings already prised apart by some other agency. In assessing the overall role of roots, two contrasting effects need to be remembered. On the one hand, they act as a stabilizing agent by binding weathered materials together and thereby retarding the exposure of fresh rock. On the other, they occasionally become a disruptive agent, as when large trees are blown over. In the same way, animals contribute little to the direct fragmentation of rocks, yet have an important role in disturbing the detritus already produced and thereby enhancing the efficacy of other weathering processes. The increasing significance being attached by geomorphologists to soil biota is discussed again on pp. 103 and 149.

In addition to the forms of mechanical weathering already

Fig. 6.3. Laboratory experiments on salt crystallization: (A) tests on the relative effectiveness of different solutions; (B) tests on the susceptibility of different rock types. Data record diminishing weight of initial sample (after Goudie et al., 1970).

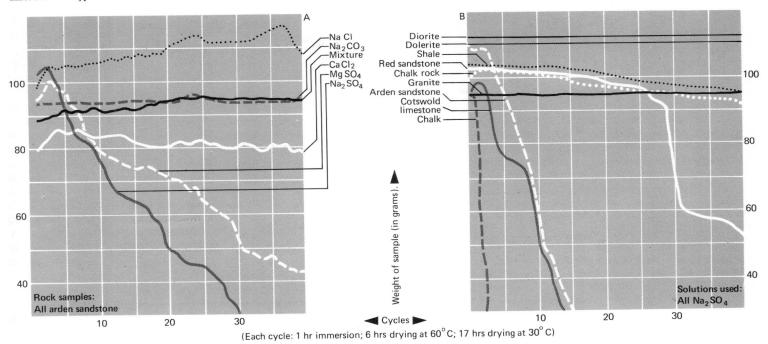

(Each cycle: 1 hr immersion; 6 hrs drying at 60°C; 17 hrs drying at 30°C)

discussed, a number of others have been suggested. One of the most interesting is colloidal plucking since it is known that gelatine drying in a tumbler can detach small flakes of glass. As an important fraction of soil organic matter exists in colloidal form, such plucking appears to be a potential means of rock disintegration, but so far proof of its effectiveness in the natural environment has not been established.

Chemical weathering

General background

In considering mechanical weathering it was possible to identify four major processes and to relate these to differing climatic environments and such rock properties as porosity and jointing. The web of chemical processes is much more complex and it is impossible to convey in a few pages the enormous wealth of chemical interactions revealed by almost any field investigation. Attention must therefore be concentrated on a few of the more basic processes. Before discussing these it will be helpful to review, firstly, some of the properties of rocks influencing their liability to chemical alteration and, secondly, the nature of the major reactant, rainwater.

Eight elements make up more than 98 per cent of the earth's crust. By far the most abundant is oxygen, comprising by weight almost half the total, followed by silicon comprising a little over a quarter. The six others are, in descending order, aluminium, iron, calcium, sodium, potassium and magnesium. On the basis of volume, nearly all the crust consists of oxygen anions bonded to metal cations in the form of oxides. By far the most important oxide occurring in mineral form is silica (SiO_2) best known as the rock material quartz; other mineral oxides include alumina (Al_2O_3) and haematite (Fe_2O_3). High proportions of the two most abundant elements are bonded together as silicate tetrahedra (SiO_4) (see Fig. 6.10) which form the basic building blocks not only of most minerals found in igneous rocks, but also of that apparently very dissimilar group, the clay minerals. The immense variety of silicate minerals derives from contrasting linkages between the tetrahedra, from different elements bonding the tetrahedra together and from replacement of some of the silicon ions by aluminium. All these factors influence the susceptibility of the minerals to chemical alteration since the major weathering processes involve the disruption of the tetrahedral lattice and removal of the bonding cations. Attempts by means of both laboratory and field observations to tabulate the relative resistance of the primary rock-forming minerals have shown that, in general, minerals crystallizing at the highest temperatures during cooling of a magma are the least resistant and those crystallizing last are the most resistant (Table 6.1). One may reasonably discern in this result the importance of thermochemical factors, and Curtis has demonstrated a consistent relationship between mineral stability and changes in what are known as the free-energy values for various weathering reactions. It is the minerals involved in reactions releasing the greatest energy that are the least stable; illustrative calculations are included in Table 6.1.

The geomorphologist must also remember that, although sedimentary rocks comprise less than 5 per cent of the total crustal volume, they outcrop over about 75 per cent of the earth's land surface. On

Table 6.1 The susceptibility of common rock-forming minerals to chemical weathering.
The basic arrangement reflects the original assessment of Goldich (J. Geol 1938), but also shown are figures indicating what are known as the Gibbs free energy values for some of the associated weathering reactions. This thermochemical approach envisages that, in any spontaneous reaction, there must be a net release of free energy ($\triangle G_r$) which can be related to the free energy of formation ($\triangle G_f$) of the reactants and their products i.e. $\triangle G_r = \triangle G_r = \Sigma \triangle G_f$ (products) $- \Sigma \triangle G_f$ (reactants). Free energy values for many different substances have been calculated and are tabulated in various works of reference: thus $\triangle G_f$ for anorthite ($CaAl_2Si_2O_8$) is -3998.2, for kaolinite ($Al_2Si_2O_5(OH)_4$) -3782.3, for water -237.2 and for the calcium ion -553.1. The weathering of anorthite, with hydrogen ions derived from dissociation of H_2CO_3, may be written $CaAl_2Si_2O_8 + 2H^+ + H_2O \rightarrow Al_2Si_2O_5(OH)_4 + Ca^{++}$. Inserting the appropriate values for $\triangle G_f$ yields $-(3998.2 + 237.2) \rightarrow -(3782.3 + 553.1)$. $\therefore \triangle G_r = -100$ kJ mol^{-1}, or, dividing by 18, -5.56 kJ g atom^{-1}
The noteworthy feature of the table is the relatively high free energy values associated with reactions involving the least stable minerals.

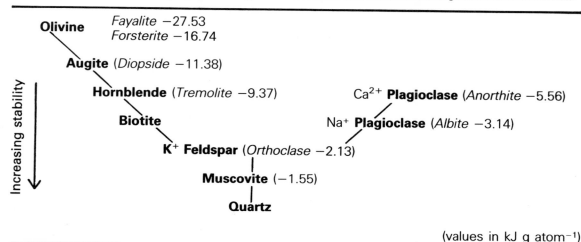

the basis of this last figure, it has been estimated that four major mineral groups constitute 85 per cent of the material exposed to weathering: feldspars; quartz; clay minerals and micas; calcite and dolomite. Yet the apparent simplicity implied by this statement is deceptive. There are many different types of feldspar and clay mineral, all altering chemically in different ways. Moreover, susceptibility to decomposition is not solely dependent on the mineralogy of a rock. Since the major reactant is percolating rainwater, accessibility for water is a highly significant property. One measure of accessibility is porosity which ranges from less than 1 per cent in most igneous rocks to over 15 per cent in some sandstones. Porosity also provides a rough guide to the surface area within a rock over which chemical activity can take place. Volume for volume, finer-grained rocks generally offer a greater surface area than coarse-grained so that they tend to weather more rapidly; other things being equal, for instance, feldspars will be altered with greater speed in a fine arkosic sandstone than in a coarse-textured granite. A second measure of water accessibility is permeability. The importance of routeways followed by percolating water is clearly seen in the development of karstic landforms in limestone regions, and is also apparent in spheroidal weathering where chemical decomposition is concentrated at the intersections of joint planes (Fig. 6.6).

Rainwater, besides acting as a carrier of dissolved atmospheric oxygen and carbon dioxide, often possesses a surprisingly complex chemical make-up. Salts derived from the sea, nitrous oxide produced by lighting and dissolved to form nitric acid, sulphur dioxide emitted by the combustion of sulphurous fuel and dissolved to form sulphuric acid – these and many other common constituents confer on precipitation the capacity to attack rock in multifarious ways. In both Europe and North America excessive sulphur dioxide has been indicted as a major cause of acidified rain that is now believed to be profoundly altering long-established weathering processes. Of course, once rainwater has percolated beneath the surface its composition is further modified by reaction with both the mineral and organic fractions of the soil.

Pure water percolating through a rock can initiate three main chemical processes: solution, hydration and hydrolysis. By virtue of dissolved carbon dioxide and oxygen, rainwater can regularly induce two further processes, carbonation and oxidation. Biological influences within the soil are so important as to require discussion as a sixth topic. Although each of these subjects will be treated separately, it must be exphasized that very rarely does one process operate alone; their overall effectiveness is largely the result of interactions between them.

Solution

Most rock materials are of only very limited solubility, the one significant exception being halite or rock salt. This does not mean that solution itself is unimportant, but its main function is probably better viewed as transporting the products of other weathering processes. If these products are not removed they can retard or radically alter the whole sequence of rock decomposition. Many solutions depend on external factors for their maintenance. For instance, the solubility of both ferrous iron and manganese rises rapidly under slightly acidic conditions, while it requires a pH of about 4 to render alumina soluble; it is mobilization of alumina and its transport into lakes and watercourses that is regarded as one of the most destructive consequences of the artificially acidified rain alluded to above. Such differential solubility also helps explain why the weathering residue from the same rock can vary widely according to local conditions; equally important, it helps account for reprecipitation as the properties of the solvent change during migration through the soil and underlying weathered horizons.

Hydration

The hydration of precipitated salts has already been discussed (p. 98), but numerous different minerals can incorporate water into their molecular structure. This includes many of the constituents not only of igneous rocks but also of sedimentary rocks; for instance, haematite may be converted to limonite, while most of the silicate clay minerals swell significantly on becoming hydrated. This property of swelling when entering into combination with water is one of the most important aspects of the process. Because of the resultant tendency to cause crumbling of the whole rock, hydration has been described as the normal precursor of most of the more profound chemical alterations.

Hydrolysis

The classic experiments in the weathering of the silicate minerals, and particularly of the feldspars, were performed by the Frenchman Daubrée in the nineteenth century. Agitating 3 kg of orthoclase feldspar in water, he found that after 200 hours the water had turned milky in appearance and the surfaces of the feldspar crystals were decomposed. Later analysis showed the water to contain 2.52 g of K_2O in solution, together with a fine suspension of clay minerals. The chemistry of the process is extremely complex but may be approximately represented by the following equation:

$$2KAlSi_3O_8 + 2H_2O \rightarrow Al_2Si_2O_5(OH)_4 + K_2O + 4SiO_2$$

Orthoclase + Water → Kaolinite + Soluble + Soluble
potassium silica
oxide

As can be seen, the reaction involves dissociation of water to produce hydroxyl ions, removal of the bonding potassium ion, and disruption of the original silicate lattice. The silicon and aluminium ions are rearranged to form the basis of the hydrous clay mineral known as kaolinite. A further part of the silica may be removed in solution. This decomposition of orthoclase illustrates a chemical reaction with water that affects many other silicate minerals. The sodium- and calcium-rich feldspars are decomposed in a very similar fashion, as are such less closely related minerals as olivine, augite and hornblende. However, in all these cases chemical changes are accelerated under natural conditions by the impurity of the water, and in particular by the presence of dissolved CO_2.

Carbonation

Dissolved atmospheric CO_2 turns rainfall into a very weak carbonic acid with an average pH of about 6. When the water percolates underground more CO_2 is rapidly dissolved from the soil air which is much richer in that gas than the free atmosphere. Experiments by Daubrée and numerous later workers have shown that orthoclase decomposition greatly accelerates if the mineral is churned in water enriched with CO_2. The enhanced reaction appears to arise from the ready ionization of weak carbonic acid into hydrogen and bicarbonate ions. The weathering of orthoclase in such circumstances can be approximately represented as follows:

$$2KalSi_3O_8 + 2H_2O + CO_2 \rightarrow Al_2Si_2O_5(OH)_4 + K_2CO_3 + 4SiO_2$$

Orthoclase + Water + Carbon → Kaolinite + Soluble + Soluble
dioxide potassium silica
carbonate

In essentials the process clearly resembles that already outlined for pure water. The silicate structure of the orthoclase is disrupted with

the release of the potassium ion, and the end-product is a clay mineral together with soluble residues. Weathering of the other main feldspar groups also results in the development of clay minerals, but with the release of Na_2CO_3 from the soda-rich and $Ca(HCO_3)_2$ from the calcium-rich minerals. Whereas the sodium and calcium is readily removed to become part of the solution load of streams, more of the potassium tends to be held by colloidal clays, possibly for ultimate use by plants.

Water enriched in CO_2 attacks many other silicate minerals. Olivine, for example, can be almost entirely dissolved in a sequence of reactions represented in much simplified fashion as follows:

$$Mg_2SiO_4 + 2H_2O + 4CO_2 \rightarrow 2Mg(HCO_3)_2 + SiO_2$$

Olivine + Water + Carbon → Magnesium + Soluble
dioxide bicarbonate silica

However, the best-known example of chemical alteration by carbonated water is the dissolution of limestone. Calcium carbonate is only slightly soluble in pure water, but undergoes the following reaction in the presence of weak carbonic acid:

$$CaCO_3 + H_2CO_3 \rightarrow Ca^{2+} + 2(HCO_3)^-$$

The calcium and bicarbonate ions may then be removed in solution, leaving behind impurities to accumulate as a superficial residuum; in many areas this contains a considerable proportion of iron which oxidizes to form the bright red material known as terra rossa. The quantity of dissolved calcium carbonate that may be transported by the water is very sensitive to changes in the amount of dissolved CO_2. Equilibration with the low CO_2 pressure of cave air, particularly by comparison with soil air, is a primary cause of the reprecipitation in the form of stalactites and stalagmites. Similar reactions at ground-level may result in the formation of tufa incrustations.

Oxidation

In chemistry, oxidation and reduction are normally defined in terms of the loss or gain of electrons by an element participating in a reaction; iron, for example, can exist in several different oxidation states and in strongly oxidizing or reducing environments transformation from Fe^{2+} to Fe^{3+} or vice versa can bring about an electrostatic imbalance that greatly weakens the surrounding crystal lattice. However, in discussion of weathering it is often the more restrictive idea of oxidation as the chemical combination with oxygen that is envisaged. In particular, oxygen dissolved in water reacts with the almost ubiquitous iron released by other weathering processes, converting it to the ferric state in the form of haematite or its hydrated equivalent, limonite. It is in the ferric state that iron tinges so many soils with characteristic red, brown or yellow hues.

Biological influences

Biological agencies affect chemical weathering both by influencing the rates of the processes already discussed, and by producing reactions that are specifically biochemical in nature. Into the first category come the various controls exercised by vegetation on the quantity and quality of percolating water. By intercepting precipitation the vegetation cover can regulate the amount of water reaching ground-level, by the production of humus it can influence the infiltration rate through the soil. The effects on water quality are of even greater significance. The most fundamental is in the supply of CO_2, already mentioned in the earlier account of carbonation. Oxidative bacteria decomposing organic residues, together with the respiration from plant roots, regularly raise the CO_2 concentration in soil air to between 0.2 and 2.0 per cent and sometimes even higher. Even the lowest of these values represents a sixfold increase by comparison with the mean atmospheric concentration. The actual amount of CO_2 in soil air varies in response to a number of factors. One is temperature since bacterial action declines rapidly as the soil temperature falls below 10 °C; another is water content since organic activity falls significantly when the moisture content sinks below about 10 per cent. Further factors include soil aeration and vegetation cover so that the overall effect is a complex pattern of seasonal and regional variations in dissolved CO_2. Although H_2CO_3 is a weak acid, this can in turn influence the solubility of such weathering products as the iron and aluminium compounds. It is scarcely surprising that geomorphologists interested in weathering are now attempting to produce maps depicting global variations in soil CO_2 concentration.

One of the most dynamic agents in changing both the quantity and quality of percolating water has been Man. By felling natural vegetation, covering large urban areas with impermeable surfaces and tile-draining agricultural land he has modified soil-water movement over vast areas of the earth's surface. Simultaneously, water composition has been altered by an almost endless list of activities. Emission of substances into the atmosphere, ploughing and harvesting of crops, application of fertilizers – all have affected the character of percolating water. Although Man's activities rarely have the dramatic impact of soil erosion in the United States during the 1930s, this must not be allowed to hide his more subtle yet extremely pervasive influence.

Two important weathering reactions that appear to be primarily biochemical are reduction and chelation. Many anaerobic bacteria obtain part of their oxygen by reducing iron from the ferric to the ferrous state, and in more extreme circumstances from the ferrous state to the metal itself. One consequence is to produce ferrous compounds that are significantly more soluble in water than the original ferric ones; this is a major way in which iron can be mobilized and removed from the soil. An analogous reduction of manganese compounds by bacterial activity has been shown to be a principal cause of the surface coating known as 'desert varnish' in arid areas. Chelation involves the union in a stable ring structure of metallic cations and complex hydrocarbon molecules. It is primarily by this mechanism that rootlets extract from the mineral soil the metals essential for the maintenance of plant life. Lichens may act in a rather similar fashion, deriving necessary nutrients by means of chelating exudates that thoroughly decompose the underlying rock surface. A third illustration is provided by decaying plant debris that may yield organic matter capable of 'complexing' metallic ions; this is now regarded as a major factor in the downward translocation of iron during podzolization. A lot remains to be learnt about biochemical weathering, but increasingly it appears that its importance has hitherto been much underestimated.

Climatic controls

In the preceding pages, frequent reference has been made to the effect of both climate and vegetation on the nature of weathering processes. Since the influence of the plant cover is often an indirect expression of climatic controls, several workers have sought a closer definition of the relationship between climate and weathering. One of the best-known schemes is due to Peltier who began with a number of simplifying assumptions. The first was that mechanical weathering is almost entirely due to freeze–thaw activity, the second that chemical weathering is so dependent on the presence of water that its intensity should vary with precipitation. Arguing that regions could be both too hot and too cold for fully effective freeze–thaw activity, he attempted to identify those climatic regimes where the temperature most frequently crosses the freezing-point. Further

Fig. 6.4. Climatic weathering regimes as defined by Peltier (1950).

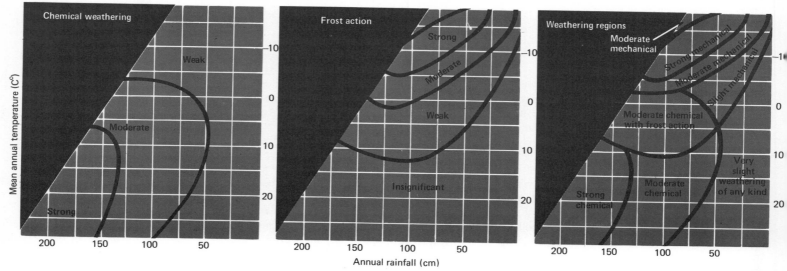

contending that chemical weathering is accelerated by high temperatures and dense vegetation, he defined regimes where he believed such conditions would prevail. By these arguments he managed to delimit climatic regimes characterized by distinctive combinations of mechanical and chemical weathering (Fig. 6.4), and most workers would probably accept that Peltier's diagram provides at least a first approximation to regional weathering patterns.

In a rather different analysis in 1967, the Russian worker, Strakhov emphasized the interaction of climate and relief controls. He argued that mechanical weathering is normally far outstripped by chemical, which reaches its acme in the tropical zone under conditions of plentiful moisture, high temperatures and abundant vegetation; an area of secondary maximum in moist temperate latitudes has a weathering rate less than one-twentieth of that in the tropics. Only in deserts and extreme northern latitudes do climatic controls encourage relatively vigorous mechanical weathering. However, an additional factor stressed by Strakhov is tectonic uplift. He argued that, with increasing relief, mechanical denudation by surface wash becomes so intense that chemical weathering is finally suppressed altogether. However, to achieve this state in the humid tropics demands excep-

tionally rapid uplift; it can be achieved much more readily in temperate latitudes.

Finally, more refined quantitative approaches essayed by a number of investigators have been confined to calculating climatic indices for individual weathering processes, such as frost action or chemical leaching in soils, rather than providing an overall assessment of climate and weathering.

Weathering forms

Bedrock outcrops may be fashioned into distinctive surface forms owing their shape primarily to weathering processes. Two groups can be recognized, those originating at ground-level and those initiated underground but suffering later exhumation. It is often difficult to distinguish between the two, and investigation of many of the features has involved a history of conflict between surface and exhumation hypotheses.

Weathering pits

Small-scale depressions known as weathering pits are very widespread. They may be observed on horizontal bedrock surfaces under a variety of climates ranging from arctic Canada to tropical Africa. They develop on many different rocks, but are particularly prominent on granite and some coarse sandstones. Commonly 1 m or so in diameter they may range in depth from 0.1 to over 0.75 m; some, but by no means all, possess obvious overflow spillways. Although pitted surfaces are often revealed when a soil is stripped back, many forms have undoubtedly developed in purely subaerial situations. An initial shallow depression becomes the site of a puddle around and beneath which the rock is regularly wetted. Hydration and hydrolysis ensue, with the soluble products evacuated by percolation or overflow. Almost all weathering pits occasionally dry out; such desiccation could permit part of the insoluble residue to be removed by wind although the flushing action of the occasional storm may also be significant.

Karren

The rocky surfaces of many limestone outcrops exhibit shallow depressions very similar to the weathering pits described above. These constitute one minor type of a whole group of limestone features best described by the German term *Karren*. This word embraces forms ranging from runnels only a few millimetres deep to prominent clefts several metres deep. All are due primarily to solutional weathering, but some appear to form at the surface and others beneath a cover of soil or superficial material. Elaborate classifications of *Karren* have been proposed, but the four types listed under the headings below appear to be particularly common.

(a) Rillenkarren

These are shallow grooves or runnels often scored in great numbers down sloping limestone faces. They are due to solution by small rills of water and can develop within a few years on freshly uncovered limestone surfaces.

(b) Rinnenkarren

These are much larger and deeper runnels, sometimes approaching 0.5 m in depth. Individual grooves may be traceable tens of metres down suitable limestone faces, and after heavy rain each carries a fast-flowing rivulet of water.

(c) Rundkarren

As the name implies, these are rounded blocky forms that develop initially beneath a soil cover. In many areas of active exhumation the rounded upper surfaces can be observed gradually appearing from beneath the overlying material. The intervening depressions are less consistent in slope and direction than in the case of *Karren* produced at the surface; they are due to percolating rather than free-flowing water.

(d) Kluftkarren

These are the features frequently described in Britain as 'grikes' from the local name used in North Yorkshire where they are very fully developed on the Carboniferous Limestone plateaux. Controlled almost entirely by rock structure, they are essentially joints widened by solutional activity until the cleft may finally penetrate to a depth of several metres. Although solution in some instances occurs directly at the surface, a two-stage origin has also been invoked since there are indications that some clefts were initiated beneath a soil cover and are currently being widened by the growth of *Rillenkarren* along their walls.

Tors

Fierce controversy has surrounded the origin of tors. Tors are small residual bedrock eminences rising steeply above their immediate surroundings. They are particularly characteristic of, but by no means confined to, areas of granite. Three main theories may be distinguished, two invoking weathering processes but the third attributing them to surface-water flow. The first hypothesis, originally advanced in its modern guise by Linton in 1955, envisages deep chemical rotting followed by exhumation (Fig. 6.5). Structure is regarded as a major influence since zones of massive rock prove much more resistant to chemical decay than intervening areas of closely jointed rock. A prolonged weathering phase is visualized, possibly under a moist tropical climate, during which rotting proceeds to a depth of at least tens of metres in the less resistant zones. Following tectonic uplift or some other cause of accelerated erosion the rotted debris is removed to leave the coarsely jointed blocks as upstanding residuals. Support for this hypothesis comes from the irregular weathering front, together with groups of large unaltered core-stones available as future tor material, occasionally seen in deep sections. A factor complicating the interpretation of some sections is hydrothermal alteration of an igneous rock during a late stage of its original emplacement: such decomposition can be

Fig. 6.5. Stages in the evolution of tors that originate from deep chemical weathering (after Linton, 1955): (A) initial form; (B) deep weathering; (C) removal of regolith.

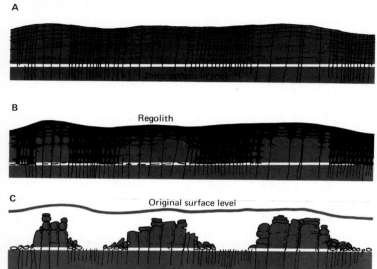

A

B

Regolith

C

Original surface level

difficult to distinguish from that due to atmospheric weathering. The second hypothesis, supported by many workers familiar with arctic environments, is that similar hilltop features can be produced by intense frost shattering, an idea discussed more fully in Chapter 14. The third viewpoint, espoused particularly by King, is that many of the features termed tors and attributed to weathering processes are no more than residuals left during the final stage in the development of subaerial erosion surfaces; he believes this is certainly the origin of numerous tor-like features in Africa.

At the moment the origin of tors appears to illustrate the problem of 'equifinality' where very different processes eventually lead to almost identical landforms. On the other hand, one may suspect that further research will reveal discriminating criteria allowing a valid genetic classification to be formulated. For instance, samples of the weathered debris from around the tors on Dartmoor in south-western England have been examined under the scanning electron microscope to see whether individual grains bear surface textures diagnostic of chemical alteration under hot, moist conditions. The results

in this particular case were inconclusive, perhaps because the problem is complicated by the great climatic fluctuations that the British Isles have experienced during Quaternary times; it is quite feasible that some tors were initiated by deep rotting during warm interglacial periods and exhumed by accelerated mass movement during glacial episodes.

Etchplains

Certain authors have suggested that, during phases of tectonic stability, chemical weathering in humid tropical regions may cause very extensive deep rotting. When the equilibrium is disturbed and a new phase of erosion initiated, the first result is the stripping of the decomposed layer to produce an 'etchplain' whose relief reflects resistance to chemical alteration as much as to the processes responsible for exhumation. These ideas are more fully examined in Chapter 19.

Physical properties of the mantle of weathered material

The immediate result of the mechanical and chemical processes outlined earlier in this chapter is a weathered mantle or regolith that veneers virtually all the continental land surfaces. Understanding the ways in which this material is set in motion and ultimately transported by the agencies of water and wind is at the very heart of any analysis of subaerial denudation. A primary concern of the geomorphologist must therefore be to examine those physical properties of the weathered layer influencing its mobility; a selection of such properties is reviewed below.

Thickness and layered structure

The thickness of the regolith varies from nil to many metres. Over much of the earth's surface it is between 1 and 2 m but occasionally totals in excess of 100 m have been recorded. The greatest values are usually found in tropical areas in conformity with the general expectation that chemical alteration will be most effective under hot, moist conditions. The character of the regolith also varies with depth.

Fig. 6.6. Deep chemical weathering producing 'spheroids' by selective alteration along intersecting joint planes. The white lines in the above photograph are quartz veins which have proved much more resistant to decomposition than the igneous rocks in which they are set.

Near the surface the mineral matter is intimately mixed with organic material to form the soil. The ratio of organic to mineral matter decreases with depth, and beneath the soil a second layer consisting very largely of weathered bedrock may be distinguished. At greater depth this in turn gives place to unaltered bedrock. The boundaries between these layers may be either abrupt or gradational. For instance it is often possible to identify an intermediate zone above the sound rock in which the material, although thoroughly decomposed, retains clear traces of its original structure (Fig. 6.6); where undisturbed these offer a guide to the limiting depth of mass movement, as opposed to weathering. In very deep regolith it is usually the basal layer of rotted rock that is exceptionally thick, and several workers have attempted further subdivision by recognizing an additional zone near the base characterized by residual core-stones. It is obvious that, in any assessment of the overall properties of the regolith, cognizance must be taken of these very significant variations with depth. Moreover, a further potential complication is that, in suitable circumstances, shallow layers in the regolith may be recemented by precipitates from circulating water; among the common cementing compounds are calcium carbonate, iron oxide and silica forming 'duricrusts' respectively known as calcrete, ferricrete and silcrete.

Infiltration capacity

The maximum rate at which the regolith permits downward percolation of water is referred to as its infiltration capacity. This extremely important property is a major determinant of whether rainwater sinks underground or flows across the surface; theoretically, when the precipitation intensity surpasses the infiltration capacity, the excess water that is unable to percolate downwards tends instead to generate overland flow. Infiltration rates of the surface layers may be measured by an infiltrometer. This normally consists of an open-ended tube, 0.1 m or more in diameter, partially sunk into the ground so as to isolate a cylindrical core of soil. The rate of percolation by water applied artificially to the surface can then be measured. Usually a container with a mechanism for maintaining a constant head of water is employed, and the rate of water loss from the container is taken as a measure of infiltration. In practice the rate is found to vary with time. Once the dusty surface is wetted, there is often an initial rapid acceptance of water as pore spaces in the upper soil horizons fill. The magnitude and duration of this early phase depends in part on the original soil-moisture content, being greatest when the soil is dry. Thereafter the rate normally diminishes until, within an hour or so, a relatively steady rate of infiltration is established. It is this constant value that is generally quoted as the infiltration capacity, although when comparison is made with precipitation intensities it is vital to recall the high initial values associated with a dry soil. Measured capacities cover a very wide range, but most tend to lie between zero and 100 mm hr^{-1}. Evaluated in this way infiltration is related less to total porosity than to what may be termed 'macro-porosity'. While total porosity refers to the gross volume of voids, macro-porosity includes only the larger pores through which water moves relatively freely. This distinction is well exemplified by sandy soils with a total porosity near the surface of 35–50 per cent, and fine-textured soils with a total porosity of 40–60 per cent. Yet so many of the voids in the latter are small that the infiltration capacity of a sandy soil is ordinarily much higher.

Field measurements have usually been confined to a shallow soil layer of 0.2 m or less. One reason is the increasing experimental problems encountered at greater depths, but even more important is the tendency for water under natural conditions to move laterally as well as vertically; so long as water is confined within a small cylinder, lateral percolation is precluded. On normal hillsides an important fraction of the infiltrating water soon moves downslope, parallel to the surface, as interflow or throughflow. Such movement can be caused by the presence of a buried horizon having a lower permeability than that of the surface layer, an extremely common situation since the B-horizon of a soil is regularly less permeable than the A-horizon. The significance of this retarded vertical movement may be assessed by inserting infiltration cylinders to different depths; those penetrating an illuviated layer will normally yield lower infiltration capacities than those confined to higher horizons. More direct measurements of throughflow have been attempted by inserting receptacles at different levels into the wall of a trench dug across a hillslope. However, on some hillsides a significant proportion of the subsurface movement is not a diffuse flow but is concentrated instead in shallow, concealed pipes. Such pipeflow is especially common on peat-covered slopes but has been reported from a wide range of environments. The large volumes of water conveyed through the pipes after heavy rainfall have been evaluated by artificially diverting the flow through suitable measuring devices.

Infiltration measurements are essentially a form of point sampling.

Experimental data, standardized as to size and depth of cylinder employed, often display large variations according to the nature of bedrock, slope and vegetation. Even where these variables appear relatively constant divergent results indicate a high sensitivity to other irregularly distributed factors. To overcome this problem of 'noise' numerous samples must be included in any systematic investigation.

Mechanical strength

The hillslope regolith is subject to stresses arising in many different ways, and its response to these determines both the pattern of movement and ultimately the form of the slope itself. Analysing the mechanical strength of the regolith poses many problems which are still far from complete solution and are beyond the scope of this book. Nevertheless, in view of the importance of the subject and the growing recognition of its significance in geomorphology, at least a brief introduction is required.

The behaviour of materials is normally considered in terms of applied stress and induced strain. Stress is a vector quantity expressed as a force per unit area. Strain is the technical term to denote deformation of a substance and is expressed as a fractional change in the original dimensions of a body. Strain may occur in different ways according to the nature of the material; for instance, air normally deforms by turbulent flow, a thin film of water by laminar flow and a solid by elastic displacement. These different behaviours under stress constitute the basis for distinguishing between fluids and solids. A fluid is a substance which cannot sustain shear forces and in which deformation is directly proportional to applied stress (Fig. 6.7). A solid is a substance which possesses both strength to resist a small applied stress and elasticity to restore its original shape when that stress is removed. Three different types of solid may be recognized. In a rigid solid small stresses produce little strain until a critical value, known as the elastic limit, is reached, whereupon abrupt fracturing occurs. In an elastic solid, on the other hand, there is considerable strain before the point of fracture is reached. In a plastic solid deformation beyond the elastic limit does not involve fracturing but an increase in strain proportional to the increase in stress. The boundaries between these three types of solid are arbitrarily defined and in different circumstances the same material may act in different ways. Thus granite would normally be considered a rigid solid, fracturing abruptly beneath the geologist's hammer; under high temperature and pressure, however, its strain pattern resembles

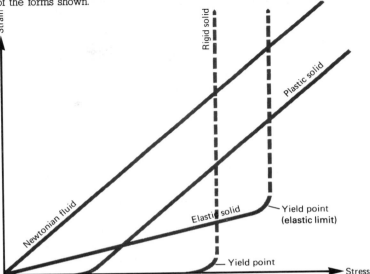

Fig. 6.7. Stress–strain relationships for four idealized materials. In practice many solid materials exhibit properties that are some complex combination of the forms shown.

that of plastic deformation. The duration of stress may also have important consequences, since a force that can be accommodated by elastic strain if applied for a short time may induce 'creep' if applied for a longer period.

These examples illustrate the complexity of apparently simple materials, and when it is remembered that the regolith consists of an intimate mixture of solids, water and air, in proportions that may vary greatly with time, the difficulty of defining even an approximate mechanical model is readily appreciated. Many analyses have treated the regolith as being in a solid state of aggregation and possessing properties conforming reasonably closely to those of a plastic solid. Yet a substantial part of true plastic deformation takes place by intracrystalline gliding, whereas in the weathered mantle movement is mainly by intergranular displacements. The physical nature of the movement is therefore very different and the plastic model by no means entirely satisfactory. A further problem arises from the fact that the mechanical properties vary greatly with depth, particularly near the surface where there are strongly contrasting soil horizons.

The problems are formidable, not to say daunting. Nevertheless, they must be tackled if a fuller understanding of regolith movement is to be obtained. Much of the initiative has come from engineers concerned with the practical difficulties of building earthen dams or digging deep motorway cuttings. For such work some knowledge of the way the materials will behave is essential, and investigation of this behaviour is the field of soil mechanics. In this context the word 'soil' is not restricted to the superficial layer in which plants grow but applies to any natural earthen material, whether regolith, superficial sediments or bedrock, in which or of which engineering structures are to be built. Modern soil mechanics is founded on pioneer work of the eighteenth-century French physicist Coulomb who expounded the general rule that the shear strength of a soil is equal to $c + \sigma \tan \phi$, in which c is cohesion, σ the stress normal to the shear plane and ϕ the angle of internal friction. In bedrock it is the cohesion that is by far the most significant value, but in certain loose, dry aggregates the value of cohesion falls to zero. Each value in the Coulomb equation merits further examination.

Coulomb regarded ϕ as the angle of repose of loose materials, but in modern theory it is designated the angle of internal friction which may be measured by the use of direct-shear apparatus. Assuming firstly that the material is non-cohesive such as dry sand or gravel, a sample is placed in a split box and the maximum force that can be transmitted through the sample to a proving ring is measured (Fig. 6.8a). By changing the load on the box, the stress normal to the shear plane can be varied, and if the results of repeated experiments are plotted as shown in Fig. 6.9a, the angle of the line joining the plotted points is the angle of internal friction; for most non-cohesive materials it lies between 30 and 40°. If a suitable cohesive soil is similarly tested in the direct-shear apparatus, the plotted data provide a measure of the cohesion of the sample (Fig. 6.9b).

As originally formulated by Coulomb, shear strength is directly related to the total stress normal to the plane of shearing. In 1923, however, the distinguished engineer Terzaghi published a modification of Coulomb's equation which afforded an explanation for certain discrepancies that had arisen between predictions of the equation and measured values. It had been found, for instance, that sand slopes which were stable when dry collapsed when wet. This was first ascribed to a substantial fall in the value of ϕ, and yet direct shear tests on wet and dry sand failed to reveal such a contrast and

Terzaghi suggested that Coulomb's equation would remain valid if total normal stress were replaced by effective normal stress. The underlying concept is that the total stress consists of two parts, effective and pore stress. The former is the fraction of the total transmitted by the solid-to-solid contacts, the latter the fraction transmitted by the pore spaces. So long as the pore spaces are filled with air the pore stress is zero and the effective and total stresses remain identical. When the pores are filled with water a variable proportion of the total stress is transmitted through the fluid and the effective stress is correspondingly diminished. In consequence the Coulomb equation in modern guise is usually written $S = c' + \sigma' \tan \phi'$, where σ' refers to effective rather than total normal stress. This modification is of crucial importance in explaining the reduced shear strength of porous materials as they become saturated; it was increased pore pressure that started the disastrous flows on the tip-heaps of Aberfan in Wales that led to the deaths of 116 schoolchildren in 1966. On natural hillslopes regolith that has remained stable for decades or even centuries may suddenly be mobilized by a fall in shear strength after exceptional rainfall.

An alternative, and ofter preferable, method of measuring the shear strength of a soil is by use of triaxial compression apparatus (Fig. 6.8b). The procedure involves capping one end of a cylindrical soil sample with a porous disc. The sample and disc are then enclosed in an impermeable membranous sleeve before being inserted into a pressure cell that can be filled with liquid. A load is then applied along the axis of the cylinder until failure of the sample occurs. The stress normal to this principal axis is varied by controlling the fluid pressure in the cell. One great advantage of triaxial over direct-shear apparatus is the ability to control and measure the pore-water pressure in the specimen. Two types of test are commonly employed. In the first, drainage is permitted so that there is no build-up of pore pressure; in the second, drainage is prevented so that the shear strength is measured under conditions of varying pore pressure.

The cohesion measured in either direct-shear or triaxial apparatus is due to binding forces between individual particles. In rocks it is normally the result of cementing and fusion, but as weathering progresses these bonds are gradually weakened. This is especially clear where the rock breaks down into debris of coarser size than silt, since such material has low cohesion but relatively high values

Fig. 6.8. Diagrammatic portrayal of the essential components in two types of equipment for measuring the shear strength of soil samples.

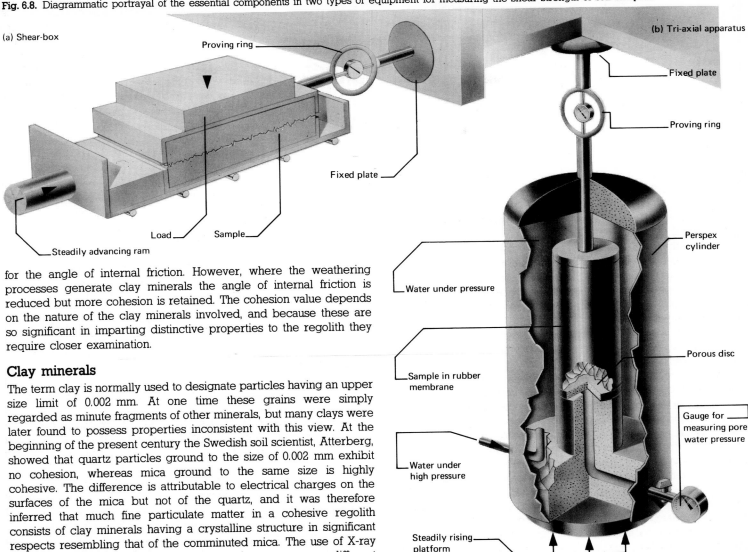

(a) Shear-box

Proving ring

Load

Sample

Steadily advancing ram

Fixed plate

(b) Tri-axial apparatus

Fixed plate

Proving ring

Water under pressure

Sample in rubber membrane

Water under high pressure

Steadily rising platform

Perspex cylinder

Porous disc

Gauge for measuring pore water pressure

for the angle of internal friction. However, where the weathering processes generate clay minerals the angle of internal friction is reduced but more cohesion is retained. The cohesion value depends on the nature of the clay minerals involved, and because these are so significant in imparting distinctive properties to the regolith they require closer examination.

Clay minerals

The term clay is normally used to designate particles having an upper size limit of 0.002 mm. At one time these grains were simply regarded as minute fragments of other minerals, but many clays were later found to possess properties inconsistent with this view. At the beginning of the present century the Swedish soil scientist, Atterberg, showed that quartz particles ground to the size of 0.002 mm exhibit no cohesion, whereas mica ground to the same size is highly cohesive. The difference is attributable to electrical charges on the surfaces of the mica but not of the quartz, and it was therefore inferred that much fine particulate matter in a cohesive regolith consists of clay minerals having a crystalline structure in significant respects resembling that of the comminuted mica. The use of X-ray diffraction equipment later showed that there are many different

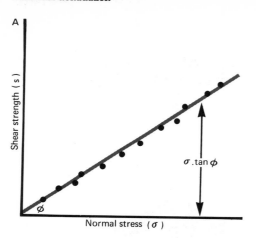

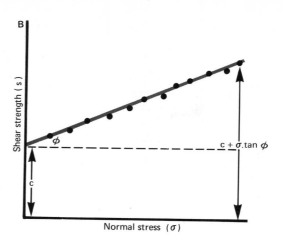

Fig. 6.9. Schematic representation of the results of multiple tests with shear-box apparatus: (A) non-cohesive material; (B) cohesive material.

types of clay mineral, each capable of endowing the regolith with unique physical and chemical properties.

The most abundant clay minerals are layered aluminosilicates, often classed as phyllosilicates owing to their sheet-like structure. In addition there are hydrous oxide clays composed chiefly of aluminium and iron oxides bonded with variable quantities of water. A good example is gibbsite ($Al_2O_3.3H_2O$), one of the principal minerals of bauxite. Of the iron-rich materials the most common is limonite ($Fe_2O_3.H_2O$), but clays of this composition occur in two distinct crystalline forms known as goethite and lepidocrocite. Besides these crystalline minerals, compounds may also occur in an amorphous state; one of the most important is allophane, a hydrous aluminosilicate of rather variable composition. This latter may represent a gel formed by the weathering of primary minerals, and be one source from which clay minerals later crystallize.

Three major groups of aluminosilicate clay minerals may be distinguished: the kaolinite, montmorillonite and hydrous mica groups. They are all phyllosilicates but differ in crystal structure, surface electrical charges and resulting mechanical properties.

Crystal structure

Minerals in the kaolinite group consist of octahedral alumina sheets alternating with tetrahedral silica sheets (Fig. 6.10). Adjacent alumina and silica layers, held together by shared oxygen atoms, form what is known as the crystal unit. Since each unit consists of one

alumina and one silica sheet, the lattice is often described as being of the 1 : 1 type. The units themselves are held together by linkages between the oxygen and hydroxyl ions, forming a relatively rigid overall structure. Minerals in the montmorillonite group are composed of crystal units in which two silica sheets enclose a single alumina sheet; on account of this arrangement the lattice is often described as being of the 2 : 1 type. The bonding between successive units is extremely weak with the result that the crystals readily expand to admit water and exchangeable cations. The hydrous mica group of clay minerals, of which illite is the commonest example, also possesses a 2 : 1 type of lattice. However, in the silica sheet a considerable proportion of the silicon is replaced by aluminium, leaving valences which are satisfied by the incorporation of potassium ions. These lie between the units and produce a more rigid structure than is normally the case with montmorillonite.

Although the phyllosilicate clays are conveniently divided into three structural groups, many intermediate forms occur under natural conditions. Individual crystals quite commonly contain several different types of structural unit interstratified with each other; these are known as mixed-layer minerals. The very existence of illite as a distinctive clay mineral has even been questioned since many so-called illites are now known to have mixed-layer lattices. Moreover, in many clays ionic substitution is so frequent that the chemical composition varies widely; in montmorillonite, for instance, up to 20 per cent of the aluminium may be replaced by magnesium.

Fig. 6.10. Diagrammatic sketches of the crystal structure of three common phyllosilicate clay minerals.

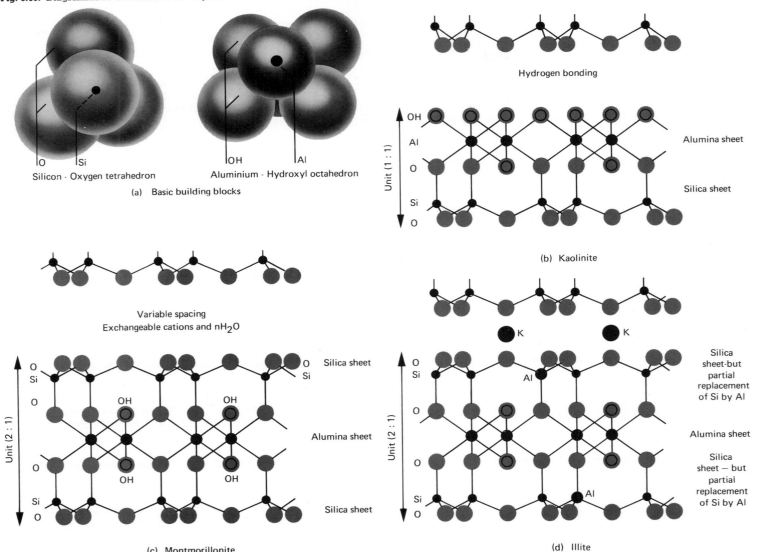

O   Si

Silicon - Oxygen tetrahedron

OH   Al

Aluminium - Hydroxyl octahedron

(a)   Basic building blocks

Hydrogen bonding

OH

Al

Alumina sheet

O

Si

Silica sheet

O

Unit (1 : 1)

(b)   Kaolinite

Variable spacing
Exchangeable cations and nH₂O

O   Silica sheet
Si

O

OH   OH

Alumina sheet

O

OH   OH

Si

O   Silica sheet

Unit (2 : 1)

(c)   Montmorillonite

K   K

O   Silica
Si   sheet-but
         partial
Al   replacement
         of Si by Al

O

Alumina sheet

O   Silica
         sheet — but
         partial
Si   replacement
         of Si by Al
O

Unit (2 : 1)

(d)   Illite

The chemically altered mantle of rock waste normally contains a varied assemblage of clay minerals. The factors leading to one particular assemblage rather than another are complex, and there is no set sequence of decomposition products. In the initial phases of weathering parent material may exercise a significant control. Granites with a high Si : Al ratio and low concentration of other metallic cations tend to alter to kaolinite; conversely, basic igneous rocks tend to weather to montmorillonite. However, the distribution of clay minerals is evidently influenced by many other factors. Acid soil conditions favour the production of kaolinite, while alkaline conduce to the development of montmorillonite. Regional patterns of dominance by certain aluminosilicate clays are discernible in mid-latitude areas. In the United States kaolinite is particularly abundant in the warm, humid south-east, the hydrous mica group becomes increasingly prominent in the cooler north-east, while montmorillonite reaches its fullest development in the drier west. Another general association is between tropical climates and abundant hydrous oxide clays, reflecting the faster leaching of silica in wet low-latitude areas than in most temperate regions. All these relationships between climate and particular clay minerals are indicative of broad trends in the weathering cycle, but at the detailed local scale many other factors are found to operate.

Surface electrical charges

All silicate clay minerals are covered with electronegative charges. In part these may be due to dissociation of the hydroxyl ion into separate oxygen and hydrogen ions. The former are negatively charged and remain part of the basic structure, whereas the latter are positively charged and are loosely held on the external surface. In a strongly acidic environment the hydrogen ions remain adsorbed on the crystal, but under more alkaline conditions may be replaced by other cations such as calcium. Another source of negative charges may be 'broken bonds' at mineral edges and corners. A third cause of electronegative conditions is ionic substitution of a high-valency cation by one of lower valency. Examples are afforded by the substitution of Mg^{2+} for Al^{3+} in such minerals as montmorillonite, and of Al^{3+} for Si^{4+} in the tetrahedral sheet of some of the hydrous micas.

The number of replaceable ions that can be adsorbed on to a clay mineral reflects its 'cation exchange capacity' which is a readily measured property. Owing to slight differences in structure and composition, samples of clays that are nominally the same yield different values for cation exchange capacity. Nevertheless, the figures for montmorillonite are often ten times those for kaolinite and four times those for a typical hydrous mica. The high value for montmorillonite stems from the widespread ionic substitution, whereas most of the charges in the case of kaolinite arise from dissociation of the hydroxyl. A further factor is the loose structure of montmorillonite which permits penetration between the crystal units so that adsorption can take place not only on the external surface but also on interior faces.

Mechanical properties

The contrasts in crystal structure and electrical charge produce important variations in mechanical properties. Some result simply from differences in particle size. For example, the rigid kaolinite lattice is associated with particles that are often about 0.002 mm in diameter, whereas montmorillonite crystals rarely exceed 0.001 mm; the hydrous micas tend to be of intermediate size. A complicating factor is the differential expansion of the minerals when wetted. The loosely structured montmorillonite tends to swell the most owing to the penetration of water between the crystal units. A regolith rich in this mineral therefore experiences correspondingly greater volumetric changes. This in turn affects water percolation owing to the tendency for deep surface cracks to develop during dry spells; these may admit large quantities of rainwater before the superficial layers are sufficiently soaked to close the cracks.

The most significant variations, however, result from the different adsorptive powers of the clay minerals. The electrostatic attraction of the typical mineral holds a layer of ions bonded to it; the concentration of ions decreases with distance from the mineral into the surrounding solution. Also bonded is a sheath of water molecules. Those next to the clay are extremely tightly held and give the molecular film properties resembling those of a solid. Slightly further away the water is still influenced by the electrical charges, but behaves more like a very viscous liquid. These adsorbed layers may be only 0.005×10^{-6} m thick, but nevertheless exercise a profound influence on such attributes as plasticity and cohesion.

Plasticity was originally investigated by Atterberg. He divided into five stages the transition from a clay so saturated with water that it behaves as a liquid to the same clay acting as a solid. The boundaries between his stages have gradually been modified and standardized, but still retain the name Atterberg limits. The three most important are:

1. The liquid limit which is the water content of a sample that allows exactly 20 mm of vertical penetration following release of a specially standardized 30° cone.
2. The plastic limit which is the water content at which a sample begins to crumble when rolled out into long threads 3 mm in diameter.
3. The shrinkage limit which is the water content below which further moisture losses cause no reduction in volume.

From these measured values can be derived the important 'plasticity index'. This defines the range of water content over which a sample remains plastic, and is simply the numerical difference between the liquid and plastic limits, both of which are conventionally expressed as percentages of the dry weight of sediment. All three Atterberg limits and the plasticity index vary according to the type of clay mineral involved. For instance, montmorillonite remains plastic over a much wider range of water content than does kaolinite, with the hydrous micas as usual occupying an intermediate position. In making such comparisons materials of the same grain size must be tested since the higher the proportion of fine particles the higher the plasticity index; this is due to the greater surface area on which adsorption can occur per unit weight.

The influence of adsorbed water on the cohesion of clay particles can be demonstrated by mixing them instead in paraffin; the paraffin molecules lack the polarizing properties of water and a drastic fall in cohesion results. As might be expected, montmorillonite clays are normally more cohesive than either kaolinite or illite clays. However, the shear strength of a clay is affected not only by the mineral type and amount of adsorbed water but also by the number and nature of adsorbed cations. The nature of the cations obviously depends upon their availability, which in turn depends upon the weathering processes and the chemical composition of the rock under attack. Moreover, certain cations are adsorbed more strongly than others, those which are loosely held being liable to replacement by others which are more tightly held. A solution of a neutral salt allowed to percolate through a clay will often emerge altered in composition owing to the loss of some of its original cations by adsorption and the addition of others by release from the clay surfaces. The actual amount of exchange depends on the strength of the solution since an equilibrium tends to be established between it and the adsorbed cations. The order of strength of adsorption when the cations are present in equivalent quantities is Al, Ca, Mg, K, Na. These reactions are fundamental for the soil scientist since they help determine soil fertility. Their significance for the geomorphologist lies in the varying physical properties they can confer on the same clay assemblage. This was demonstrated many years ago by Sullivan who treated nine identical clay specimens with different solutions each containing only one type of cation. Tests on the specimens showed that, with the same water content, their strengths varied greatly. Of the common cations, sodium and calcium produced relatively weak clays, potassium and hydrogen relatively strong. At certain water values the potassium clay was found to have a shear strength double that of the sodium clay. On the other hand, sodium-rich clays often have a high liquid limit and only become semifluid with an unusually large water content.

It is evident that the clay minerals are a very significant but very variable component of weathering residues. Although the foregoing account touches upon only a few of their properties, it demonstrates that many factors influence their behaviour as constituents of a mobile regolith. In reviewing the regolith as a whole mention must also be made of the organic component. Concentrated in the soil horizons the complex organic compounds known collectively as humus endow the superficial layers with qualities not found at greater depth. Humus is a colloidal substance having an adsorptive capacity much higher than that of montmorillonite. Its cation exchange capacity is more than twice that of most mineral clays and it has the ability to hold exceptionally large quantities of water. The range of swelling and contraction is correspondingly large, and the presence of humus often confers on the surface horizons a distinctive crumb structure which, in turn, affects their infiltration capacity and general erodibility.

References

Brook, G. A., Folkoff, M. E. and **Box, E. O.** (1983) 'A world model of soil carbon dioxide', *Earth Surf. Processes Landf.* **8**, 79–88.

Colman, S. M. (1982) 'Chemical weathering of basalts and andesites: evidence from weathering rinds', *U.S. Geol. Surv. Prof. Pap.* 1246.

Cooke, R. U. (1979) 'Laboratory simulation of salt weathering processes in arid environments', *Earth Surf. Processes* **4**, 347–59.

Curtis, C. B. (1976) 'Stability of minerals in surface weathering reactions', *Earth Surf. Processes,* **1**, 63–70.

Doornkamp, J. C. (1974) 'Tropical weathering and the ultra-microscopic characteristics of regolith quartz on Dartmoor', *Geogr. Annlr.* **56A**, 73–82.

Dorn, R. I. and **Oberlander, T. M.** (1982) 'Rock varnish' *Progr. Phys. Geogr.* **6**, 317–67.

Evans, I. S. (1969–70) 'Salt crystallization and rock weathering: a review', *Rev. Geomorph. Dyn.* **19**, 153–77.

Gerrard, A. J. W. (1974) 'The geomorphological importance of jointing in the Dartmoor granite', *Inst. Brit. Geogr. Spec. Pub.* **7**, 39–51.

Goudie, A. S. (1974) 'Further experimental investigation of rock weathering by salt and other mechanical processes', *Z. Geomorph. Supplementband* **21**, 1–12.

Goudie, A. S. *et al.* (1970) 'Experimental investigation of rock weathering by salts', *Area* **1** (4), 42–8.

Goudie, A. S. and **Watson, A.** (1984) 'Rock block monitoring of rapid salt weathering in southern Tunisia', *Earth Surf. Processes Landf.* **9**, 95–8.

Hekinian, R., Chaigneau, M. and **Cheminee, J. L.** (1973) 'Popping rocks and lava tubes from the Mid-Atlantic rift valley at 36°N', *Nature, London* **245**, 371–3.

Hills, R. C. (1970) 'The determination of the infiltration capacity of field soils using the cylinder infiltrometer', *Brit. Geomorph. Res. Group Tech. Bull.* 3.

Kerr, A. *et al.* (1984) 'Rock temperatures from southeast Morocco and their significance for experimental rock-weathering studies', *Geology* **12**, 306–9.

King, L. C. (1958) 'The problem of tors', *Geogr. J.* **124**, 289–92.

King, L. C. (1975) 'Bornhardt landforms and what they teach', *Z. Geomorph.* **19**, 299–318.

Lautridou, J. P. and **Ozouf, J. C.** (1982) 'Experimental frost shattering: 15 years of research at the Centre de Géomorphologie du CNRS', *Prog. Phys. Geogr.* **6**, 215–32.

Linton, D. L. (1955) 'The problem of tors', *Geogr. J.* **121**, 470–87.

Peel, R. F. (1974) 'Insolation weathering: some measurements of diurnal temperature changes in exposed rocks in the Tibesti region, central Sahara', *Z. Geomorph. Supplementband* **21**, 19–28.

Peltier, L. C. (1950) 'The geographic cycle in periglacial regions as it is related to climatic geomorphology', *Ann. Ass. Am. Geogr.* **40**, 214–36.

Rice, A. (1976) 'Insolation warmed over', *Geology* **4**, 61–2.

Sharp, A. D. *et al.* (1982) 'Weathering of the balustrade on St Paul's Cathedral, London', *Earth Surf. Processes Landf.* **7**, 387–9.

Skempton, A. W. and **Northey, R. D.** (1952) 'The sensitivity of clays', *Geotechnique* **3**, 30–53.

Smith, B. J. (1977) 'Rock temperature measurements from the northwest Sahara and their implications for rock weathering', *Catena* **4**, 41–64.

Smith, B. J. and **McGreevy, J. P.** (1983) 'A simulation study of salt weathering in hot deserts', *Geogr. Annlr.* **65A**, 127–33.

Strakhov, N. M. (1967) *Principles of Lithogenesis* (trans. J. P. Fitzsimmons), Oliver and Boyd.

Sweeting, M. M. (1972) *Karst Landforms,* Macmillan.

Thorn, C. E. (1979) 'Bedrock freeze–thaw weathering regime in an Alpine environment, Colorado Front Range', *Earth Surf. Processes* **4**, 211–28.

Winkler, E. M. and **Wilhelm, E. J.** (1970) 'Salt burst by hydration pressures in architectural stone in urban atmosphere', *Geol. Soc. Am. Bull.* **81**, 567–72.

Selected bibliography

Three texts that provide, in rather different ways, a good insight into weathering processes are: F. C. Loughnan, *Chemical Weathering,* Elsevier, 1969; C. D. Ollier, *Weathering,* Longman, 1975 and R. C. L. Wilson, (ed.), *Residual Deposits: surface related weathering processes and materials,* Blackwell, 1983.

There are numerous introductory books on soil mechanics that outline the methods of assessing the physical properties of rock materials. One such popular text is G. N. Smith, *Elements of Soil Mechanics for Civil and Mining Engineers* (5th edn), Granada, 1982.

Part 2 SUBAERIAL DENUDATION

Chapter 7
The geometry and energetics of fluvially eroded terrains

This and the next two chapters are concerned with the ways in which the continental surfaces are fashioned by the natural application of rainwater into the everyday pattern of ridges and valleys. Examination of almost any topographic map will reveal contrasts in the form of such fluvially eroded terrains, and if this spatial variability is to be explained it is essential that some method of describing the relief, in reasonably precise and objective terms, should be devised. Yet the three-dimensional geometry of ridge and valley patterns has proved bewilderingly complex. Two different strategies for solving the problem may be distinguished. The first is that which essays a quantitative description of the general relief properties of an area. At the global scale the difficulties that this encounters have already been alluded to on p. 5 and it is an approach that will not be pursued further here. The second strategy has been to focus attention on the basic units, the building blocks, of which a fluvially eroded landscape is composed. The component unit is the drainage basin, and analysis of basin form has proved a relatively fertile approach to the problem of quantitative description; an additional advantage is that the drainage basin also constitutes the natural unit within a stream-eroded topography for investigating the dynamics of the geomorphic system. However, a prerequisite of any attempt to measure and compare the morphological properties of drainage basins is some scheme of classification. The key has been found in the procedure known as stream ordering.

Morphometry of drainage systems

Methods of ordering drainage networks

During the first half of the present century several workers experimented with the notion of analysing drainage networks with the aim of devising a rational classificatory system for both streams and the basins that they drain. However, the acknowledged father of most recent network studies is the American engineer, Horton. The basis of the particular method he advocated in 1945 is the procedure known as stream ordering (Fig. 7.1). This involves assigning each stream a number dependent upon its relative position in the drainage network. For example, all headstreams above the first confluence points are designated first-order streams. Where two first-order streams join, they combine to form a second-order stream; where two second-order streams join, they combine to form a third-order stream, and so on. A high-order stream may receive a low-order tributary without its designation altering; a change only occurs at the confluence of two streams of like order. In the scheme originally proposed by Horton, the initial ordering was followed by a process of re-designation in which the higher-order streams were traced back to what might be regarded as their sources. Although rules for carrying out this operation were suggested, the whole process seemed rather arbitrary and lacking in real justification, and in 1952 Strahler recommended that this second part of the Horton procedure should be abandoned. Since that time most work has employed the so-called Strahler system of ordering, even though certain inherent weaknesses have been the subject of continuing discussion. Particular attention has focused on the failure of the Strahler system to take any account of those confluences that involve streams of different order. Shreve suggested an alternative scheme in which each link of a network is assigned a magnitude equal to the sum of the values for the tributary links immediately upstream. In effect this means that each link has a magnitude equal to the number of undivided headstreams that feed into it. Although the logic of Shreve's approach has much to commend it, neither his scheme nor a number of more complex variants has yet formed the basis of many morphometric studies.

Once stream ordering is complete, corresponding values may be assigned to the drainage basins. Thus the area drained by a first-order stream is designated a first-order basin, that drained by a

Fig. 7.1. Systems of stream ordering.

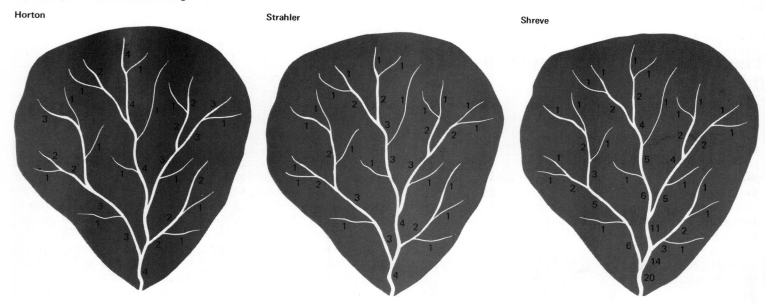

Horton

Strahler

Shreve

second-order stream a second-order basin, and so on. It should be noted that the whole area tributary to a particular stream link is included in the catchment; a third-order basin therefore includes not only the slopes shedding water directly to the third-order stream, but also the area draining into that stream via first- and second-order tributaries.

The apparent rigour of procedural rules for ordering streams and drainage basins should not be allowed to obscure the practical difficulties that are often encountered. The best way to map a drainage network has been much discussed. Although the blue lines on ordinary topographic maps have frequently been used, field checking has revealed serious inconsistencies, even on the same map, in the way minor headstreams are depicted. In some instances virtually all stream channels are shown, whereas in others only a selected proportion are portrayed. Consistent recording of small headstreams is crucial to the proper ordering of a drainage net, and many geomorphologists have therefore undertaken their own surveys

by means of either field-mapping or aerial photography. Recourse has occasionally been had to contour crenulations as the best cartographic guide to the positioning of headstreams, but this seems only marginally better than reliance on watercourses shown in blue. Delimitation of drainage basins may be attempted from the contours on topographic maps, but for accurate results field surveys or careful plotting from air photographs must be undertaken.

Analysis of data

Through the introduction of stream ordering Horton equipped the geomorphologist with a logical system for comparing the form of water-eroded landscapes. The field of enquiry thereby opened is so immense that it is beyond the scope of a few paragraphs to indicate the full range of possibilities, but little imagination is needed to visualize some of the more significant potentialities. In his original work Horton investigated three relationships which have since come to be known as the component parts of a 'Horton analysis'. The first concerns the number of streams of different orders within a single network. Studies have repeatedly shown that the total number of streams of successively lower orders increases as a simple geometrical progression. This may be demonstrated by plotting order on an arithmetic scale against the number of streams on a logarithmic scale (Fig. 7.2). The ratio of the number of streams of one order to that of the next higher order is known as the bifurcation ratio; its value commonly lies between 3 and 5. The second relationship concerns the average length of streams of different orders within a single network. Many studies have revealed a geometric increase in the mean channel length of successively higher orders, again made evident by plotting the relevant data on semilogarithmic graph paper. The third and final relationship concerns the average size of drainage basins within a single catchment unit. It is found once more that the mean size for successively higher orders increases as a geometrical progression. Following a Horton analysis of different regions, it is possible to construct hypothetical drainage nets incorporating the observed values of stream number, channel length and basin area; these show schematically the sharp contrasts that can exist between different drainage systems (Fig. 7.3).

Within the framework offered by an ordering system many other morphological properties can be measured and compared. Of course, operational procedures need to be carefully specified to ensure true comparability, and in many instances there has been prolonged debate about the best techniques to adopt. Reference

Fig. 7.2. An example of a 'Horton analysis' applied to an individual drainage basin (data from Brush U.S.G.S. Prof. Pap. 282-F, 1961). Susquehanna River basin, USA (Horton ordering system).

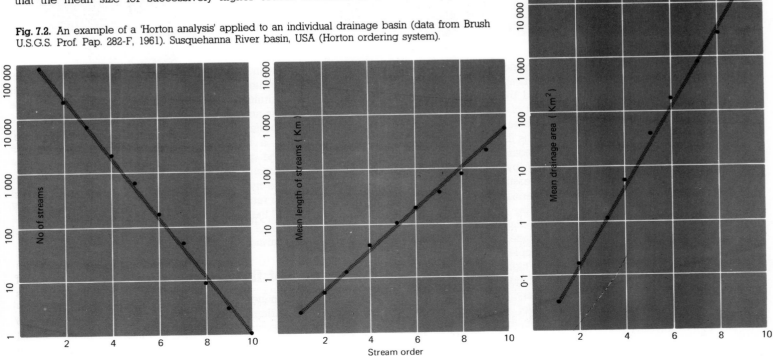

Fig. 7.3. Hypothetical fourth-order drainage basins in three contrasting environments (after Selby, 1967).

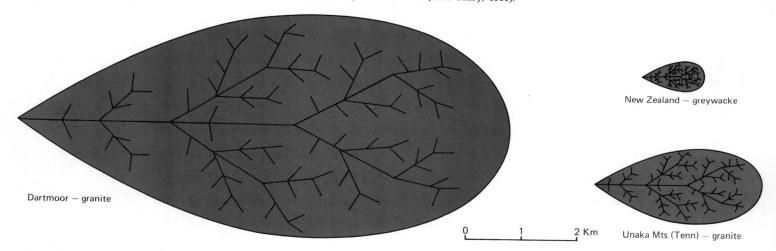

Dartmoor — granite

New Zealand — greywacke

Unaka Mts (Tenn) — granite

0   1   2 Km

may first be made to two additional planimetric properties, drainage density and basin shape. Drainage density is normally defined as the total channel length within a basin divided by the basin area. The inverse relationship, known as the constant of channel maintenance, is of interest as a guide to the catchment area required to sustain a unit length of channel. A further derivative is the value known as the distance of overland flow, by convention regarded as the reciprocal of twice the drainage density. Basin shape, by contrast, has proved much less amenable to simple analysis; measures of elongation and circularity have been devised by comparing, respectively, basin lengths and sizes with those that would be possessed by circles having either the same area or the same perimeter.

Introduction of the third dimension, altitude, affords many further properties worthy of investigation. The simplest is relief amplitude, or the difference in elevation between the highest and lowest points within a basin. A second derivative of altitude is stream slope. In many instances this has been shown to vary inversely with stream order; as in the cases discussed above, a law of geometric progression often seems to apply. So far attempts to investigate the angle of valley-side slopes within the framework of basin ordering have not proved very fruitful, but one technique that has aroused

interest is computation of the value known as the hypsometric integral. This involves measuring the area above successively lower contours within a drainage basin. Instead of dealing with absolute figures, both heights and areas are then converted into proportional values of the total basin relief and area (Fig. 7.4). Plotted on a graph these values yield a line known as the hypsometric curve. The complete area of the graph may be visualized as the gross volume of rock in the catchment before erosion began; the area beneath the curve, known as the hypsometric integral, represents the volume of rock that still awaits removal. Both rock type and geomorphic history seem likely to affect the integral; the computation tends to be tedious, but it remains one of several indices that may appropriately be calculated in comparing drainage basins.

Energetics of fluvial denudation

Shaping of the continental surfaces into the intricate geometrical forms described earlier in this chapter is the work of what has aptly been termed the geomorphology machine. Its primary driving force is gravity. Both the rocks raised to form mountain chains and the water evaporated from the oceans into the atmosphere possess

Fig. 7.4. Calculation of the hypsometric integral. The integrals for four contrasting landscapes are shown, although there is a strong tendency for most fluvially dissected areas to cluster between curves (b) and (c) with integrals between 40 and 60 per cent.

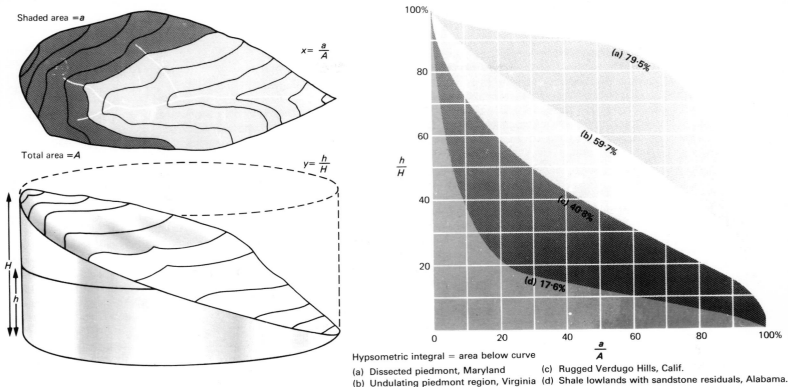

Hypsometric integral = area below curve

(a) Dissected piedmont, Maryland   (c) Rugged Verdugo Hills, Calif.
(b) Undulating piedmont region, Virginia (d) Shale lowlands with sandstone residuals, Alabama.

potential energy by virtue of the gravitational attraction of the earth. The energy needed to elevate these materials comes from two sources. In the first instance it is internal terrestrial heat, in the second solar radiation. The latter is by far the more important in the fashioning of fluvially eroded landscapes; a major goal of geomorphology, it has wisely been claimed, is to understand how the energy of solar radiation is converted into the mechanical work responsible for the moulding of landforms.

Solar radiation

The radiant energy emitted by the sun is enormous. At a distance of 149 million km, equivalent to the mean radius of the earth's orbit, the energy falling on a surface normal to the sun's rays is just under 1 360 W m^{-2}; this means that every minute the earth intercepts over 5×10^{18} J of radiant energy. However, of this total only a small fraction actually penetrates to the surface of the globe to become potentially available for geomorphic work. The Russian climatologist

Budyko has compiled a map (Fig. 7.5A) depicting the total solar radiation received at the earth's surface in a year. The distribution reflects the interaction of two major factors, the geometry of the earth's orbit round the sun and the shielding effect of the atmosphere. Although interesting as a statement of the energy directly available at the earth's surface, Budyko's map has important limitations. Studies of the global heat budget show that in latitudes below about 38 °N and S there is an excess of incoming solar radiation over outgoing terrestrial radiation, whereas in higher latitudes the relationship is reversed. To compensate this imbalance which would otherwise cause a steady heating of the tropics and cooling of the polar regions, there must be massive poleward transfers of heat energy. This is achieved by the movement of relatively warm air and water from low to high latitudes, and is a fundamental feature of the atmospheric and oceanic circulations of the planet.

The solar energy reaching the earth's surface is responsible for a complex web of processes. For example, part is used in sustaining the biosphere, and it has been estimated that each year photosynthesis by the continental vegetation cover converts 6.7×10^{20} J from radiant into chemical energy. A further part is absorbed by the ground with resultant heating and enhanced rates of chemical and biochemical change. A yet further part falling on wet surfaces is responsible for the evaporation of moisture into the atmosphere. The proportion of the incoming energy devoted to the different processes obviously varies from place to place and also from time to time. In a desert area, for instance, surface heating normally predominates, whereas in a forested region the vegetation shields the soil from direct radiant energy. Seasonal variations are well illustrated by deciduous forests which not only consume less energy in photosynthesis and transpiration during winter but also provide less effective shading. Great contrasts are also found in the amount of energy devoted to evaporation. The latent heat required to evaporate 1 g of water ranges from 2 260 J at 100 °C to 2 512 J at 0 °C, so that a mean annual precipitation over the globe of 857 mm demands on average about 2.1×10^5 J cm^{-2} yr^{-1} or some 30 per cent of available energy. Budyko has compiled a map (Fig. 7.5B) portraying the regional variations in the energy expended in evaporation; as might be anticipated, the highest values occur over the oceans where more than 50 per cent of the energy is commonly expended in evaporation compared with under 15 per cent for most desert areas.

The hydrological cycle

The energy expended in evaporation may be viewed in another way, that is, as the primary source of energy for the hydrological cycle. In essence the cycle is a transfer of water between four major reservoirs, the ocean basins, the atmosphere, the continental surfaces and underground storage. Estimates of the contents of each reservoir vary because of the difficulty in making reliable global measurements, but the amounts quoted in Fig. 7.6 are believed accurate to ± 10 per cent. They reveal that over 97 per cent of all the water is contained in the oceans, over 2 per cent on the land surfaces, about 0.6 per cent underground and less than 0.001 per cent in the atmosphere; of the water on the continental surfaces 99 per cent is held in the form of glacier ice, less than 1 per cent constitutes the lakes and rivers and a minute fraction is stored in the biosphere. Estimated rates of transfer between the various reservoirs are also incorporated in Fig. 7.6. Combined annual evaporation from the continents and oceans at 423×10^3 km^3 is sufficient to replenish the water in the atmosphere once every 11 days. By contrast it would require almost three centuries for precipitation to replenish the continental reservoir, the main reason being the very large volumes of water stored in the major ice-sheets; the water in the lakes and rivers could be replaced in little over 2 years. However, by far the longest residence period for water is in the oceans since it would require almost 4 000 years of evaporation at the current rate to completely desiccate the ocean basins.

Defined as weight times head, potential energy is generated on a tremendous scale by the hydrological cycle. The annual runoff from the continents is 37×10^3 km^3 with a total weight of 37×10^{15} kg. If it be assumed that this huge body of water descends from the mean continental height of 870 m the potential energy amounts to over 34×10^{19} J. It must be appreciated, however, that a very high proportion of this energy, once converted into kinetic form as the water moves downhill, is dissipated as frictional heat and so becomes unavailable for the work of erosion. The above figure may also be compared with that for the rock materials elevated above sea-level by tectonic processes. Conversion of this second source of potential energy into kinetic form is controlled by the rapidity of weathering and disaggregation of material at the earth's surface. It will be shown in Chapter 18 that, in areas of moderate relief, lowering of the land surface appears to be proceeding at a rate of approximately

Fig. 7.5. Solar energy at the earth's surface: (A) total energy received; (B) amount of energy expended in evaporation (after Budyko, 1974).

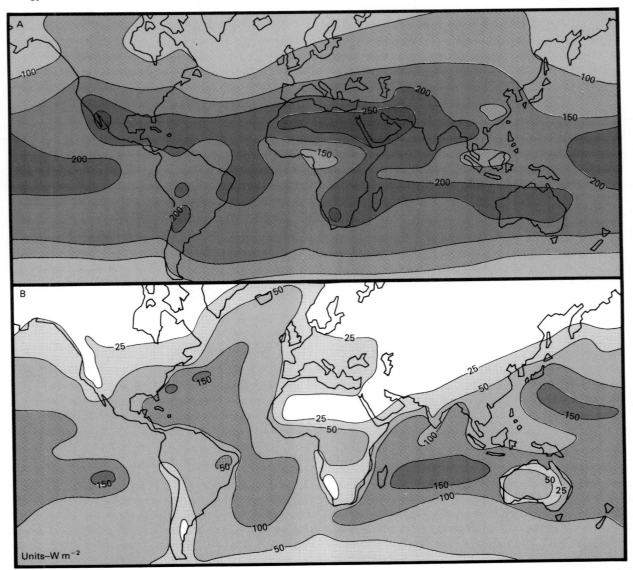

Fig. 7.6. Schematic representation of the hydrological cycle on a global scale.

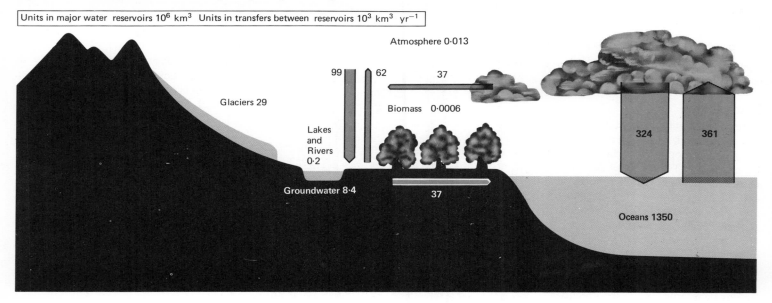

Units in major water reservoirs 10^6 km^3 Units in transfers between reservoirs 10^3 km^3 yr^{-1}

Atmosphere 0·013

Glaciers 29

99   62   37

Biomass 0·0006

Lakes and Rivers 0·2

324   361

Groundwater 8·4   37

Oceans 1350

0.05 mm yr^{-1}. The total weight of sediment released in this way each year is about 20×10^{12} kg; its potential energy, assuming it descends from the mean continental height, is a little under 18×10^{16} J, or less than one-thousandth part of the energy available from the hydrological cycle.

The drainage basin as a unit of study

Although fundamental to a full appreciation of the erosional system fashioning the earth's surface, the global energy values quoted above remain remote from everyday experience. Moreover, the variable ways in which solar radiation is converted into forms of energy means that world averages, even if they could be accurately computed, would have little practical significance for purely local studies. In consequence research has tended to concentrate on small catchments where the geomorphic system can be examined in much closer detail. Such 'catchment studies' permit the investigator to select drainage basins according to some preconceived plan, thereby simplifying interpretation of any data that may be collected. For example, he may choose as a unit of study a basin underlain by rocks of just one lithology and supporting a homogeneous vegetation cover. Alternatively, for purposes of comparison, two basins may be chosen that differ radically in only one respect. Thus contiguous catchments with similar geological but contrasting vegetational characteristics might be selected. A third common procedure is to study temporal changes that occur when one of the controls is radically altered, say, by deforestation.

A great advantage of the drainage basin as a unit of study is that it can be analysed as an open system, that is, a system in which there is a constant throughput of both mass and energy. The rest of this chapter is devoted to the techniques by which the two principal inputs and outputs of mass, namely water and rock waste, can be measured and their interrelations thereby elucidated.

The hydrology of individual catchments

A basic requirement for understanding water input and output is the ability to put figures into the hydrological equation:

Precipitation = Runoff + Evapotranspiration ± Storage

Each of the four components must be separately assessed in order to compile a complete water budget.

Precipitation

Normally most of the moisture input can be measured by means of rain-gauges, but where such forms as snow, dew or frost constitute a significant proportion of the total other methods of assessment may be necessary. Two types of rain-gauge are in regular use, the non-recording which is usually read by an observer at 24-hour intervals, and the recording which automatically provides a continuous trace on a rotating chart. The latter has the great advantage that it records rainfall intensity which, for the geomorphologist, may be at least as important as the rainfall total. A further facility is the easy conversion of the instruments to electronic recording and telemetry, so that even in remote areas data from an array of instruments can be rapidly received and analysed.

However, advances in the electronics used should not be allowed to obscure inherent limitations in accuracy. Precipitation measurement is fundamentally a sampling procedure and two sources of uncertainty are immediately identifiable. The first is instrumental and the second statistical. One problem with the conventional rain-gauge is that it creates a pattern of turbulent air flow that deflects a variable amount of the precipitation away from the orifice. It is difficult to apply any correction factor because the deficiency varies with wind speed and size of raindrop; experimental work suggests it commonly amounts to 5 per cent and in adverse conditions may exceed 15 per cent. Different designs of shield have been tried, and gauges have been sunk in pits so that their orifices are flush with the ground. The problem in the latter case is to prevent 'insplash' from the surrounding surface, but slats styled to divert the excess water away from the gauge have proved successful in experimental studies. Where snow constitutes more than, say, 5 per cent of the precipitation, accurate assessment requires the adoption of some form of snow-measuring device. Instruments employed include modified rain-gauges incorporating heating elements or antifreeze mixtures to melt the snow, and special weighing gauges which record the weight of precipitation falling in a pan. In areas of regular heavy snowfall sampling with snow corers may be substituted, but whichever method is adopted a high proportion of snow undoubtedly reduces the reliability of measured precipitation totals.

The statistical uncertainty is not easily assessed, especially as gauge density can be of critical importance where a high proportion of the rainfall comes from localized storm centres. Various procedures have been devised for calculating, from the point samples, total precipitation input. They range from the manual drawing of isohyets to geometrical constructions, known as Thiessen polygons, so designed that the value for each gauge is assigned to an area determined solely by the proximity of other gauges. The latter technique is appropriately employed where the gauge network is unevenly distributed and little supplementary information is available to guide the drawing of isohyets. The advent of the computer permitted more elaborate statistical techniques to be used, particularly where data are abundant. It is possible, for instance, to compute regional trends and by regression to assess the influence of such local factors as altitude, aspect and exposure. This ultimately allows prediction of the precipitation values for all points within the study area and the insertion of isohyets by the computer. Nevertheless, the intrinsic unreliability of the raw data must always be remembered, and it is doubtful whether total precipitation even in a small catchment can be estimated more precisely than ± 5 per cent of the true value.

Runoff

Measurement of stream flow is a relatively simple procedure, and with reasonable precautions accuracy should at least equal that normally attained in the assessment of precipitation. Two basic methods are in common use, one relying upon the natural channel and the other involving installation of a short section of artificial channel. Gauging procedures in a natural channel can be divided into three phases: firstly measurement of water-level, secondly determination of discharge and thirdly definition of the relationship between water-level and discharge.

The water-level or stage of a river can be measured by periodic observation with a manual gauge or by a continuous recording instrument. The simplest form of manual gauge is the vertical calibrated staff either set in the river bed or attached to a fixture such as a bridge pier. The value of the gauge is obviously dependent upon the frequency with which it is read. The interval between readings is commonly 24 hours, and although adequate for a large river of relatively stable flow, this is quite unsuitable for a small stream with a very 'flashy' regime. If the interest of the investigator lies in flood discharges these regular readings can be supplemented by installation of a peak gauge. This simple device records the maximum water-level reached between successive readings. It consists of a tube set vertically into the stream and either coated internally with a water-sensitive dye or containing some form of float that will remain at its maximum height even when the water-level falls. The principle is akin to that of estimating a flood-level from the 'tide mark' on buildings or from driftwood trapped in bushes along the river bank. These latter methods are often employed where an exceptional flood exceeds the capacity of the installed gauges, or where a catchment lacks any instruments at all.

The severe limitations of manual gauges require that, for detailed studies, some form of recording gauge be installed. The commonest type is that in which a float, free to move vertically with changes in water-level, is coupled to a pen that plots the variations on a rotating drum. In order to protect the float and damp the turbulent water movement, a stilling well is usually built and connected to the river by a submerged pipe. The resulting continuous trace of river stage indicates not only the levels of peak and low flow but also the rates of change during the passage of a flood. A second type of continuous recorder is actuated by pressure changes. As the water-level rises or falls pressure alterations at the stream bed are monitored by compression of air in a vessel fitted with a flexible diaphragm. A tube transmits the pressure variations to a sensitive aneroid which in turn is coupled to a recording pen. Both this and the float recorder can be readily adapted to electronic recording and telemetry so that in remote areas the instruments may be left unattended for prolonged periods.

Instantaneous discharge is usually measured by surveying the cross-sectional area of a stream and then determining the velocity of flow perpendicular to that section. The velocity can be measured by means of a current meter fitted with either a propeller or rotating cups like the familiar anemometer. The rate of rotation of propeller or cups is proportional to the speed of water flow and is normally computed by an electrically operated counting device. The greatest problem is the variation in velocity at different points in the cross-section since it is well known that flow close to the banks and bed is retarded by friction leaving a central thread of faster moving water (see Fig. 9.4, p. 164). The standard procedure to overcome this difficulty is to partition the cross-section into ten or more vertical segments depending on the width of the stream. Each segment is then treated as a separate unit within which the water velocity is measured at depths equivalent to 0.2 and 0.8 of the total depth; the mean of these two readings has been found to correspond closely to the mean for the whole segment. More speedy but slightly less reliable results can be obtained by one measurement at a depth equivalent to 0.6 of the total depth. The discharge of each segment is then calculated by multiplying its area by its mean velocity, and the total for the whole stream obtained by simple summation.

An alternative means of assessing instantaneous discharge is known as the salt-dilution method. For a period of about an hour a suitable concentrated salt solution is injected into a stream at a constant rate. Owing to turbulence the solution soon thoroughly mixes with the stream and by measuring the salt concentration in water samples a short distance below the point of injection the discharge can be calculated. The salt-dilution technique is particularly useful on small mountain streams where other methods are difficult to apply.

The instantaneous discharge at a gauging station must be measured at a variety of river stages in order to define its relationship to water-level. Once completed the results are plotted on a graph and a smooth curve drawn through the points (Fig. 7.7). This is known as a rating curve and may be used in conjunction with a continuous trace of river-level to provide a complete record of discharge. Certain precautions need to be observed in the construction and use of rating curves. The site for gauging should consist of a straight length of channel with a symmetrical cross-section. The banks and bed should preferably be smooth, and must in any case be stable; it is obviously impossible to use a rating curve where the channel is constantly changing its shape, and if necessary the banks and bed must be artificially stabilized.

An alternative procedure avoiding dependence on a suitable reach of natural channel is to install either a weir or a flume

Fig. 7.7. Examples of two types of rating curve: (A) the stage-discharge relationship of a cross-section on the Willamette River, Oregon, USA; (B) the sediment–water discharge relationship at a cross-section on the Rio Puerco, New Mexico, USA.

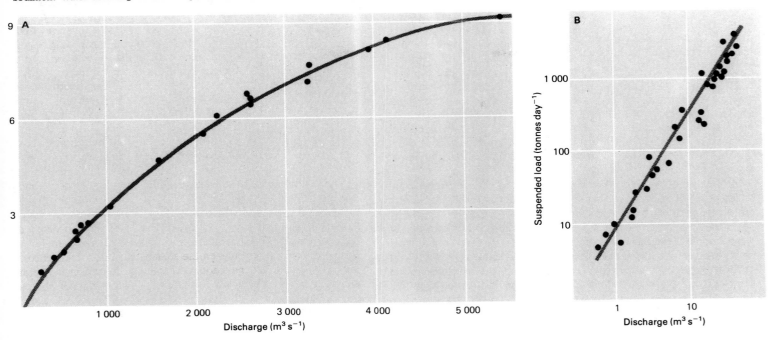

(Fig. 7.8). Often impracticable in the case of a large river, these so-called 'control structures' are much used on small streams. They have the great advantage that both types can be so built and calibrated that discharges are directly calculable, by simple mathematical formulae, from the values for the head, that is, the height of the water surface above the crest of the weir or the throat of the flume; it is this head that is normally measured by float recorder. On large rivers discharge measurements are regularly made as an adjunct of engineering structures built for other purposes. For instance, navigation works commonly involve the construction of locks and weirs which can be used to provide stream-flow records, while the spillways of dams and turbines of generating plants can both be calibrated to yield data about discharge.

Evapotranspiration

Evapotranspiration presents such formidable problems of measurement that it has often been estimated indirectly. If the hydrological equation is applied to an adequate time-interval, the change in storage becomes proportionately so small that the equation may justifiably be rewritten in the form:

Evapotranspiration = Precipitation − Runoff

and evapotranspiration evaluated by means of the techniques already described. However, for full understanding of water movement within a catchment direct measurement is essential. The very name evapotranspiration indicates part of the problem, for it is derived from two separate terms – evaporation and transpiration. The former operates alone over large water bodies, but is supplemented over the land by the transpiration from plant stomata of moisture originally absorbed through the roots.

Evaporation from water surfaces is usually assessed by means of pans in which either the fall in water-level is observed or the volume of water required to maintain a constant level is measured. However, evaporation from a small pan tends to surpass that from a large water body, partly because of heat exchange with the atmosphere through the pan walls and partly because of the 'oasis effect' in which there is a constant supply of relatively dry air to enhance the evaporation rate. As a consequence the various designs of instrument are assigned a 'pan coefficient' by which the observed values need to be multiplied in order to derive a truer indication of natural evaporation.

An additional problem when considering land surfaces is the availability of water. Clearly, neither evaporation nor transpiration will be the same under natural conditions as where water is constantly replenished by artificial means. This difference led to formulation of the concepts of potential evaporation and potential transpiration, both much used in climatological and hydrological studies. However, for present purposes it is better to concentrate on the measurement of actual values. Evaporation from an unvegetated soil surface may be assessed by a percolation gauge consisting of a metal cylinder some 40 cm in diameter filled with an undisturbed column of soil. The cylinder is then set in the ground above a receptacle to catch all the downward percolating water. The difference between the measured precipitation and the volume of water caught in the receptacle is the amount lost by evaporation. Such an instrument gives reasonable results over a long time-interval, but for day-to-day changes variations in soil-water storage must also be evaluated. This is most readily achieved through a mechanism for continually weighing the soil column, as is sometimes done in more elaborate installations designed to measure the combined effects of evaporation and transpiration. Termed lysimeters, these instruments consist of a much larger soil monolith set in a tank and supporting a vegetation cover. Owing to their greater size they are expensive to build and maintain,

Fig. 7.8. A typical permanent gauging station on a small mountain stream. The stilling well and equipment for recording the water-level are housed in the structure on the left.

and are not generally suited for use in small catchment studies. Nevertheless, they are of considerable scientific interest as one of the few sources of data about a vital phase in the hydrological cycle. The most famous early lysimeters were built at Coshocton in Ohio where the tanks holding the monoliths are about 2.5 m deep and have a surface area of 9 m²; despite loads of the order of 50×10^3 kg the modern hydraulic weighing mechanisms are sensitive to the nearest 0.3 kg or the equivalent of about 0.03 mm of precipitation. Such sophisticated instruments remain rare, but the number of more basic lysimeters has steadily grown. In Britain, for instance, an experimental monolith was installed during the late 1970s in Cambridgeshire with a superficial area of 25 m² and depth of 5 m, while an array of sixty-four smaller lysimeters was constructed in Oxfordshire as part of an agricultural research programme. Nevertheless, doubts persist about the relative effects on evapotranspiration of climate, soil and vegetation, and this remains a weak link in many attempts to compile full catchment budgets.

Storage

Between its initial point of impact as precipitation and its appearance as channelled flow at a gauging station water may follow many different routes. Some may flow over the ground surface, some may pass downslope at shallow depth along various horizons within the soil, and some may sink downwards to join the groundwater that fills the voids in the zone of saturation. Whereas rain falling on an interfluve may reach a stream channel in a matter of minutes by surface flow, it may take weeks, months or even years to reach the same point by an underground route. Studies by means of radioactive isotopes have shown that some groundwater now in storage was precipitated several thousand years ago. Of course this is not a direct measure of the response time between heavy precipitation and increased flow from springs and seepage lines. Accession of more groundwater tends to steepen the upper surface of the zone of saturation and so produce a greater hydraulic gradient. According to the law formulated by Darcy over a century ago but still forming the corner-stone of modern groundwater studies, the velocity of flow through a rock of given permeability is proportional to the hydraulic gradient; consequently, addition of water from renewed precipitation soon increases outflow at the point of issue to the ground surface. Conversely, as the zone of saturation subsides during a dry spell, the rate of groundwater discharge tends to decline exponentially. The overall effect of water sinking underground is to reduce the flashiness of the stream régime.

By convention, storage changes are often deemed to be restricted to variations in groundwater volume. This is because so many attempts to balance water budgets are concerned with periods ranging from a few months to a year or more, and at this time-scale only groundwater changes are of any real significance. Measurements are normally made by observing fluctuations in the height of the water-table. The level of the water surface in suitable wells and boreholes may be monitored by either periodic manual inspection or the installation of continuous recording devices. In temperate latitudes these generally reveal a prominent seasonal rhythm, with high winter levels followed by a pronounced decline during spring and summer. Superimposed on this annual cycle are shorter-term changes reflecting spells of wet or dry weather. In addition, where the water-table is sufficiently shallow to be tapped by tree roots, there is often a diurnal rhythm with a falling level during the daylight hours of active transpiration, and a recovery at night when abstraction declines below the rate of recharge. The irregular spacing of observation points can make volumetric calculations difficult, particularly in areas of complex geological structure, but the crucial importance of water resorces in many parts of the world is encouraging much attention to this problem.

If attempts are made to balance a water budget for much shorter periods, extra storage locations need to be taken into account. These include watercourses, sites of surface detention and the soil. The term detention refers to entrapment of precipitation that is either intercepted before reaching ground-level or held temporarily in the numerous shallow depressions that characterize almost any ground surface. Interception is mainly due to vegetation, the effectiveness of which is well recognized in the common practice of sheltering from a storm beneath a tree. The proportion intercepted varies widely according to the nature of both the precipitation and the vegetation. In showery summer weather it may be as high as 90 per cent, and even with heavier storm rainfall values up to 40 per cent have been recorded beneath a Douglas fir cover. These figures underline the desirability of siting gauges to measure not only total precipitation but also the effective ground-level precipitation; it is worth noting that the latter should include both the 'throughfall' and also the 'stemflow' trickling down the trunks of trees, which can be diverted by special collars into suitable measuring containers.

For budgetary purposes soil water ought theoretically to encompass moisture held both in the soil *sensu stricto* and in the vadose zone of unsaturated rocks above the water-table. In practice, however, most measurements are concentrated in the soil layer owing to the difficulty of making observations at greater depth. The simplest way of measuring moisture content is to weigh soil samples before and after drying in a laboratory oven, the difference between the two readings being an indication of water content. This, however, is not a very practical approach to the problem of continuous monitoring and several methods have been devised that attempt an *in situ* evaluation of soil-moisture content. Various types of porous block have been inserted into the soil, the principle being that as soil moisture fluctuates so will the water content and electrical conductivity of the blocks; this latter property can be measured without removal of the blocks. A second technique has been to determine the tension with which water is held on the soil particles since, as indicated on p. 114, the thinner the film of water the greater the tension; this can be achieved by using a vacuum gauge to monitor the tendency for water to be sucked from or returned to a porous ceramic pot buried in the soil. In both these methods the instruments will require calibration against known moisture contents for the soil under study. A third and more recent technique has involved the use of a neutron probe. This consists of two parts, a radioactive source of fast neutrons and a slow neutron counter. When the probe is lowered into the ground the fast neutrons are scattered by the hydrogen in the soil and the resulting slow neutrons are counted by the detector. Since almost all the hydrogen atoms in the soil are in the water, the number of slow neutrons is a measure of the moisture content within about 25 cm of the probe. Once a series of holes has been drilled and lined with aluminium tubing, the actual process of lowering the probe to the required depth and taking a reading can be completed in a matter of minutes.

The output of rock waste from individual catchments

The yield of rock waste from a catchment can be evaluated by monitoring the total load carried by a stream at its point of exit. The material in transit along a stream channel is normally divisible into three parts: the suspension load, solution load and bed load. The contribution to the total made by each of these parts varies greatly, and each poses a special problem of assessment requiring individual discussion.

Suspension load

Material is carried in suspension by the turbulent motion of flowing water in which the tendency of the particles to settle is offset by the residual upward momentum of the transporting medium. The water close to the river bed will normally be more heavily laden than that near the surface, so that samples taken from different parts of a stream cross-section will vary in the amount of sediment they contain. Since the water velocity also varies with depth, calculation of the total suspension load passing through a particular cross-section must take account of both variables.

In early investigations many hundreds of samples were often collected and analysed in order to derive reasonably accurate results. However, design of an instrument known as a depth-integrating sampler came as a vital technical innovation. The equipment consists essentially of a metal body shaped so as to direct the intake end upstream, a projecting nozzle to permit the ingress of the water, an internal sampling bottle, and a duct to allow escape of the air as the bottle fills. An electrically controlled valve permits remote opening and closing of the intake and also equalizes pressure in the sampling bottle so as to avoid an initial surge. The operating principle is simply that the amount of water entering the nozzle is proportional to the velocity of flow. This means that if the sampler is lowered at a constant rate from the surface to the bed of the stream, each depth in that particular vertical section will be represented by a volume proportional to its velocity; in other words, the sample is automatically compensated for variations in velocity and load, so that its sediment concentration need only be multiplied by the appropriate discharge to obtain the total suspension load passing through the vertical section in question. The depth-integrating sampler may conveniently be lowered in each of the vertical segments used for water-gauging purposes, as outlined on p. 126, and the gross value for the stream calculated by summation. An alternative procedure, known as equal-transit-rate sampling, is to employ exactly the same rate of lowering and raising the instrument at all verticals; the composite sample then requires no more than a single analysis to yield a sediment concentration figure which, if multiplied by the stream discharge, gives the total suspension load.

The depth-integrating sampler has greatly simplified the work of measuring the suspension load passing through a particular cross-section at any one instant. There remains the problem, however, of deriving a continuous record to take account of the variations with time. One possible approach is to construct a sediment-rating curve that directly relates suspension load to discharge. When observed values for some streams are plotted on logarithmic scales a close correlation is found (Fig. 7.7B), but for others a much wider scatter of points is apparent. This is because the suspension load is not simply a function of discharge but is affected by a wide range of other factors. Some of these are seasonal so that the amount of material carried at a specified discharge may differ in summer and winter; others are related to the character of the precipitation, particularly its intensity and duration. Deviation from a consistent suspension load–discharge relationship is well illustrated by the San Juan River (Fig. 7.9) where the amount of material carried on the rising leg of a seasonal peak may be ten times that transported by the same discharge on the falling leg. Other rivers show the reverse relationship with the maximum sediment concentration occurring as the flood waters recede. It is obvious that sediment-rating curves must be used with some caution, and for reliable results it is often necessary to pursue a sustained programme of regular sediment sampling.

Attempts are increasingly being made to devise mathematical models that will allow prediction of the total suspension load from a limited number of point samples. Such point samples can be collected either by installing a fixed array of bottles capable of being opened remotely so as to allow ingress of the water, or by arranging for water to be pumped electrically from specified points in the cross-section into bottles stored on the river bank; in both these cases sampling can be carried out automatically either at a pre-set time or at a predetermined river stage. The ultimate goal would be the ability to calculate the total suspension load from a few locations in a cross-section that are being continuously monitored. Experimental continuous recording has involved such techniques as measuring the scattering of light from a laser beam by the concentration of suspended particles, but at the present time further refinement appears to be necessary and for reasonable accuracy traditional but rather laborious methods appear to offer the best solution.

Solution load

The variations with depth that characterize suspension load do not apply to most dissolved material. The turbulent flow tends to ensure relatively complete mixing and for many purposes a 'bottle on a string' will provide a representative sample. Nevertheless, for a complete chemical analysis, and particularly where a stream flows through deep quiet pools, it is preferable to collect a depth-integrated sample. A complete analysis covers a very wide range of dissolved solids and gases, and it is not appropriate to discuss here all the complex laboratory techniques that may be required. For the geomorphologist most interest focuses on those dissolved solids, released by weathering, from which the rate of chemical denudation may be estimated. The simplest and most widely studied example is the calcium carbonate content of streams draining limestone catchments. Approximate determinations of calcium hardness may be made in the field by adding specially prepared tablets to a water sample and observing the ensuing colour changes, but laboratory titration is normally used for the more precise determinations needed in a research programme. The results of each analysis, initially expressed in parts per million, are convertible into values of limestone dissolved. However, in order to ascertain the total amount of limestone being removed from a catchment per unit of time it is is obviously necessary to monitor both stream discharge and the changing calcium hardness of the water; account may also need to be taken of the magnesium in the water since much limestone is composed of both calcium and magnesium carbonate.

In general, with rising river-level the total amount of the solution load increases but the concentration of individual solutes falls (see, for instance, Fig. 7.10). However, different elements behave rather differently so that the overall pattern is one of considerable complexity. Individual solute-rating curves may be constructed, but the inconsistent relationships between load and water discharge alluded to in discussion of suspended material apply equally to the solutes. This is scarcely surprising when it is remembered that stream flow is a mix of waters that have followed many varied pathways, each imprinting the water with its own particular chemistry, and that the mix can be quite different even when the discharge is identical. Furthermore, the strong biological influence on many chemical

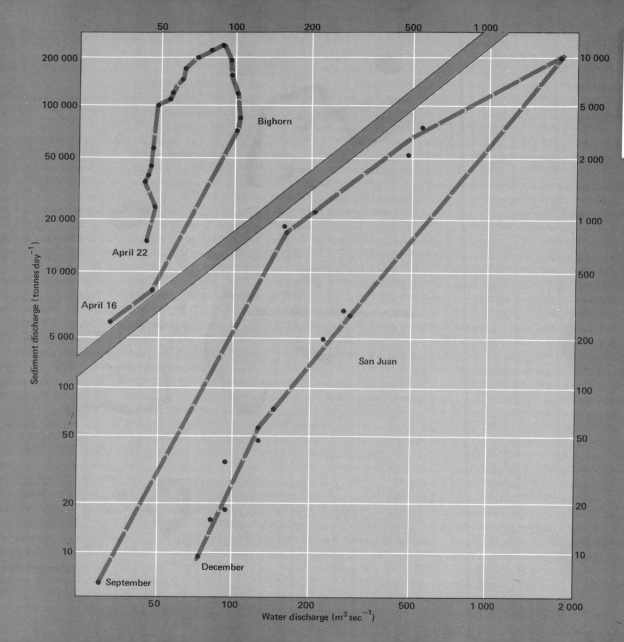

Fig. 7.9. Short-term variations in the suspended sediment concentration on the Bighorn and San Juan Rivers in the USA. For the same water discharge the 4-month record on the San Juan illustrates a substantially greater sediment load on the rising stage than on the falling stage; the week-long record of the Bighorn illustrates the reverse relationship (after Guy, 1964).

Fig. 7.10. Two illustrations of the way the solution load of a stream varies with discharge. Note the tendency for the solute concentration to fall with rising discharge, although there may sometimes be an initial 'flush' as accumulated salts are washed out of the catchment. Individual basins can show considerable differences in their response to flood events (after Walling, 1975).

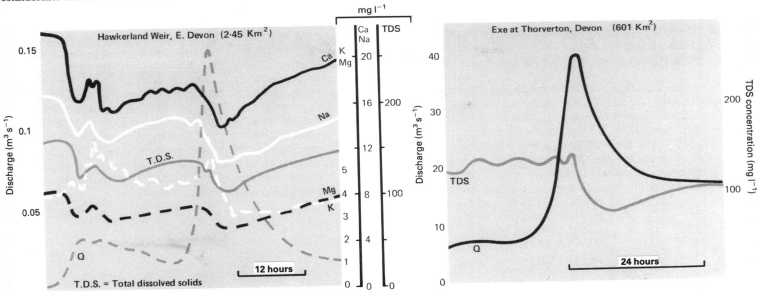

processes means that pronounced seasonal contrasts may exist. As always, the ideal solution would be continuous monitoring but this poses many technical problems. For some purposes the electrical conductivity of the water can be regarded as a satisfactory surrogate for the total quantity of dissolved solids, and this can be measured continuously without undue difficulty. However, data regarding individual solutes are much harder to obtain. Various 'ion-specific electrodes' that will measure the concentration of individual ions are being developed, and in future these may provide valuable accessory information. In the meantime it must be recognized that details of the solution load in transit along a stream channel often have significant shortcomings. A final noteworthy point is that a small but variable fraction of the dissolved material enters a catchment as a component of the precipitation (see p. 101). This part of the solution load needs to be substracted if that part due to rock weathering is to be correctly evaluated.

Bed load

This third component of the total load, sometimes also called the traction load, is the most difficult to measure. It may be defined as that part of the sediment in transit that forms the mobile bed of the stream. Displacement of individual particles is by rolling, sliding or occasional saltation into the basal layers of the water. As with many other geomorphological investigations, the problem is to make measurements without disturbing the very processes one is attempting to measure. One device that has been widely tried is the basket sampler anchored to the stream bed. This consists of a box made of mesh screening but open at its upstream end. If the mesh is too large part of the load passes through, whereas if it is too small it easily clogs and accentuates the intrinsic problem of any basket sampler, the diversion of flow away from the inlet. To counter the latter difficulty the sampler is best shaped so that the walls near the entrance flare outwards in a downstream direction. This creates a pressure drop which can be adjusted to ensure that the inlet velocity approximates that of the surrounding stream. A second method that has been tried involves the excavation of a sampling pit. This comprises an elongated pit, dug across the full width of a stream, in which are placed receptacles that can be removed at intervals to measure the amount of trapped debris. Although more efficient than basket samplers, these pits demand elaborate preparation and if incorrectly designed can cause eddying that disturbs the natural flow of the water.

Many of the difficulties in making short-term measurements are compounded when it comes to assessing total bed-load movement over a relatively long time interval. Careful work with basket samplers has revealed that, even at constant discharge, the bed load moves in surges with an interval of a few minutes between each peak. As a result of problems in obtaining reliable field measurements, recourse has been had to developing mathematical formulae to describe bed-load transport. These must rely on readily measured properties such as velocity, depth, slope, width and particle size of the bed material. Numerous such formulae have been proposed, often predicting quite different transport rates, but none has proved sufficiently satisfactory to command really widespread support.

When values for total sediment load are quoted, the limitations indicated above need to be borne in mind. The proportions being carried as suspension and traction load appear to vary considerably, although a figure of 5–10 per cent as traction load is often regarded as characteristic. On the other hand, some measurements suggest it may occasionally exceed 50 per cent where most of the debris is very coarse. It should also be remembered that suspension-load samplers do not normally collect material closer to the bed than about 0.1 m, so that an important fraction of the material in suspension is left unsampled. Increasingly the problem of measuring material in transit at, and close to, the bed of a stream is being recognized by a realistic division of the detrital load into measured and unmeasured fractions; the latter is then simply estimated by what appears to be the most appropriate transportation formula.

References

Budyko, M. I. (1974) *Climate and Life* (ed. D. H Miller), Academic Press.

Cannell, R. Q. and **Burford, J. R.** (1979) 'The use of lysimeters at ARC Letcombe Laboratory', *Inst. Geol. Sci. Rep.* **79/6**, 17–19.

Edwards, A. M. C., McDonald, A. T. and **Petch, J. R.** (1975) 'The use of electrode instrumentation for water analysis', *Brit. Geomorph. Res. Group Tech. Bull.* 15.

Guy, H. P. (1964) 'An analysis of some storm-period variables affecting stream-sediment transport', *U.S, Geol. Surv. Prof. Pap.* 462-E.

Horton, R. E. (1945) 'Erosional development of streams and their drainage basins: hydrophysical approach to quantitative morphology', *Geol. Soc. Am. Bull.* **56**, 275–370.

Kennedy, B. A. (1978) 'After Horton', *Earth Surf, Processes* 3, 219–31.

Kitching, R. and **Day, J. B. W.** (eds) (1979) 'Two-day meeting on lysimeters', *Inst. Geol. Sci. Rep.* **79/6**.

Murphy, P. J. and **Amin, M. I.** (1979) 'Compartmented sediment trap', *Am. Soc. Civ. Engr. J. Hydr. Div.* **105** (HY5), 489–500.

Selby, M. J. (1967) 'Aspects of the geomorphology of the greywacke ranges bordering the lower and middle Waikato basins', *Earth Sci. J.* **1**, 1–22.

Shreve, R. L. (1967) 'Infinite topologically random channel networks', *J. Geol.* **75**, 178–86.

Strahler, A. N. (1952) 'Hypsometric analysis of erosional topography', *Geol. Soc. Am. Bull.* **63**, 923–38.

Walling, D. E. (1975) 'Solute variation in small catchment streams: some comments', *Trans. Inst. Brit. Geogr.* **64**, 141–7.

Selected bibliography

A valuable review of morphometric techniques, accompanied by an extensive bibliography, is provided by V. Gardiner, *Drainage Basin Morphometry*, Brit. Geomorph. Res. Group Tech. Bull. 14, 1974. Standard works on catchment hydrology include R. C. Ward, *Principles of Hydrology* (2nd end), McGraw-Hill, 1975, and J. C. Rodda, R. A. Downing and F. M. Law *Systematic Hydrology*, Newnes-Butterworth, 1976. The more geomorphological aspects of catchment studies are covered in K. J. Gregory and D. E Walling, *Drainage Basin Form and Process*, Arnold, 1973.

Chapter 8
Hillslopes

In Chapter 6 the preparation of rock material for transport by subaerial geomorphic agencies was outlined, and in the last chapter attention was turned to the basic energetics of water-eroded landscapes. However, an individual catchment was treated largely as a 'black box' with inputs and outputs of mass and energy, but with very little reference to actual processes operating within it. The present chapter and that which follows take a much closer look at the internal workings of fluvially eroded landscapes. Such areas are, in essence, assemblages of hillslopes and an appreciation of the mechanisms responsible for shaping the slopes is therefore vital to an understanding of the landscape as a whole.

In the general realm of slope studies three particularly significant spheres of investigation may be identified. The first concerns the physical properties of slope materials, a subject that has already been outlined in Chapter 6. The second concerns the measurement and quantitative description of slope form, a topic that is discussed as the first part of the present chapter on pp. 136–140. The third sphere of investigation concerns the nature of hillslope processes. A major difficulty arising in this context is the multiplicity of such processes. Some are chemical, others mechanical; some are confined to the surface layers, others penetrate much more deeply; some go on almost continuously, others are highly intermittent; some are triggered by temperature changes, others by humidity changes. So many different processes operate simultaneously that the slope form cannot be ascribed to any one acting in isolation. The range of processes that assist in the shaping of hillslopes is reviewed as the second part of the present chapter under the individual heads of

particulate movement, mass movement and solutional movement.

The intricate web of relationships between material, form and process cannot be stressed too strongly even though, as a matter of convenience, each is treated under a separate heading. Equally, the dangers of assuming that present-day morphology is due entirely to current processes should be noted. There is abundant evidence that the form of many hillsides was roughed out by processes no longer operative, and it needs to be remembered that most areas of the globe have experienced major environmental changes during the last 10 000 years or so.

Measurement of form

Morphological mapping

The conventional method of depicting relief is by means of contours. Although these have immense advantages for the cartographer in conveying the maximum information in the minimum space, for the geomorphologist they have severe limitations. This is especially true of contours surveyed as part of a national mapping system. The interval between contours is ordinarily too large for them to delineate other than the broad pattern of relief. Details of slope form are inevitably lost. That is one reason why, for many decades, individual geomorphologists have devised special schemes of morphological mapping aimed at depicting the surface form of the ground in much greater detail than is possible by widely spaced contours. Originally each worker tended to be highly selective, choosing to portray only those morphological features he believed to have significance for the geomorphic history of the region under study. Later it was argued that the primary aim of morphological mapping should be different, namely to provide an accurate and full two-dimensional representation of the three-dimensional form of the ground surface. Yet this apparently simple objective has proved remarkably difficult to achieve.

Morphological mapping is founded on the premise that the complex shape of hillslopes and valley sides can be resolved into a finite number of units capable of delimitation in the field. Most of the proposed schemes involve recognition of two basic types of unit. On the one hand are facets over which the profile gradient remains constant, on the other are curved units over which the profile

Fig. 8.1. Morphological mapping as advocated by Savigear. The sketch on the left shows how the forms represented in the block diagram would be depicted. The map on the right records the morphology of an area near Ibadan, Nigeria (redrawn from Savigear, 1965).

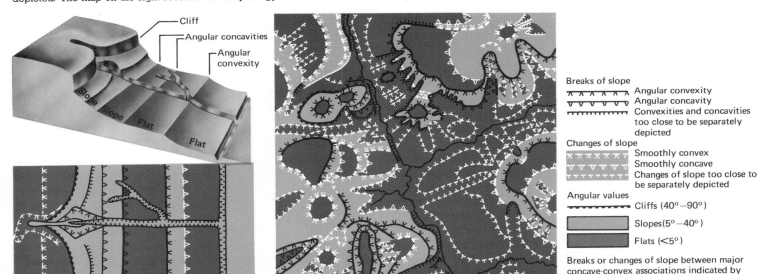

gradient steadily changes. Curved units are further subdivided into concave and convex categories. Two types of boundary separating these morphological units are usually distinguished. A break of slope is an angular discontinuity, a change of slope is a more gradual transition from one unit to another. One such scheme employing this general approach is illustrated in Fig. 8.1. Certain terrains lend themselves much more readily to portrayal in this way than do others. Satisfactory results have been obtained in the southern Pennines where many of the bold escarpments associated with alternating gritstone and shale outcrops can be resolved into facets separated by breaks of slope, but much less success has attended efforts to employ the same technique in clay lowlands dominated by subdued concavo-convex slopes.

In evaluating any mapping scheme a major criterion must be the extent to which the results can be replicated by groups or individuals working independently. In theory, for instance, contour maps produced by different survey parties should be identical. However,

in the case of morphological mapping schemes, no set of field instructions has yet been devised that will guarantee closely comparable maps being produced. In order to achieve uniformity more and more complex rules have been suggested, but these have both failed to attain the desired end and also added greatly to the range of field measurements that need to be made; the increase in objectivity has not appeared commensurate with the extra effort involved. In the meantime morphological mapping undoubtedly remains a very useful technique for the rapid recording of hillslope form in approximate terms. It can greatly supplement the information provided by normal contours, and has important applications in such fields as soil survey and terrain analysis. Yet its value in detailed quantitative slope studies appears at present to be severely circumscribed.

Determination of slope-angle frequencies

A totally different approach to hillslope studies has been adopted in a number of methods that seek to analyse the frequency with which given slope angles recur. One procedure is simply a cartographic exercise employing the contours shown on a normal topographic map. The inter-contour distances equivalent to selected critical angles are first calculated and a template scale drawn. Using the scale, shading is then applied to areas where the contours are closer than the selected values; with denser shading on steeper slopes a good visual impression of the relief is obtained. By measuring the total area within each slope-angle class, an estimate of the frequency of each class can be obtained. Because manipulation of the template scale is extremely tedious an alternative photographic technique has sometimes been tried. This fuses contours at a predetermined critical spacing into a single blurred image. By repeating the photography at different scales a map can eventually be compiled showing the distribution of chosen slope angles. The major limitation of these 'isoclinal maps', whether constructed manually or by some other means, is that the contours on which they are based contain a strictly limited amount of information about actual surface form. It has to be assumed, firstly, that the ground between contours slopes at a uniform angle and, secondly, that changes in angle occur only along published contours. Patently, neither of these assumptions is totally valid and the procedure can only yield a first approximation of the true frequency distribution of slope angles.

In theory similar results may be obtained rather faster by a sampling procedure. The gradient can be calculated at randomly selected points by presuming a constant gradient between the nearest higher and lower contours. Strahler, who has pioneered many methods of analysing slope forms, employed this technique in studies of water-eroded landscapes in the Appalachians and the mountains of southern California. An important possible modification is to make the measurements in the field rather than derive them from topographic maps. Admittedly this is more time-consuming, but the increased precision often makes it a desirable innovation. The return on the extra effort depends in part on the nature of the relief. In an accidented area dominated by linear slope profiles the differences between field and map measurements may be quite small. On the other hand, in a lowland region with concavo-convex profiles, sharply contrasting results can emerge; this is especially true where the contour interval is large relative to the amplitude of relief.

Profiling

A third approach to the analysis of hillslope forms is by measurement of representative profiles. Before this method can be employed two decisions have to be taken: how are the lines of the profiles to be selected, and what procedure is to be used for the actual surveying. The choice of profile lines has often been made subjectively with a particular purpose in mind, but this is open to the criticism that the investigator may subconsciously be making a selection that will confirm some preconceived idea. On the other hand, randomly chosen survey lines may prove difficult of access and yield profiles which the geomorphologist regards as not strictly comparable owing to the operation of localized factors such as active stream undercutting. Much depends on the objective of the study and the nature of the profile analysis that it is hoped to undertake. Profiles might, for example, be chosen by surveying upslope and downslope from randomly selected points. In 1972 Young advocated a procedure by which mid-slope lines are first drawn around each catchment halfway between stream and watershed (Fig. 8.2). These mid-slope lines can then be used for sampling, with points selected along them according to a suitable plan. Through the points profiles are surveyed from stream to watershed.

Several different methods have been used for the actual survey. The most popular equipment consists of ranging poles set up at appropriate points, a tape to measure the distances between them and an Abney level to determine the corresponding gradients. Normal practice in early studies was to site the ranging poles at obvious changes of slope, ignoring the considerable differences in ground distance this might involve. Later work has tended to employ a constant ground distance, largely on the score that choosing significant changes of slope can be highly subjective. If the uniform measured lengths are too long significant breaks of slope can be smoothed out, whereas if they are too short there appears to be an unnecessary waste of time in surveying lengthy segments of constant gradient. Despite the tediousness, profiles are probably best surveyed by means of short uniform distances, although much must depend on the final type of analysis it is proposed to employ.

Several instruments have been designed specifically for profile surveying. The simplest is the slope pantometer composed of a four-sided equilateral wooden frame with adjustable corners. A spirit-level incorporated into the frame allows the operator to position two of the sides vertically, and an angular scale then indicates the slope of the remaining two sides; since the lower is resting on the ground this is the gradient at that particular point. A convenient size for the

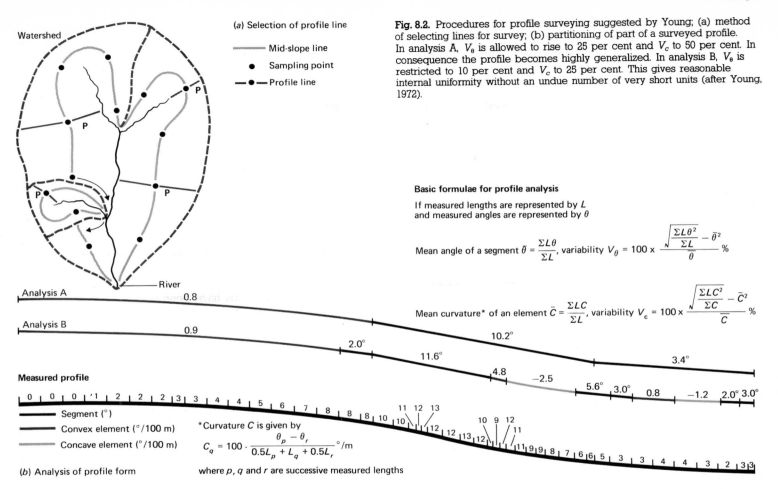

(a) Selection of profile line

Watershed

- - - - Mid-slope line
● Sampling point
—●— Profile line

P

P

P

P

River

Fig. 8.2. Procedures for profile surveying suggested by Young; (a) method of selecting lines for survey; (b) partitioning of part of a surveyed profile. In analysis A, V_θ is allowed to rise to 25 per cent and V_c to 50 per cent. In consequence the profile becomes highly generalized. In analysis B, V_θ is restricted to 10 per cent and V_c to 25 per cent. This gives reasonable internal uniformity without an undue number of very short units (after Young, 1972).

Basic formulae for profile analysis

If measured lengths are represented by L and measured angles are represented by θ

Mean angle of a segment $\bar{\theta} = \dfrac{\Sigma L \theta}{\Sigma L}$, variability $V_\theta = 100 \times \dfrac{\sqrt{\dfrac{\Sigma L \theta^2}{\Sigma L} - \bar{\theta}^2}}{\bar{\theta}}$ %

Mean curvature* of an element $\bar{C} = \dfrac{\Sigma L C}{\Sigma L}$, variability $V_c = 100 \times \dfrac{\sqrt{\dfrac{\Sigma L C^2}{\Sigma C} - \bar{C}^2}}{\bar{C}}$ %

Analysis A

0.8

Analysis B

0.9

2.0°

11.6°

10.2°

3.4°

4.8

−2.5

5.6° 3.0° 0.8 −1.2 2.0° 3.0°

Measured profile

— Segment (°)
— Convex element (°/100 m)
— Concave element (°/100 m)

*Curvature C is given by

$$C_q = 100 \cdot \dfrac{\theta_p - \theta_r}{0.5L_p + L_q + 0.5L_r} \;°/m$$

where p, q and r are successive measured lengths

(b) Analysis of profile form

frame is 1.5 m square so that the measured length is much shorter than is normal with ranging poles and a tape. Speed of operation is high but vegetation and minor surface bumps can cause difficulties. A second and much more complex instrument, known as a profile recorder, has as its essential components a set of four wheels coupled to devices for recording ground distance and slope angle. The distance is measured by a mechanism counting revolutions of the wheels, while slope angle is monitored by sensing changes in the relative elevations of the two axles. When the recorder is pushed up or down a slope, internal linkages ensure that a continuous profile is automatically drawn on a rotating drum. In the hands of a skilled operator tests have shown that very accurate profiles can be obtained; however, the machine is expensive and only works well on unvegetated or grassy slopes.

Analysis of measured profiles normally involves division of the slope into rectilinear, convex and concave portions. This may be done subjectively, but different workers are liable to vary in their interpretation of both the number of slope units to be recognized and the positions of the intervening boundaries. It is preferable, therefore, for the division to be made in accordance with a set of rules that will eliminate operator bias and, if desired, permit the calculations to be performed on a computer. In 1972 Young proposed an elaborate scheme, known as the best units system, based upon the variability of angles to be allowed within a single profile unit (Fig. 8.2). He recognized two basic types of unit, a segment over which the angle remains approximately constant, and an element over which the curvature remains approximately constant. The need is to define in quantitative terms what is meant by 'approximately'. For this purpose Young employs the value known as the coefficient of variation, which effectively expresses the deviation from the mean as a proportion of the mean value itself. This is chosen since it is believed, for instance, that a deviation of 1° is often unimportant on steep slopes but may be highly significant on gentle slopes. A segment is consequently defined as a portion of a slope profile within which the coefficient of variation of angle does not exceed a specified value; an element is defined as a portion of a slope profile within which the coefficient of variation of curvature does not exceed a specified value. At first sight the system, only given in outline here, may seem unduly complicated but rigour demands that all possible circumstances be covered. Even so, the system has been criticized because in certain cases different results may be obtained if calculations begin at the base of the slope and work upwards, rather than starting from the crest and working downwards.

Provided methods of both surveying and partitioning are the same, there are several different ways in which profiles can be compared. Given that the normal valley-side slope is composed of an upper convex element, a lower concave element and an intermediate segment, the relative lengths of each of these units may be analysed. It has been shown that rectilinear segments are much more common than was at one time supposed, often occupying over 10 per cent of the total profile length and sometimes surpassing 50 per cent; this is true even in areas such as the chalklands that have traditionally been described as composed of concavo-convex slopes. Concave elements rarely occupy as much as half of a profile and in some situ-ations are completely absent. Convex elements, on the other hand, are virtually always present and, according to Young, frequently occupy over 40 per cent and occasionally over 80 per cent of British profiles. The same author has described slopes in Brazil where more than 95 per cent of the total length is composed of the upper convexity. Although significant contrasts evidently exist in the proportion of profiles occupied by the different basic units, agreement about the underlying causes has yet to emerge. Another aspect of profiles that is frequently the subject of comparison is the length and angle of the rectilinear segment. One reason for the interest in this property is the belief that it may represent a limiting gradient under given conditions of process and material. Finally, mention should be made of efforts to analyse the form of concave and convex elements, normally by fitting simple polynomial curves to discover whether they exhibit close mathematical similarities. The results so far have not proved very illuminating, with controversy surrounding the extent to which apparently curved units are actually composed of several short rectilinear segments.

Particulate movement

Particulate movement refers to the downslope transfer of individual particles moving independently of the rest of the slope materials. It differs from mass movement, to be discussed below, in that surface debris alone is displaced and the material in transit is disaggregated. Many different processes give rise to particulate movement, but two important groupings can be recognized. The first gravity-driven grouping encompasses those processes which induce instability on a free rock face and lead eventually to detachment of segments that may range in size from sand grains to huge boulders tumbling down to accumulate as basal scree. The second water-driven grouping includes those processes that operate when heavy precipitation bombards a slope with large raindrops and ultimately generates a surface run-off capable of carrying fine loose debris further downslope. This twofold division of the activities responsible for particulate movement will be used as the basis for further discussion under the general heads of rockfall and surface wash.

Rockfall

Rockfalls are normally limited to free faces that slope at angles of over

Fig. 8.3. Schematic representation of the effect of rock structure on the angle of a cliff face. In the case of stratified rocks dipping into the cliff face a critical factor is the relative spacing of the joints (*c*) and the bedding planes (*d*). White lines indicate the angle of friction at which slip along joints and bedding planes is assumed to start.

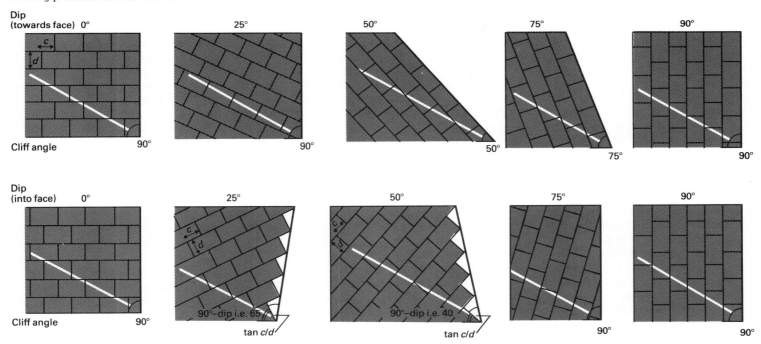

40°; this limiting gradient may be specified because granular debris can sustain maximum angles of repose of about 40°, and only above that steepness does dislodged material automatically tend to roll or slide further downhill. In nearly all cases the initial detachment takes place along a prominent joint system. From an analysis of the stresses in a vertical face Terzaghi early showed that moderately strong unjointed rock should be able to sustain vertical cliffs 1 500 m high. Because such cliffs do not occur in nature and many slopes of much lower height have collapsed, Terzaghi concluded that limiting heights must be determined by mechanical 'defects' such as joints and faults, rather than by the strength of the rock itself. Failure occurs when the spread of joints through the material so reduces its overall strength that it can no longer withstand the shearing stresses.

The pattern of joints may be as significant as their frequency. In massive igneous rocks sheeting structures commonly develop parallel to the rock face, with cross-joints dividing the sheets into the smaller blocks that constitute the typical rockfall debris. In stratified rocks, on the other hand, the tilt of the bedding and joint planes relative to the free face may be crucial (Fig. 8.3). If an absence of cohesion across these planes be assumed, the strata may be treated as little more than a body of dry masonry. Then, with an outward dip, the component blocks will remain stable so long as their bases incline less steeply than the relevant angle of friction, and the maximum slope angle will be vertical; where their bases incline more steeply than the angle of friction, the unstable blocks will tend to slip downslope so that the maximum slope angle will equate with the dip. With an inward dip the maximum possible slope angle will normally be vertical. However, if there are master joints traversing all the strata, slipping along these will reduce the gradient towards the angle of friction; if the joints are slightly offset in the various strata, a crenulate face may develop with the overall slope angle governed by the ratio of the offset distance to the distance between adjacent bedding planes. This theoretical analysis is obviously very idealized. It assumes zero cohesion across the bedding planes and an unnatural regularity of the joint systems. Moreover, in most stratified rocks there are significant lithological changes to be taken into account. Nevertheless, it does serve to emphasize the critical role played by jointing in the development of most free faces.

The frequency and magnitude of rockfalls may be assessed in a variety of ways. One of the most common is to place some form of trap at the foot of a free face and to measure the amount of debris caught in a particular time interval. In a celebrated study in 1960 of erosional processes at Karkevagge in northern Scandinavia, Rapp employed hessian carpets and wire netting of 10 mm mesh to catch the falling fragments. He emphasized, in addition, the value of simple observations on the volume of rock debris lying on snow banks during early summer; under conditions like those of northern Lapland a high proportion of the rockfall appears to be concentrated in late spring and early summer. An extra inventory may be made in late summer, the freshly fallen boulders then being identified by the crushing of vegetation. Rapp estimated that in an 8-year period of observation the mean annual volume of rockfall from faces with a vertical area of 900 000 m^2 amounted to some 50 m^3, equivalent to a retreat of 0.06 mm yr^{-1}. The techniques so far described are suitable for sites where numerous small-scale falls occur each year. More difficult to quantify are large but infrequent collapses. By collating observations from residents in the National Parks and Monuments of the south-western United States, Schumm and Chorley at least managed to demonstrate that large-scale falls in this arid region are currently a significant mechanism of rockwall retreat. In theory relatively recent falls can also be located through the contrasting appearance of the newly exposed rock faces; for example, in dry regions they may lack the desert varnish that coats adjacent older faces. However, the actual age of falls cannot be defined in this way and care needs to be exercised since, prior to collapse, there are often gaping joints down which weathering agents can penetrate and alter the rock surfaces.

At very active sites the incidence of rockfall often displays a striking seasonality. Many investigators have recorded maxima in either autumn or spring, or in some cases in both periods. One of the best-documented examples comes from Norway where almost 60 per cent of the rockfalls have been found to occur in the 4 months of April, May, October and November; in a study in Northern Ireland pronounced maxima for relatively large events, but not for small, were identified in November–December and again in February–March. At these high latitudes most workers have associated the peak frequencies with a large number of freeze–thaw cycles, but another potentially important factor is water occupying the joint system. This can generate what Terzaghi termed cleft-water pressure which, like the analogous porewater pressure, reduces the mechanical strength of the rock. Terzaghi argued that the Norwegian data are best explained as due to high cleft-water pressure; in October this might be ascribed to heavy autumnal rain, in April to snowmelt while the exits from the joint system are still plugged by ice. Another illustrative case, although not involving immediate rock-fall, is provided by measurements made during the 1930s on a sand-stone monolith known as Threatening Rock in Chaco Canyon, New Mexico. Resting on a bed of shale, this huge vertical slab, 50 m long, 30 m high and 10 m thick, was slowly moving away from the cliff of which it was once a part. A steel bar was cemented to the top of the slab, and another to the cliff face. For 5 years monthly measurements of the gap were made. When cumulative movement is plotted as in Fig. 8.4 five periods of rapid movement and five periods of relative stability are discernible. The intervals of rapid displacement all span the winter season. This is a relatively dry period in Chaco Canyon so that it is difficult to attribute the accelerated movement to wetting of the underlying shale by rain. On the other hand, much of the winter precipitation is in the form of snow and it is conceivable that this is a more effective wetting agent than the brief summer storm. The obvious alternative is that freeze–thaw action during the winter plays a significant role in weakening the structure of the shale. Eventually in January 1941, after the gap had widened by over 0.5 m the 30 000-tonne monolith toppled over and was destroyed.

Finally in this discussion of rockfall, reference must be made to the scree slopes that develop as a result of debris accumulation. Most scree profiles are either rectilinear or gently concave, with a maximum angle near the apex of between 30 and 40°. The largest fragments may be concentrated near the lower end of the scree

Fig. 8.4. The movement of Threatening Rock, New Mexico, USA. The steps in the curve represent the summer periods, while the risers represent the winter periods (i.e. maximum movement but minimum precipitation) (after Schumm and Chorley, 1964).

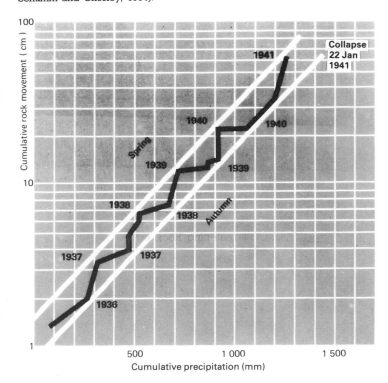

Fig. 8.5. A model of cliff development where scree accumulates at the foot of a vertical slope. The model assumes no loss of material and no change in bulk density.

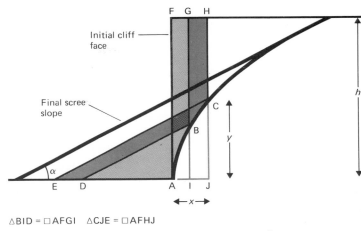

$\triangle BID = \square AFGI \quad \triangle CJE = \square AFHJ$

$\frac{1}{2}y^2 \cdot \cot \alpha = h \cdot x \quad y^2 = 2h \cdot x \cdot \tan \alpha \quad y = (2h \cdot x \cdot \tan \alpha)^{\frac{1}{2}}$

owing to their extra momentum as they fall, but the pattern is often made complex by some particles tending to roll at high velocity while others merely slide. Moreover, it must not be assumed that all scree material is added by simple rockfall. In high latitudes at least an equal volume may be contributed by snow and slush avalanches carrying with them considerable quantities of both coarse and fine debris. In 1969 Caine described how, in the South Island of New Zealand, the upper part of the scree is nourished primarily by rock-fall, while the lower part is supplied mainly by spring avalanching. The passage of an avalanche may also assist in redistributing material

that had earlier lodged near the top of the scree. Following a 7-year study in the Canadian Rockies Gardner emphasized the temporal and spatial variability of such diverse accretionary processes.

The form of the rock face concealed beneath a scree is generally unknown, although several theoretical models have been proposed that provide some guidance as to the likely shape. The simplest case was considered over a century ago by Fisher (Fig. 8.5). He assumed a vertical cliff retreating without loss of angle, together with a constant scree slope. He further simplified the model by assuming no loss of material and the same bulk density for *in situ* bedrock and accumulated scree. Granted these conditions it can be shown that the buried rock surface should be parabolic in form.

Later workers have progressively relaxed some of the constraints imposed by Fisher. For example, cases have been considered where the initial face is not vertical or where there are volumetric changes associated with scree accumulation. These changes may be expressed as the ratio of the scree volume to the volume of the corresponding rock, and are conventionally designated by the constant c. In Fisher's original model $c = 1$, but several factors can significantly alter this value. The most obvious is the need to take account of the volume of talus voids, thereby raising the value of c. On the other hand, some material may be lost by weathering and removal at the base of the scree, thereby reducing the value of c. The limiting case occurs where virtually all the debris is removed as rapidly as it falls, in which case c approaches zero. It is not too difficult to demonstrate that, so long as c has a positive value, the buried rock face will be convex upwards with its lowest point tangential to the ultimate scree surface. If $c = 0$, however, the rock face becomes rectilinear and slopes at an angle equal to that at which the scree would stand if it were present. This is true whether or not the cliff face maintains its steepness during retreat. The rather surprising prediction is that rockfall with rapid basal removal should tend to produce bedrock slopes inclined at angles corresponding to those of screes. There is observational evidence that this prediction may be fulfilled in suitable mountain environments, although local factors such as lithological variations often complicate the picture. It also needs to be remembered that rockfall rarely acts alone and there is usually a simultaneous functioning of many other processes.

Surface wash

Surface wash is a term denoting the downhill displacement of superficial slope materials by moving water. Two distinct processes may contribute to the displacement, namely raindrop impact and overland flow. Normal raindrops vary in diameter from 0.5 to 7 mm and attain terminal velocities ranging from under 1 m to over 9 m s^{-1}. From these figures it is clear that the kinetic energy of individual drops differs enormously, and that a brief thunderstorm can be a vastly more effective erosional agent than a prolonged period of drizzle. Rainsplash erosion has been the subject of much experimentation. The first requirement in the laboratory was a rainfall simulator capable of delivering raindrops of a controlled size and velocity. Once such equipment had been designed, most workers applied the artificial rain to a pan covered with fine but non-cohesive sediment. Particles splashed from the pan were generally caught in a surrounding tray, but in one of the more elaborate experiments tissue screens were used that could be reweighed in order to determine the amount of displaced sediment; by tilting the whole apparatus, the effect of slope angle could also be examined in the laboratory. Such investigations have shown that fine sand grains can occasionally be splashed as much as 1 m in a single leap, and that at moderate gradients the net downslope transport varies as a function of the sine of the slope angle. Some rainfall simulators were adaptable to outdoor operation and therefore permitted controlled field experiments, but many workers have relied on natural precipitation in their study of splash activity. In the English Midlands, for instance, Morgan installed twelve 'splash cups' on unvegetated slopes with gradients between 6 and 11°. Each cup consisted of a central circle of unprotected soil, 100 mm in diameter, surrounded by a 300 mm collecting tray that was subdivided in such a way that the splash fractions moving upslope and downslope were kept separate. After 900 days of observation, Morgan calculated that in a year the maximum net downslope transport recorded by any of the splash cups was just over 8 g for each centimetre of slope width.

It is true, of course, that the effectiveness of raindrop impact must depend very much on the nature of the local vegetation. Not only can a complete foliage cover absorb the entire energy of the raindrops, but the extra cohesion of organically rich soils also acts as a brake on movement. Splash erosion reaches its peak effectiveness on cultivated soils where no protective measures have been taken, and secondly in arid regions where there are considerable areas of bare ground between individual plants. Yet it may also be significant on patches of bare ground beneath a forest cover owing to water dripping from leaves and branches; large drops formed in this way can approach their maximum theoretical velocity after falling distances of some 7 or 8 m. In a beech forest in New Zealand specially treated filter papers were used to sample the number and size of drops in both the direct precipitation and in the throughfall. Calculations showed the kinetic energy of the throughfall was actually greater than that of the rainfall, and cups of sand placed beneath the forest canopy showed greater splash displacement than similar cups placed in the open.

Overland flow has been the object of study by workers in such diverse fields as soil conservation, hydrology, engineering and geomorphology, with the consequence that a large body of data has now been collected. An early analyst of the theoretical basis for surface runoff was the American engineer Horton. He contended that surface discharge should primarily be a function of rainfall intensity, infiltration capacity and slope position. So long as the soil can absorb all the precipitation no overland flow is to be expected; on the other hand, if the precipitation rate exceeds the maximum infiltration rate, the excess volume will be discharged across the surface. As seen in Chapter 6, methods for the field measurement of infiltration capacity have been devised and these show that values normally lie between zero and about 100 mm hr^{-1}. As heavy storms may produce precipitation intensities in the range 20–50 mm hr^{-1}, the implication seems to be that on many slopes direct surface runoff will be of rare occurrence. The third factor emphasized by Horton was position on the slope relative to the nearest divide. He argued that, given the same excess volume at all points on a slope, the actual discharge, involving both depth and velocity, will increase as a function of distance from the watershed. If surface particles are accorded a uniform resistance to displacement, it follows that near the crest they may remain undisturbed in what Horton terms a 'belt of no erosion', while lower down the slope they are picked up and transported. While this model of the generation of overland flow is logical and straightforward, later observations suggested that it did not encompass all conditions leading to surface runoff. In 1963 Hewlett and Hibbert suggested that throughflow passing down a hillside will tend to saturate the soil near the base of the slope and greatly reduce its capacity to absorb more rainwater. This saturated zone may build up to the point where there is seepage back to the surface and any water that is expelled joins with the direct precipitation to produce overland flow. The Hewlett and Hibbert model envisages surface runoff being concentrated at the foot of hillslopes close to a stream channel; the area affected will depend primarily upon the intensity and duration of the rainfall. It should be emphasized that the Hortonian and Hewlett and Hibbert models are not mutually exclusive and both may occur in suitable circumstances; however, current opinion seems to regard pure Hortonian flow as restricted to rains of exceptional intensity.

The term sheetwash refers to one particular form of overland flow in which the water moves as a thin and relatively uniform film. As discharge increases the flow concentrates into a series of slightly deeper pathways and is known as rillwash. This is turn grades into gully flow and ultimately into channelled stream flow. The nature of fluid movement in sheetwash has occasioned much discussion, some authors contending that it is primarily laminar, others that it is entirely turbulent (see pp. 162–4). In a series of elegantly designed experiments in 1970, Emmett examined the whole question using both laboratory flumes and specially prepared hillslope plots. In both cases artificial rain was applied at known rates and detailed measurements made of depth and velocity of flow. By varying the rate of water application changes in depth and velocity were induced, and from the data amassed in this way an attempt was made to decide whether flow was laminar, turbulent or a mixture of the two. Theoretically, as discharge rises, depth increases more slowly in laminar than in turbulent flow. The relevant relationships are $D \propto Q^{0.33}$ for laminar flow and $D \propto Q^{0.6}$ for turbulent flow, where D is depth and Q is discharge. In the laboratory flumes the transition from laminar to turbulent flow was clearly identified as the exponent in the above formulae jumped from the lower to the higher value. On the field plots the change was much more difficult to detect, probably owing to surface roughness, but Emmett concluded that Hortonian overland flow would be primarily laminar near the watershed and progressively more turbulent as discharge increases down the slope.

The erosional and transporting activity of overland flow has been studied in a variety of ways. Small experimental plots have been instrumented so that the input of rainfall, either natural or artificial, can be gauged and the runoff, together with its rock waste, can be measured by diversion into a small gutter at the foot of the plot. On natural hillslopes specially designed troughs have been sunk into the ground, the lip of each trough being so positioned that it catches surface water and debris, but excludes water moving through the soil; a fitted cover keeps out precipitation and any extraneous rainsplash sediment. Sometimes known as a Gerlach trough, such a device may be equipped with a collecting bottle into which the trapped water and debris are diverted. One weakness of an installation of this type is that it gives no indication of the slope area from which the trapped material comes, and for closer study it is often desirable to insert a whole series of troughs at intervals up a hillside. An alternative technique for assessing the effect of surface wash *sensu lato* is to measure directly any changes in slope elevation. This is most often done by driving long rods into the ground, but leaving protruding ends whose lengths can be monitored as an indication of the rate of surface change. In some studies a loose washer has been slipped over the top of each rod so as to provide a firm base on which to make the measurements and also allow ready identification of any site where deposition has occurred. Another possible procedure is to use a specially constructed frame supported at each corner by a deeply

Fig. 8.6 Under the influence of surface runoff in semi-arid regions, badland slopes such as those pictured opposite recede with exceptional rapidity. The contrast between the rounded crest and gullied margin of each individual ridge underlines the way in which the dominant process may change with passage down the slope.

buried leg; the changing elevation of the ground may then be periodically measured by means of light-weight rods slotted through the frame. This has the advantage of causing less disturbance at the actual point of measurement.

The incidence and destructiveness of Hortonian overland flow has been enhanced in many areas by Man's activities, but under natural conditions it almost certainly attains its maximum effectiveness in arid and semi-arid regions (Fig. 8.6). Contributory factors include the sparse vegetation cover, permitting a relatively uniform and unimpeded flow of water, and the very intense precipitation of local convectional storms. Several studies have shown the potency of runoff to depend as well on the surface being rendered suitably friable. By comparing the form of clay slopes in Colorado during different seasons Schumm identified an annual cycle. Winter frost weakens the surface structure so that individual particles are readily dislodged by raindrop impact in spring and early summer. At the same time the precipitation compacts the surface, thereby enhancing runoff which, during late summer, cuts shallow rills that persist until the return of winter frosts. Measurements with stakes have shown that certain clay and shale slopes in the arid south-western United States are retreating with remarkable rapidity under such an annual cycle. Surface lowering of over 2 mm yr^{-1}, and occasionally over 5 mm yr^{-1}, has been recorded in a number of separate investigations; these figures are also consistent with those from an unusually large-scale experiment with artificial precipitation on two small catchments in Colorado where an average lowering of about 0.15 mm followed the application of 40 mm of water in an hour.

The magnitude of the erosion measured on bare arid and semi-arid slopes dwarfs that recorded on vegetated temperate slopes. Stakes have invariably proved too crude a means of determining the minute surface lowering in the latter areas, and the only information comes from the use of sediment traps. Rarely have values as great as 0.1 mm yr^{-1} been recorded and a more representative figure would appear to be 0.01 mm yr^{-1}. It is evident that there must be huge contrasts in the effectiveness of sheetwash erosion. The dominant factor seems to be the protection afforded by the plant cover. This can act by intercepting the precipitation, shielding from raindrop impact, increasing the surface roughness and inducing an absorbent crumb structure in the soil. It might be anticipated that the importance of surface wash would also diminish under a tropical forest cover, but current evidence on this point seems rather equivocal. The major proponent of overland flow in tropical humid conditions has been Rougerie. By means of stakes he recorded rates of surface lowering in the Ivory Coast comparable to those in arid regions. Potentially favourable factors include high-intensity rainfall, thin cover of leaf litter and low permeability among many latosols. Yet it needs to be remembered that most of the work by Rougerie was concentrated in areas of semi-evergreen seasonal forest and his results may not be applicable to all tropical environments. At one time surface wash under equatorial evergreen forests was dismissed as unimportant, but during the last few decades there has been a gradual shift of opinion favouring it as a possibly significant process. What is indisputable is that the heavy precipitation of low latitude regions can be highly destructive if the vegetation cover is removed.

Several workers have formulated theoretical models of slope development through the action of surface wash. Emmett for example, contended that laminar flow will produce a downhill increase in gradient and turbulent flow a downhill decrease. If he is correct in supposing that, under Hortonian overland flow, there is a progressive downslope change from laminar to turbulent flow, the resultant profile should be concavo-convex in form. It is widely agreed that turbulent sheet-floods by themselves will produce a concave profile. Carson and Kirkby have argued that different circumstances will induce different degrees of concavity. They maintain, for instance, that slopes formed from coarse-grained bedrock will tend to be more strongly concave than those formed from fine-grained material. An effective vegetation cover will generally reduce the concavity so that grass-covered clay slopes, if moulded mainly by sheetwash, will be relatively rectilinear. If a factor for rainsplash is introduced, which it is widely agreed tends to generate convex hill crests, their model predicts that surface wash acting over a long period of time should ultimately produce concavo-convex profiles.

Mass movement

Mass movement, sometimes referred to as mass wasting, may be defined as the downhill transference of slope materials moving as a

Fig. 8.7. Schematic representation of the forces operating on any slope material, and of the three types of mass movement most commonly induced.

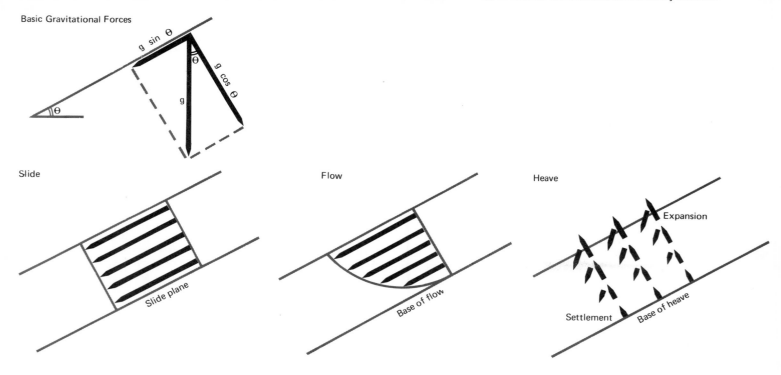

coherent body. The layer moving in this way may range in thickness from a few centimetres to many metres and in velocity from under 1 mm yr^{-1} to over 1 km hr^{-1}. Given these and other contrasts, it is small wonder that many different classificatory schemes have been suggested. Here the classification proposed by Carson and Kirkby will be followed. This distinguishes three main types of movement: slide, flow and heave (Fig. 8.7). In slide all differential movement is concentrated at the base of the mobile layer; a clearly defined shear plane separates the undeformed moving mass from the stable material beneath. In flow the velocity is greatest at the surface and diminishes with depth; this implies internal shear which normally reaches its maximum close to the base of the moving layer. In heave

the surface layers expand and contract in a direction normal to the slope; by itself this will not produce lateral movement, but repeated heaving followed by gravitational settling will generate downslope transport. Since most mass movements involve more than one of these three types acting concurrently, the classification is conveniently depicted in the form of a triangular diagram (Fig. 8.8).

Soil creep

Soil creep is the slow downhill movement of the regolith that results from constant minor rearrangements of the constituent particles. It is the product of at least three separate mechanisms, continuous creep, heaving and biological activity. Continuous creep refers to the

Fig. 8.8. Classification of mass movement processes (after Carson and Kirkby 'Hillslope Form and Process').

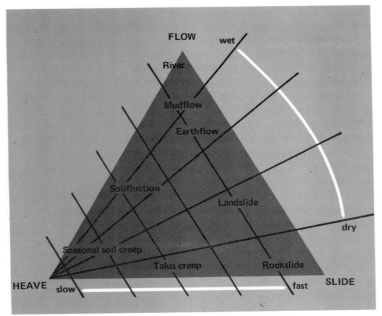

behaviour of fine moist materials under sustained gravitational shear stresses; although very small, these can cause progressive deformation and migration downslope without actual heaving. True heaving, on the other hand, is occasioned near the surface by cycles of wetting and drying or freezing and thawing. Both the frequency and depth of penetration of such cycles is primarily a function of climate. Since, under almost any climatic regime, the number of operative cycles will be greatest at the surface, and the effect of each will diminish with depth, the overall result should be to induce the fastest creep in the uppermost soil layers. The actual downhill velocity must be expected to vary with both slope and climate, but another factor is likely to be the nature of the materials involved. Here it is necessary to distinguish between moisture and temperature fluctuations. Wetting and drying will produce greater volumetric

changes where the regolith contains a high proportion of clay; moreover, as explained in Chapter 6, certain clay minerals are much more prone to swelling than others so that the mineralogical composition is a further important factor. Freezing and thawing is also affected by grain size, but in this case a relatively high proportion of silt promotes the greatest degree of expansion and contraction (see p. 280). The third mechanism mentioned above, biological activity, is one of the most difficult to evaluate, but current research suggests its importance may have been seriously underestimated. The growth and decay of plant roots, the burrowing activities of the soil fauna and the trampling of animals all contribute to disturbances leading to downhill creep. The number of roots in a small volume of soil can be quite startling; beneath a grass sward it has been shown that roots and root hairs in each cubic centimetre may have a total length of over 100 m. Equally the number and activity of worms is often not fully appreciated. In some soils a density of 2.5×10^6 per hectare has been found, and it has been estimated that they annually 'digest' as much as 30 tonnes of dry earth. Measurements beneath a forest cover in Luxembourg have shown that in a year the production of surface worm casts amounts to 15 tonnes per hectare; in addition to causing soil disturbance, this activity also renders the excreted material very liable to rainsplash erosion. When to the above figure is added a further 2.5 tonnes excavated in the same forest area by moles and voles, the scale of biological activity can begin to be appreciated. In tropical areas termites become extremely important, and even in desert areas the effects of porcupines and woodlice have been shown to be far from negligible. It must also be remembered that the above data relate to only a few species, not the total soil fauna of a region.

The field measurement of soil creep poses a difficult problem to which no entirely satisfactory solution has yet been devised. Methods tried can be divided into four groups. The first involves the use of rigid pegs. Sometimes these have simply been inserted to a shallow depth and the positions of the protruding ends resurveyed at intervals by reference to some fixed point. The fact that the pegs tend to tilt renders the measured displacements open to several different interpretations. Variants of the simple peg include the T-bar, in which a crosspiece fitted to the top of the main shaft is so designed that its angular tilt can be very accurately measured by spirit-level, and the rod inclinometer, in which a device that magnifies the necessary angular measurements, and can be read very precisely, is fitted over the top of each peg. Apart from the practical problem of detecting disturbance by animals and birds, a more fundamental objection to any rigid stake is the indeterminacy of the differential movement that must occur between it and the deforming soil.

The second group of methods, sometimes referred to as Rudberg pillars from their use by an early Scandinavian investigator, aims to surmount this basic problem by drilling a small hole and inserting an easily identifiable material that will deform during downhill creep. Rudberg used wooden dowelling sawn into short pieces, but other workers have employed lengths of plastic, specially dyed sands, glass beads and even radioactive tracers. Because of the need to re-excavate the hole to ascertain how much movement has occurred, continuous monitoring is virtually impossible.

This has encouraged exploration of a third group of methods in which the hole that has been drilled is lined with some form of tubing that, although very flexible, retains its circular cross-section. It may then be feasible to make repeated measurements and relate observed changes to meteorological events. Some workers have employed a specially designed inclinometer that records electrically the progressive deviation of the tube from the vertical. An alternative procedure is to fix several cross-wires in the tubing before it is placed in position and then to monitor their relative positions by means of a vertically mounted telescope. A less sophisticated but cheaper method may be to measure with callipers the distance of the tubing from a long datum stake permanently installed down the centre of the hole. In all this group of experiments the insert is assumed to be sufficiently flexible to move and deform with the soil and not be bypassed.

The final group of methods involves tracing various types of buried marker. Young initiated a technique that has been adopted in many subsequent studies in many different countries. A shallow pit is first dug and into a vertical face aligned downslope a grid of horizontal pins, a few centimetres long, is inserted. The positions of the ends of the pins are surveyed with respect to some fixed point and then the pit is carefully infilled. After an appropriate interval the hole is excavated again and the positions of the pins resurveyed. With care it may be possible to repeat the procedure several times using the same set of pins, but each time the pit is dug out there is an obvious risk of accidental disturbance. One consequent limitation of Young-pits, as

they have been termed, is that measurements can only be made at infrequent intervals, and it is correspondingly difficult to relate creep to such transient soil conditions as moisture content and temperature. Alternative forms of buried marker include small cones, sometimes connected by means of fine wires to a reference shaft above ground-level. Although the pull on the wires theoretically permits regular monitoring, individual markers behave so unpredictably that any movements are very difficult to interpret.

All the measuring methods so far devised suffer from the shortcoming that the means employed are liable to affect the very processes it is desired to observe. Insertion of stakes and augering of holes will disrupt the original soil structure; thermal relations in the soil will be affected by metal rods and hollow plastic tubes; most important of all, the passage of water through the regolith will be prone to alteration by almost any of the techniques described. Despite these limitations the results from many parts of the world are beginning to yield a tolerably consistent pattern. On moderately steep slopes the uppermost soil layers commonly creep downhill at a mean rate of $1-2$ mm yr^{-1} in humid temperate regions, $3-6$ mm yr^{-1} in humid tropical regions and $5-10$ mm yr^{-1} in semi-arid regions with cold winters. An alternative way of expressing creep measurements that takes into account velocity variations with depth is by the volume of material that passes downslope through a unit length of contour in unit time. Less consistency is apparent in these figures (Table 8.1), and in a few instances movements have been recorded which do not diminish with depth but exhibit two distinct peaks, one at the surface and the other normally just below the densest part of the root zone. The assumption in such cases is that the roots are a restraining influence, and there seems little doubt that further research will emphasize the close relationship between vegetation and creep movement.

Almost all investigators have concluded that physical creep acting alone would tend to produce convex slope forms. Various arguments have been adduced to substantiate this view, although it is often expressed in the simple statement that the downslope increase in the volume of regolith to be displaced requires a progressively steeper gradient to ensure its disposal. If it is postulated on theoretical grounds that the rate of transport by creep should be proportional to the sine of the slope angle, it can be shown by a simple mathematical model that a rectilinear profile will gradually be altered to a convex form. A reduced gradient will first appear at the slope crest and then extend

Table 8.1 Selected volumetric measurements of the rate of soil creep

Area		Slope angle (degrees)	Downhill creep (cm³ cm⁻¹ yr⁻¹)
British Isles	*S. England	0–25	4.0
	*N. England	25	0.6
	†Scotland	17	2.1
	*Wales	27	1.0–3.0
North America	‡Washington D.C.	18–28	0.2
	†Maryland	17	1.3
	†Ohio	20	6.0
	‡New Mexico	45	5.6
	*California	7–8	150.0
	*Puerto Rico	17–20	7.5–9.1
Other regions	*Volgaland (USSR)	5–28	5.0–15.0
	‡Tatar (USSR)	22	5.7–8.4
	‡New South Wales	15	1.9–3.2
	*New Zealand	7–33	25.0–50.0
	‡Malaysia	10	12.4

* Compilation by Saunders and Young (1983).
† Compilation by Carson and Kirby 'Hillslope form and process'
‡ Compilation by Young (1974).

towards the base. Subsequent evolution will include a diminishing maximum angle and slope curvature, but with no tendency for a concave element to develop. It is worth noting, however, that the primacy formerly accorded to creep in the fashioning of hillslopes has been questioned, and its true importance *vis-à-vis* other processes has yet to be fully established.

Solifluction

As a process characteristic of periglacial regions solifluction will be discussed in greater detail in Chapter 14. Here it will suffice to note that accelerated movement of the arctic regolith is ascribable both to increased freeze–thaw activity and to exceptionally high seasonal values for soil-moisture content. The frost action contributes to faster creep, the soil water to periods of flowage. In addition, water during winter may be preferentially concentrated into ground-ice lenses that substantially lift the overlying soil. In theory this elevation is likely to be normal to surface while settlement during melting will be vertical. The process may be regarded as a seasonal heave of exceptional magnitude, supplementary to any displacement induced by freeze–thaw cycles of much shorter duration. The faster movement means that solifluction is more easily measured than normal soil creep. Surface stones in transit on a mobile solifluction sheet have been marked with paint and their changing distances from a fixed point regularly surveyed. Short stakes have been driven into the ground and their downslope movements monitored. To determine the velocity distribution at depth, plastic tubing has been employed in the same fashion as for soil creep. These procedures have all served to show a highly seasonal movement, with mean annual surface rates between 10 and 50 mm and downslope volumetric transport as much as 50 cm³ cm⁻¹ yr⁻¹. One characteristic of solifluction emerging in many studies is the wide range of movement even on the same section of slope. This is probably associated with marked variations in the composition, and therefore in the water content of the mobile layer.

In theory solifluction might be expected to produce convex slopes similar to those predicted for creep. However, several studies have found little correlation between slope angles and rates of movement, so that there must be some hesitation before applying the same arguments as were earlier adduced for creep. It may well be that variations in lithology, particularly in terms of frost susceptibility, assume such importance in periglacial regions that simple models predicated on a homogeneous bedrock have little practical application.

Talus creep

The fragments comprising open-work scree are subject to small intermittent displacements which together constitute the movement known as talus creep. Dislodgement of individual particles may be initiated by ice wedging, temperature oscillations, wetting and drying, avalanching, the impact of additional falling debris, faunal activity and a variety of other factors. The relative importance of each differs from locality to locality. The rate of talus creep has been assessed by laying lines of painted stones across the surface of a scree and repeatedly surveying their positions with respect to a datum on nearby bedrock. In practice this procedure usually reveals wide variations in downslope movement. Some stones may remain completely stable for several years while adjacent ones travel 1 m or more. This erratic behaviour arises from constant minor readjustments, sometimes triggered by virtually random factors; however, in arctic and alpine environments the single most important trigger is almost certainly ice wedging.

Rockslide

The term rockslide denotes a large-scale slope failure in hard, jointed rock where there is some disintegration of the moving body as it speeds downslope. On the one hand the rockslide grades into rockfall, with some authors distinguishing an intermediate category of rock avalanche; on the other it grades into landslide where the whole mass moves as a slower, more coherent unit. Terzaghi made a theoretical analysis of conditions conducive to rocksliding. As with rockfall, he emphasized the critical importance of the joint system, contending that it is the spread of continuous joints through a massive rock that ultimately weakens the structure and leads to sliding. Contributory factors include variations in the cleft-water pressure and gradual chemical weathering of the sound rock between joint planes. Once the stresses exceed the shear strength along any plane in the rock, movement will ensue. The failure will often be progressive, since weakening in one part will throw additional stresses on others. If water in the joint system is a major factor, failure will frequently spread upwards from an initial fracture near the base since it is here that hydrostatic pressure reaches its maximum. Terzaghi estimated that, with a random joint pattern, hard massive rock should be able to sustain slope angles of up to about 70°. This figure, derived primarily from the measured shear strength of crushed rock aggregate, receives strong support from observations made during the design of cuttings for the Panama Canal. It was then reported that in Central America rocks such as massive granite and quartzite sustain high natural faces of at least 70° without loss of stability.

In practice many large rockslides have occurred in consequence of local structural peculiarities, especially master joint or bedding planes dipping steeply towards the axis of a mountain valley. One such example is illustrated in Fig. 8.9, and an even more disastrous instance is provided by the Vaiont rockslide of 1963 in the Italian Alps. This involved a body of rock 1.8 km long, 1.6 km wide and

Fig. 8.9. A rockslide in Montana triggered by the same earthquake as that responsible for the floodplain scarp illustrated in Fig. 4.8. Note how part of the 28 million m³ of debris, possibly supported on a cushion of air, rode up the opposite valley side to a height of 130 m. In the foreground the lake dammed by the slide reached a depth of over 55 m. (Photo J. R. Stacey, USGS.)

with an estimated volume of 250×10^6 m³ plunging into an artificial reservoir and creating a wave of water below the dam that killed over 2 000 people in the town of Longarone. Once set in motion these large rock masses may attain remarkably high velocities. Speeds in excess of 200 km hr^{-1} have been estimated in a number of cases, and a noteworthy feature of many rockslides in the great elevation to which the toe rises on the opposite side of the valley. Such velocities and mobility are inconsistent with simple frictional sliding, and it has long been held that a cushion of compressed air trapped beneath the moving mass may help explain its observed properties. This view has itself been challenged, partly because analogous features have been found on the moon where there is no air, and other more complex forms of fluidization have been invoked. The trigger of most large rockslides is either an earthquake or prolonged heavy rain leading to high cleft-water pressures; the initial sliding action is rapidly converted into other forms of movement that enormously increase the mobility of the descending mass.

Landslide

Landslides assume many different forms depending on the nature of the bedrock and regolith involved. One possible classificatory scheme suggested by Skempton in 1953 is based upon the shape of the sliding mass. Using the ratio of maximum thickness to hillslope length, he distinguished two major classes: surface slides with a ratio of between 2 and 5 per cent, and deep slides with a ratio of between 10 and 30 per cent. In the former case the failure plane is so shallow that most of the slide material normally consists of regolith, whereas in the latter it plunges so deeply that a high proportion of the moving body is usually composed of bedrock.

Analysis of the stresses associated with shallow slides is reasonably simple (Fig. 8.10). It is rendered easier if the mobile layer is assumed to be uniform in thickness, and to rest on a slope of constant angle and infinite extent. This dispenses with the need to consider side- and end-effects, and may be justified on the grounds that a

Fig. 8.10 Diagrammatic representation of the stresses that operate in a shallow slide when side- and end-effects are ignored.

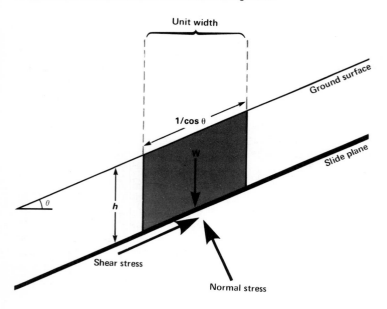

If w is unit weight of soil, total weight (W) of column = $h.w.$ Acting over an area $1/\cos\theta$ the vertical stress = $h.w.\cos\theta$. This may be resolved into a shear stress equal to $h.w.\cos\theta.\sin\theta$ and a normal stress equal to $h.w.\cos^2\theta$.

shallow slide has very great length in comparison with its depth. The computed shear stress acting along the plane of the slide can be compared with the shear strength of the material as measured in a triaxial test. At the time of failure the two values should be approximately equal. It will be remembered that the shear strength is reduced to a minimum when the regolith is saturated and porewater pressure attains its greatest value (see p. 110); this is one reason why most shallow sliding occurs after periods of heavy rain. The actual porewater pressure in the field may be measured by installation of piezometers. These vary in complexity but preferably consist of a small porous pot buried at the depth of the failure plane and connected by tubing to some form of gauge at the surface that allows monitoring of the local water pressure.

In practice shallow sliding often occurs on slopes of lower angle than the foregoing analysis would suggest. Two reasons for this seeming anomaly may be adduced, both related to the special properties of materials composed of clay minerals. The first arises from the fact that many investigations are concerned with over-consolidated clays. These are deposits which, during their history, have been subject to much higher overburden pressures than currently exist. This is most often due to erosion having stripped later geological strata from above the clay, although some tills are over-consolidated owing to their original accumulation beneath a thick ice-sheet. The physical properties of the material do not return to those that existed prior to the consolidation since the particles remain more tightly packed and the water content lower. Removal of the super-incumbent weight leads to fissuring and jointing, and as with hard rock penetrated by joints the overall strength of the clay will be less than that of the average small laboratory sample. The second reason arises from an important distinction between the properties known as peak and residual strength. During laboratory testing, the original strength of a sample is often high but once shearing has started the strength diminishes before finally settling to a more constant residual value. The change is associated with a reorientation of the flaky clay particles parallel to the direction of shear and a simultaneous fall in cohesion. This means that a clay, once sheared past its peak strength, will thereafter be a much weaker material. It follows that if one section of a clay slope is forced past its peak, additional stress is liable to be thrown on some other section, possibly causing the peak to be surpassed there also. Such progressive failure may reduce the shear strength to its residual value along the entire slip surface, after which the more significant property for long-term stability will be the residual rather than the peak strength.

Illustrative of the importance of residual strength are studies of slopes on the London Clay in south-eastern England. It has been found that these slopes are only stable at angles of less than about 10°. This figure is the same as that predicted from measurements of the residual strength of the clay and leads to the conclusion that it is the limiting value for long-term stability. Of course, this does not mean that all slopes will lie at 10° since other processes may operate to lower them; however, if stream undercutting steepens a slope beyond its critical gradient, sliding will eventually function to reduce its angle to the stable value. A further noteworthy aspect of these studies on the London Clay, and equally on Jurassic clays in the English Midlands, is the widespread occurrence of landsliding. This may not always be obvious from surface morphology, but in many excavations is clearly attested by shallow slip planes that in some cases bear prominent slickensides.

Another factor that can influence shallow sliding, but in the reverse direction to those just quoted, is the additional strength given to the regolith by plant roots. This can operate to support much steeper slopes than laboratory testing would indicate as having long-term stability. In a painstaking investigation in Ohio it was estimated from the residual strength of the clay regolith that sliding might be expected to occur at slope angles of no more than 14°, whereas buttressing by plant roots enables hillsides as steep as 35° to resist sliding; of the total shear strength of the hillslope materials, almost 80 per cent was contributed by the roots.

Analysis of the stresses in deep slides is more complex, and

Fig. 8.11. Basic principles of analysing stresses on an arcuate shear plane by dividing the overlying mass into vertical slices.

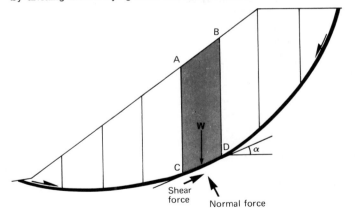

For column ABCD   shear force = $W \sin \alpha$
normal force = $W \cos \alpha$
Total shear force along arc = $\Sigma W \sin \alpha$
(but note the effect of toe where shear force operates in reverse direction)

consideration here will be limited to the basic principles involved. Observation shows that the slip planes beneath deep slides tend to approximate an arcuate form. The moving masses in consequence often exhibit a degree of rotation, and the term rotational slip has appropriately been applied to such cases. Deep slides along arcuate failure planes are especially characteristic of thick clay strata, and one of the earliest scientific investigations was undertaken by the Royal Swedish Geotechnical Commission in 1922 following a number of slides that disrupted their national railway system. The basic concepts then formulated are still employed in most methods of stability analysis. The initial premise is that the slide surface conforms to a circular arc, with the centre of the imaginary circle defining the axis of rotation (Fig. 8.11). If the axis lies outside the intersections of the slip plane with the ground surface, all the relevant weight is contributing to the potential failure; otherwise the rotational tendency is partially resisted by the weight of the toe. For purposes of computation the whole mass is divided into vertical slices and the stress contribution of each is separately assessed. The aggregate stress may then be compared with the shearing resistance along the arc due to the strength of the materials involved. Allowance must again be made for porewater pressure and the particular properties of over-consolidated clays. The greatest practical difficulty used to lie in determining along which of the many possible slip planes movement was most likely, but these calculations are now normally performed by computer.

Rotational slips are commonly associated with permeable caprocks overlying thick clay or shale strata. They are therefore especially characteristic of scarp faces where undercutting at the foot of the slope steepens the gradient until the cuesta summit becomes unstable. The precise form of the slip plane is much influenced by structural details, and is by no means always a simple circular arc. Nevertheless, backward tilting of the resistant caprock is commonplace, sometimes resulting in an ill-drained hollow in which organic matter begins to accumulate. Radio-carbon dating may then be employed to give an estimate of the minimum age of the slide. Detailed mapping in the British Isles has shown that many of the scarps of central and southern England have been severely affected by landsliding, much of it probably induced by late Devensian periglacial conditions when the strength of materials was significantly reduced by their high water content.

Earthflow

Although many landslides, particularly along the saturated toe, involve a minor flow component, other forms of displacement occur predominantly by flowage and it is to these that the term earthflow is normally applied. They vary greatly in nature, from minor disturbances in which an original turf cover can remain almost intact apart from small marginal ruptures, to spectacular events in which massive bodies of water-saturated material progress rapidly downslope. Rates of movement range from under a metre per month to several kilometres per hour, with the primary controls being the nature of the materials involved and their water content.

An example of relatively slow flowage is provided by long clay tongues descending the edge of the Antrim plateau in north-eastern Ireland. The material comes from weathering of Liassic shales and includes varying proportions of montmorillonite, illite and kaolinite. The rate of movement has been assessed by surveying the positions of specially installed pegs at monthly intervals. To provide more complete records an instrument was devised to monitor continuously the displacement of an individual peg. By these techniques a close association between rainfall and velocity was demonstrated (Fig. 8.12); a second association was found between patterns of instability within the tongues and local concentrations of montmorillonite. In a 14-year period of monitoring in New Zealand a comparable lagged response to both short-term and long-term precipitation cycles was demonstrated.

The most startling illustrations of rapid earthflow occur in certain post-glacial marine clays found in such areas as eastern Canada and Scandinavia. Sometimes known as quickclays, these sediments have the remarkable property known as high sensitivity. Sensitivity refers to the shear strength of a remoulded sample compared with its strength in the undisturbed state. The ratio of these two values may exceed 15 in a highly sensitive clay, and in extreme cases surpasses 100. In effect, the material can be transformed from a weak solid into a viscous fluid without any change in water content. Once this alteration has taken place, rapid flow can occur down very gentle gradients. Disastrous earthflows have affected a number of settlements along the valley of the St Lawrence and its major tributaries. In 1971, for instance, the village of St Jean Vianney suffered over thirty

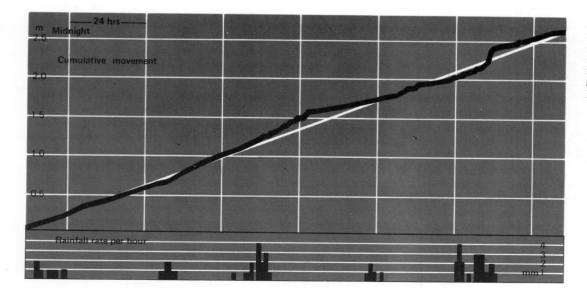

Fig. 8.12 The relationship between rainfall and short-term movements on a continuously monitored earthflow in Northern Ireland. Note how movement regularly accelerated a few hours after significant precipitation had fallen (after Prior and Stephens, 1972).

deaths and bad structural damage when some 8×10^6 m^3 of clay, underlying an area of about 30 hectares, flowed into the local valley at speeds of up to 25 km hr^{-1}. The cause of the rapid liquefaction has occasioned much dispute but appears to relate to a combination of factors. The sediments were laid down as flocculated marine silty clays in which much of the material was a finely comminuted rock flour rather than true clay minerals. Connate water trapped in the pore spaces was highly saline, but leaching has gradually reduced this salinity and it is in areas where leaching has progressed furthest that most quickclay activity is encountered. Localized internal shearing or basal sliding probably triggers a collapse of the flocculent honeycomb structure, with speedy liquefaction uninhibited by the low porewater salinity.

Mudflow

Loose debris mantling a hillslope or mountain gully can be mobilized by either heavy rain or melting snow to produce a waterlogged slurry capable of moving at high speed down quite gentle gradients. There is a continuum between this form of mass movement and a very heavily debris-laden stream, with any distinction depending on an arbitrary ratio of sediment to water. The term mudflow may seem to imply a restriction to fine-grained material, but one of the features of many mudflows is the diversity of their load, with cobbles and even boulders being carried along by the high-density fluid. Where much of the transported material is of relatively coarse grade the term debris flow is sometimes employed. In many instances mudflows lose significant quantities of water by percolation into the surface across which they pass; this can lead to a downstream increase in both density and viscosity until finally the flow is brought to a halt.

Some of the mountain regions of southern California, with their semi-arid climate and winter snowfall, provide favourable conditions for intermittent mudflow activity. One damaging example engulfed part of the settlement of Wrightwood on the slopes of the San Gabriel Mountains. Here a mudflow with an estimated density of just over 2 000 kg m^{-3} ran for some 24 km along a course that, at its lower end, had a gradient of no more than 1°. Velocities were estimated to have achieved at least 18 km hr^{-1}, a figure comparable to that obtained from the use of specially installed cameras above the routes of regular debris flows in Japan.

Movement of material in solution

As will be indicated in Chapter 18, many studies of the solution load carried by streams have indicated the important role played by chemical weathering in the denudation of individual catchments. Yet for many years geomorphologists neglected the potential significance of solutional activity in the development of hillslope forms. One reason for this was an inadequate appreciation of the diverse pathways followed by precipitation as it journeyed to the stream channels. However, as models of hillslope hydrology have improved, so there has been more incentive to measure the chemical characteristics of water following those different pathways since this should ultimately bring a better understanding of the way the whole denudation system works.

Four components of stream runoff are worthy of investigation. As might be expected, samples of overland flow are normally low in their concentration of solutes since the water has had only brief contact with the ground; one exception may occur when throughflow is forced to the surface to become part of the overland discharge. The other main source of rapid runoff, pipeflow, is generally found to be much more heavily charged with dissolved material, possibly because of the shunting effect in which old soil water is forced out into the pipes by the advent of the new precipitation. Throughflow has been monitored in a number of studies by installing troughs at different elevations in the upslope face of a pit and diverting the water into devices for measuring its volume and solute concentration. By this technique it has been shown that throughflow after a storm is often heavily charged with solutes, and in some cases there may even be a flushing effect with displaced soil moisture causing the concentration to rise at the very time the flow is increasing. Such a relationship has not been found universally, however, and more studies are needed to establish the spatial and temporal variability that may exist. Groundwater, owing to its relatively long residence period, is usually much enriched in dissolved mineral matter. The chemical composition tends to an equilibrium with the host rocks, so that the groundwater contribution to stream discharge normally exhibits a greater constancy in solute concentration than most other inputs.

As the previous paragraph indicates, significant progress has been made in understanding the chemical role of precipitation falling on hillslopes. However, this should not be allowed to obscure the need for much more information if a proper comprehension is to be achieved. Such questions as the source of the dissolved load being transported by pipeflow and throughflow, or the extent of reprecipitation of some of the soluble salts, have yet to be fully investigated. Of critical importance is the rapidity with which percolating water equilibrates with the local soil environment. Experimental evidence suggests that approximate equilibrium may be established in a matter of a few hours or at most a few days. In certain limestone areas, for instance, virtual saturation with $CaCO_3$ has been demonstrated within 1 m of the surface and Williams has estimated that as much as 25 per cent of the total solution may take place actually at the surface. Elsewhere the concentration of silica has been found to be remarkably constant, even during flood discharge, with the implication that this solute, too, is picked up very rapidly.

To investigate local spatial variation in the rate of chemical activity, Crabtree and Burt buried specially prepared rock tablets, each about 3 cm in diameter and 7 mm thick, at twenty points on a hillside in south-western England. After 15 months the tablets were recovered and, following suitable cleansing, their precise weight losses were determined. In conformity with the previous hypotheses of a number of workers, the tablets higher on the slope were found to have suffered the greatest weight loss; it is presumably in this sort of topographic location that solutional denudation reaches its maximum. At the same site specially prepared spheres, about 20 mm in diameter and made from plaster of Paris, were also buried. Solutional loss from these spheres exhibited a very different pattern, with maximum values near a seepage zone at the base of the slope. Since the dissolution of gypsum is largely independent of soil pH, on many bedrock types it will not mimic chemical denudation and it was argued that the primary control of weight loss from the spheres was the quantity of local throughflow. From this experimental work it can be seen that the use of specially prepared inserts into the soil may greatly assist our comprehension of chemical processes in hillslope evolution.

References

Anderson, E. W. and **Cox, J.** (1978) 'A comparison of different instruments for measuring soil creep', *Catena* **5**, 81–93.

Caine, N. (1969) 'A model for alpine talus slope development by slush avalanching', *J. Geol.* **77**, 92–100.

Campbell, I. A. (1974) 'Measurement of erosion on badland surfaces', *Z. Geomorph. Supplementband* **21**, 122–37.

Carson, M. A. (1981) 'Influence of porefluid salinity on instability of sensitive marine clays: a new approach to an old problem', *Earth Surf. Processes Landf.* **6**, 499–515.

Carson, M. A. and **Kirby, M. J.** (1972) *Hillslope Form and Process* CUP.

Cleaves, E. T. et al. (1970) 'Geochemical balance of a small watershed and its geomorphic implications', *Geol. Soc. Am. Bull.* **81**, 3015–3032.

Cox, N. J. (1978) 'Hillslope profile analysis', *Area* **10**, 131–3.

Crabtree, R. W. and **Burt, T. P.** (1983) 'Spatial variation in solutional denudation and soil moisture over a hillslope hollow', *Earth Surf. Processes Landf.* **8**, 151–60.

Crabtree, R. W and **Trudgill, S. T.** (1984) 'The use of gypsum spheres for identifying water flow routes in soils', *Earth Surf. Processes Landf.* **9**, 25–34.

Dedkov, A. P., Moszherin, V. I. and **Tchasovnikova, E. A.** (1978) 'Field station study of soil creep in the central Volgaland', *Z. Geomorph. Supplementband* **29**, 111–16.

De Ploey J. (ed.) (1983) 'Rainfall simulation, runoff and soil erosion', *Catena* Supplement **4**.

Dunne, T. and **Dietrich, W. E.** (1980) 'Experimental investigation of Horton overland flow on tropical hillslopes', *Z. Geomorph. Supplementband* **35**, 60–80.

Emmett W. W. (1970) 'The hydraulics of overland flow on hillslopes', *U.S. Geol. Surv. Prof. Pap.* 622–A.

Finlayson, B. L. (1981) 'Field measurements of soil creep', *Earth Surf. Processes Landf.* **6**, 35–48.

Fullen, M. A. (1982) 'Laboratory and field studies in the use of the isotope ^{59}Fe for tracing soil particle movement', *Earth Surf. Processes Landf.* **7**, 285–93.

Gardner, J. S. (1983) 'Accretion rates on some debris slopes in the Mt. Rae area, Canadian Rocky Mountains', *Earth Surf. Processes Landf.* **8**, 347–55.

Hazelhoff, L. et al. (1981) 'The exposure of forest soil to erosion by earthworms', *Earth Surf. Processes Landf.* **6**, 235–50.

Hewlett, J. D. and **Hibbert, A. R.** (1963) 'Moisture and energy conditions within a sloping soil mass during drainage', *J. Geophys. Res.* **68**, 1081–7.

Horton, R. E. (1945) 'Erosional development of streams and their drainage basins: hydrophysical approach to quantitative morphology', *Geol. Soc. Am. Bull.* **56**, 275–370.

Imeson, A. C. (1976) 'Some effects of burrowing animals in slope processes in the Luxembourg Ardennes', *Geogr. Annlr.* **58A**, 115–25.

Kiersch, G. A. (1965) 'Vaiont reservoir disaster', *Geotimes* **9**, 9–12.

Lewis, L. A. (1974) 'Slow movement of earth under tropical rainforest conditions', *Geology* **2**, 9–10.

Lewis, L. A. (1975) 'Slow slope movement in the dry tropics: La Paguera, Puerto Rico', *Z. Geomorph.* **19**, 334–9.

Lusby, G. C. (1977) 'Determination of runoff and sediment yield by rainfall simulation' in *Erosion – research techniques, erodibility and sediment delivery* (ed. T. J. Toy), Geobooks.

McGregor, D. F. M. (1980) 'An investigation of soil erosion in the Colombian rainforest zone', *Catena* **7**, 265–73.

Morgan, R. P. C. (1978) 'Field study of rainsplash erosion', *Earth Surf. Processes* **3**, 295–9.

Morton, D. H. and **Campbell, R. H.** (1974) 'Spring mudflows at Wrightwood', *Q. J. Eng. Geol.* **7**, 377–84.

Mosley, M. P. (1973) 'Rainsplash and convexity of badland divides', *Z. Geomorph. Supplementband* **18**, 10–25.

Mosley, M. P. (1982) 'The effect of a New Zealand beech forest canopy on the kinetic energy of water drops and on surface erosion', *Earth Surf. Processes Landf.* **7**, 103–7.

Okuda S. et al. (1980) 'Observations on the motion of a debris flow and its geomorphological effects', *Z. Geomorph. Supplementband* **35**, 142–63.

Prior, D. B. and **Stephens, N.** (1972) 'Some movement patterns of temperate mudflows: examples from north-eastern Ireland', *Geol. Soc. Am. Bull.* **83**, 2533–44.

Rapp, A. (1960) 'Recent development of mountain slopes in Karkevagge and surroundings, northern Scandinavia', *Geogr. Annlr.* **42**, 71–200.

Riestenberg, M. M. and **Sovonick-Dunford, S.** (1983) 'The role of woody vegetation in stabilizing slopes in the Cincinnati area, Ohio', *Geol. Soc. Am. Bull.* **94**, 506–18.

Rougerie, G. (1956) 'Etudes des modes d'erosion et du faconnement des versants en Cote d'Ivoire equatoriale', *Slopes Comm. Rep.* **1**, 136–41.

Saunders, I. and **Young, A.** (1983) 'Rates of surface processes on slopes, slope retreat and denudation', *Earth Surf. Processes Landf.* **8**, 473–501.

Savat, A. (1981) 'Work done by splash', *Earth Surf. Processes Landf.* **6**, 275–83.

Savigear, R. A. G. (1965) 'A technique of morphological mapping', *Ann. Ass. Am. Geogr.* **55**, 514–38.

Schumm, S. A. (1964) 'Seasonal variations of erosion rates and processes on hillslopes in western Colorado', *Z. Geomorph. Supplementband* **5**, 215–38.

Schumm, S. A. and **Chorley, R. J.** (1964) 'The fall of Threatening Rock', *Am. J. Sci.* **262**, 1041–54.

Schumm, S. A. and **Chorley, R. J.** (1966) 'Talus weathering and scarp recession in the Colorado plateaus', *Z. Geomorph.* **10**, 11–36.

Skempton, A. W. (1953) 'Soil mechanics in relation to geology', *Proc. Yorks. Geol. Soc.* **29**, 33–62.

Skempton, A. W. (1964) 'Long-term stability of clay slopes', *Geotechnique* **14**, 77–102.

Strahler, A. N. (1950) 'Equilibrium theory of slopes approached by frequency distribution analysis', *Am. J. Sci.* **248**, 800–14.

Terzaghi, K. (1962) 'Stability of steep slopes on hard unweathered rock', *Geotechnique* **12**, 251–71.

Wasson, R. J. and **Hall, G.** (1982) 'A long record of mudslide movement at Waerenga-O-Kuri, New Zealand', *Z. Geomorph.* **26**, 73–85.

References

Williams, P. W. (1968) 'An evaluation of the rate and distribution of limestone solution and deposition in the River Fergus basin, western Ireland', in *Contributions to the Study of Karst* (ed. P. W. Williams and J. N. Jennings), Aust. Nat. Univ. Canberra.

Yair, A. and **Rubin, J.** (1981) 'Some aspects of the regional variation in the amount of available sediment produced by isopods and porcupines, Northern Negev, Israel', *Earth Surf. Processes Landf.* **6**, 221–34.

Young, A. (1972) *Slopes*, Longman.

Young, A. (1974) 'The rate of slope retreat', *Inst. Brit. Geogr. Spec. Pub.* **7**, 65–78.

Young, A. (1978) 'A twelve-year record of soil movement on a slope', *Z. Geomorph. Supplementband* **29**, 104–10.

Selected bibliography

The modern growth of interest in hillslope studies has been reflected in the publication of a wide range of valuable texts. Of particular note are: A. Young, *Slopes*, Longman, 1972; M. A. Carson and M. J. Kirkby, *Hillslope Form and Process*, CUP, 1972; M. J. Kirkby (ed.) *Hillslope Hydrology*, Wiley, 1978; and M. J. Selby, *Hillslope Materials and Processes*, OUP, 1982.

Chapter 9
Streams

Whether as surface runoff or after temporary residence underground, the discharge from hillslopes is normally gathered into stream channels whose forms are shaped by the downstream passage of the water. This fashioning may be achieved either by erosion of bedrock or more commonly by picking up and redepositing the fragmental materials that form most channel floors and walls. All these processes require energy which the streams possess by virtue of their elevation above sea-level. This potential energy is converted into kinetic energy as the water flows down the channel, the amount of kinetic energy available for modifying the stream bed varying greatly at different river stages. As the water-level rises during a flood, increasing stresses are placed upon the channel material which, if loose, may be set in motion when a critical stress is reached. The detritus entrained in this way is transported downstream, being either removed from the river basin entirely or redeposited when the stream lacks the power to carry it further.

In the following pages attention will be directed in turn to the erosional and transporting activities of streams, to the forms and patterns assumed by stream channels and finally to the landforms that result from fluvial deposition. However, as an essential preliminary, the basic principles of stream flow will first be reviewed.

Stream hydraulics

Basic concepts

The flow of fluids presents a subject of immense range and

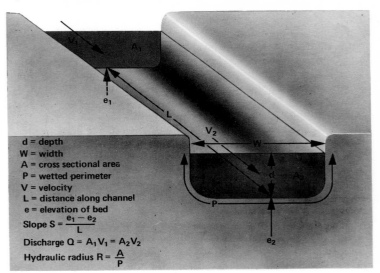

Fig. 9.1. Conventional terminology in describing stream channels.

d = depth
W = width
A = cross sectional area
P = wetted perimeter
V = velocity
L = distance along channel
e = elevation of bed
Slope $S = \dfrac{e_1 - e_2}{L}$
Discharge $Q = A_1 V_1 = A_2 V_2$
Hydraulic radius $R = \dfrac{A}{P}$

complexity. Much of it lies beyond the scope of this book but is adequately covered in numerous texts on fluid mechanics. For present purposes only those basic concepts required for a proper understanding of the geomorphological activity of streams need be outlined. Figure 9.1 illustrates the conventional terminology employed in the description of streams and the channels they occupy.

As a Newtonian fluid, water is unable to resist shearing stresses when in static equilibrium, and any deformation it suffers is directly proportional to the applied stress (see Fig. 6.7). The rate of deformation or flow is a function of its internal friction or viscosity, which may therefore be defined as the ratio of shear stress to shear strain. The viscosity of water at 0 °C is 0.018, at 10 °C 0.013 and at 20 °C 0.010 P. Strictly speaking the value thus defined should be designated the dynamic viscosity to distinguish it from another frequently used measure known as the kinematic viscosity. The latter is the dynamic viscosity divided by the density, and for stream water normally has a value of about 0.000 0015 m^2 s^{-1}. Both the dynamic and kinematic viscosity are calculated on the assumption that flow is laminar. Where flow is turbulent a third and rather different form of viscosity becomes important. Known as the eddy viscosity, this is usually much greater than the dynamic viscosity and is significant in sharply increasing the shear stress that a flowing stream can exert upon the materials composing its channel. It should be noted, however, that it is not an intrinsic property of the fluid but varies with the turbulent flow of the water.

Fluids may exert both static and dynamic pressure. Static pressure, as experienced when diving, is proportional to the height of the overlying water column. Dynamic pressure, which results from movement, is due to the kinetic energy possessed by flowing water and equal to $\frac{1}{2}mV^2$, where m is mass and V is velocity. The total energy of a small body of moving water may be computed by adding the energy equivalents of the static and dynamic pressures to the potential energy. Expressed in symbolic terms for a single point this may be written

$$p + \frac{V^2}{2g} + h = \text{total energy} = \text{constant}$$

where p is pressure, V is velocity, g is gravity and h is height above an arbitrary datum. Known as the Bernoulli equation, this is strictly applicable only to an ideal fluid with zero viscosity; in a stream potential energy is converted into kinetic energy which in turn is dissipated in overcoming the frictional resistance represented by the viscosity of the water. The conversion of the energy to heat is exactly analogous to the more familiar case of frictional heat generated between two solids. In this way the available energy diminishes downstream. Figure 9.2 illustrates the application of the Bernoulli equation to stream flow. The line depicting total energy less the frictional loss is called the energy grade line, and the distance between it and the channel bed indicates the quantity known as specific energy.

Nature of stream flow

Almost all stream flow is turbulent, that is, characterized by irregular eddying superimposed on the overall downstream movement. Only at exceedingly low velocities in a shallow straight channel could the flow be laminar. The best-known expression for defining the transitional velocity from laminar to turbulent flow is the Reynolds number

(Fig. 9.3). This dimensionless figure is computed by dividing the product of the velocity and depth by the kinematic viscosity. With values for the latter of around 0.000 0015 m² s⁻¹, insertion of typical values for velocity and depth (or hydraulic radius) yields Reynolds

Fig. 9.2. Application of the Bernoulli equation to stream flow and derivation of the subcritical and supercritical flow states.

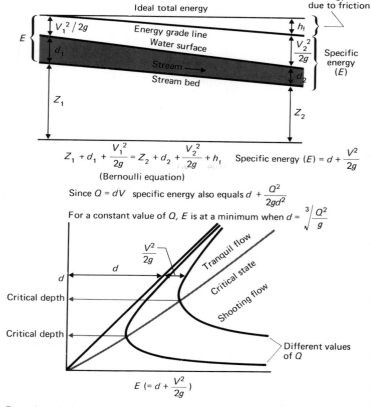

$$Z_1 + d_1 + \frac{V_1^2}{2g} = Z_2 + d_2 + \frac{V_2^2}{2g} + h_1 \quad \text{Specific energy } (E) = d + \frac{V^2}{2g}$$

(Bernoulli equation)

Since $Q = dV$ specific energy also equals $d + \frac{Q^2}{2gd^2}$

For a constant value of Q, E is at a minimum when $d = \sqrt[3]{\frac{Q^2}{g}}$

For a given discharge, there are both supercritical and subcritical depths and velocities

At the critical depth $d^3 = \frac{Q^2}{g}$. Since $Q = dV$, $d^3 = \frac{d^2 V^2}{g}$, $d = \frac{V^2}{g}$, $V = \sqrt{gd}$

i.e. Froude number 1.

Fig. 9.3. Criteria for the classification of different types of stream flow.

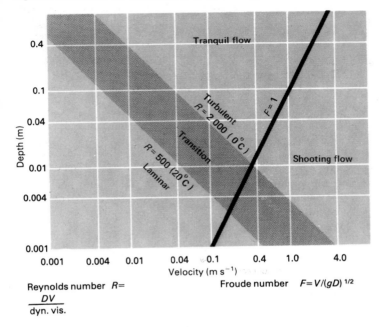

Reynolds number $R=$
$$\frac{DV}{\text{dyn. vis.}}$$

Froude number $F = V/(gD)^{1/2}$

by a second dimensionless value known as the Froude number. This is the quotient of the velocity divided by $\sqrt{gD}$, where g is gravity and D is depth. When the Froude number is below unity the flow is tranquil, but when above it is shooting. Since the value $\sqrt{gD}$ also defines the speed at which surface ripples traverse shallow water, this explains why disturbances are not propagated upstream when the Froude number exceeds unity. The existence of different types of turbulent flow results from the fact that, for given values of specific energy and discharge, there are two alternative regimes, either a shallow high-velocity form or a deeper low-velocity form (Fig. 9.2). The regime may change quite rapidly from one form to the other. The transition from tranquil to shooting flow is accompanied by a sharp drop in water-level, whereas the opposite change generates the feature known as a hydraulic jump. This is the prominent reversed gradient, often foaming at the surface, where the water reverts to slow, deep flow.

In both types of turbulent flow random eddies cause velocity at a point to fluctuate. Velocity as measured with a normal current meter is time-averaged, and instantaneous maximum and minimum values can differ from the mean by 50 per cent or more. The degree of turbulence at any point may be assessed with sensitive instruments that record the changes in velocity along three separate coordinates, one parallel to the general flow direction and the other two at right angles. The variability, commonly expressed in the form of the standard deviation, is then employed as a measure of the intensity of turbulence (Fig. 9.4). In a straight segment of channel, the intensity is least near the surface and increases to a maximum near the bed; in a sinuous segment the zone of greatest intensity tends to be displaced towards the outer and deeper side of each curve.

Two types of shear layer are found within a flowing stream. The more important occurs along the channel boundary. Although turbulence increases towards the stream bed, there may be a thin film actually at the contact zone in which movement is entirely laminar, followed outwards by a buffer zone in which there is mixed laminar and turbulent motion. Details of the laminar sublayer are much influenced by the roughness of the channel materials. In general the sublayer can only develop fully where it is thicker than the diameter of the boundary grains; if the grains are too large they protrude directly into the zone of turbulent flow. The second type of shear layer is associated with the phenomenon known as separation. This

numbers far in excess of the 2 000 usually regarded as about the upper limit for laminar flow.

Two types of turbulent flow may be distinguished. By far the more common is subcritical or tranquil flow, but in certain circumstances this may be transformed into supercritical or shooting flow. As the name implies, shooting flow involves marked acceleration of the water and occurs where there is an abrupt constriction of the channel. This is normally confined to rapids where the stream is flowing either on bedrock or around large boulders, since a channel in finer detrital material tends to have a relatively uniform cross-section. The visual characteristics of shooting flow are a fast, streamlined appearance with little disturbance on the upstream side of any obstacle and stationary oblique rolls on the downstream side. In tranquil flow, on the other hand, obstacles tend to produce disturbed water on the upstream side and standing cross-waves on the downstream side. The transition velocity from tranquil to shooting flow is defined

Fig. 9.4. Schematic diagram illustrating variations in mean velocity, sediment concentration and intensity of turbulence (i.e. deviation of velocity from local mean value).

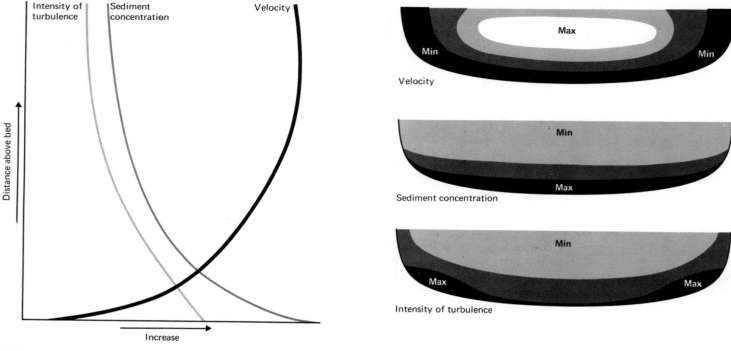

(a) Vertical distribution

(b) Cross-sectional distribution

occurs where an abrupt change in channel alignment causes the water to break away from the solid boundary. Immediately downstream from large obstacles or sharp bends, separation generates a linear zone of intense turbulence known as the free shear layer; systematic eddying patterns are a common accompaniment.

Relationship of flow velocity to channel characteristics

Engineers concerned with such practical tasks as the regulation of river courses and the construction of canals have enunciated a series of empirical formulae that relate flow velocity to such variables as channel slope, shape and roughness. The earliest important formula was introduced by the French engineer Chézy in 1768 when he designed a canal for the Paris water supply. He argued that flow velocity would be proportional to the square root of the product of the slope and hydraulic radius; the actual velocity would also depend upon a coefficient, now often called the Chézy coefficient, incorporating an allowance for channel roughness. During the nineteenth century elaboration of the relationships in the Chézy equation led to formulation of the now widely employed Manning equation $V = R^{2/3}S^{1/2}/n$. In this equation n is a coefficient indicative of channel roughness. The term 'roughness' here refers not only to the size of bed material but also to the sinuosity of the channel and the presence of such obstructions as reeds and fallen trees. With experience the likely value of n can be estimated with considerable precision, and since it may be as low as 0.025 for a clean straight channel but over 0.075 for a winding overgrown channel, the flow velocity in the same size and slope of stream course can vary by a factor of over three.

Erosion

Bedrock channels

Although erosion of channel bedrock must presumably be a fundamental mechanism in subaerial denudation, it has been the subject of relatively little detailed investigation. Three separate but inter-related processes are commonly invoked: corrosion, corrasion and hydraulic action. Corrosion refers to the chemical activity of the flowing water, and is therefore closely related to chemical weathering. Changing river stage regularly subjects part of the bare rock surface to wetting and drying cycles, and these can maintain their effectiveness because any loosened grains are periodically swept away to expose fresh material. Of course, if the water is already saturated with respect to certain salts it may be chemically less aggressive than the precipitation from which it is derived; this suggests that the maintenance, by changing river stage, of a bedrock surface open to the atmosphere may be as significant as direct fluvial corrosion. In either case there seems little doubt that chemical decomposition is a vital precursor to much corrasional and hydraulic activity.

Corrasion refers to the wearing away of bedrock by the scraping and gouging effect of rock debris in transit along the stream channel. The size, shape, composition, number and velocity of the fragments all influence the corrasive power of the debris. In general the bedload sliding and rolling along the channel floor will be the most effective, and as will be shown below (p. 166) the size of the largest entrained fragment increases rapidly with rising discharge; as long ago as 1753 Brahms expounded the so-called sixth-power law stating that the weight of the heaviest mobile fragment varies as the sixth power of the velocity. As the calibre of the load grows in this way, so also does the speed of individual particles with the result that corrasion is enormously enhanced during periods of spate. The clearest testimony to this abrasion is the presence of smoothed and rounded rock surfaces, often moulded into deep potholes. Unlike mobile channel sediments, bedrock forms impose persistent flow patterns within the stream. An initial hollow can induce a swirling motion by which rock fragments are repeatedly made to impinge on the walls and floor; ultimately the depression will be transformed into a prominent pothole, in some cases with internal spiral flutings. High-velocity flows are required for such action to be fully effective, and the best examples often lie so far above normal river-level that they can only have been formed during periods of spate. Most of the time the hollows remain dry, their floors sometimes littered with the well-rounded pebbles sporadically used as abrasive tools.

Hydraulic action refers to the mechanical work of the water alone. The momentum of the water can undoubtedly dislodge large joint-bounded blocks, especially if already loosened by weathering, but its capacity to erode virgin rock in other ways is still open to question. However, one process which may locally be important is the so-called cavitation that arises from abrupt changes in pressure as water accelerates and decelerates. According to the Bernoulli equation (p. 162) any increase in velocity must involve a reduction in pressure. If the acceleration is sufficient, the pressure can drop to the point where numerous small bubbles form and create the impression that the water is foaming. As the pressure rises again the bubbles implode and emit powerful shock waves which hammer adjacent solid surfaces. The effect of cavitation has long been known to engineers concerned with wear on ships' propellers and turbine blades where extremely high velocities are involved. It has also been invoked to explain pits gouged into solid concrete spillways on certain dams, one such example attaining a depth of almost 0.5 m in a single day. It is not easy to prove a particular erosional form as due to cavitation, and few rock surfaces exhibit the fretted appearance characteristic of a badly eroded propeller; the difference could be due to cavitation acting alone on propellers but in conjunction with a sediment load on stream channels. Although the importance of cavitation is difficult to assess directly, the conditions under which it is a potential means of erosion can be derived from the Bernoulli equation. If localized acceleration is assumed to exceed mean stream velocity by a factor of two, water 1 m deep is capable of inducing cavitation when flowing at an average speed of about 9 m s^{-1}. To attain this speed in an open channel a sustained gradient of over 5°, or a vertical fall of about 4 m is necessary. Although these calculations indicate no more than a general order of magnitude, they suggest that cavitation may be effective along certain steep rapids and at the foot of substantial waterfalls.

Alluvial channels

The concept of erosion in alluvial channels is rather different from that in bedrock. In the latter case erosion is an irreversible process since any later aggradation will involve replacement of the bedrock by detrital sediments. Erosion in alluvial channels, on the other hand, normally occurs during periods of spate with any mobilized debris being replaced during the falling river stage by similar material from higher upstream. In the vertical plane one may distinguish for any reach of channel three possible long-term conditions: aggradation, degradation and equilibrium; all three will represent the net effect of a large number of much shorter cycles, each involving both erosion and deposition. In the horizontal plane the amount of channel erosion will determine the potential rate of lateral stream migration.

Research has concentrated on the factors that conduce to short-term erosion of the channel floor and sides. The general field is known as loose boundary hydraulics, and because conditions in streams are so complicated much of the experimental work has been conducted in laboratory flumes where far greater control of the many variables is possible. The underlying principle is that the passage of the water exerts a force on the channel perimeter. It can be shown that the mean bed shear stress is equal to the product of the unit weight of water, the hydraulic radius and the water surface slope, normally written as the DuBoys equation $\tau_0 = \gamma\, RS$. The problem is then to understand how that stress may displace particles composing the bed of the channel; a simplified theoretical analysis of this problem is illustrated in Fig. 9.5. Work on the initial entrainment of sediment has normally concentrated on one of two values, known as the critical erosion velocity and the critical shear stress, the word 'critical' in each case connoting the minimum value required to set a particle of specified size in motion. In an empirical evaluation of critical erosion velocities, the Swedish geomorphologist Hjulstrom in 1935 collected the results of over thirty experiments and plotted the data in the form of a graph. In this way he showed that the most readily moved material has a diameter of 0.25 mm and is set in motion by water flowing at 0.2 m s^{-1} (Fig. 9.6). Both above and below this size the critical erosion velocity rises. For large particles this is due to the extra force required for lifting and rolling; for very small particles it is partly explained by the cohesion of the material.

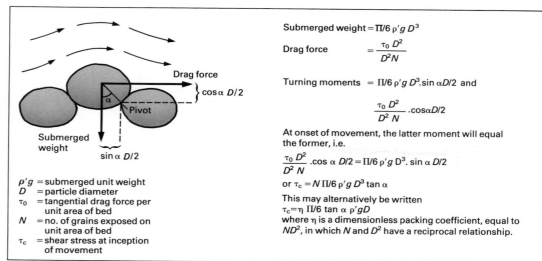

Submerged weight $= \Pi/6\, \rho'g\, D^3$

Drag force $\qquad = \dfrac{\tau_0\, D^2}{D^2 N}$

Turning moments $= \Pi/6\, \rho'g\, D^3.\sin\alpha D/2$ and

$$\dfrac{\tau_0\, D^2}{D^2\, N}.\cos\alpha D/2$$

At onset of movement, the latter moment will equal the former, i.e.

$$\dfrac{\tau_0\, D^2}{D^2\, N}.\cos\alpha\, D/2 = \Pi/6\, \rho'g\, D^3.\sin\alpha\, D/2$$

or $\tau_c = N\, \Pi/6\, \rho'g\, D^3 \tan\alpha$

This may alternatively be written
$\tau_c = \eta\, \Pi/6 \tan\alpha\, \rho'gD$
where η is a dimensionless packing coefficient, equal to ND^2, in which N and D^2 have a reciprocal relationship.

$\rho'g$ = submerged unit weight
D = particle diameter
τ_0 = tangential drag force per unit area of bed
N = no. of grains exposed on unit area of bed
τ_c = shear stress at inception of movement

Fig. 9.5. Theoretical analysis of conditions at onset of particle movement on a stream bed. The conditions are idealized, and experimental data exhibit substantial deviations from the forecast values. One factor not considered is a 'lift force' acting upwards away from the fluid boundary. Predicted by the Bernoulli equation from the faster flow across the top of the grain, this has been shown in some circumstances to compare in magnitude with the drag force. Under natural conditions, such additional factors as the non-sphericity of the particles, variation in the value of the angle α, irregular packing and the impact of previously mobilized grains need to be borne in mind.

A year after Hjulstrom's work was published, a German engineer named Shields described his experimental work on the flows required to set cohesionless particulate material in motion. Concentrating on the idea of critical shear stress, he emphasized the crucial importance for fine material of the thickness of the laminar sublayer. As their particle sizes diminish, fine sand and silt become progressively more submerged within the sublayer with the result that higher shear stresses, and therefore higher velocities, are required to set them in motion. On the other hand, above a particle diameter of about 0.2 mm at which the grains begin to project well into the turbulent flow, any increase in size also requires an increase in shear stress to initiate movement. Thus the basic forms of the Hjulstrom and Shields curves closely resemble each other.

Although conceptually very important, it should be emphasized that the practical application of a diagram such as Fig. 9.6 to ordinary stream channels is fraught with difficulty. The diagram refers to smooth but not irregular beds and if, for instance, there are sand ripples on the stream floor the grains tend to move before the predicted critical velocity is reached. Fine cohesive material tends to break away in lumps rather than be entrained grain by grain. Perhaps most important is the relatively unsorted character of bed material when compared with the uniform grains employed in the majority of flume experiments. As early as 1914 Gilbert had already recorded that the addition of fine material to uniform coarse debris increases not only the total transported load but also the amount of course debris that is moved. Later it was noted that, in mixed sediments, a few of the larger fragments may actually begin to move before the fine. Sundborg in 1956 argued that some of these apparent anomalies might be due to the particle diversity affecting flow patterns near the stream bed. From an analysis of published data he distinguished three situations: in coarse mixed material over 6–8 mm in diameter, it is the finer fraction that is first entrained; in mixed sediments between 6–8 and 0.8–1.0 mm, it is the medium grain size; and in sediments under 0.8–1.0 mm, it is usually the coarsest. He concluded that, although data for well-sorted sediments indicate the 0.2–0.5 mm size range as the most readily entrained, for more poorly sorted sediments it is the 1–7 mm size range.

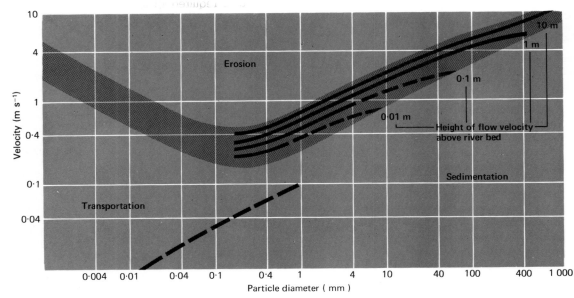

Fig. 9.6. Stream velocities required to set channel material in motion. For larger particles the data are from Sundborg (1956) and are based on the work of Shields; for finer particles the data are from Hjulstrom (*Geogr. Annlr.* 1935).

Tests of critical velocities and stresses in the field have generally involved ascertaining the largest fragments moved in the course of some particular flood. These have been identified in several different ways. Boulders and cobbles of a range of sizes have been marked with paint and their before-and-after positions accurately surveyed; cameras have been installed so that repeated pictures of the channel bed can be taken; the largest rocks trapped in a bedload pit sampler have been used; and in at least one case selected cobbles have been placed upon small floats designed to rise to the surface when the stone starts to move. It is worth emphasizing that the above experiments have covered the incipient movement of boulders weighing up to more than 100 kg and therefore of very different dimensions from those used in most laboratory experiments. The results have not produced a consistent pattern. For example, a review of American data led one group of workers to suggest a relationship of the form $d = 65\tau^{0.34}$, whereas a study by means of pit samplers in the English Pennines was held to support a relationship of $d = 18.3\tau^{1.58}$ for wide streams and $d = 1.01\tau^{1.97}$ for narrow streams; in each case d is particle diameter in mm and τ is shear stress in kg m^{-2}. Such divergences clearly imply that we still have much to learn about the threshold of particle movement.

The foregoing arguments refer solely to the calibre of displaced material and not to the amount of erosion that will occur. As far as the latter is concerned a self-regulating mechanism may be envisaged since channel enlargement will itself reduce the water velocity. Several early workers claimed that severe vertical scour occurs during spate discharge, quoting as an example the deepening of the Colorado channel at Yuma by as much as 8 m during peak flow. However, in a review of the evidence available in 1954 Lane and Borland emphasized the patchy distribution of such overdeepening, a conclusion supported by later observations which serve to confirm abrupt channel constrictions as the only locations of regular deep scour.

Lateral erosion of stream banks, particularly on the outside of meanders, has been evaluated by the insertion of long stakes whose protruding ends can be measured at intervals and the rate of channel migration calculated. Many experiments carried out in this way have underlined the importance of weathering in preparing fine bank material for rapid removal. A bank may thus be eroded much more swiftly in winter after freeze–thaw cycles have rendered the surface friable than at the same discharge in summer. Where erosion is particularly active, repeated surveying or even the comparison of successive topographic maps may provide suitable data on the long-term rate of channel migration; such approaches may be essential where channel erosion occurs mainly by bank caving.

Sediment transport

Basic ideas

The material in transit along a stream channel may partly derive from erosion of the bed and banks, and partly be contributed as so-called washload directly from surface runoff. When in motion the sediment is divisible into two fractions, the bedload and suspension load. The distinction between these two is often blurred by an exchange of material between them, but it remains an important one because of the different principles underlying their respective modes of transport. If one first imagines a bed of fine but non-cohesive material, water flowing at below the critical erosion velocity will leave all the particles undisturbed. As the velocity increases a few particles will slide or roll a short distance and then come to rest again; meanwhile others will be disturbed and make similar downstream movements. Further acceleration will lead to more and more grains being set in motion until a complete layer, a few particle diameters thick, becomes mobile. These grains roll, slide or skip down the channel with their immersed weight borne mainly by contact with the bed. Meanwhile other dislodged particles, particularly the finer ones, are lifted into the water where their immersed weights are supported by turbulent upward currents. These latter particles constitute the suspension load and may be transported long distances without any contact with the bed. Substantial differences exist between the bedload and suspension load in terms of grain size and downstream velocity. The finer particles in the suspension load may move at approximately the same speed as the water, whereas the bedload as a whole travels much more slowly.

The above account is an idealized picture, and natural conditions are often much more complex. For example, with heterogeneous material the fines may be removed at low velocity to leave a surface armoured with a lag of coarse debris that can only be entrained with a major velocity increment. Even with homogeneous sand, the mobile layer does not necessarily move as a uniform sheet. Transport at velocities just above the threshold for movement leads to the development of ripples. As bed stresses increase, the ripples are replaced by dunes which, in the biggest rivers, can be as much as 10 m high and 250 m long. In both ripples and dunes the movement of the individual grains is by sliding and rolling up the gentle upstream face and avalanching down the lee face; this type of movement is characteristic of flow with low Froude numbers. With further acceleration to Froude numbers around 0.4 the dunes progressively flatten and are washed out to produce a mobile 'plane bed'. This is transitional to the development of antidunes at Froude numbers of about 0.7; these bedforms have an approximately sinusoidal form and, unlike ordinary dunes, migrate upstream because they are eroded on the downstream face while growing by deposition on the upstream side. When antidunes are fully developed, the position of each is marked by a prominent standing wave on the surface of the water.

The precise mechanism by which the available power represented by the moving water is converted into the work of transportation still defies full understanding. One cause of the difficulty is the complex nature of the variables involved. For instance, two of the most fundamental appear intuitively to be stream velocity and sediment supply. Yet, in considering the former, attention must be directed not only to a mean value but also to the velocity profile and degree of turbulence; similarly the sediment must be described in terms of size, shape, density and cohesion. The interdependence of many of the variables further complicates the subject to the point where it is difficult to derive laws of any great generality from experimental work. On the other hand, a deductive approach from fundamental physics needs to make so many simplifying assumptions that the results can rarely be applied directly to natural streams.

Theoretical models

Several workers have endeavoured to compute sediment transport rates from fundamental physical principles. Bagnold, for example, argued in 1966 that the stream may be regarded as a transporting machine and that the concept of efficiency, long familiar to engineers dealing with machines, can appropriately be applied to sediment transportation. The basic idea is formulated in the expression 'Rate of doing work = Available power × Efficiency'. The available power in the case of a stream is clearly the rate of liberation of kinetic energy as the stream flows down its channel. This energy is available for the two separate mechanisms of bedload and suspension load transport. Space does not permit a full recapitulation of Bagnold's arguments and it must suffice to note that, from a few fundamental axioms, he derived for the total transport rate the relatively simple expression

$$\omega \left(\frac{e_b}{\tan \alpha} + 0.01 \frac{U_s}{V} \right) \text{ (For further elaboration, see Fig. 9.7).}$$

This formula was then tested against data from both flume experiments and natural streams. Bagnold found the predicted values corresponded closely to observed flume values over a wide range of flow conditions and grain sizes, the only pronounced discrepancy occurring with a very fine suspension load. The shortage of reliable bedload measurements for natural streams meant that he had to compare the data from sediment-gauging stations with predicted suspension loads. Again a reasonable correspondence was found although some of the discrepancies proved much larger than those associated with flumes. Observed values lower than predicted were explained in several ways, such as a shortage of suspendable material or flow temporarily below the critical level needed to lift available particles into suspension. More puzzling were instances of the actual load greatly exceeding the predicted capacity of the river. No entirely satisfactory explanation of these cases could be offered, but Bagnold suggested the discrepancy might sometimes result from sampling procedures in which part of the true bedload has been counted as suspension load.

Bagnold's analysis has been selected as an outstanding example of the deductive approach in fluvial geomorphology. Based upon the application of a few simple physical principles, its formulation was entirely independent of observational data. Its comparative success in forecasting transport rates under a wide range of artificial and natural conditions is eloquent testimony to its general soundness. Nevertheless it also pinpoints certain limitations in our present knowledge. The relationships that apply for sand are not necessarily true for finer materials; in natural streams unlimited availability of transportable solids cannot be assumed; the constant discharge employed in many laboratory studies is uncharacteristic of natural streams where variations due to changing stage can be of prime importance;

Fig. 9.7. A brief synopsis of the main arguments deployed by Bagnold (1966) in predicting likely sediment discharge from commonly measured streamflow properties. No single diagram can do justice to Bagnold's elegantly argued hypothesis and, for further details, reference should be made to the original work.

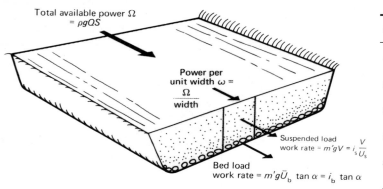

Total available power Ω $= \rho gQS$

Power per unit width $\omega = \dfrac{\Omega}{\text{width}}$

Suspended load work rate $= m'gV = i_s \dfrac{V}{\bar{U}_s}$

Bed load work rate $= m'g\bar{U}_b \tan \alpha = i_b \tan \alpha$

Symbols used

ρ	= fluid density	$m'g$	= immersed weight of load
g	= gravity acceleration	V	= fall velocity of suspended solids
Q	= discharge	i	= transport rate
S	= slope	$\bar{U}$	= mean transport velocity ($\bar{U}_s$ assumed to equal water velocity)
ω	= stream power per unit width	e	= efficiency
$\tan \alpha$	= friction coefficient ($\alpha \approx 33°$)		

For both types of load, work rate = available power x efficiency, **therefore**

$$i_b \tan \alpha = e_b \omega \text{ or } i_b = \frac{e_b \omega}{\tan \alpha}$$

and, assuming available power for suspended load is now $\omega(1 - e_b)$,

$$i_s \frac{V}{\bar{U}_s} = e_s \omega (1 - e_b) \text{ or } i_s = \omega \frac{e_s \bar{U}_s}{V} (1 - e_b)$$

By summation

$$\text{total load} = \omega \left(\frac{e_b}{\tan \alpha} + \frac{e_s \bar{U}_s}{V} (1 - e_b) \right)$$

Assuming $e_s = 0.015$ and $(1 - e_b) = \frac{2}{3}$, the latter equation simplifies to

$$\omega \left(\frac{e_b}{\tan \alpha} + 0.01 \frac{\bar{U}_s}{V} \right)$$

For purposes of comparison with gauging-station records (**see** bar graph) only the suspended material was considered

$$i_s = \omega \left(0.01 \frac{\bar{U}_s}{V} \right)$$

Subscripts s and b refer to suspended load and bedload respectively

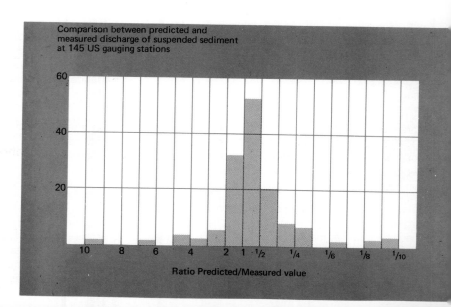

Comparison between predicted and measured discharge of suspended sediment at 145 US gauging stations

Ratio Predicted/Measured value

channel roughness attributable to ripples and dunes may itself affect the energy relationships close to the stream bed; and doubts still persist regarding the precision of certain gauging techniques.

Field investigations

For field investigations proceeding beyond the routine gauging techniques described in Chapter 7, a prime requirement is some means of labelling sediment so that its actual movement either on the stream bed or suspended in the fluid can be monitored. Most methods in common use involve either fluorescent or radioactive tracers. The former have the advantage of being cheap and safe, but normally demand the recovery of samples so that the counting of fluorescent grains can take place under ultraviolet light. The passage of radioactive tracers can be monitored by drawing detectors across the stream bed without the need for removing samples, but on the other hand the equipment is relatively expensive and stringent health safeguards need to be observed. In a study of the slow-flowing River Idle in Nottinghamshire sand-sized tracers incorporating the radionuclide Sc^{46} were released and their downstream movement plotted by lowering detectors to the stream bed. During a 3-week experiment the mean velocity of the water varied between 0.3 and 0.6 m s^{-1}, while the tracers moved at an average rate of 0.064 m hr^{-1}. A flood led to pronounced acceleration of the tracers with their velocity rising to over 2 m hr^{-1} for a period of $4\frac{1}{2}$ days. In a similar 13-day study of the North Loup River in Nebraska when the water velocity averaged 0.7 m s^{-1}, sand-sized tracers were found to move at a mean rate of about 1.15 m hr^{-1}.

Sands coated with fluorescent dyes were employed in a study of suspended sediment transport along a gravel-bed river in Colorado. Three distinctive colours were used for labelling three grain sizes, 0.15–0.3, 0.3–0.52 and 0.52–1.29 mm. Some 800 m below the point of injection the suspension load was analysed by means of a depth-integrated sampler, while a pump was also used to abstract material from just above the channel bed. The travel rate for the fastest particles was found to be inversely proportional to the square of their diameters. Although the mean velocity was 1.4 m s^{-1}, particles of 0.86 mm diameter travelled at 0.015 m s^{-1}, particles of 0.3 mm at 0.135 m s^{-1}, and only the smallest grains of 0.15 mm approached the velocity of the transporting fluid. It seems clear that the coarser particles spent a considerable proportion of their time on the river bed.

Some workers have experimented with alternative tracers, such as pebbles tagged with magnetic inserts, but others have endeavoured to develop more novel techniques. These have included the use of movie cameras, acoustic apparatus and highly sensitive echo-sounders. Photography is often ruled out by turbidity, and devices which monitor and amplify the acoustic energy generated by particle impacts, although potentially very useful, have so far proved difficult to calibrate. Modern ultrasonic echo-sounders are capable of detecting changes in bed elevation of less than 10 mm, and an array of such instruments mounted above a sand-bed channel can trace the downstream migration of both ripples and dunes. However, any assumption that this bedform movement can be directly translated into a transportation rate ignores two possible sources of error. If numerous particles leap several ripples at a time, the true transportation rate may greatly exceed that calculated from bedform migration. The same would be true if the mobile layer were significantly thicker than the height of the bedforms. The retrieval of cores from streams seeded with tracers offers the potential of testing this latter point; in the experiments mentioned above tracers on the River Idle were found to extend 0.12 m, and on the North Loup 0.44 m, below the channel bed.

Theoretical computations, sediment-gauging techniques, experiments with labelled particles and measurements of bedform migration, all provide estimates of transportation rates. However, even replicated experiments using the same measuring method occasionally provide data varying by 20 per cent or more, and discrepancies of that order are commonplace when comparisons are made between different methods. Much further work is evidently needed before sediment transportation by natural streams is fully understood.

Attrition

The downstream attrition suffered by the transported load has proved remarkably difficult to measure, largely because its effect can rarely be differentiated from that of sorting. The progressive rounding of pebbles from an identifiable source leaves little doubt concerning its impact on coarse debris, but much more uncertainty surrounds the fine fraction. Most information has come from laboratory experiments. Kuenen, for instance, employed a large circular dish coated with concrete. An electrically driven rotating paddle impelled the water and sediment round the dish and the distance travelled by individual particles was estimated by coating grains with fluorescent paint and measuring their velocities under ultraviolet light. Kuenen concluded that the attrition of fine quartz material is extremely slight since grains 1.5 mm in diameter that had travelled 120 km at 0.45 m s^{-1} suffered a weight loss averaging only 0.008 per cent. This may be compared with a more recent field experiment with gravels where a mean reduction of 6.1 per cent in the long-axis length of pebbles was recorded after travel equivalent to only 26 km. And yet the investigators in the latter case still concluded that attrition was responsible for less than one-tenth of the observed down-valley diminution in particle size, the rest being attributable to sorting. One may perhaps infer that particle abrasion, although difficult to quantify, is a significant process in gravel-bed streams; it is undoubtedly much influenced by the hardness of the materials involved and is also abetted by any weathering the debris may suffer while stationary on channel bars.

Channels in cross-section and long profile

Stream channel forms are immensely varied. In addition to obvious contrasts in size, they may be single- or multiple-thread, deep or shallow, straight or meandering, steep or gentle. They are, of course, fashioned by the water that is discharged down them, and the streams may be steady or flashy, perennial or intermittent. A primary aim of the fluvial geomorphologist must be to relate the spatial variability of form to the temporal variability of flow. Channel morphology may be regarded as reflecting the interaction between fluid stress and boundary resistance. Fluid stress varies with stream stage, whereas boundary resistance remains relatively constant. Intuitively, fashioning of the channel might be expected to occur during bankfull discharge when fluid stress attains its maximum value. For this reason the dominant or formative discharge is often equated with the bankfull discharge, although the possibility that the increased frequency of a slightly lower stage might be equally effective should not be overlooked. The interval at which bankfull discharge recurs appears to vary a little from area to area, but most studies in humid regions have suggested a periodicity of between 1 and 2 years; viewed in an alternative way, data from British river channels show them to be full to capacity about 0.6 per cent of the time.

For purposes of further discussion, two approaches in attempts to relate form to discharge may be distinguished. The first is devoted to temporal changes occurring at a single point on the long profile, where 'at-a-station' relationships may be analysed, and the second is concerned with examining downstream changes in channel morphology.

At-a-station relationships

Leopold and Maddock used the term hydraulic geometry in 1953 to denote the mutual adjustment of channel form and discharge. For instance, as discharge rises at a station there is normally an increase in stream width, depth and velocity. If the data for these three variables are plotted on logarithmic graph paper (Fig. 9.8) the points are found to lie close to a straight line indicating that each tends to be a power function of discharge. One may therefore express the relationships by three equations: $w = aQ^b$; $d = cQ^f$; $v = kQ^m$ in which a, b, c, f, k and m are empirical constants. Because Q is itself the product of w, d and v, it follows that both $a \times c \times k$ and $b + f + m$ must equal unity. As might be expected the actual values of the constants differ from one site to another. In 1977 Park investigated the value of exponents for 139 stations in different parts of the world. Much diversity was discovered with values of b ranging from zero to 0.59, f from 0.06 to 0.73 and m from 0.07 to 0.71. Since the exponents must sum to unity, he plotted their values in the form of a triangular diagram to search for possible regional patterns related to climate. Although the exponents for width tended to be higher in semi-arid than in humid temperate regions, and those for depth and velocity lower, there was such a wide area of overlap that it appeared other factors must be masking any strong association with climate. In practice the cohesive nature of the bed material appears to be the major determinant of whether width or depth increases more rapidly. When the perimeter sediments contain a high proportion of silt or clay, a steep-sided channel tends to develop with values of b less than 0.05; in coarse, non-cohesive materials the banks slope much more gently with the result that b is commonly within the range of 0.2–0.4.

The three properties just considered are so clearly interrelated as to form a natural grouping for at-a-station studies. However, it is possible to enquire how certain other properties change as discharge increases. For instance, Knighton examined variations in channel fric-

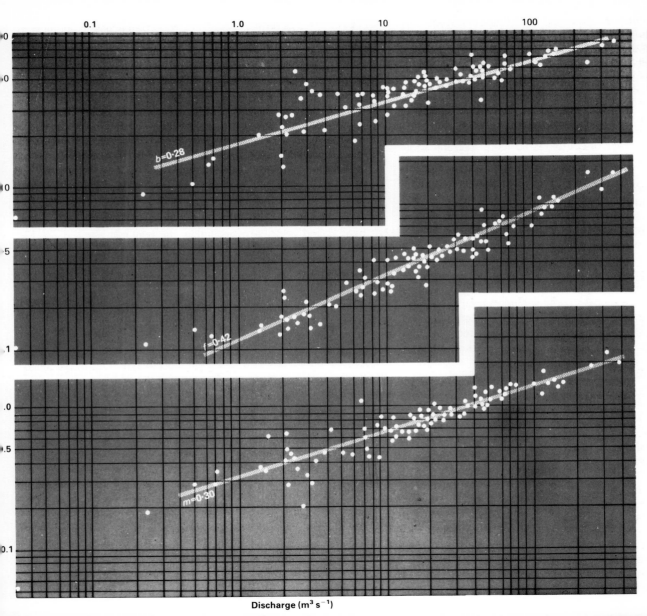

Discharge (m³ s⁻¹)

Fig. 9.8. 'At-a-station' relationships on the Powder River at Locate, Montana, USA (data from Leopold and Maddock, 1953).

tion and found, as might be anticipated, that this value tends to decrease rapidly in a stream strewn with coarse debris where there is a concurrent swift rise in velocity; on the other hand, in a sand-bed stream the development of ripples and dunes may cause the change in velocity to take place very slowly. Other workers have plotted suspension load against discharge and have suggested that, at least as a first approximation, one may write $G = pQ^j$, where G is sediment load and p and j are empirical constants; values of j commonly lie between 2.0 and 3.0. This, of course, is no more than the sediment rating curve with the shortcomings already discussed (p. 131).

Downstream variations in channel properties

In 1953 Leopold and Maddock published a series of graphs designed to show how width, depth and velocity alter in response to changing discharge along the length of an individual stream channel. Using logarithmic scales they plotted values for each against mean annual discharge; the latter was employed in place of the theoretically preferable bankfull discharge because its value was more readily available at most gauging stations. Despite some scatter among the points, the authors generally found a simple straight-line relationship to characterize each stream (Fig. 9.9). This suggests a series of equations may be formulated analogous to those for at-a-station records. The exponents, however, are found to differ from those of a single cross-section, the calculated values commonly approximating to $b = 0.5$, $f = 0.4$ and $m = 0.1$. These figures indicate that a channel accommodates the downstream increase in discharge mainly by an increase in width, with a rather smaller change in depth. Surprise is sometimes expressed that velocity also increases in a downstream direction, partly because of a perception of upland streams as cascading down a series of rapids. However, careful collation of records shows that this is far from a universal picture, if only because the friction of coarse bed material often acts as a major restraint on water velocity. If flow at comparable stages is analysed, many streams exhibit a slight downstream acceleration. However, the increase in speed is generally small and the scatter of values rather wide owing to the influence of other factors.

One further aspect of the changes in the three properties so far discussed is their cyclical regularity along certain limited reaches. In its most fully developed form this gives rise to the feature known as a pool-and-riffle sequence. Shallow gravel bars or riffles alternate with long deep pools in which the bed material is noticeably finer; at low flow the stream traverses the riffles with broken surface and then passes more slowly through the pools. Many workers have noted that riffles are regularly spaced at intervals of five to seven times the channel width. One explanation for this spacing was proposed by Yalin in 1971 who argued that turbulence along the channel banks, even along a straight uniform reach, will generate large-scale eddies which will in turn produce alternating longitudinal zones of accelerating and decelerating flow spaced at intervals equal to the channel width multiplied by 2π. To even out velocity values along the length of the reach, erosion may be necessary in the zones of fast flow and deposition in zones of slow flow. It is inferred that all these changes occur at bankfull stage and that the resultant bedforms are inherited as the river-level falls. This can lead to what has been called velocity reversal; thus, the pool that was scoured by the zone of fastest flow now becomes the site of slow flow, while the opposite relationship applies to the riffle. The reality of velocity reversal was very clearly established in an investigation in Wyoming when eleven cross-sections on a 400 m length of channel were monitored over a wide range of flows; as an example, at one site velocities changed from 0.22 to 1.70 m s^{-1}, while near by they were changing from 0.49 to 1.31 m s^{-1}. It appears that the fine material on the floor of the typical pool reflects the reduced transporting capacity of the stream as a whole during low flow, and in particular its reduced capacity when passing through the pools.

A fourth channel property of very considerable interest is the downstream change in gradient. Although the seaward decline in slope has elicited comment for several centuries, mathematical analysis is a more recent development. Two separate but related approaches may be discerned. The first has been concerned with the link between channel slope and other stream properties, the second with the overall form of river long profiles. Into the first category comes research in which measured slope has been plotted against discharge and size of bed material. Often catchment area has had to be employed as a surrogate for discharge, but with this modification several workers have claimed a relationship of the form $S = tQ^z$. The scatter of points, however, has generally proved rather wide, with the calculated exponent z varying considerably from one catchment to another. This clearly implies the intervention of other factors, of which the calibre of the bed material is often regarded as

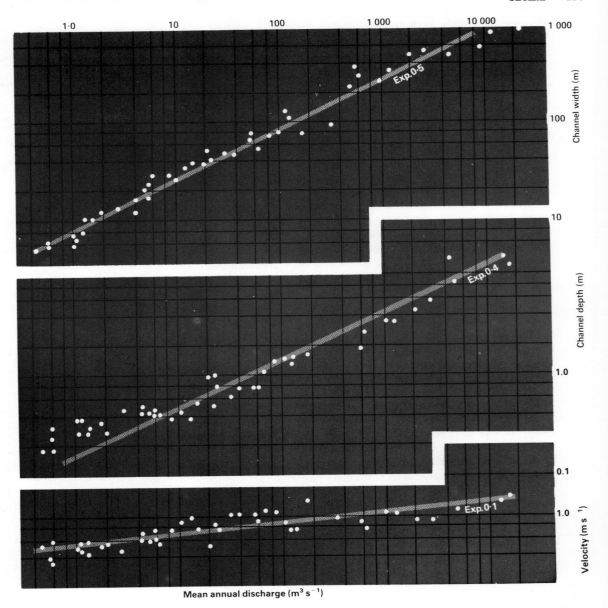

Fig. 9.9. The relationship of variations in width, depth and velocity to downstream increases in discharge within the Missouri and Mississippi basins, USA (data from Leopold and Maddock, 1953).

the most important. In a study of fifteen streams in Virginia and Maryland, Hack in 1957 first plotted channel slope against catchment area and found a poor correlation (Fig. 9.10). However, the scatter of points was considerably reduced when they were classified according to geology. Further analysis suggested that geology was actually acting through the calibre of the load supplied to the stream, and Hack concluded that the relationships are best represented by the expression $S = k\,(M/A)^{0.6}$ which incorporates values for both the median size of bed material, M, and the contributory drainage area, A. Of course, this expression by itself does not specify the form of long profiles since the rate of downstream change in particle size and in catchment growth may vary. Nevertheless, employing arguments analogous to those of Hack, Brush analysed sixteen stream profiles in central Pennsylvania. Graphing channel slope against stream length, he managed to distinguish three groups of plotted points, each identified with a specific bedrock lithology and each conforming to a relationship defined by the formula $S = kL^h$ (Fig. 9.11). Although the exponents in the equations do not differ greatly, statistical analysis confirmed the variation of slope with length to be significantly different for the three rock types, and Brush calculated hypothetical long profiles for each lithology. He also examined the possibility of geological control being exercised through the calibre of bed material, but found that the introduction of such a factor only slightly reduced the scatter of plotted points.

The above investigations reflect one approach to the study of stream gradients. The underlying idea is that of a mutual adjustment between discharge, channel cross-section, bed materials and slope. The down-valley increase in discharge, supplemented by the diminishing particle size that results from both sorting and attrition, is theorized to lead to concave profiles defined by some form of power function. Even within this conceptual framework complications will be introduced by streams crossing geological boundaries and by abrupt changes that may occur at tributary junctions.

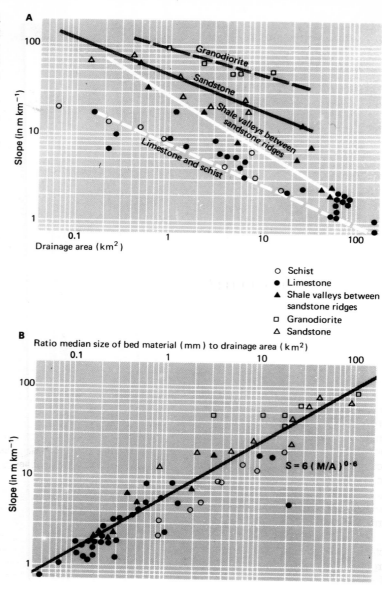

Fig. 9.10. Channel gradients in parts of Virginia and Maryland USA; (A) plot of channel slope against drainage area, showing a wide overall scatter of points that appears to be due, at least in part, to contrasts in geology; (B) plot of channel slope against the ratio of the size of bed material to drainage area (M/A). This appears to lead to a substantial reduction in the scatter of the points (data from Hack, 1957).

Fig. 9.11. Analysis by Brush (1961) of channel gradients in central Pennsylvania USA; (A) plot of channel slope against stream length for sandstone, shale and limestone-dolomite streams; (B) hypothetical long profiles, calculated from regression lines in (A) and related to arbitrary datum, H_o.

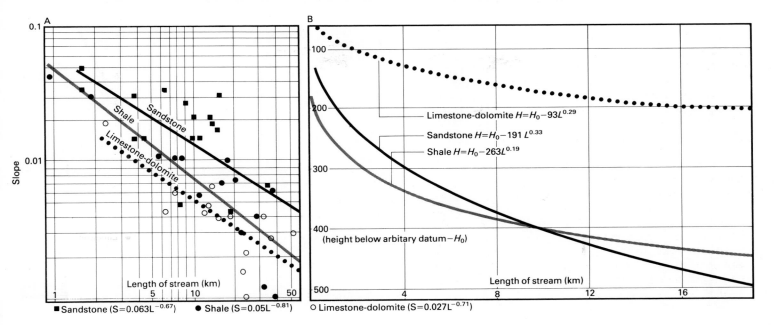

In examining lengthy stream profiles Hack in 1973 advocated calculation of a 'gradient index' obtained from the ratio of fall within a reach to the difference between the natural logarithms of the river lengths at each end of the reach as measured from the stream source. Sudden changes in this index along a stream profile mark localities where the idealized smooth curve is perturbed by other factors. Hack contended from work in the Appalachians that it may be possible by this means to identify areas where recent tectonic movements, as distinct from lithological variations, are influencing the form of river profiles.

Although this approach was relatively new, the idea of tectonic activity finding some reflection in stream profiles has a long history, largely because they are seen as one of the least transient aspects of river geometry. In south-western England, for instance, Kidson showed that the profile of the Exe can be described by a series of intersecting logarithmic curves of the general form $y = a - k \log x$, where y is elevation, x the distance from a theoretical source and a and k are constants (Fig. 9.12). Each concave segment was treated separately and the constants yielding the closest fit to the measured points were independently computed. This procedure was followed in the belief that each segment represents the product of a stable erosional phase ultimately terminated by uplift and initiation of a new knickpoint that worked its way headwards up the valley; by extrapolation of the fitted curves it was hoped to reconstruct the earlier profiles. Such a technique was often employed with little regard to the underlying cause of the concavity, yet it did serve to confirm some form of logarithmic relationship between elevation and downstream distance. What remains in dispute is the cause of the

Fig. 9.12. The long profile of the middle and upper Exe as surveyed and interpreted by Kidson (1962). It has been partitioned into nine concave segments, each treated as part of a former profile that may be reconstructed by extrapolating mathematically the short reach that survives. Also shown for one such reconstructed stage (the Westermill) is a succession of benches believed to have been eroded at or close to the contemporaneous sea-level.

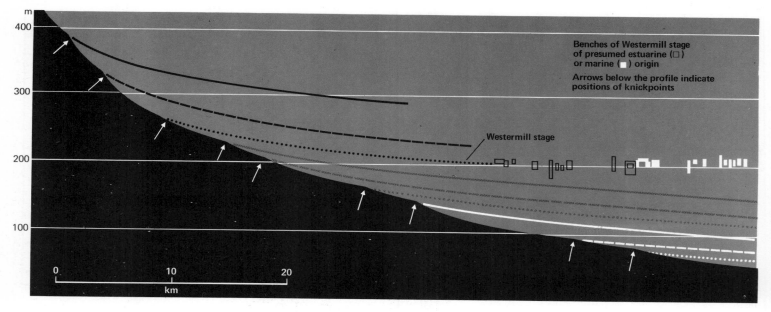

frequently observed temporary steepening in stream gradients. The various explanations that have been offered are not mutually exclusive, and there can be no doubt that in examining individual river profiles geological history has to be carefully borne in mind. For example, Wheeler showed from an analysis of 115 British rivers that, although the degree of concavity exhibited by the profiles is partly a function of the rate of growth in discharge, there are also many aberrant values. 'Over-concave' rivers occur where there has been substantial sedimentation along the lower reaches of the valley, or where elongation by glacial interference has taken place; 'under-concave' rivers can equally result from glacial diversion, or may be associated in at least one case with recent tectonic movement. If interruption by glacial or tectonic activity results in a substantial steepening of a river profile, some flume experiments and computer

modelling suggest the resultant knick-point can eventually migrate long distances upstream, but other work indicates this can only happen in very special cases.

Channel patterns

In some ways the great variety of channel patterns to be observed on the earth's surface is rather surprising. As Scheidegger has remarked, were water to flow in a straight channel from high to low point we would probably regard it as virtually self-explanatory. The fact that some streams are highly sinuous and that others constantly subdivide to form anastomosing systems obviously requires some explanation. In order to analyse the patterns they must be measured and classified, but this in itself presents problems. Although there is often a clear distinction between a single-thread and a braided stream, or between a straight and a meandering channel, these are end-members of spectra that include numerous intermediate forms. These latter should not be forgotten in the course of the following discussion where emphasis will naturally tend to focus on the more striking examples of the phenomena. The regional distribution of channel patterns can also beget an unfortunate tendency to regard one type as the norm and any other as exceptional. Thus, an inhabitant of north-western Europe might well regard the single-thread channel as typical and the braided channel as unusual, but this is scarcely the view that would be adopted by a resident of, say, Iceland or the Great Plains of the United States where braided channels are commonplace (Fig. 9.13). This regional differentiation is itself significant since it points to common causal factors of widespread extent.

Braided channels

Among factors that have been invoked to explain channel braiding are steep gradient, large sediment load, non-cohesive bed material, widely fluctuating discharge and rapid aggradation. While each may be present in a particular case, none appears to offer an adequate explanation by itself, and usually a combination of factors is responsible. Early workers emphasized the importance of channel steepness in relation to the discharge, and this relationship was further stressed when experiments by Schumm and Khan demonstrated that increasing the gradient of a flume can induce a threshold at which a single-thread flow is converted to a multi-thread form (Fig. 9.14). In natural channels, however, another factor conducing to braiding is the calibre of the bed materials, and several workers have endeavoured to incorporate this into equations that will predict the critical slope for the onset of braiding. However, it is not only the average particle size that is significant but also the silt and clay content which, as it rises, gives the bank material greater cohesion and therefore a greater resistance to splitting into subsidiary channels. Another factor that is often quoted as favouring a multi-thread channel is a widely fluctuating discharge. This may operate in part through its ability to suppress soil and vegetation cover which would otherwise tend to

Fig. 9.13. An aerial view of the braided channels characterizing meltwater streams issuing from Vatnajökull in south-eastern Iceland.

Fig. 9.14. A plot of experimental data obtained by Schumm and Khan (1972) as they progressively increased the gradient of a laboratory flume under conditions of constant discharge. An initially straight channel became more and more sinuous until, at a slope of 0.016, a braided form developed.

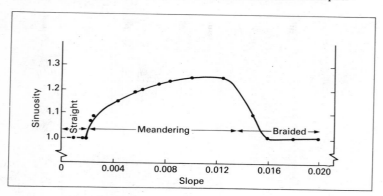

ington the channel alterations were so rapid that Fahnestock used time-lapse photography as part of his study. Shifts of whole stream courses are especially common where rivers emerge from mountains on to gentler plainlands; the River Kosi in northern India, renowned for its devastating flooding at the foot of the Himalayas, has migrated more than 100 km during the last two centuries. And yet to emphasize the point made earlier about the diversity within any simple classification, it is worth noting that the braided Knik River in Alaska appears to have experienced pronounced channel stability over a period of at least 30 years. The Knik River is unusual because, until recently, for about 1 week each summer it carried up to 9 800 m³ s⁻¹ of outflow from an ice-dammed lake compared with only 560 m³ s⁻¹ during the rest of the melt season. Another abnormal aspect of this particular case is the development by such a gravel-bed river of a gradient averaging no more than 1 m km⁻¹.

stabilize the channel floor; it is the essence of most braiding that bedforms are subject to virtually constant reworking. It was once a common view that braiding indicates an overload of the stream leading to aggradation, but it now seems that such a pattern has prevailed for substantial periods on some streams with little or no deposition.

During falling discharge on many braided rivers, the gravel bedload begins to accumulate in, and gradually choke, the active channels. Field observation suggests this material then constitutes the nucleus of bars which grow by later accretion at their downstream end. In braided rivers transporting sand rather than gravel the channel forms assume a somewhat different pattern. Instead of relatively high-standing longitudinal bars, the characteristic feature is the linguoid bar which is rhombic in shape and has a gentle upstream face and much steeper downstream face. Linguoid bars are initiated beneath the water surface at high flow and then progressively exposed as the river-level falls.

Frequent oscillations between erosion and deposition can lead to remarkably rapid shifting of either individual braids or the entire river bed. The former case is particularly well exemplified by glacial outwash streams, and in one well-known investigation in the valley below the Emmons Glacier on the flanks of Mount Rainier in Wash-

Sinuous and meandering channels

Few single-thread streams follow a straight course, observed sinuosities ranging between slightly winding and intricately meandering (Fig. 9.15). Of various criteria proposed for measuring channel sinuosity, the most widely employed is the ratio between stream length and valley length. The ratio of straight or winding streams ranges from unity to about 1.5; the latter figure has conventionally been adopted as the lower limit for meandering, although in a fully developed meander belt the ratio may exceed 4. Problems of definition can arise with incised meanders where there may be no obvious valley length to measure; the length of the meander belt axis is apropriately substituted in such cases. Richards in 1977 argued for the same statistical procedure to be extended to braided rivers, so that the index would then become a measure of the ratio of active channel length to valley length; observation shows there to be an overlap in such an index between a highly tortuous meander belt and a modestly braided river system. One claimed advantage for this innovation would be to emphasize the continuum in the rate of potential energy expenditure per unit valley length as the channel pattern

Fig. 9.15. An intricate series of meanders traced by the River Jordan across its floodplain north of the Dead Sea. The mobility of the river channel is indicated by the large number of abandoned loops, several of which are quite fresh, although others are now marked merely by the pattern of vegetation.

changes from straight through sinuous to meandering and ultimately to braided (Fig. 9.14).

In some straight or slightly winding streams it may be difficult to discern any clear geometrical pattern. However, in others the line of maximum depth may be found to oscillate from one side of the channel to the other in a relatively regular fashion. This is seen, for instance, in some pool-and-riffle sequences where the riffles lie alternately on opposite sides of the bed diverting the main current against each bank in turn. These transverse movements have been invoked by some workers as the prime cause of meandering and there certainly appear to be analogies between aspects of the pool-and-riffle sequence and some of the regularities exhibited by meander belts. However, an alternative way of viewing the sequence is not as a developmental one, but as a reflection of an increasing ability of the flowing water to impose regular channel patterns, presumably as a result of a changing power-to-resistance ratio. It is natural that, as the most distinctive of all stream patterns, meanders have attracted much attention and two approaches in particular have yielded significant results.

The first approach has been concerned with the geometry of the pattern and has evolved a widely used terminology (Fig. 9.16). Analysis of active meander belts on alluvium has demonstrated a number of simple mathematical relationships, particularly between channel width and meander wave length, amplitude and radius of curvature. These are best expressed in the form $L = aW^b$ $A = cW^d$ and $R = eW^f$, but each exponent is so close to unity that simple linear equations are almost equally appropriate. Since channel width is related to stream discharge (see p. 172) many meander belt dimensions are themselves a function of discharge. This fact has been used by several workers in computing the volume of water responsible for shaping abandoned meander traces. Dury made the most elaborate calculations in a study of streams in many parts of the world that have apparently shrunk to a fraction of their former size. The basis for recognizing these streams was that the

Fig. 9.16. (A) Terminology employed in describing meanders; (B) examples of geometrical relationships of meanders in alluvium.

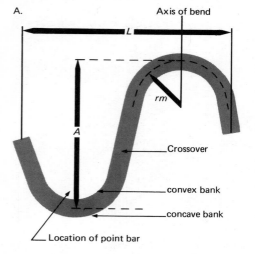

A.

L = Meander length
A = Amplitude
rm = Mean radius of curvature

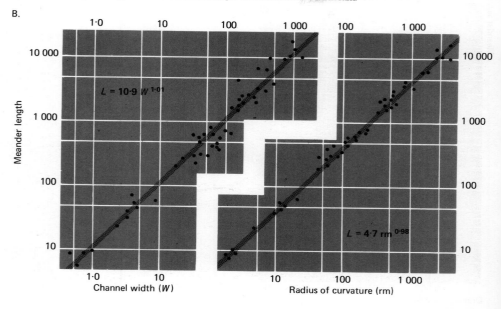

B.

$L = 10.9\,W^{1.01}$

$L = 4.7\,rm^{0.98}$

meanders they now describe are much smaller than those described by the bedrock valleys in which they flow. Two categories of meander were identified, stream meanders adapted to present conditions of discharge, and valley meanders shaped during former periods of much greater discharge (Fig. 9.17). Dury argued that if channel width is a power function of discharge with an exponent of 0.5, and wave length a linear function of width, then wave length and discharge are related by the expression $l \propto q^{0.5}$. If the same arguments apply to valley meanders so that $L \propto Q^{0.5}$, it follows that $Q/q = (L/l)^2$. In many examples the ratio between the wave length of stream and valley meanders proved to be about 5, so that the inferred reduction in formative discharge was as much as twenty-five times. Possible causes of reduction of this magnitude were discussed by Dury, with preference being accorded to decreased precipitation, increased temperature and diminished storminess.

Although it is unnecessary to envisage former mean annual runoff even approaching twenty-five times its present value, the changes required by Dury's analysis remain extremely formidable and often lack supporting evidence from the fields of pollen analysis and sedimentology. In consequence many workers have queried his basic assumptions, with attention focusing in particular on whether stream and valley meanders are strictly analogous. Inasmuch as meander dimensions reflect the interaction between flowing water and channel materials the contrasting properties of bedrock and alluvial deposits might be expected to produce dissimilar patterns. The idea that bedrock and alluvial meanders have different form relationships is a

Fig. 9.17. A mass plot of the wavelength of valley and river meanders against drainage area (after Dury, 1965).

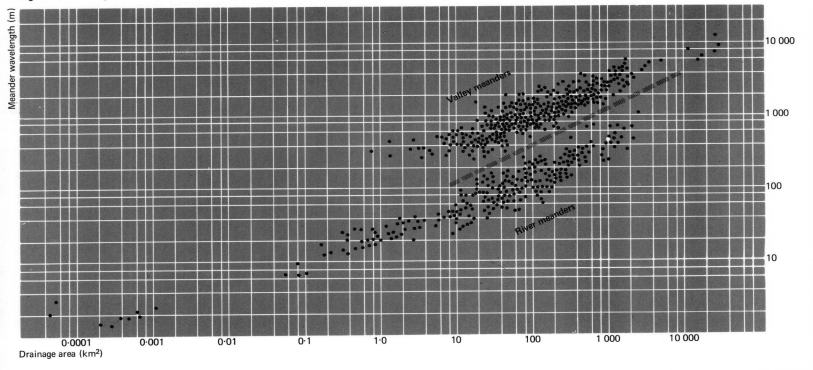

very old one, since it was used over 50 years ago to refute the view of W M Davis that all entrenched meanders are simply floodplain meanders that have been vertically incised. More recently Hack has shown that the wave length of a meander train in Michigan alters as the stream passes from bedrock on to alluvium, while in 1983 Braun demonstrated from an analysis of 1 089 meander loops in the Appalachians that their lengths in shaly lithologies are about twice those in non-shaly lithologies for the same discharge. One may perhaps conclude that hypotheses founded purely upon dimensional similarities need to be treated with caution so long as underlying causes remain poorly understood.

The second approach to meander studies has been concerned with the nature of the flow round the curves. Many investigators have shown that, along the outer bank of each curve, a superelevation of the water surface, directly correlated with velocity and tightness of the bend, generates return currents at depth so that the flow assumes a helical form. Simultaneously, downstream from the inner convex bank a flow separation may develop, with a free-shear layer characterized by spiral vortices. This can reduce the effective width of the main current and concentrate flow against the outer bank. It is also a common observation that the thalweg or line of maximum depth hugs the outer side of each meander and passes from one side of the channel to the other near the cross-over point. There is, however, an important lack of congruence between the channel and flow patterns. The thalweg approaches closest to the bank at the downstream end of each meander, and crosses the centre-line of the channel downstream from the point of inflection. Erosion is thus concentrated on the outer bank below the axis of the meander, while deposition occurs in the form of a point bar at the corresponding position on the inner bank. Such observations help explain the down-valley migration of individual channel bends, but not the beautifully regular pattern they together constitute in many meander trains.

Bagnold identified one possible factor contributing to the dimensional relationships between channel width and meander curvature when, in 1960, he pointed out that the total flow resistance round a 90° bend in a pipe is minimized when the ratio of the radius of curvature to pipe diameter lies between 2 and 3; analogous relationships in open-channel flow might account for meander wave length commonly tending towards eight to twelve times the channel width. Yet it is not immediately apparent why a river should minimize the flow resistance, and more recently Davies and Sutherland have actually suggested that meander curvature could be related to a local peak of resistance in pipe flow when the above ratio has a value of 2.5. Moreover, even if the analogy with pipes explains certain details of form, it still cannot account for the overall pattern. It is the latter on which Langbein and Leopold focused attention in 1966. They theorized that, if hydraulic considerations are assumed to dictate a certain length of channel between two points on the valley floor, the question resolves itself into the property possessed by meanders that would not be possessed by any other pattern. They argued that meanders tend towards a form which may be described as a sine-generated curve in which the direction at any point is a sine function of distance along the meander (Fig. 9.18). Comparison of this family of curves with actual meanders reveals a close fit to examples in both alluvium and bedrock, and ranging in size from laboratory flumes to the Mississippi River. The sine-generated curve has the property of minimizing the variance of the deflection angles along each successive unit length and therefore the amount of work done in turning.

The arguments deployed by Langbein and Leopold constitute one part of a more general thesis that there is a tendency towards a minimization of the variance of a range of hydraulic properties including, for example, energy expenditure per unit area of channel. While the planimetric form of meanders may appear consistent with this general thesis, to test its wider application the authors selected five streams in Wyoming with adjacent meandering and straight reaches of channel. Slope, depth and velocity were measured at closely spaced stations during periods of near-bankfull discharge. As might be expected the variance for these three values proved to be much greater on the curved than on the straight reaches, but when the data were combined to yield values for bed shear and a friction factor DS/v^2 the variance along the meanders proved to be appreciably lower than along the straight reaches. If the minimum variance theory is accepted, therefore, this field investigation might be held to support the inference that a meandering pattern is more stable than a straight or sinuous one.

It should be stressed that the symmetry of meanders can easily be exaggerated, and several workers have advocated techniques for a more detailed planform analysis. Brice, for instance, proposed a fourfold classification of loops into simple and compound variants of

Fig. 9.18. The geometry of meanders: (A) the relationship between a hypothetical meander and the sine curve from which it is derived; (B)–(D) comparisons of actual river meanders with matching sine-generated curves. Above, for each segment of river the sine-generated curve is shown by a dashed line. Below, the corresponding sine curve is shown by a continuous line while the actual river is depicted by plotted points representing the direction of flow as a function of distance along the channel (in part after Langbein and Leopold, 1966).

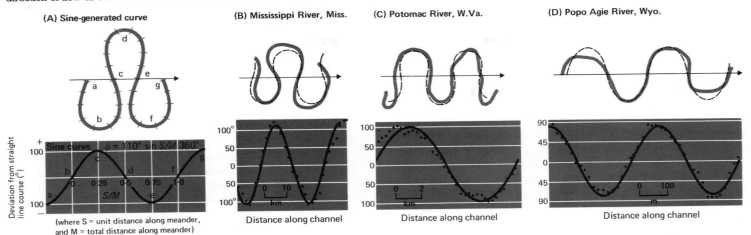

(A) Sine-generated curve

(B) Mississippi River, Miss.

(C) Potomac River, W.Va.

(D) Popo Agie River, Wyo.

(where S = unit distance along meander, and M = total distance along meander)

Distance along channel

basic symmetrical and asymmetrical styles. More recently Carson and Lapointe have contended that the great majority of meanders are asymmetrical since a down-channel displacement of the cross-over point regularly produces a down-valley convexity; where this is very pronounced, curved gooseneck loops are the result (see, for example, Fig. 9.15). As a consequence of the lack of congruence between flow pattern and channel form already alluded to, some degree of asymmetry is not surprising, and indeed it can be argued that strict conformity to a sine-generated curve is inherently improbable where active meander migration is taking place.

Landforms resulting from fluvial deposition

Of the many landforms produced by stream deposition, three that are illustrative of the range of processes involved will first be discussed, namely floodplains, alluvial fans and deltas. Attention will then be turned to river terraces in which active accumulation has ceased and the previously deposited sediments are suffering dissection.

Floodplains

Defined as areas of alluvial land periodically inundated by the streams which they fringe, floodplains have too often been dismissed as flat monotonous features unworthy of much attention. Yet in prac-tice detailed survey will often reveal subdued relief forms providing important clues to the processes of development. Two major groups of processes may be distinguished. The first includes deposition on the floodplain surface by waters escaping from the confines of the channel during periods of spate. As they spread out across the flood-plain, such waters lose speed and deposit as a thin sheet the fine sediment previously carried in suspension. The second group of processes derives from the shifting of stream channels. As a channel loop migrates it leaves behind it point-bar deposits approximately equal in volume to the material eroded from the concave bank. Since these deposits consist primarily of the traction load of the stream they are often significantly coarser than the overbank sediments. The importance of channel migration is clearly attested on many flood-plains by the numerous meander scrolls visible on aerial photographs. On the other hand, where lateral stream shifting is less active, fine overbank deposits may provide a relatively uniform and featureless cover. The balance between overbank sedimentation and channel migration is thus a highly significant control of floodplain form.

Among distinctive floodplain features are levees, sandsplays, floodbasins and oxbows. Levees are ridges trending parallel to the stream channel and may, in the case of a large river, rise 5 m or more above the general floodplain level. Composed of the coarser fraction of the suspension load deposited as soon as the water escapes from its channel, they normally consist of fine sand and silt. Even where relatively continuous their line is interrupted at intervals by transverse depressions, called crevasses, through which the water first spills as the river-level rises during a flood. The fast-flowing current passing through the crevasse can carry comparatively coarse sediment which is spread out across the adjacent floodbasin as a lobate sandsplay. Once they have spilled across the levees, floodwaters may be diverted many kilometres downstream before being able to regain the main channel. Re-entry will often be deferred until the confluence of a tributary which may itself have been diverted by levee development along the trunk river; tributaries occupying such a lateral position have been termed yazoos after the stream of that name on the Mississippi floodplain. It is implicit in such cases that the main meander belt does not occupy the full width of the floodplain. With aggradation concentrated close to the meander belt the position of the trunk river may ultimately become unstable. A lateral shift of the whole meander train then occurs and the river comes to occupy one of its former floodbasins. Known as avulsion, this process has characterized the growth of many of the great floodplains of the world. It is also possible that a form of avulsion in which a rapidly aggrading river spills from its original valley into a neighbouring lower one, offers a better explanation for certain river diversions than the traditional concept of capture.

Within the confines of a meander belt a uniform down-valley progression of channel loops may generate relatively few instances of meander cut-off, but in other cases overrunning of one loop by another is so common that the floodplain becomes littered with oxbow lakes in various stages of infilling. When a cut-off occurs, the abandoned channel segment is first isolated by deposition of bed material across its ends and then infilled with fine overbank sediments,

possibly augmented by organic residues from any plants growing in the resultant lake. Former oxbows may so diversify the composition of an alluvial sheet that the orderly down-valley progression of meander loops is interrupted and many new instances of cut-off induced. It has been pointed out that two types of cut-off can be distinguished. The first is the neck cut-off where the new channel develops across the narrowest point of the meander core, and the second the chute cut-off where the stream assumes a fresh course along the depression between successive scroll bars. In general the former type is characteristic of clay and silt floodplains, the latter of sand and gravel floodplains where overbank sediments are relatively insignificant.

In a study of the Missouri floodplain near Kansas City, Schmudde illustrated the value of watching the sequence of events during a rising flood stage. Natural levees are virtually absent and each bottom, the local term for floodplain fragments isolated by loops of the river, slopes quite steeply down-valley and has a corrugated transverse profile. The steepness arises from the fact that, where the river flows obliquely across the floodplain, the down-valley bank consistently stands 1 m or more above the up-valley bank. In consequence it is the lower end of each bottom that is first inundated by sluggish backwater during a flood. Only when the river rises substantially higher does water begin to spill directly down the valley. Separate zones of rapid current may then attain the velocity required to scour lengthy furrows, while in between there is simultaneous floodwater deposition. In this way the floodplain surface becomes corrugated with a relief of 1–2 m. The forms and processes described by Schmudde probably have their counterparts in many smaller floodplains, and it is certainly true that individual floodplain segments often slope more steeply than the valley floor as a whole; a simple model of regular and even overbank accretion is clearly inappropriate to such cases.

Sections through floodplains in temperate regions often display 1 m or so of fine sediments resting on coarser material. This succession is not necessarily indicative of a long-term change in fluvial regime; the contrast in deposits may simply reflect accumulation by the separate mechanisms of overbank sedimentation and channel migration. It is even possible to visualize a steady-state condition under which a thick layer of sand and gravel could be built up beneath a constant capping of clay. Nevertheless, the upward change from coarse to fine sediments beneath many modern flood-plains does seem to be associated with major climatic fluctuations. The basal sands and gravels often contain structural and organic evidence of aggradation under periglacial conditions. They also display minor abandoned channels indicative of deposition by braided rather than single-thread streams. It is believed, for instance, that the Mississippi assumed its meandering habit less than 5 000 years ago, and that the great thickness of coarse sediments beneath the modern floodplain is attributable to a formerly braided river (Fig. 9.19). Thus the familiar streams of cool temperate latitudes meandering across their floodplains are often recent additions to the landscape. In Australia Schumm detected a rather different sequence of accumulation on the Murrumbidgee riverine plain in New South Wales. The modern river occupies a deep, narrow and meandering channel, but traces of at least two earlier channel systems are discernible. The oldest comprises a series of shallow, wide and straight channels attributable to rivers carrying a coarse sandy bedload during a period of relative aridity. Of intermediate age are channels resembling those of today but with much larger dimensions; these rivers are believed to have been transporting fine clayey sediment, but with discharges computed from meander sizes and application of the Manning equation (p. 164) up to five times that of the modern drainage. Schumm ascribed these large meandering channels to a period in which the total rainfall greatly exceeded that of today.

These examples illustrate the importance of considering not only the current processes but also the evolutionary history of a floodplain. In some cases the actual surface is reworked with such frequency by the migrating river that no features of great antiquity can survive. Schmudde records that one-third of the entire alluvial surface of the lower Missouri floodplain has been worked over in 75 years, while in Maryland it has been estimated that a much smaller floodplain will be completely reworked in about 1 000 years; substantial local variations, however, are implied by the observation of Alexander and Nunnally that the lower Ohio channel has shifted very little during the

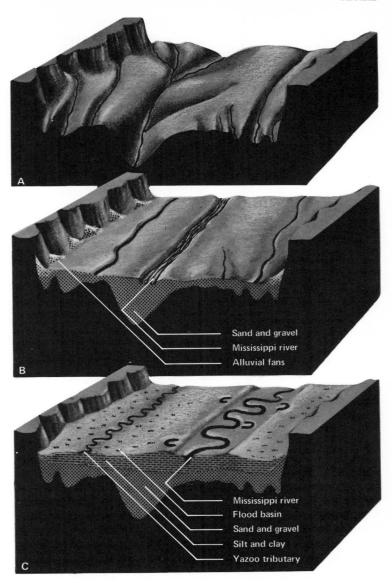

Sand and gravel
Mississippi river
Alluvial fans

Mississippi river
Flood basin
Sand and gravel
Silt and clay
Yazoo tributary

Fig. 9.19. Stages in the recent evolution of the lower Mississippi floodplain (A) entrenchment during lowered sea-level *c.* 20 000 years BP; (B) rapid aggradation by heavily laden braided river; tributaries building up alluvial fans *c.* 10 000 years BP; (C) slow aggradation by modern meandering river: levees, backswamps, yazoos (after Fisk, 1947).

last millennium. In Britain the evidence of archaeological sites on the Thames floodplain has been employed by Robinson and Lambrick to show that that river channel has remained relatively stable for most of the Holocene epoch. The same authors, moreover, are able to relate phases of fine-grained overbank sedimentation to episodes of more intense agricultural activity in Roman and medieval times. They argue that in this case alluviation has been more a response to Man's activities than to climatic change, an inference that has also been drawn from renewed overbank accretion that seems to have begun as late as the nineteenth century in parts of Tennessee.

Alluvial fans

Composed of detrital sediment and shaped like a segment of a low-angle cone, alluvial fans are found in almost all climatic zones. Although best known from arid and semi-arid areas, the single most important factor in their initiation is the abrupt juxtaposition of mountain and lowland. Where a stream emerging from the upland loses its transporting capacity all or part of its load is deposited; at first the loss may be due to reduced gradient, but later it is more likely to be associated with the change from confined channel within the mountains to unconfined channel on the fan. The size of the resulting fan is often directly related to the area of the upland catchment. Along a continuous mountain front coalescent fans may form a compound feature termed a 'bajada' in arid regions. Alluvial sediments sometimes accumulate to total thicknesses in excess of 300 m, the greatest values being attained a short distance below the apex of the fan. The steepest surface gradients also occur near the apex where they may locally surpass 10°, but over most large fans slopes are rather lower and angles of under 5° more normal.

Many individual fans are constructed not only by normal stream flow but also by viscous debris flows or mudflows in which sediment concentrations are sometimes such that the overall density exceeds 2 000 kg m^{-3}. Debris flows, unlike streams, do not generally give rise to selective deposition but result in ill-sorted accumulations that terminate on the fan surface as slightly elevated lobes; cobbles and even boulders can be transported and often appear to become concentrated around the margins of a lobe. The area over which normal stream deposition occurs is determined in part by percolation of water into the fan and in part by diminishing surface gradients. The location of accumulation frequently shifts as old courses are abandoned and new ones adopted. The surface of the fan may consequently be divisible into active and disused washes. The former have a micro-relief of 1 m or so, but after abandonment this gradually diminishes and in an arid environment the undisturbed gravel acquires a coating of desert varnish: thereafter weathering and soil-forming processes are as important as mode of deposition in explaining the character of the surface materials. From this character, or from the nature of soil and vegetation development in humid regions, relative ages may be assigned to different segments of a fan. The actual speed at which the surface of a typical fan is reworked by migrating washes is more difficult to assess. Many fans are certainly very active, and in California one-third of a new 3 km length of road across a fan was affected within 50 years.

The cyclical evolution of fans has provoked much discussion. Whereas many modern examples possess an apex grading smoothly into the upland valley, others are deeply trenched by the main watercourse. These fanhead trenches appear to indicate that aggradation has ceased and a phase of erosion has supervened. After faulted uplift of the mountain front, the initial response is likely to be fan accumulation but, with further erosion of the upland, declining stream gradients may engender dissection of the previously deposited materials. Such a long-term cycle of growth and decay seems plausible, but identification of the stage reached by an individual fan poses many problems. It is tempting to associate a smooth surface lacking a fanhead trench with vigorous aggradation, but in some instances it seems that as much material as is added at the apex by the main feeder is removed from the toe by numerous smaller watercourses. This apparently implies an equilibrium stage with neither net growth nor net loss but a form resembling that of a young aggrading fan. Similar care may be needed in the interpretation of fanhead trenches. Debris flows tend to build steeper surfaces than stream deposition so that if both processes are operative the debris-flow material may be subject to almost immediate stream dissection; trenching on this interpretation has little to do with the age of a fan. However, the depth of dissection arising in this way is probably limited, and deep trenches seem more likely to result from changed environmental controls. Since many fans owe their initiation to faulting, they are particularly prone to continued tectonic movement and many aspects of their morphology, including fanhead trenches, have been ascribed to this cause. A further significant factor may be climatic change. This is especially true in arid regions where even modest changes can induce fluctuations in stream activity. Alluvial fans are evidently complex features subject to a variety of controls and it is often difficult to relate details of form to one particular control. Indeed, laboratory simulation has suggested there may well be threshold gradients capable of triggering alternate phases of aggradation and trenching without any radical environmental alteration.

Deltas

Deltas, like fans, are sedimentary accumulations sited where streams lose part of their transporting capability. However, because deltas are built out into water the factors influencing their growth are rather different. The initial requirement is that the volume of material supplied by the river should exceed that removed by marine or lacustrine agencies. Given that prerequisite, at least four major factors influence the form of the resulting delta: the density and salinity of the water body, the character of the river load, the nature of the coastal processes, and the growth habits of colonizing vegetation. Although deltas have traditionally been subdivided into arcuate and bird's-foot types, this diversity of factors controlling sedimentation actually results in a multitude of different forms.

Thus rivers with a coarse traction load are likely to deposit much of their debris at the point of debouchment and build up relatively steeply sloping structures known as fan-deltas; these are especially common in high latitudes and in glaciated regions. By contrast, the rate at which incoming fine load is deposited varies with the density and salinity of the receiving water. If a muddy river flows into highly saline water, the clay fraction of the suspension load flocculates into larger aggregates which rapidly settle to the bed; the outstanding illustration is the Terek delta which at one time prograded into the Caspian Sea at about 300 m yr^{-1}. On the other hand, dense, heavily charged rivers may plunge beneath the surface of freshwater lakes and carry virtually all their sediment to the deeper parts of the basin. In this way the proportion of fluvial sediment incorporated into a delta varies widely, irrespective of such factors as wave action and tidal range. This is not to deny the importance of coastal processes which often exercise a critical influence on form. In a low-energy environment where the water body is incapable of redistributing the sediment, long embankments or levees may grow outwards from the land on both sides of the major river channel. The process resembles an inflowing jet of water decelerated along its margins with attendant deposition of the load. Bars may also develop on the channel bed and this appears to be a primary cause of bifurcation into separate distributaries. During floods weaknesses in the levees may be revealed, the overflow and enlargement of any breaches being known as crevassing. In this way a composite bird's-foot delta evolves with relatively slow infilling of the inter-channel regions.

In an environment with coastal processes capable of redistributing all the sediment supplied by a river, long isolated levees are precluded and delta progradation takes place along a much broader front. The varied depositional processes include not only overbank fluvial sedimentation but also wave and current action along the coastal margin. Much of the actual prograding takes place by the construction of delicate beach ridges or the growth of offshore bars with enclosed lagoons. The basins surrounded by levees and barrier beaches then become the sites for accumulation of flood deposits. Fluvial, lacustrine and marine sediments frequently interdigitate. Additional to normal river floods may be occasional large-scale marine inundations during storm surges: these can be highly destructive of all earlier depositional forms.

An early view that deltas are confined to coasts of small tidal range is patently untrue, with the Ganges prograding into the Bay of Bengal where the tidal range exceeds 3 m and the Irrawaddy advancing seawards where it is over 5 m. In fact, tidal scouring can have an important effect on surface detail, since the induced currents may be able to mobilize the fluvial sediments accumulating along the distributaries. As on the Ganges or Mekong delta, deep channels then become characteristic, together with complex flaring seawards shaped to accommodate the regularly reversing ebb and flow currents. An important secondary influence is the nature of the vegetation cover. In low latitudes mangroves tolerant of periodic inundation by salt or brackish water trap sediments with their long stilt roots. The rich organic mud becomes the habitat for a varied burrowing fauna that promotes rapid assimilation of any new sediment. Mangrove growth may so encroach on minor waterways that eventually flow is concentrated into a few relatively large channels. The surface of the intervening islands is concurrently built up until they can be colonized by freshwater swamp vegetation. On deltas of higher latitudes the wetland pioneers are usually grasses

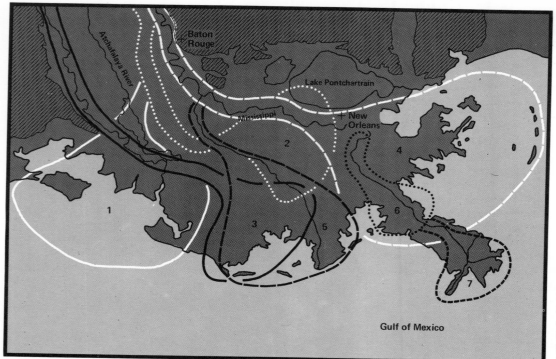

Fig. 9.20. The component deltas of the Mississippi delta system, together with an indication of their individual ages.

Delta	Years B.P.
1 Salé-Cypremort	5 400 – 4 400
2 Cocodrie	4 600 – 3 600
3 Teche	3 900 – 2 750
4 St. Bernard	2 850 – 1 650
5 Lafourche	1 900 – 700
6 Plaquemines	1 150 – 500
7 Balize	500 –

such as *Spartina* or rushes of the *Juncus* family; these also trap sediment during the periodic marine or fluvial inundation. By contrast, drier habitats on the levees and sandy beach ridges or cheniers may soon be occupied by trees; the derivation of the word 'chenier' from the French for oak itself testifies to the characteristic vegetation on the Mississippi delta. Ultimately woodland may spread across the former basins as the land level is raised and the various environmental zones migrate seawards.

It is not unusual to find one segment of a delta stationary or even retreating while another is prograding rapidly. This often arises from periodic floodplain avulsion and associated changes in distributary pattern, many examples of which are now well documented. The best-known case is that of the Mississippi which has constructed five

separate deltas and two smaller lobes in the last 5 000 years (Fig. 9.20); moreover without strenuous efforts at artificial regulation, an increasing proportion of the discharge would now be using the Atchafalaya channel and starting to build a new delta. Prior to development of the Sale – Cypremort delta, the rapid post-glacial rise in sea-level, aggravated by long-term subsidence of the Gulf Coast (p. 65), ensured that the deltaic landforms were submerged almost as quickly as they were formed; presumably most of the surface features of the other great marine deltas of the world are similarly youthful.

River terraces

Fluvial sediments deposited along the valley floor may later be dissected so as to form flat or gently shelving terraces raised above the level of contemporary flooding. Terraces so formed vary in morphology, structure and origin. One consequence is that many different criteria have been employed in attempts at classification. A much-used division is into paired and unpaired terraces. The former result from rapid vertical incision of the river, the latter from concomitant vertical and lateral displacements leaving a terrace first on one side and then on the other at slightly different levels. A second frequent division is into rock and aggradational terraces. A rock terrace consists of a bedrock platform strewn with fluvial deposits no thicker than might be moved during a single migration of the river, while an aggradational terrace is composed of river materials so thick as to imply an important period of accumulation prior to the start of downcutting. A third classificatory system is based upon the presumed initiating factor. This normally distinguishes two primary causes of terrace formation, changes in base level and changes in climate. A variation in either of these is capable of inducing river incision and the abandonment of an old floodplain. By combining the three classifications, which appear to be largely independent of each other, eight different types of terrace might immediately be distinguished.

Probably because of their insular location British workers have undoubtedly tended to emphasize base-level changes at the expense of other initiating factors. Both eustatic falls in sea-level and localized tectonic uplift have regularly been invoked to explain the dissection of former floodplains. The assumption in such models is that a terrace diverges downstream from the modern river, attaining its maximum relative altitude close to the old coastline (e.g. Fig. 9.12). Although such terraces doubtless exist, it is becoming increasingly apparent that they represent only a fraction of the total number. Dating by means of organic materials often shows an aggradation to have been contemporaneous with a low sea-level. In such cases it seems clear that the fundamental control is less likely to have been a change in base level than a change in climate.

The rapid climatic fluctuations of the Pleistocene epoch were especially favourable for terrace development. The growth and decay of ice-sheets induced sea-level oscillations with the same effect on lower river courses in all latitudes. Simultaneously, variations in sediment supply and transporting capacity modified fluvial activity nearer the headwaters of many catchments. The changes in the

upper reaches are likely to have varied according to the particular climates experienced. In temperate regions beset with periglacial conditions, accelerated hillslope movements and highly seasonal stream regimes probably led to fluvial aggradation; in modern desert areas, periodic cooler and wetter phases may have induced downcutting by the better nourished streams; in equatorial areas, less extreme environmental fluctuations may have had a more limited influence on fluvial activity. It is evident that no single model of terrace development can cover all such circumstances, and instead attention here will be focused on two illustrative examples.

The first example concerns the evolution, since the Ipswichian interglacial, of the Avon valley in the English Midlands (Fig. 9.21). Four terraces of this age are normally identified, No. 4 being the highest and No. 1 the lowest. The gravels of all but No. 3 have yielded organic materials indicative of intensely cold conditions. On the other hand, No. 3 has yielded remains of *Hippopotamus* and other mammals that seem to imply an interglacial aggradation. The relationship between terraces Nos. 3 and 4 has been a matter of dispute, and underlines the need to draw a distinction between terrace form and the age of the underlying gravels. Most workers have held that, after an initial phase of valley deepening, a major aggradation began below the level of No. 3 terrace and culminated at the height of No. 4 terrace; the total thickness of the deposits exceeded 15 m. The ensuing episode of downcutting was interrupted by a brief interlude of lateral erosion during which No. 3 terrace was fashioned from the earlier fluvial sediments. On this interpretation the aggradation began during Ipswichian and terminated in early Devensian times. After the fashioning of No. 3 terrace, renewed downcutting ended with the accumulation of the gravels of No. 2 terrace. Radio-carbon assays indicate this terrace deposit to be of mid-Devensian age, with dates ranging from approximately 38 000 to 28 000 years BP. On two further occasions during late Devensian times the Avon deepened its valley as a prelude to brief aggradational episodes; the first ended with the deposition of No. 1 terrace, the second with accumulation of the modern floodplain.

The Avon terraces bear no consistent relationship to the modern floodplain. For instance, No. 4 terrace lies some 12 m above present river-level at Rugby but 24 m above it near Tewkesbury; on the other hand, No. 2 terrace between the same two points maintains a more or less constant altitude of about 6 m. Presumed correlatives of the

Fig. 9.21. The terraces of the Avon–Lower Severn. Each terrace fragment is presumed to belong to one of four former river-levels whose approximate downstream fall in height is indicated by white bands (compiled from Shotton, *Phil. Trans. R. Soc.* 1953; Tomlinson, *Quart. J. Geol. Soc.* 1925; and Wills, *Quart. J. Geol. Soc.* 1938).

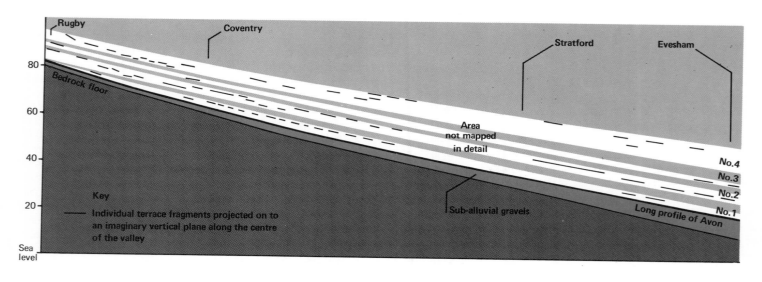

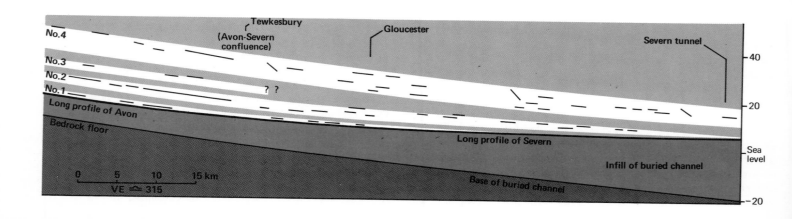

Avon succession have been traced to the Severn estuary 60 km below Tewskesbury. The equivalent of No. 4 terrace converges on the modern alluvial flats until no more than 12 m above them at the head of the estuary. The equivalent of No. 2 terrace similarly declines and is virtually at the level of the flats where the river debouches into the estuary. The correlative of No. 1 terrace plunges below the level of modern alluvium in the vicinity of Gloucester. A deep, sediment-filled channel has been traced from near Tewkesbury downstream to the estuary where it descends to a depth of at least −22 m. Further west the exact line of the channel is imperfectly known, but it clearly falls to at least −30 m within a further 20 or 30 km.

The passage of many of the terraces beneath alluvium, and ultimately below sea-level, conceals much of the evidence necessary to relate them to former marine levels. However, certain inferences can be drawn from the apparent ages of the terraces and known glacio-eustatic oscillations (see pp. 322 et seq.). During the last interglacial, when the gravels of No. 3 terrace were accumulating, sea-level around the Severn estuary probably differed little from that of today. By the time the aggradation responsible for No. 4 terrace was complete, a marked refrigeration of the climate had occurred and presumably sea-level had fallen. It could even be speculated that the initial result of the climatic change was gravel accumulation, but this was superseded by incision as the effect of the eustatic change worked upstream. Later aggradational phases undoubtedly coincided with low sea-levels, but evidence is as yet lacking to relate them directly to either climatic or eustatic fluctuations.

The second example to be discussed is the remarkable succession of cut-and-fill phases in the arid and semi-arid areas of the southwestern United States. Considerations of both geography and dating mean that in this case the influence of sea-level change can be safely discounted. Within the last few thousand years there have been numerous oscillations between aggradation and the cutting of steepwalled ravines or arroyos. Most workers have invoked climatic change as the primary cause, but have failed to agree about the precise mechanism involved. Some have argued for arroyo-cutting during wetter interludes, others during drier interludes. Neither hypothesis receives unequivocal support from palynological data, and several workers have maintained that channel incision begins during periods of greater summer storminess. The most recent episode of arroyo-cutting dates from about 1870 and has converted many shallow washes, formerly crossed by pioneer settlers without difficulty, into steep-walled channels incised over 10 m below the original valley floor. Climatic statistics do not entirely support the ascription of this phase of arroyo-cutting to increased storminess, and overgrazing by stock and other forms of human interference have been claimed as important contributory factors. However, that could not apply to earlier episodes and it must be admitted that the decisive controls are still far from clear. It has even been suggested that cut-and-fill sequences do not reflect climatic oscillations but are an expectable consequence of random meteorological events. Arroyo-cutting would then be caused by 'super-floods' with a very long recurrence interval; aggradation, on the other hand, would take place during intervening periods of less extreme conditions. One result of this cataclysmic hypothesis might be different cut-and-fill sequences in different valley systems, since 'super-floods' could be due to highly localized storms.

Two final points of very general application are worthy of mention. Deposition in one area obviously implies erosion in another and vice versa, so that there will often be sharply contrasting conditions within a single catchment; elaboration of this idea led to enunciation of the principle of 'complex response' (see p. 404). Secondly, once certain threshold values have been exceeded, a longstanding equilibrium may be destroyed and only restored after many consequential changes. For example, if a protective soil cover is gullied during an exceptional storm, the underlying material may be highly susceptible to removal by the same rainfall that had previously been quite impotent; any soil remnants on the inter-gully areas may be continually undercut so that their protective role is effectively ended.

References

Alexander, C. S. and Nunnally, N. R. (1972) 'Channel stability on the Lower Ohio River', *Ann. Ass. Am. Geogr.* **62**, 411–17.

Andrews, E. D. (1979) 'Scour and fill in a stream channel, East Fork River, Western Wyoming', *U.S. Geol. Surv. Prof. Pap.* 1117.

Bagnold, R. A. (1960) 'Some aspects of the shape of river meanders', *U.S. Geol. Surv. Prof. Pap.* 282-E.

Bagnold, R. A. (1966) 'An approach to the sediment transport problem from general physics', *U.S. Geol. Surv. Prof. Pap.* 422-I.

Baker, V. R. and Ritter, D. F. (1975) 'Competence of rivers to transport coarse bedload material', *Geol. Soc. Am. Bull.* **86**, 975–8.

Beaty, C. B. (1974) 'Debris flows, alluvial fans and a revitalized catastrophism', *Z. Geomorph. Supplementband* **21**, 39–51.

Bradley, W. C.., Fahnestock, R. K. and Rowekamp, E. T. (1972) 'Coarse sediment transport by flood flows on Knik River, Alaska', *Geol. Soc. Am. Bull.* **83**, 1261–84.

Brackenridge, G. R. (1984) 'Alluvial stratigraphy and radiocarbon dating along the Duck River Tennessee: implications regarding floodplain origin', *Geol. Soc. Am. Bull.* **95**, 9–25.

Braun, D. D. (1983) 'Lithologic control of bedrock meander dimensions in the Appalachian Valley and Ridge Province', *Earth Surf. Processes Landf.* **8**, 227–37.

Brice, J. C. (1974) 'Evolution of meander loops', *Geol. Soc. Am. Bull.* **85**, 581–6.

Brush, L. M. (1961) 'Drainage basins, channels and flow characteristics of selected streams in central Pennsylvania', *U.S. Geol. Surv. Prof. Pap.* 282-F.

Bull, W. B. (1977) 'The alluvial-fan environment', *Progr. Phys. Geogr.* **1**, 222–70

Carling, P. A. (1983) 'Threshold of coarse sediment transport in broad and narrow natural streams', *Earth Surf. Processes Landf.* **8**, 1–18.

Carson, M. A. and Lapointe, M. F. (1983) 'The inherent asymmetry of river meander planform', *J. Geol.* **91**, 41–55.

Cooke, R. U. and Reeves, R. W. (1976) *Arroyos and Environmental Change in the American South-West*, OUP.

Crickmore, M. J. (1967) 'Measurement of sand transport in rivers with special reference to tracer methods', *Sedimentology* **8**, 175–228.

Davies, T.R. and Sutherland, A. J. (1980) 'Resistance to flow past deformable boundaries', *Earth Surf. Processes* **5**, 175–9.

Denny, C. S. (1967) 'Fans and sediments', *Am. J. Sci.* **265**, 81–105.

Dury, G. H. (1965) 'Theoretical implications of underfit streams', *U.S. Geol. Surv. Prof. Pap.* 452-C.

Ergenzinger, P. and Conrady, J. (1982) 'A new technique for measuring bedload in natural channels', *Catena* **9**, 77–80.

Fahnestock, R.K. (1963) 'Morphology and hydrology of a glacial stream – White River, Mount Rainier, Washington', *U.S. Geol. Surv. Prof. Pap.* 422- A.

Ferguson, R. I. (1986) 'Hydraulics and hydraulic geometry', *Prog. Phyr. Geog.* **10**, 1–31.

Fisk, H. N. (1947) 'Fine-grained alluvial deposits and their effects on Mississippi river activity', *Mississippi River Comm. Waterways Experimental Station*.

Gardner, T. W. (1983) 'Experimental study of knickpoint and longitudinal profile in cohesive, homogeneous material', *Geol. Soc. Am. Bull.* **94**, 664–72 (see also comment by C. G. Higgins, **95**, 122–3).

Gomez, B. (1983) 'Temporal variations in bedload transport rates: the effect, of progressive bed armouring', *Earth Surf. Processes Landf.* **6**, 235–50.

Hack, J. T. (1957) 'Studies of longitudinal stream profiles in Virginia and Maryland' *U.S. Geol. Surv. Prof. Pap.* 294-B.

Hack, J. T. (1965) 'Post-glacial drainage evolution and stream geometry in the Ontonagon area, Michigan' *U.S. Geol. Surv. Prof. Pap.* 504-B.

Hack, J. T. (1973) 'Stream profile analysis and stream-gradient index', *U.S. Geol. Surv. J. Res.* **1**, 421–9.

Hack, J. T. (1982) 'Physiographic divisions and differential uplift in the Piedmont and Blue Ridge', *U.S. Geol. Surv. Prof. Pap.* 1265.

Helley, E. J. (1969) 'Field measurement of the initiation of large bed particle motion in Blue Creek near Klamath, California', *U.S. Geol. Surv. Prof. Pap.* 562-G.

Hooke, R. LeB. (1968) 'Steady state relationships on arid region alluvial fans in enclosed basins', *Am. J. Sci.* **266**, 609–29.

Keller, E. A. (1971) 'Area sorting of bed material: the hypothesis of velocity reversal', *Geol. Soc. Am. Bull.* **82**, 753–6.

Kennedy, V. C. and Kouba D. L. (1970) 'Fluorescent sand as a tracer of fluvial sediments', *U.S. Geol. Surv. Prof. Pap.* 562-E.

Kidson, C. (1962) 'The denudation chronology of the River Exe', *Trans. Inst. Brit. Geogr.* **31**, 43–66.

Knighton, A. D. (1975) 'Variations in at-a-station hydraulic geometry', *Am. J. Sci.* **275**, 186–218.

Kuenen, Ph. K. (1959) 'Experimental abrasion 3. Fluviatile action on sand', *Am. J. Sci.* **257**, 172–90.

Lane, E. W. and Borland W. M. (1954) 'River-bed scour during floods', *Trans. Am. Soc. Civ. Engrs.* **119**, 1069–80.

Langbein, W. B. and Leopold, L. B. (1966) 'River meanders – theory of minimum variance', *U.S. Geol. Surv. Prof. Pap.* 422-H.

Leopold, L. B. and Maddock, T. (1953) 'The hydraulic geometry of stream channels and some physiographic implications', *U.S. Geol. Surv. Prof. Pap.* 252.

Lewin, J. (1978) 'Floodplain geomorphology', *Prog. Phys. Geogr.* **2**, 408–37.

Miall, A. D. (1977) 'A review of the braided-river depositional environment', *Earth Sci. Rev.* **13**, 1–62.

Park, C. C. (1977) 'World-wide variations in hydraulic geometry exponents of stream channels: an analysis and some observations', *J. Hydrol.* **33**, 133–46.

Richards, K. S. (1976) 'Channel width and the riffle-pool sequence', *Geol. Soc. Am. Bull.* **87**, 883–90.

Richards, K. S. (1976) 'The morphology of riffle-pool sequences' *Earth Surf. Processes* **1**, 71–88.

Richards, K. S. (1977) 'Channel and flow geometry: a geomorphological perspective', *Prog. Phys. Geogr.* **1**, 65–102.

Richards, K. S. (1982) *Rivers*, Methuen.

References

Richards K. S. and **Milne L.** (1979) 'Problems in the calibration of an acoustic device for the observation of bedload transport', *Earth Surf. Processes* **4**, 335–46.

Robinson, M. A. and **Lambrick, G. H.** (1984) 'Holocene alluviation and hydrology in the upper Thames basin', *Nature, London* **308**, 809–14.

Sayre, W. W. (1965) 'Transport and dispersion of labeled bed material, North Loup River, Nebraska', *U.S. Geol. Surv. Prof. Pap.* 433-C.

Schick, A. P. (1974) 'Formation and obliteration of desert stream terraces – a conceptual analysis', *Z. Geomorph. Supplementband* **21**, 88–105.

Schmudde, T. H. (1963) 'Some aspects of landforms of the Lower Missouri floodplain', *Ann. Ass. Am. Geogr.* **53**, 60–73.

Schumm, S. A. (1968) 'River adjustment to altered hydrologic regimen – Murrumbidgee River and paleochannels, Australia', *U.S. Geol. Surv. Prof. Pap.* 598.

Schumm, S. A. and **Khan, H. R.** (1972) 'Experimental study of channel patterns', *Geol. Soc. Am. Bull.* **83**, 1755–70.

Smith, N. D. and **Smith, D. G.** (1984) 'William River: an outstanding example of channel widening and braiding caused by bed-load addition', *Geology* **12**, 72–82.

Sundborg, A. (1956) 'The river Klaralven. A study of fluvial processes', *Geogr. Annlr.* **38**, 125–316.

Wheeler, D. A. (1979) 'The overall shape of longitudinal profiles of streams', in *Geographical Approaches to Fluvial Processes* (ed. A. F. Pitty), Geobooks.

Yalin, M. S. (1971) 'On the formation of dunes and meanders', *Proc. 14th Int, Congr, Hydr. Res. Ass.* **3**, C 13, 1–8.

Yalin, M. S. (1977) *Mechanics of Sediment Transport* (2nd edn), Pergamon.

Selected bibliography

Excellent modern texts on the geomorphic activity of rivers are provided by S A. Schumm, *The Fluvial System*, Wiley-Interscience, 1977, and K. S. Richards, *Rivers: form and process in alluvial channels*, Methuen, 1982.

Chapter 10
Aeolian activity

The three preceding chapters have been concerned with the multifarious ways in which water moulds the continental land surfaces into the distinctive pattern of valleys and interfluves. There is, however, one other subaerial agency capable of redistributing a selected fraction of the material originally released from the bedrock by the processes outlined in Chapter 6. That agency is the wind, and although on any global view it is manifestly subordinate to water, its significance is increasingly being recognized even outside the traditional subject of desert sand-dunes. The present chapter therefore reviews the general principles underlying aeolian transport and then seeks to examine the consequences of wind-blown sediments in the contrasted fields of erosion and deposition.

Principles of aeolian transport

To be fully effective the wind requires a range of favourable circumstances to coexist. The most obvious needs are a supply of non-cohesive sediments of an appropriate grain size; the absence of a vegetation cover that would otherwise bind the surface materials together and obstruct the free flow of the air; and, thirdly, atmospheric conditions to create wind velocities adequate to set the available material in motion. Such circumstances are most likely to occur under extreme arid or frigid climates, or, at a more local scale, where other geomorphological processes leave spreads of sediment temporarily open to the atmosphere. The range of particle sizes in aeolian sediments is strictly limited. In wind-blown dunes the diameters of the sand grains rarely exceed 3 mm and yet infrequently fall below

0.05 mm. The latter figure does not, of course, represent the lower limit of wind-blown sediments since finer materials, commonly known as dust in arid regions, may be transported long distances while suspended in the atmosphere. The upper size limit for full suspension is believed to be about 0.2 mm. This means that dune sands often incorporate a size fraction theoretically capable of being lifted into suspension but which has not been removed by aeolian action; this residue of fine particles may be attributable in part to variations in the local wind speed.

The distinction between sand and dust in aeolian activity is a vital one since it controls the distances over which material may be dispersed. In effect the finest dust, once in the upper atmosphere, is unlimited in its potential travel since it can circulate right round the globe. This possibility is perhaps best illustrated by the well-known phenomenon of fine volcanic dust, once ejected high into the atmosphere, acting as a veil to incoming solar radiation at all points on the earth's surface. Intercontinental transport is exemplified by the Saharan dust plume which regularly propagates very fine aeolian material from West Africa to the West Indies; it has been estimated that each year between 60 and 260×10^6 tonnes are carried out over the Atlantic Ocean. Much more rarely Saharan dust has been identified over the British Isles when, after heavy precipitation, it has left such objects as cars covered with an extremely thin red film. By contrast, owing to their lack of transport in suspension, dune sands cannot traverse even narrow bodies of water, although unlike streams they can move upslope and so are less constrained by relief.

Thresholds of particle movement

The preceding paragraphs indicate some of the natural conditions conducive to aeolian activity. However, the topic of sediment transport by the wind is one which lends itself to laboratory investigation by means of wind tunnels. The pioneering studies in this field were undertaken by Bagnold in the 1930s and it is on his classic work *The Physics of Blown Sand and Desert Dunes* that the following recapitulation is based. He initially demonstrated, from calculation of the appropriate Reynolds numbers (see p. 162), that the relevant air flows are invariably turbulent. Plotting measured mean velocity against height above the surface, he showed a logarithmic relationship to apply, with the zero intercept just above the surface (Fig. 10.1). The thickness of the underlying stationary sublayer is

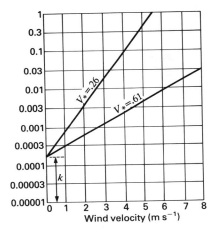

0 1 2 3 4 5 6 7 8
Wind velocity (m s^{-1})

0 1 2 3 4 5 6 7 8
Wind velocity (m s^{-1})

Fig. 10.1 Measurements of two wind velocity profiles immediately above ground-level, on the left plotted with height on a linear scale, and on the right plotted with height on a logarithmic scale. Note that, in the latter case, the intercepts with the line of zero velocity occur at a height of just over 0.000 1 m; this latter value is a function of surface roughness. Each of the lines on the right is assigned a value, known as the drag velocity, indicated by the symbol V_*. This latter is proportional to the rate of increase of wind velocity with the log-height, i.e. to the tangent of the angle between the velocity ray and the height ordinate on the above figure; a proportionality factor of 5.75 is found to apply. Thus the fluid velocity, V, as measured at height z, the gradient V_*, and the roughness constant k are all related by the equation $V_z = 5.75 \ V_* log(z/k)$.

independent of velocity but is related to the surface roughness in such a way that it is approximately equal to 0.03d, where d is the diameter of the particles composing the bed. One consequence is that the drag stress increases as a direct function of velocity: $\tau = \rho V_*^2$, where ρ is the density of the air and V_* is a value known as the drag velocity. The last-named is found to be proportional to the rate of increase of the wind velocity with the logarithm of the height or, in other words, to the tangent of the angle between the velocity ray and the vertical ordinate in Fig. 10.1.

If the drag stress can be calculated in this way, it needs to be compared with the force required to dislodge an individual grain resting on the sand bed. At the instant of displacement the grain is turned about its leeward point of contact with adjacent grains (Fig. 10.2). Movement will occur when the magnitude of the drag force equates with that required for dislodgement. Bagnold calculated the threshold drag velocity as equal to

$$A(\frac{\sigma - \rho}{\rho} \ gd)^{0.5}$$

where A is a constant, g is gravity, σ and ρ are the densities of the grain and air respectively and d is grain diameter. Experimental evidence indicated A as having a value of 0.1 for air which yielded fluid threshold values as plotted in Fig. 10.3. However, the above equation only applies to particles with a diameter in excess of 0.1 mm. Below that figure the surface becomes so smooth that the coefficient A is no longer constant but increases as the grains become smaller and smaller. Bagnold described how a layer of fine Portland cement powder did not move even when subject to a wind velocity that should theoretically be capable of displacing material up to 4.6 mm in diameter (cf. p. 167 for the corresponding effect in water). Of course, this does not mean that the air is incapable of transporting such material, but simply that it may be unable to initiate motion. If fine dust is disturbed by some other agency, the wind may then carry it long distances. This is well illustrated by the powdery material disturbed by a vehicle or a flock of animals being blown into a large dust cloud where previously the same wind had appeared totally impotent.

It will be noted that on Fig. 10.3, in addition to the line for the fluid threshold already discussed, there is a second line denoting an impact threshold. Thus far the arguments deployed have been closely analogous to those for fluvial sediment, but the extra line for an impact threshold signifies an important difference between air and water transport. By means of high-speed photography in his wind tunnel Bagnold was able to plot the trajectory of displaced sand grains. He showed that, once in motion, a sand particle striking a stationary grain may either itself bounce upwards or else eject the

Fig. 10.2. Principles of aeolian sand transport. On the left the fundamental considerations in initial grain movement are outlined. The analysis closely follows that for non-cohesive fluvial sediments on p. 166, but relates incipient motion to the drag velocity of the wind as outlined in Fig. 10.1. On the right the nature of sand saltation is examined and its relationship to the drag velocity once particle movement has started is outlined.

Initiation of movement

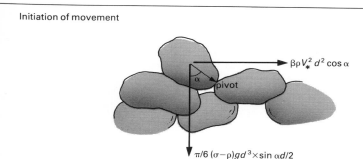

Saltation

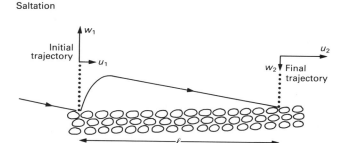

Referring back to Fig. 9.5, it can be seen that the turning moment about the point of pivot is $\pi/6\,(\sigma-\rho)gd^3\times\sin\alpha d/2$

where σ and ρ are the densities of the grain and air respectively. This has to be equated at the onset of movement by the drag stress which, according to Fig. 10.1, is ρV_*^2. Bagnold argues that the proportion of this acting on the projected area of a grain will vary as a function of the square of the grain diameter, i.e. $\beta\rho V_*^2\,d^2$, where β is a coefficient reflecting such factors as temporary turbulent velocities, position of the grain relative to the surface and the height at which the drag force acts. The turning moment is therefore given by $\beta\rho V_*^2\,d^2\cos\alpha$.

Equating these two moments yields for the critical value of the drag velocity,

$$V_{*t}=A\sqrt{\frac{\sigma-\rho}{\rho}\,gd}$$

and for the critical value of the velocity as measured at height z (from Fig. 10.1)

$$V_t=5.75\,A\,\sqrt{\frac{\sigma-\rho}{\rho}\,gd}\times\log\frac{z}{k}$$

In these expressions the coefficient A is found to have a value of 0.1 for grains with diameters exceeding 0.1 mm (see Fig. 10.3 which is essentially a graphical plot of the first of the above expressions).

The above sketch depicts the characteristic path of a saltating sand grain as recorded photographically. Initial movement is vertically upward, and the grain acquires most of its forward momentum from the wind while travelling near the top of its path. The momentum extracted from the air will be $m(u_2-u_1)$, where m is the mass of the grain, and this may be regarded as a loss per unit length of $m(u_2-u_1)/\ell$. Since u_1 is extremely small, this simplifies to mu_2/ℓ. For a large number of saltating grains of mass q passing over a unit width of bed in a second, the wind will lose momentum equal to $q(\bar{u}/\ell)$. This, of course, is a measure of the drag exerted on the wind by the moving sand, and is conventionally designated as τ'. This is equal to $\rho V_*'^2$, where V_*' refers to the altered wind velocities above the level of constant velocity (k') (see Fig. 10.4). One may therefore write $\tau'=\rho V_*'^2=q(\bar{u}/\ell)$. The ratio ($\bar{u}/\ell$) is found to approximate to g/w_1 so that $q=(\rho/g)\,V_*'^2 w_1$; similarly, w_1 is found to be related to the drag velocity V_* with a proportionality factor of 0.8. This yields transport by **saltation** equal to $0.8\rho V_*'^3/g$; measurements show that such saltation accounts for only three-quarters of the total transport rate, the other fraction being contributed by surface creep.

immobile grain into the moving air. In both cases the uplifted grain attains a forward speed of the same order as that of the surrounding air. On descending, normally at an angle of between 10 and 16°, it will strike the surface with sufficient momentum either to bounce again or to eject another grain to a comparable height. This saltation process is thus self-sustaining, deriving its energy from the moving carpet of air. The resultant drag is very large and exercises a major control on the velocity profile. Under steady conditions an equilib-

rium is established so that, whatever the velocity profile, at a certain height above the surface, generally between 0.3 and 1.0 cm, the velocity remains constant. This means that the wind speed very close to the surface is actually reduced as the wind above is strengthened. In this way the velocity at the surface is kept less than the fluid threshold so that the wind is unable to pick up any more grains by its own direct action. By experimenting Bagnold found that, for grains of a diameter below 0.25 mm, the impact threshold is given by the

Fig. 10.3. Variation of threshold drag velocity with grain size. The upper line is a plot of the velocities required to initiate motion by virtue of the induced drag stresses; the lower line is a plot of the velocities required to sustain motion under the impact of saltating grains.

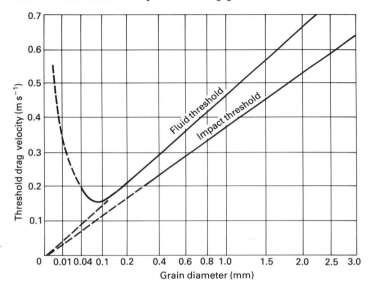

Fig. 10.4. Field measurement of the change in wind velocity with height, after initiation of sand movement. Note the similarity in form with Fig. 10.1, although the value k' designates a level of constant rather than zero velocity. Similarly, the velocity rays designate values for V_*', rather than for V_* as on Fig. 10.1 where no sand movement is involved.

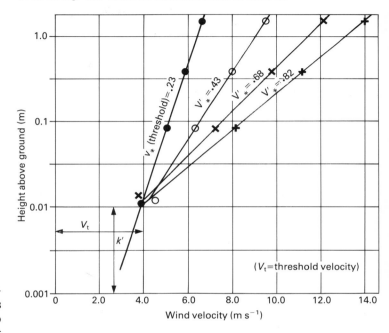

same equation as quoted in the previous paragraph, but with the coefficient A having a value of 0.08 instead of 0.1. The consequences in terms of measured wind velocities are shown in Fig. 10.4. Two further aspects of grain impact deserve mention. The first is that sand grains may disturb dusty material and thereby allow it to be picked up and transported by suspension. The second is that the constant bombardment by saltating grains induces a surface creep of particles that may be far too large to be moved by the wind alone; material moved in this fashion may have a diameter up to six times that of the saltating sand.

Rates of sediment transport

Whereas streamflow is essentially unidirectional, wind may blow from any point of the compass and the associated movement of sand grains is similarly unconstrained. Both the wind speed and its duration from different quarters need to be taken into account if the net amount and

direction of aeolian transport are to be understood. One of the primary objectives of Bagnold's wind-tunnel experiments was to determine the sand transport rate under different wind velocities. From a sequence of theoretical calculations, based primarily upon the probable trajectories of saltating sand grains, and from a range of experimental data, he derived an equation relating the transport rate to particle size and wind velocity as measured at a particular height. One such graphical expression of this equation is shown in Fig. 10.5 where it can be seen that there is an exceedingly sharp rise in total sand movement with only modest increases in wind velocity. As Bagnold himself points out, a wind blowing at 16 m s⁻¹, or 58 km hr⁻¹, will move as much sand in 24 hours as would be moved

Fig. 10.5. The variation in sand transport rate with changing wind velocity. The material is assumed to be average dune sand and the wind velocity is that occurring 1 m above the ground.

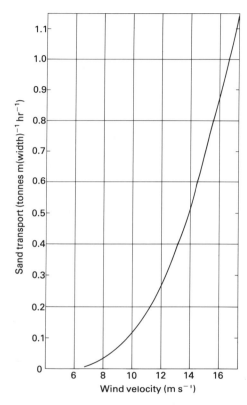

in 3 weeks by a wind blowing at 8 m s^{-1}, or 29 km hr^{-1}.

Although later workers have modified and refined some of Bagnold's methods, his calculations have proved remarkably enduring and are still used as the basis for many modern studies. Bagnold himself in 1953 suggested a simplified formula that might be used to predict relative sand movements from standard meteorological measurements:

$$Q = \frac{1.0 \times 10^{-4}}{\ln(100z)^3}\ t(u-16)^3$$

where Q is tonnes of sand per m width, z is the height of the velocity measurement and t is the number of hours that the wind of u km hr^{-1} blows. The value of 16 in the above equation reflects the belief that a figure of about 16 km hr^{-1}, as measured in standard meteorological observations, reflects the threshold velocity for most desert sands. At a later date Hsu proposed as an alternative formulation

$$Q = (4.97d - 0.47)\ \frac{0.4(V-2.75)^3}{\ln[z(gd)^{0.5}]}$$

where d is the mean grain diameter in mm and V is the hourly averaged annual wind velocity in one direction. Recent tests with fluorescent tracers have cast some doubt on the accuracy of both Bagnold's and Hsu's equations, but if a satisfactory relationship can be defined Q can then be summed for winds from each direction and the results plotted as a so-called 'sand rose'. Since the 1970s this procedure of constructing sand roses has been adopted in attempts to compare the presumed direction and rate of sand transport with dune morphology as revealed on satellite imagery.

Wind erosion by abrasion and deflation

Ideas among geomorphologists concerning the erosive capabilities of the wind have undergone violent shifts in the past. Some early workers claimed that the wind might be capable of the planation of vast continental areas, but partly as a reaction against that view there was a later period when wind erosion was almost entirely ignored as an effective geomorphological process. Modern studies have resurrected the idea that wind, in certain limited circumstances, can be a significant agent of erosion, and there remains scope for debate about its possible role in the evolution of a number of features still defying explanation by other agencies. Two different processes have been invoked to explain aeolian erosion, abrasion and deflation, and each requires separate consideration.

Abrasion

It is almost universally acknowledged that strong winds, armed with small, hard sand grains, are capable of abrading exposed surfaces, but most workers envisage such action as restricted to polishing and streamlining bedrock forms initially shaped by other means. Sand-blasting is, of course, an accepted industrial process, but it is difficult to translate its effectiveness in such highly artificial circumstances to the diverse conditions of the natural environment. Wind-tunnel investigations have provided a closer laboratory simulation and several workers have demonstrated in this way the potentiality of abrasive action on relatively soft rocks. Controlled field investigations have also been attempted, and in 1952 Sharp set out a large experimental plot near Palm Springs in California where both natural granite boulders and artificial blocks were subjected to the abrasive action of the wind. Small rubber strips cemented to the upwind faces of the boulders protected part of the original surface which could then be used as a datum from which the cutting rate could be measured. However, although abrasion was undoubtedly occurring, field monitoring proved difficult because, even after 10 years, the results were too small to be measurable. On the other hand, a common brick after 11 years had had its upwind face cut back by a maximum of 57 mm, and its top fluted and grooved. An experimental leucite rod in approximately the same period had experienced a maximum wear of 0.97 mm, the point of greatest abrasion being situated 23 cm above the ground.

It can be theorized that sand-blasting is likely to be most effective where there is an adequate supply of hard, angular grains, transported by a wind of high velocity, and bombarding a relatively soft rock surface. The work on sand transport described earlier in this chapter has assumed a virtually infinite supply of sand, and as seen this has the effect of severely restricting the wind velocity close to the ground. One consequence is that the rate of abrasion will reach its maximum a short height above the surface, and small stones actually resting on the ground may be relatively immune. The overall effectiveness of sand-blasting could well increase where there is only a limited supply of abrasive material since wind speeds close to the ground will not then be reduced to the same extent. The consequences of sand-blasting may be seen in a range of small-scale features that have been described not only from desert regions but also from high-latitude periglacial areas. One of the most common is a polishing of exposed rock surfaces. This may be seen, for instance, on pebbles which develop a distinctive sheen quite unlike anything resulting from water transport. In its extreme form this polishing can confer on large flat areas of bedrock a reflectivity that, in a low-angle sun, makes them appear almost mirror-like. The cutting of new faces in rock materials is best exemplified by the production of ventifacts. These pebbles, on which at least one fresh facet has been imposed by prolonged wind abrasion, can assume several different shapes. The best known is that of the dreikanter, rather like a three-sided pyramid, but other forms such as the einkanter with just one sloping facet can also occur. The conditions needed to produce a dreikanter have attracted much attention, especially whether the form indicates winds blowing from several directions or whether the pebble itself has been periodically disturbed. On some gravel-covered surfaces it is possible to see thousands of einkanters so aligned that their facets all face into the dominant wind. Such cases are instructive because they imply that the cutting of the facets has occurred since the last time the pebbles were moved by intermittent desert floods.

On a slightly larger scale, wind abrasion may flute and groove bedrock surfaces, especially where the wind is funnelled through a restricted opening. In southern California vertical fluting up to 1 m in length has been described from steep faces of gneiss and schist, while in Egypt some widened joint faces near the summits of limestone scarps carry prominent horizontal grooving several metres long. Such features, however, are still mere modifications of existing forms, and it is not difficult to envisage their production by wind abrasion. It is when the scale of the grooving becomes much larger that the ascription to sand-blasting on its own becomes more dubious. For that reason, such features will be reviewed only after the subject of deflation has been introduced.

Deflation

Literally translated as the act of blowing away, the term deflation can be applied to a very wide range of aeolian activity. However, it is used here primarily with reference to the blowing away of material either from bedrock or from sediments earlier deposited by some agency other than the wind. An example is provided by the removal of the fines from water-laid gravels, leaving behind a surface layer enriched in pebbles and helping to form what is known as a desert pavement in North America, a gibber plain in Australia and a reg or serir in North Africa. The coarse fraction is acting here as a protective armour that prevents further deflation taking place, but in other deposits that protection is not present and deflation may eventually remove a substantial thickness of material. An illustration of this latter situation is provided by playas in which dried surface sediments are scoured by the wind to produce a deflation hollow whose base may extend to, but not below, the local water-table.

It is often, though not invariably, from old lake sediments that the forms known as yardangs are fashioned. These are salient ridges, from a few metres to a kilometre or more in length, elongated parallel to the dominant wind direction, and having high upwind faces and long pointed tails. Until recently they had been the subject of comparatively little research, and the factors controlling such aspects as their location and form are still poorly documented. In essence they are streamlined residuals left by the removal of the sediment in the intervening hollows. The process by which that initial removal occurred could well be deflation, although many workers have seen in the streamlined form the later imprint of abrasion by wind-driven sand. Until further work has been completed it may be better to regard yardangs as wind-eroded forms without attempting to specify too closely the detailed processes involved; some modern experimental work points to abrasion at the windward end being accompanied by deflation at the downstream end. Among the finest yardangs are those in the Ica valley of southern Peru where individual examples are up to several kilometres in length and about 100 m high. They are here fashioned from fine-bedded white siltstones of upper Oligocene to upper Miocene age.

It is, of course, possible to view an area covered with yardangs as being grooved on a very large scale. Until recently most grooving of such dimensions had been described either from glaciated areas where non-aeolian processes might be invoked (see p. 228), or from a limited number of periglacial and subhumid regions where the pattern might conceivably be due to stabilized linear dunes. However, the advent of satellite imagery has now revealed dramatic illustrations of huge grooving in parts of the central Sahara. Eroded in Palaeozoic sandstones, the grooves are between 500 and 2 000 m wide, up to 30 km in length and cover an area of at least 90 000 km². The orientation of the grooves parallel to the direction of other aeolian features leaves little doubt about their wind-eroded origin. Disaggregation of the sandstones under the present severe climatic regime may increase the erodibility of the bedrock, but it is difficult to envisage sand abrasion as not contributing significantly to the scouring action of the wind. Yet the reasons for the spacing and dimensions of the grooves still remain to be determined.

One other puzzling feature of desert morphology that has at times been ascribed to wind erosion is the presence of huge enclosed basins for which no other satisfactory explanation has ever been found. The outstanding example is the Qattara Depression in the Western Desert of Egypt which covers an area of some 18 000 km² and descends 130 m below sea-level, but many smaller basins fall into the same category. To suggest that wind by itself was responsible for excavating such enormous basins is almost certainly incorrect, but that it played a significant role in their long-term evolution now seems increasingly likely.

Wind deposition

Dust

Material fine enough to be carried long distances suspended in the atmosphere may be deposited over a region as a rain of extremely small dust particles. However, over much of the earth's surface the quantities of material exported by deflation are similar to those brought in from outside so that a balance is established between what are, in any case, diminutive amounts. Closer to potential source regions of abundant dust, and particularly on the leeward side where there is a strong prevailing wind direction, there may be a net accumulation. If this positive balance is small, the additional material may be absorbed into the surface layers, for instance by soil-forming processes, without constituting an identifiable separate stratum. Where the net balance is larger, the wind-blown dust will accumulate as a distinct superficial layer that may, over a prolonged period, build up to a substantial thickness. An illustration of a modern area of slow accretion is provided by measurements at Kano in Nigeria where, by means of special dust traps, it was shown during the 1970s that the net deposition rate due to harmattan winds from the Sahara amounted to just over 0.1 mm yr^{-1}. Since larger particles will normally travel shorter distances before being deposited, the accumulated sediment will tend not only to thicken as the source region is approached but also to become appreciably coarser. In areas of high net accumulation most of the material is likely to be of silt size and then constitutes the deposit known as loess. As pointed out in the earlier discussion of threshold velocities, these fine-grained particles, although capable of transport in suspension, are not readily set in motion once they have settled to the ground. Moreover, in areas of deposition wind speeds close to the ground could well be reduced by a vegetation cover. The consequence is that loess normally accumulates as a relatively uniform blanket across the underlying terrain and lacks the distinctive bedforms associated with dune sands. Since much of the research on loess has been concentrated in former periglacial areas, fuller discussion of its distribution and properties is deferred until Chapter 14.

Sand

It has been estimated that at the present time about one-tenth of the continental areas between 30 °N and S are covered by active sand deserts. The wind-driven sand, travelling by saltation and creep, is regularly built into bedforms of a range of different sizes. At the smaller end of the scale are ripples, while at the upper end are huge wind-shaped dunes sometimes separately classified as 'draas'. The smaller features are, of course, relatively transient and the constituent sand grains may remain 'deposited' for only a matter of days or even hours; in the case of the draas, on the other hand, an individual grain may remain buried within the sand accumulation for centuries or even millennia. Most aeolian ripples are ballistic in origin, that is, they result from the impact of saltating sand grains. Bagnold in his wind-tunnel experiments found a close correspondence between ripple wavelength and the mean distance travelled by saltating grains. He argued that if initial random eddying produces a slight hollow this will tend to be enlarged because the upwind side of the depression will receive little bombardment by saltating sand compared with the downwind side. From this downwind side, therefore, more grains will be ejected and they will fall back to the surface at a distance controlled by their mean flight path. They in turn will eject a fresh set of grains so that the pattern is extended in a downwind direction. A limit to the upward growth of ripples is placed by their crests rising into a zone of faster flowing air where any grains would be subject to rapid removal and deposition in the adjacent trough; this would help to account for the largest and therefore most stable grains generally being found on the crests of the ripples. Two aspects of this so-called ballistic explanation of aeolian ripples deserve mention. The first is that the processes of wind and water bedform ripples are fundamentally different since the impact of saltating grains in water is minimal. The second is that the size of aeolian ripples is strictly limited so that it is not physically possible for them to grow into dunes as was at one time suggested.

Wind-blown dunes may assume a vast variety of forms, and although they have been the subject of many studies no widely agreed system of classification has yet emerged. Some workers have sought to categorize them simply by their morphology, but others have endeavoured to establish a genetic classification by relating form to dominant wind patterns. A primary distinction, recognized by most investigators, is between tied dunes in which the position and shape of the sand accumulation is closely connected with an obstruction in the path of the wind, and free dunes in which the surface form reflects almost exclusively the interaction between mobile sand and fluid atmosphere, with the underlying topography playing, at most, a very subordinate role. Extensive dune-covered areas of the latter type are often referred to as sand seas. In the following account the dune forms encountered in sand seas will be discussed under heads adapted from a scheme proposed in 1979 by a group of American workers (Fig. 10.6). Since they rely heavily in their analysis on satellite imagery it is natural that the planform shape should form a major basis for their classification. However, they also stress the value of identifying the location of all slip faces, that is, the steep leeside slopes down which the sand avalanches as the dune migrates, since these are important indicators of the direction of dune movement. It is also worth noting that there is a simple theoretical relationship between the height of a slip face, the speed of dune migration and the rate of wind-induced sand transport. The advance of a slip face of unit width requires a volume of sand traversing the crest equal to $d \times h$, where d is the distance advanced and h is the height of the slip face. The rate of advance is therefore controlled by the value of Q, the transport rate, normally expressed as a weight per unit width per unit time, modulated by a factor to reflect the packing density of the sand avalanching down the slip face. Expressed mathematically $d = Q/\gamma h$, where γ is the packing factor.

Tied dunes

Any obstruction to the regular flow of the air, whether it be a bush, a small hill or man-made object, is liable to lead to a regular eddying pattern that can locally reduce the carrying capacity of the wind. Of the associated range of dune forms the most common is the simple lee dune trailing downwind from the obstruction, while on the upwind side there may be either a climbing dune resting directly against the

Fig. 10.6. Diagrammatic sketches illustrating some of the major dune forms.

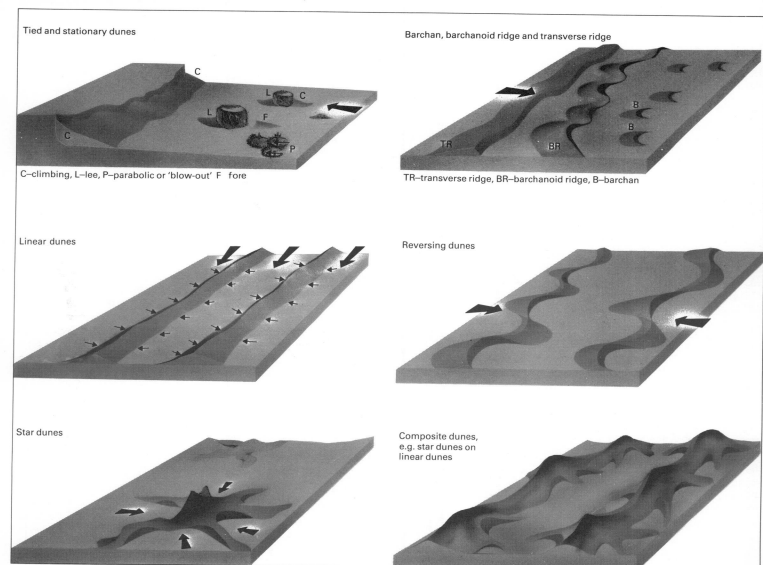

Tied and stationary dunes

C—climbing, L—lee, P—parabolic or 'blow-out' F fore

Barchan, barchanoid ridge and transverse ridge

TR—transverse ridge, BR—barchanoid ridge, B—barchan

Linear dunes

Reversing dunes

Star dunes

Composite dunes, e.g. star dunes on linear dunes

Fig. 10.7 Star and barchanoid dunes superimposed on a linear dune in the Namib Desert illustrate the complex form of many aeolian accumulations. Note the sand-free corridors on both sides of the main ridge. (*Photo E. Tad Nichols*)

obstruction or a foredune separated from it by a sand-free corridor. Lee, fore and climbing dunes can be observed at all scales from minor sand patches around an individual boulder to major accumulations flanking a desert butte.

Another common type of non-migrating dune is the parabolic, so-called from its U-shape. The arms of the U lie parallel to the dominant wind direction, and the dune in this case is tied because of its dependence on a local source of sand. It occurs most frequently in semi-arid regions where parts of a dune complex are stabilized by a vegetation cover. If that cover is locally breached the wind will scour the unprotected patches and carry the sand downwind to construct a parabolic dune. This origin explains why the form is sometimes known as a blow-out dune. Another common location is as a crescent on the leeward side of a playa basin. Examples of this type are especially characteristic of semi-arid areas of Australia where they are known as lunettes.

Barchans, barchanoid ridges and transverse ridges

Each of these forms is characterized by having just one slip face and is therefore associated with unidirectional wind movement. The isolated barchan with its horns trailing downwind is one of the best known and most studied of desert landforms. It occurs where supplies of sand are sparse and is essentially the form assumed by a large mound of sand migrating across a hard desert surface. If one assumes an initial circular mound beginning to move under the influence of a unidirectional wind, the slip face that develops near the margins will be lower than that at the centre, and therefore in conformity with the formula $d = Q/\gamma h$ will advance more rapidly to produce the crescentic shape. Having advanced beyond a certain distance, however, the horns will pass into the eddying effect of the main crest of the dune and so an equilibrium will be established by which the barchan can maintain its basic form as it migrates across the desert surface. Trenches have been dug in small barchans to observe the internal structure and, in conformity with what one might expect, the whole dune, apart from a thin layer on the upwind slope, exhibits a steeply dipping downwind stratification approximately parallel to the modern slip face. Several workers have also endeavoured to measure the rate at which barchans move. Characteristic figures range between 10 and 50 m yr^{-1}, with the latter figure generally applying to the smaller examples.

Barchanoid ridges consist in essence of coalescent barchans arranged in rows. The ridge crest is sinuous and asymmetrical, and between each row there is often a dune-free corridor, although in other cases linking dunes may be present to form a reticulate pattern. The lengths of the crescentic segments along the winding crests of barchanoid ridges have been analysed, and a correlation with ridge width confirms a dimensional similarity with individual barchans. The stratification as revealed in large artificial sections also consists of numerous sets of cross-bedded layers dipping steeply downwind. The sinuosity of the summit crests can vary, but as they become straighter the whole dune becomes more uniform in appearance. In this way the barchanoid character of the ridge gradually diminishes until eventually the feature becomes what is known as a transverse ridge. These morphological changes may well be accompanied by a more consistent sand movement, since the units of cross-bedded strata appear to enlarge with the regularity of the transverse ridge. Several cases have been described which imply that, with a unidirectional wind, the transverse ridge forms where the sand supply is greatest and the individual barchan where it is least.

Linear and reversing dunes

This pair of dune types, although contrasting in many respects, has as a common characteristic two slip faces orientated in different directions. The implication is that the formative winds blow from more than one quarter. Linear dunes, also known as longitudinal or seif dunes, are especially common and in some regions, such as the Kalahari Desert, may occupy over 80 per cent of the sand-covered areas. The largest individual lines may be as much as 150 m high, 1 500 m wide and traceable for distances of several hundred kilometres, but even much smaller dunes can be remarkably persistent. The form of the ridge tops can vary from sharp-crested to smoothly undulating, and there have been suggestions that the diversity in size and shape may reflect at least two separate origins for features broadly classed as linear dunes.

An early proponent of a bimodal wind direction for linear dunes was Bagnold who argued that even a barchan might be converted into an elongated form by periodic winds preferentially lengthening one of the horns. Bagnold stressed that he did not think all linear dunes originated in this way, but he did believe that many exhibited the influence of two powerful winds blowing from slightly different quarters. The main ridge would then trend approximately parallel to the resultant of these two formative winds. In support of this view are observations that slip faces along the crest of some linear dunes do in fact change their orientation at different times of the year to conform with the dominant wind direction. Moreover, in the few trenches so far dug across linear dunes, sets of strata dipping steeply in opposed directions have usually been uncovered. Another group of workers, however, has placed greater emphasis on secondary flow patterns within the moving body of air rather than on regional wind circulation. It is argued that regular spiral vortices develop, with their horizontal axes parallel to the general flow direction. The vortices occur in pairs, alternating between left- and right-handed spirals. Adjacent vortices sweep sand from a central corridor obliquely towards the flanking ridges and this could explain the opposed directions of stratification observed in trench sections. It has proved difficult to demonstrate in the field the existence of the large-scale spirals, but such helical motion is well established in many other areas of fluid mechanics and the general hypothesis has the support of many modern authorities. As already pointed out, there are indications that there may be more than one type of linear dune, and it is therefore quite conceivable that more than one process is involved in their development.

Reversing dunes owe their pair of slip faces to a bimodal wind distribution that drives the sand first in one direction and then later in almost the diametrically opposite direction. As a consequence the dune may grow to a substantial height but lateral migration tends to be very limited. This form of dune was noted at an early date in central Asia and has since been described from several other deserts of the world and also from cold regions such as Antarctica. Since they require a virtual 180° shift in the wind direction they are often best developed in confined mountain valleys where the local topography tends to funnel the wind in just two possible directions. In California the changes in a reversing dune were monitored over a period of 12 years during which it was found that although the total net migration in all that time was only 3.6 m, the position of the crestline shifted backwards and forwards over a zone at least 9 m wide. Sections in reversing dunes have generally revealed very complex stratification; in some cases steep lee-face bedding units that dip in opposite directions have been preserved, but elsewhere these seem to have been eliminated and contorted structures indicative of erratic slumping become relatively common.

Star dunes

These dunes are morphologically unique in possessing at least three distinct slipfaces orientated in different directions. Being the product of complex wind systems they show little tendency to lateral migration, but normally develop a very high central peak with three, four or even more radial arms. Large examples may be several kilometres in diameter, but a more characteristic figure would be about 1 000 m. They have been reported from many different desert regions under a variety of conditions. They may occur as separate mounds apparently randomly distributed, but in other cases they align themselves in parallel rows almost as 'beads' on a linear dune. There has been little trenching of star dunes, but the shallow sections that have been dug have generally shown a very complex stratigraphy, at least near the central nexus.

Sand sheets and dome dunes

Areas of almost flat, featureless aeolian sand have been termed sand sheets by a number of authors. They may cover areas of 1 000 km² or more and are distinguished on the basis that they possess no identifiable slipfaces so that the direction of any sand transport is not apparent in their morphology. They are found in virtually all desert regions, sometimes as exceptionally large interdune areas, but in other cases occupying the whole floor of an intermontane basin. Bagnold in 1941 described trench sections in a sand sheet that showed horizontal laminated bedding. He attributed the failure to develop even surface ripples to the presence of grit or very fine pebbles driven across the surface by the bombardment of saltating sand grains. This admixture of coarse and therefore less mobile particles interferes with the normal sand movement and prevents the growth of the bedforms that would otherwise be expected.

Dome dunes are low circular or elliptical mounds lacking a slipface. They may occur where powerful winds inhibit the upward growth of other forms of dune. They also often consist of coarser-grained sand left on the upwind margin of a dunefield. They vary in size from those having a diameter of only 20–30 m to large examples that may be over 1 km across. Trenches that have been dug often reveal bedding that dips steeply in one direction reminiscent of that observed in barchans. However, across the top of the dome there may also be a layer some 50 cm thick in which the stratification is parallel to the surface. Another structure that has been observed is an unusually large number of cut-and-fill scour channels parallel to the dominant wind direction. These characteristics have led a number of workers to conclude that many dome dunes were originally barchans that have been flattened and destroyed by the action of exceptionally strong winds.

To conclude this section on wind deposition, it is worth emphasizing that the foregoing classification of dunes represents a considerable simplification of the myriad of forms encountered in the field. For example, the range of aeolian bedforms is greatly enlarged by cases where one large dune either carries on its surface smaller dunes of the same type or has superimposed upon it a dune of a different type. Illustrative of the first situation are huge barchans that carry on their windward slopes a group of much smaller barchans, or major linear ridges that are covered with a series of minor such ridges; the second situation is well illustrated by the common occurrence of star dunes aligned along the crest of a linear ridge.

References

Anderson, R. S. and **Hallet, B.** (1986) 'Sediment transport by wind: towards a general model' *Bull. Geol. Soc. Am.* **97**, 523–35.

Bagnold, R. A. (1941) *The Physics of Blown Sand and Desert Dunes*, Methuen.

Bagnold, R. A. (1953) 'The surface movement of blown sand in relation to meteorology', in *Desert Research*, Res. Council Israel Spec. Pub. **2**, 89–93.

Berg, N. H. (1983) 'Field evaluation of some sand transport models', *Earth Surf. Processes Landf.* **8**, 101–14.

Carlson, T. N. and **Prospero, J. M.** (1972) 'The large-scale movement of Saharan air outbreaks over the northern equatorial Atlantic', *J. Meterology* **11**, 283–97.

Goudie, A. S. (1983) 'Dust storms in space and time', *Prog. Phys. Geogr.* **7**, 502–30.

Hsu, S. A. (1973) 'Computing aeolian sand transport from shear velocity measurements', *J. Geol.* **81**, 739–43.

Lancaster, N. (1984) 'Characteristics and occurrence of wind erosion features in the Namib desert', *Earth Surf. Processes Landf.* **9**, 469–78.

Mainguet, M. (1970) 'Un étonnant paysage: les cannelures greseuses du Bembeche (N. du Tchad)', *Ann. Geogr.* **79**, 58–66.

Mainguet, M. (1983) 'Tentative mega-morphological study of the Sahara', in *Mega-geomorphology* (ed. R. Gardner and H. Scoging), OUP.

McCauley, J. F., Grolier, M. J. and **Breed, C. S.** (1977) 'Yardangs of Peru and other desert regions', *Interagency Rep.: Astrogeology.* **81**, U.S. Geol. Surv.

McKee, E. D. (ed.) (1979) 'A study of global sand seas', *U.S. Geol. Surv. Prof. Pap.* 1052.

McTainsh, G. H. and **Walker, P. H.** (1982) 'Nature and distribution of Harmattan dust', *Z. Geomorph.* **26**, 417–35.

Morales, C. (ed.) (1977) *Saharan Dust*, Wiley.

Pye, K. (1984) 'Loess', *Prog. Phys. Geogr.* **8**, 176–217.

Robock, A. (1983) 'El Chichon eruption – the dust cloud of the century', *Nature, London* **301**, 373–4.

Sharp, R. P. (1966) 'Kelso dunes, Mohave desert, California', *Geol. Soc. Am. Bull.* **77**, 1045–73.

Suzuki, T. and **Takahashi, K.** (1981) 'An experimental study of wind abrasion', *J. Geol.* **89**, 509–22.

Vernon, P. D. and **Reville, W. J.** (1983) 'The dust fall of November 1979', *J. Earth Sci. R. Dubl. Soc.* **5**, 135–43.

Ward, A. W. and **Greeley, R.** (1984) 'Evolution of the yardangs at Rogers Lake, California', *Geol. Soc. Am. Bull.* **95**, 829–37.

Selected bibliography

Anyone interested in aeolian activity cannot do better than start with the classic work of R. A. Bagnold, *The Physics of Blown Sand and Desert Dunes*, Methuen, 1941. Among more recent texts that indicate the thrust of recent research are E. D. McKee (ed.) '*A Study of global Sand Seas*', *U.S. Geol. Surv. Prof. pap.* 1052, 1979; M. E. Brookfield and T. S. Ahlbrandt (eds) *Eolian Sediments and Processes*, Elsevier, 1983; and *Dunes: continental and coastal*, Z. Geomorph. Supplementband **45**, 1983.

Like other geomorphic features, glacial landforms can be studied either as they are developing at the present day or in the 'fossil' state after the processes responsible for their formation have ceased to operate. Direct observation of landform development at the sole of a glacier presents obvious problems and many features are known only in fossil form. However, from a study of modern glaciers important inferences may be drawn about the fashioning of the underlying land surface, and many advances in glacial geomorphology have been stimulated by glaciological research.

The area of the continents currently mantled by ice approaches 15 million km². Of this total by far the greatest part is concentrated in two regions, Antarctica with a cover of 12.5 million km² and Greenland with a cover of 1.7 million km². In addition to these two huge ice-bodies there are countless smaller masses ranging in size from those nestling in minor niches on steep mountain slopes to others blanketing thousands of square kilometres of lowland. The variety is immense and offers for study an extremely wide range of glacial environments. The purpose of the present chapter is to examine those characteristics of modern ice-bodies having particular significance for the moulding of the subglacial land surface.

Basic morphology of glaciers and ice-sheets

Surface form

As already emphasized, permanent ice-masses assume a wide diversity of shapes and in consequence it has often proved advantageous to devise a morphological classification even though the boundaries between different categories are sometimes rather blurred. The traditional twofold division is between valley glaciers in which the ice is laterally confined by rock walls, and ice-sheets in which the ice is so much thicker that its surface form can display considerable independence of topographic control. Four common variants of the first class are: (a) cirque glaciers; (b) Alpine valley glaciers; (c) outlet glaciers; and (d) piedmont glaciers. Cirque glaciers are relatively small with many examples little more than 1 km in length; they are distinguished by occupying uniquely shaped rock-walled basins and by a very characteristic pattern of ice movement. Alpine valley glaciers, on the other hand, may attain lengths in excess of 100 km and are normally fed by ice issuing from one or more cirques; their surfaces exhibit many contrasting forms ranging from gentle longitudinal gradients to precipitous ice-falls composed of seracs, and from smooth to deeply crevassed marginal slopes. Outlet glaciers differ from the Alpine type in being nourished by an ice-sheet, and although this may have important consequences for the glacier regime, in terms of pure morphology the distinction is relatively slight. Piedmont glaciers are differentiated by having greatly distended snouts where they protrude beyond any constraining valley walls so that the ice can spread freely over an adjacent lowland; the rapid change in thickness and direction of ice movement often produces an exceptionally intricate crevasse system.

Ice-sheets are divisible into two major classes based upon size. The smaller are known as ice-caps and the larger as true ice-sheets. Ice-caps are relatively thin, and both upland and lowland types may be recognized. In an upland location the ice generally rests upon an undulating plateau and feeds a group of peripheral outlet glaciers. By contrast, a lowland ice-cap occupies an area of subdued relief little above sea-level, and owing to its topographic situation does not normally support major outlet glaciers. Both types share a domed form comparable to that displayed on a much larger scale by true ice-sheets. These latter are of adequate thickness and extent to bury completely topography ranging from plainlands to substantial mountain ranges. Their upper surfaces resemble vast flattened domes in which gradients steepen towards the edges. Analyses of marginal slopes along both the Antarctic and Greenland ice-sheets have revealed a noteworthy degree of conformity. In the absence of gross disturb-ance by partially buried mountain ranges, it is found that the ice surface rises to 1 000 m in the first 50 km from the edge, to 2 000 m in the next 150 km, and to 3 000 m in a further 350 km. These figures provide one possible guide to likely gradients around the edges of active European and North American ice-sheets during Pleistocene times, although in the latter case there is some field evidence that the marginal slopes of the Laurentide ice-sheet were slightly lower.

Subglacial relief

Whereas the surface form of most ice-masses can be mapped by traditional surveying methods, and is often shown on topographic maps by means of contours, determination of the subglacial relief poses special problems. The form of the bedrock surface beneath a small valley glacier can sometimes be estimated with little risk of gross error, but a major ice-sheet so conceals the subglacial topo-graphy that elaborate techniques are required to determine the position of the ice–rock interface. Seismic surveying on a large scale was pioneered by the Wegener expedition to Greenland during the years 1929–31. In this method an array of instruments is used to measure the local travel time of compressional waves emanating from a charge detonated in the upper layers of the ice and reflected from the bedrock surface. After due allowance for the effects on wave velocity of density and temperature variations, plots of the subglacial surface should be accurate to ±5 m. Although suitable for most ice-masses, this method has proved less satisfactory in Antarc-tica where superficial dry snow so attenuates the compressional waves that the record is difficult to interpret. The problem can be partially overcome by detonating charges at depths of 50 m or more, but in a hostile environment the inevitable delays and extra equip-ment are a high price to pay.

Gravimetry has also been employed as a means of mapping subglacial relief. As explained in Chapter 1, it is possible to eliminate one by one the sources of variation in the local gravity field. Since there is no isostatic compensation for purely local topographic features, any residual variation on the gently sloping surface of an ice-sheet will be due primarily to unevenness in the bedrock relief. The rapidity with which a gravimetric survey can be carried out makes the method especially useful for filling in detail between seismic stations. On a smooth ice-sheet, away from the complications intro-duced by nunatak mountain ranges, the technique should be capable of yielding ice thicknesses accurate to ±20 m.

In the 1960s a new survey method was devised. It had previously been noted that planes fitted with radar altimeters were obtaining reflections not only from the ice surface but also from deep within the ice. It was soon realized that this second reflection was coming from the ice–rock interface and afforded a potential means of surveying bedrock relief. Special echo-sounding equipment was developed capable of operating either from aircraft or from surface vehicles and utilizing frequencies that would ensure maximum penetration of the radio waves. The method proved particularly suited to work in Antarctica where the cold ice absorbs little of the transmitted energy; here frequencies in the range 30–300 MHz have been successfully employed. Absorption increases as the temperature of the ice rises, and difficulties have been encountered on some relatively warm glaciers where higher frequencies have been needed in order to obtain satisfactory results. Certain valley glaciers have also posed problems of reflection from steep rock walls, although even here surface vehicles can be fitted with strongly directional antennae and consistent measurements obtained within 50 m of the glacier edge. The method generally has the great advantage of giving a continuous profile, and although allowance must be made for the effect of ice density on the velocity of radio waves results accurate to ±10 m can be obtained.

International cooperation in the study of ice-sheets, combined with the technical advances outlined above, has facilitated mapping of the subglacial topography in both Antarctica and Greenland. Investigations in the Antarctic continent (Fig. 11.3) have disclosed a bedrock relief consisting of two contrasting parts, an undulating plain in the eastern section and a mountain range in the smaller western section. Much of the plain is blanketed beneath 3 000 m of ice (with the thickness possibly exceeding 4 000 m in places), but over the mountains more variable values are found. Were the ice-sheet to melt and isostatic recovery occur, western Antarctica would become a mountainous archipelago, while the eastern part of the continent would be transformed into a elevated plateau; the hypsographic curve for the new land surface would not differ fundamentally from those of the currently unglaciated continents. Greenland exhibits subglacial relief consisting of an extensive central plain near sea-level, where ice thicknesses commonly exceed 2 500 m, flanked by mountains to both east and west. It is characteristic of both Antarctica and Greenland that the main ice-mass rests in a broad, shallow basin fringed by more or less continuous hills and mountain ranges; a similar form presumably characterized many of the areas depressed beneath the ice-sheets of North America and northern Europe during Pleistocene times.

Mass balance studies

For any ice-mass in equilibrium with its environment it is a necessary condition that the gains through accumulation should equal the losses through ablation. In the compilation of a budget for such an ice-mass, accumulation must encompass all those processes by which material is added to the feature; the most important is that of direct precipitation, but others that may locally be effective include avalanching, wind blowing and refreezing of runoff from surrounding land surfaces. Ablation must encompass all processes by which material is removed; the most obvious is simple melting but others include sublimation and iceberg calving. The time-period to which measurements of accumulation and ablation refer must also be specified. This is commonly the balance year which begins in late summer after ablation has reached its greatest extent and the ice surface is at its lowest point, and continues to the time of the corresponding state in the next calendar year. Owing to slight annual variations each balance year will not necessarily be of 365 days' duration, although this should be the average over a long period.

Field measurements

In practice the gross accumulation rate presents such formidable problems of measurement that most investigations concentrate instead on the more easily calculated net value. The net annual accumulation refers to that part of the material added during the balance year that is still present at the end of the ablation season. It can be assessed by measuring the annual incremental volume over that part of the ice-mass known as the accumulation area. On any steady-state ice-mass there must by definition be an accumulation area balanced by an ablation area; the boundary between the two is known as the equilibrium line (Fig. 11.1). Within the accumulation area further subdivisions are often present. In the coldest localities, summer temperatures may remain below freezing so that no meltwater is produced; such a region is referred to as the dry snow zone. It is commonly bordered by the percolation zone in which

Fig. 11.1. Schematic representation of factors in the nourishment of an ice-mass. The same basic principles apply to all ice-masses from a small cirque glacier to a large ice sheet.

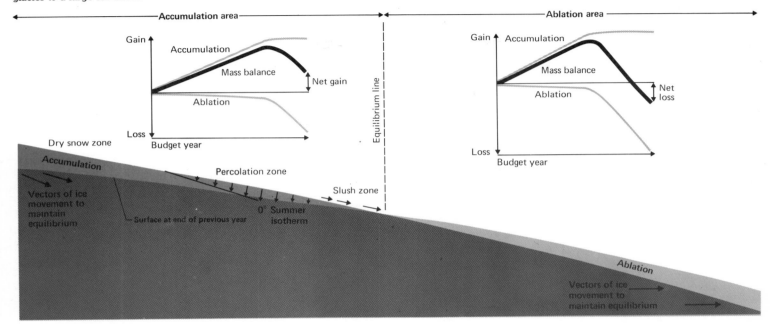

slightly less extreme conditions generate summer meltwater that percolates downwards and refreezes. Adjacent to the dry snow zone the quantity of water will be small and the depth of percolation shallow, but with increased summer warmth the whole thickness of the annual accumulation layer may be permeated with meltwater. Saturation often leads to slush avalanching in a region designated the slush zone; this is the highest location where loss of material occurs by surface runoff.

Such varied conditions clearly demand careful selection of sampling sites if an accurate assessment of net annual accumulation is to be obtained. The usual procedure is to dig pits or take cores at the chosen sites and then to identify the annual incremental layers. This is normally facilitated by thin dirt bands that mark the snow surface at the end of each balance year; these are particularly conspicuous close to rock outcrops which can supply the necessary

fine debris. Another valuable marker horizon is the compact dense ice that results from partial melting and refreezing during the summer season. On the other hand, annual layers in the dry snow zone may prove much more difficult to identify. In central Antarctica, for instance, the most useful marker is often a layer of coarsely crystalline snow known as 'depth hoar' which appears to form each autumn just below the surface. In all net accumulation studies measurement of the annual layers must be followed by determination of snow density, since the thicknesses need to be converted into water-equivalent values.

To strike a balance with net accumulation, ablation must be measured over the entire area below the equilibrium line. The aim is to determine the amount of material lost from the surface of dense ice while ignoring the melting of any superficial snow cover. During spring stakes are normally drilled through the snow deep into the underlying ice. The distances between the tops of the stakes and the surface of the ice can thereafter be monitored at intervals throughout the summer until fresh snow begins to accumulate in the autumn. Many such stakes will be needed for an accurate assessment of overall surface ablation, but the figure obtained, once converted into a water-equivalent value, can then be compared directly with that for net accumulation.

Analyses of mass budgets

If an ice-mass is in equilibrium, its accumulation and ablation totals should theoretically be identical. Yet exact equality can rarely be demonstrated. Several possible reasons may be advanced. The first relates to doubt concerning field measurements which often involve inadequate sampling and neglect of certain elements in a complete budget. A second stems from minor year-to-year variations in both accumulation and ablation; while the former can sometimes be evaluated from the study of the incremental layers, the latter can only be established by protracted field observations. A third possible reason is intrinsic to the ice-masses themselves. An ice-mass adjusts only slowly to changed climatic circumstances so that complete equilibrium may be a relatively rare occurrence. The actual rate of adjustment differs significantly from one glacier to another, generally being much longer for a large ice-sheet than for a small valley glacier.

Given methods of sufficient accuracy it should be possible to categorize ice-masses as having either a positive net balance when accumulation exceeds ablation, or a negative net balance when the reverse relationship holds. The two most important climatic controls are precipitation and temperature, and a change in either may disturb an ice-mass from its equilibrium state. For example, refrigeration will induce a positive net balance occasioned primarily by a reduction in summer ablation. Immediate effects will include the raising of the surface level over both the accumulation and ablation areas, together with a lowering of the equilibrium line. Equilibrium will only be re-established when the additional net accumulation is compensated by increased loss, and this will normally be achieved by advance of the ice margin to produce a larger ablation area. A similar pattern of changes will attend an increase in snowfall without any modification in temperature. In an early review of the theoretical effect of a positive net balance, Nye showed that a 'kinematic wave' of increased discharge will be propagated across the ablation area.

This wave travels approximately four times as fast as the ice itself, and on reaching the terminus produces first thickening and then advance. A negative net balance normally causes the whole ice-mass to thin, with the largest change occurring near the terminus. Theory predicts, and observation confirms, that minor changes in the net balance can result in large shifts in the position of the ice-front.

The intricate nature of the climatic controls, combined with the sensitivity of glacier response, can result in complex patterns of ice advance and withdrawal. During the first half of this century, when most Alaskan valley glaciers were shrinking, a few were actually growing slightly. Such variations in behaviour may arise, for instance, from different temperature and precipitation relationships on opposite sides of a watershed, especially where allied to contrasts in the relative sizes of accumulation and ablation areas. Although for short-term studies these deviations from strict synchroneity can be very important, it remains true that the major advances and retreats of the Pleistocene ice-sheets seem to have been broadly contemporaneous, even on a global scale.

Before leaving the subject of mass balance studies it is worth examining in a little more detail the case of Antarctica. Particular interest attaches to this ice-sheet in view of its sheer size. Ignoring the effect of isostatic compensation, it is estimated that release of all the water locked up in Antarctica would raise sea-level by well over 50 m; conversely, an increase in mean ice thickness of just over 100 m would cause sea-level to drop 4 m, enough to disrupt shipping and radically alter many coastlines of the world. It has even been suggested that cyclical growth and decay of the ice-sheet could exercise a primary control over global climate. Considerable practical importance therefore attaches to the present mass budget.

Unfortunately measurement presents exceptional difficulties. Much of the continent is a dry snow accumulation area with poorly defined incremental layers; it is generally agreed, however, that the interior is one of the driest areas on earth, with an annual precipitation equating to a water depth of no more than 50 mm. Near the margins of the ice-sheet the corresponding value is about 500 mm, yielding a continental mean of approximately 150 mm. The most formidable problems of assessment attach to ablation which occurs mainly by loss from floating shelves where the ice-sheet projects into the Antarctic Ocean. Iceberg calving is the most prolific source of ablation, accounting for as much as 90 per cent of the total loss.

Individual bergs usually measure a few hundred metres across, but occasionally gigantic fragments over 100 km long detach themselves from a shelf. Calving can be monitored by satellite and aerial photography, but less readily assessed is any melting, or freezing-on of new ice, that may occur at the underside of a shelf. Another potential source of error is the loss due to blowing of unconsolidated snow into the sea. Aeolian redistribution of snow is a particular feature of Antarctica, and it has been estimated that each year 10^{11} tonnes are swept into the surrounding ocean by powerful katabatic winds blowing down the flanks of the ice-sheet. In the circumstances conclusions about the mass balance must remain tentative, but most workers believe that the ice volume is either stable or growing very slightly. It should be stressed that an ice-mass like that of Antarctica responds extremely slowly to changed environmental controls, so that it might take several thousand years after a climatic alteration for equilibrium to be fully restored. It is also worth noting that the distinctive forms of ablation around Antarctica detract from its value as a model of the ice-sheets that developed in the Northern Hemisphere during the Pleistocene epoch; in this respect Greenland, with a conventional ablation zone locally 150 km broad, may provide a better analogue than does Antarctica.

Ice temperatures

The importance of ice temperature as a property of glaciers and ice-sheets can scarcely be doubted. Among vital temperature-dependent properties are the rate of strain and the interaction with meltwater. Under the same stress ice will deform much more rapidly near its melting-point than at lower temperatures; this increased strain rate must affect both the velocity of an ice-mass and ultimately its surface form. A very cold ice-mass will neither generate meltwater at depth, nor will it allow any liquid produced by surface thawing to percolate deeply before refreezing; on the other hand, where the ice is close to melting-point water may penetrate far below the surface and even enlarge tunnels by the thawing of the ice-walls. Because of their demonstrated significance, ice temperatures have been the subject of many field investigations, while much effort has also gone into the theoretical modelling of the thermal regimes of ice-masses. It is these two aspects that form the basis of the following discussion.

Field measurements

Measurements show that in a dry snow area where the thermometer remains constantly below 0 °C the temperature of the compacted snow approximates to the mean annual air temperature. At the Soviet Vostok station in Antarctica, for instance, the average air temperature is −56 °C and the compacted snow or firn at a depth of 10 m registers −57 °C. At very shallow depths there is a significant seasonal fluctuation in temperature, but below about 10 m this effect becomes negligible and below 15 m virtually undetectable. As the firn, normally defined as having a density between 400 and 830 kg m^{-3}, is further compacted by burial, its porosity declines, interconnecting air passages are sealed off, and by a process known as sintering there is a gradual conversion into true glacier ice. On high-Arctic glaciers this end-product retains a temperature still very close to the mean annual air temperature.

Under a less extreme climate, a very different sequence of changes can be observed. Each spring, as the temperature climbs and surface thawing begins, meltwater percolates down through the cold snow of the preceding winter. Eventually this water will refreeze, but in doing so it will release latent heat. Since 1 g of water can produce enough latent heat to raise the temperature of 160 g of snow by 1 °C, the effect is often dramatic. On many glaciers the total thickness of freshly fallen snow is warmed until its temperature approaches 0 °C. Such ice-masses are commonly designated temperate, and on some Alpine glaciers the firn has been shown to be 7 or 8 °C warmer than the mean annual air temperature. The presence of refreezing meltwater also accelerates the conversion of snow to ice; on temperate glaciers with abundant meltwater this transformation may occur within 10 years, whereas on some very cold glaciers the process takes several centuries.

The foregoing paragraphs describe two common extremes under which the temperature of the ice produced in the accumulation area either approximates to the mean annual air temperature or lies close to melting-point. Intermediate cases undoubtedly exist where the percolating water warms the snow but fails to bring it fully to the melting temperature. For that reason if for no other, the traditional twofold division into temperate ice-masses, in which the temperature below a depth of 15 m is everywhere at melting-point, and cold ice-masses, in which only submelting temperatures are encountered, has limited utility. The position is further complicated by the contrasts that often exist in different parts of a large ice-mass so that it may be misleading to categorize the whole body as either temperate or cold. This is particularly true when temperatures at great depth within an ice-body are taken into account.

Although there had been earlier theoretical predictions about the likely temperatures at the base of the Greenland and Antarctic ice-sheets, it was only in the 1960s that deep boreholes were sunk that allowed the opportunity for direct measurement. In 1966 at Camp Century in northern Greenland a 1 387 m borehole penetrated the full thickness of the ice and showed that, from −24 °C at a depth of 10 m, the temperature fell slightly to −24.6 °C at 154 m, but then rose steadily until at the base of the hole it was −13 °C (Fig. 11.2). Two years later a borehole at Byrd Station in Antarctica penetrated 2 164 m of ice. From surface values of −28 °C the temperature at first declined to −28.8 °C at 800 m but then rose to −13 °C at 1 800 m. Below this depth it was found impossible to take direct readings, but extrapolation of the measured curve suggested the basal ice would prove to be at its pressure melting-point of −1.6 °C. This view was validated by the presence of a thin film of water at the bottom of the hole, and it has since been inferred from radio-echo-sounding data that there are even subglacial lakes beneath parts of the Antarctic ice-sheet. In the face of this evidence it is obviously unsatisfactory to adhere to a rigid classification of complete ice-sheets as either temperate or cold. Instead, geomorphologists have tended to adopt such phrases as 'warm-based' and 'cold-based' to indicate whether, in a purely local sense, the ice at the rock inter-face is at or below its pressure melting-point; the latter value is used in preference to 0 °C since the transition from ice to water is depressed by approximately 1 °C for every increase in pressure of 13.5 MN m^{-2}.

Modelling the thermal regimes of ice-sheets

The controls on the temperature of newly formed ice within the accumulation area have already been outlined in the preceding section. Thus the primary concern here is with any changes that may occur after that material has sunk deeper within the main body of the ice-mass. As a general rule, ice temperatures cannot fall so that ice that is already close to its pressure melting-point will retain that characteristic throughout its life; it may melt but it will not get substan-

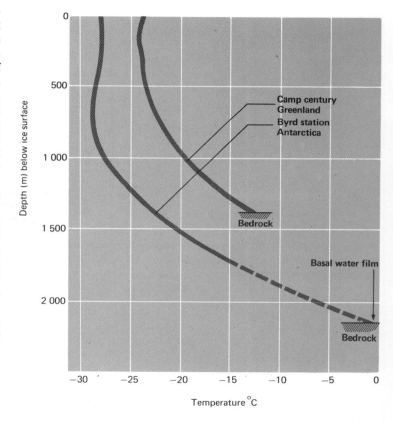

Fig. 11.2. Temperature profiles of the first two boreholes to penetrate the total thickness of the Greenland and Antarctic ice-sheets. In 1981 another very deep borehole, known as Dye 3, was sunk through the Greenland ice-sheet; this reached bedrock at 2 037 m, and recorded a basal temperature of −12 °C (−20 °C at the surface).

tially colder. Cold ice, on the other hand, such as that originating in a dry snow zone, may be progressively warmed during its passage through an ice-mass. Three sources of heat can be identified, the geothermal flux, friction with the underlying bedrock and friction due to shearing within the ice itself. Since most of the internal shearing

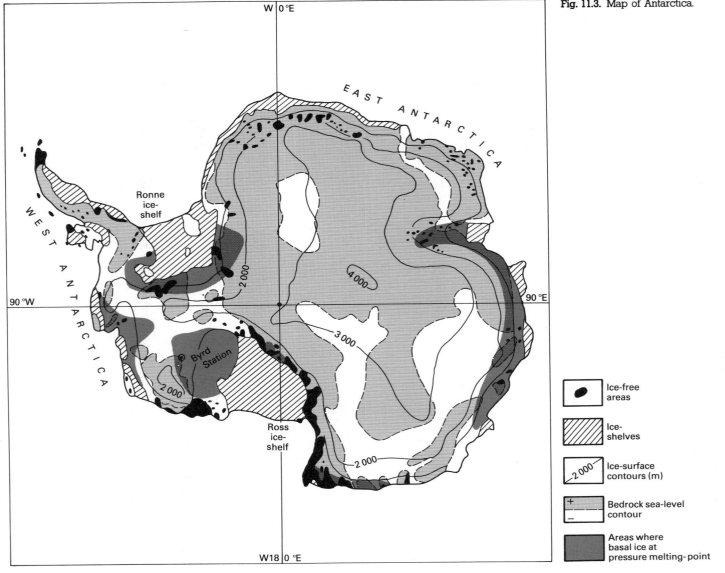

Fig. 11.3. Map of Antarctica.

occurs in the lower layers of an ice-mass, it is evident that all the heating is concentrated near the base. This explains why the upper parts of both the Greenland and Antarctic profiles featured in Fig. 11.2 are virtually isothermal.

Although the geothermal flux can locally be very high, as for example beneath parts of the main Icelandic ice-cap at the present day, over wide areas it is permissible to regard it as a constant. Calculations can also be made of the frictional heat generated by different ice velocities, so that, provided the rate of movement is known, the main heat sources can be roughly evaluated. Yet there remain two other variables influencing thermal regimes that have not so far been mentioned, namely ice thickness and the annual rate of ice accumulation. These are important because, in essence, they control the time available for the warming process to operate. For instance, if accumulation is very fast, cold ice may be carried down and outwards so rapidly that it is never warmed sufficiently to attain its melting-point; conversely, if accumulation is very slow, residence times in the ice-sheet may be such as to ensure that the basal layers everywhere attain their pressure melting temperature.

As early as 1955 Robin had already formulated a simple model of temperature changes at the centre of an ice-sheet based upon the foregoing principles. Subsequent workers have elaborated this model, endeavouring to incorporate the effects of the radial outward trajectories of the ice. These complex three-dimensional models, based upon presumed flow lines, have been used to predict the zones in Antarctica and Greenland where the basal ice will prove to be either at or below the pressure melting-point (Fig. 11.3). An important extension of this work has involved attempts to reconstruct the thermal regimes of now-vanished Pleistocene ice-sheets. As will be emphasized in Chapter 12 the nature of the erosional and depositional processes at the base of an ice-mass are profoundly influenced by the local ice temperatures. Thus for a full understanding of the geomorphic impact of the Pleistocene ice-sheets it is essential to reconstruct their thermal properties. To achieve this aim requires, as a minimum, that realistic estimates be available for the three-dimensional shape of the ice-sheet, the pattern of net accumulation, the directions of ice-flow and the surface temperature distribution. Given these exacting demands, it cannot be claimed that other than very tentative steps have so far been taken in this work and that many problems remain to be overcome; yet the underlying

ideas are extremely important and define a goal that is likely to be more nearly approached as better data become available in the future. Examples of the approach are afforded by the work of Boulton and associates on the British ice-sheet, and by Sugden on the Laurentide ice sheet.

Ice movement

The problem of ice movement can be approached from two independent but related viewpoints. Laboratory experiments may first be performed on ice to determine some of its physical properties, particularly the rate at which it deforms under a given stress. This

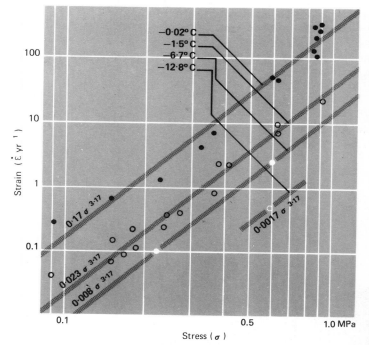

Fig. 11.4. The results of laboratory experiments to measure the strain rate of ice when subject to varying shear stresses. Ice was tested at four different temperatures. The plotted points conform to the general relationship $\varepsilon = k\sigma^n$ known as Glen's power law (after Glen, 1955).

approach has the advantage that such variables as temperature and pressure can be strictly controlled and information may be obtained regarding the true physical nature of ice deformation. Nevertheless, any predictions from laboratory experiments require testing on actual glaciers and ice-sheets, and here conditions are usually much more complex than those envisaged in the rather idealized laboratory models. The greatest difficulty in the field lies in obtaining a full three-dimensional picture since it is only the surface layer that is readily observed.

Before examining the laboratory and field methods in more detail, it is worth making some general observations that may seem obvious on reflection but can easily be overlooked. The movement of any ice-mass consists of a transfer of material from the accumulation to the ablation area. The maximum transfer must occur at the equilibrium line and in the case of a steady-state ice-mass must equal the net accumulation. By the same token discharge through any cross-section in the accumulation area must equal the net addition above that section; in the ablation area it must equal the net loss below that section. On a steady-state ice-mass the constant form of the accumulation area can only be maintained by a downward inclination of the velocity vectors; similarly in the ablation area the vectors must normally be inclined upwards relative to the ice surface.

Laboratory studies

Ice is a crystalline material that behaves neither as a very viscous fluid nor as a brittle solid. Instead, a given stress induces deformation which first increases in rate and then settles down to a relatively steady value. In a single crystal this plastic flow or creep can be seen to be taking place by the slipping or gliding of layers, one over the other, parallel to the basal plane. It is possible to induce gliding on other planes but the required stress is very much greater. It follows that the orientation of crystals within an ice-mass is an important determinant of its overall physical properties. Observation shows that crystals are not randomly aligned in glacier ice but exhibit varying degrees of preferred orientation with the c-axes perpendicular to any planes of shearing. Since crystals in the firn and shallow ice-layers do not display this preferential alignment it is presumably induced by movement taking place at depth. A similar effect can be observed in the laboratory when a block of polycrystalline ice is subjected to simple shear stress. From an initial random distribution a preferred alignment gradually develops by intergranular movement and recrystallization. As it does so the strain rate increases, apparently because a higher proportion of the movement is taking place by gliding parallel to the basal planes.

Fig. 11.5. Basic analysis of the shear stresses operating on a parallel-sided slab of ice resting on a slope of angle α. By application of Glen's power law it is relatively easy to compute the velocity profile which, in such a case, is essentially a function of slope angle and ice thickness.

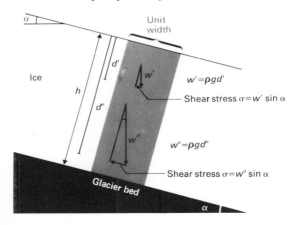

$w' = \rho g d'$

Shear stress $\sigma = w' \sin \alpha$

$w'' = \rho g d''$

Shear stress $\sigma = w' \sin \alpha$

If shear strain $\varepsilon = k\sigma^n$ (Fig. 11.4) then, by integration, surface velocity

$$V_s = \frac{k(\rho g)^n}{n+1} \sin^n \alpha . h^{n+1}$$

Velocities at intermediate depths may be similarly computed; n is commonly assumed to have a value of 3.

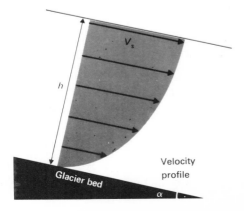

Velocity profile

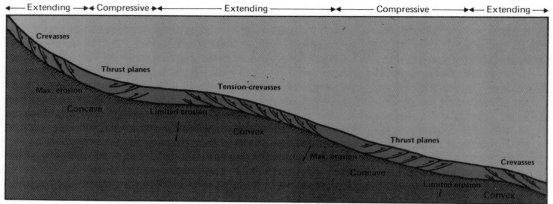

Fig. 11.6. Simplified representation of Nye's idea of extending and compressive flow related to convexities and concavities of the bedrock profile. One factor not depicted is the accumulation/ablation ratio; for the same long profile extending flow will theoretically be more common in the accumulation area, compressive flow in the ablation area.

Plotting the values of strain rate against shear stress from repeated laboratory experiments reveals a power relationship (Fig. 11.4). This is conventionally expressed in the form known as Glen's power law: $\varepsilon = k\sigma^n$, where ε is the strain rate, σ the stress, and k and n are constants. The value of k varies with temperature, but n is relatively independent of temperature. In most studies n is found to approximate 2.5 or 3, although a higher figure may better fit the data under large stresses where a high proportion of the crystals are preferentially aligned; similarly, a lower figure may be appropriate where only small stresses are involved. In order to apply Glen's flow law to an idealized glacier or ice-sheet, it is first necessary to consider the stresses that will then operate. It is easiest to begin with the example of a simple parallel-sided slab of ice resting on a uniform slope (Fig. 11.5). Shear stress is then clearly a function of two readily measured properties, slope and ice thickness. However, it needs to be remembered that its value will reach a maximum at the base of the ice and fall to zero at the surface. Yet the upper layers of ice are carried forward on the lower so that the actual movement will be greatest at the surface and decline with depth. It can then be shown by integration that in the elementary example considered, and when n in Glen's flow law has a value of 3, the surface velocity is proportional to the third power of the sine of the slope angle and to the

fourth power of the thickness. Where the slab is not of uniform thickness, so that the slopes at the base and surface are unequal, analysis of the situation is more complex. Yet it can still be shown that, provided the angles are small, the basal shear stress remains proportional to the sine of the surface slope, just as in the case of the parallel-sided slab. One corollary is that movement should always be in the direction of the surface slope and not the basal slope; this provides the theoretical explanation for the observed ability of ice to scour deep rock basins that necessitate uphill movement.

A further complication, and a closer approach to reality, is introduced by considering the effect of an irregular valley long profile. In an analysis of the changes occasioned by steeper and gentler reaches, Nye showed that they are liable to induce two contrasting types of flow which he termed compressive and extending (Fig. 11.6). Compressive flow with thickening of the ice will occur where the bedrock slope is concave; deep shear planes will tend to develop curving upwards from the base towards the surface. Extending flow will be found where the ice accelerates down a convexity; superficial tensional structures may here curve downwards from the surface towards the base. Even down a constant bedrock slope, changes in discharge must theoretically be expected to induce extending flow over the accumulation area and compressive flow over the ablation area; in practice these basic tendencies interact with variations in the bedrock gradient to produce the complex patterns of movement exhibited by nearly all glaciers and ice-sheets.

Field investigations

In field investigations (Fig. 11.7) it is essential to realize that any observed motion is normally composed of two elements. The first derives from internal deformation of the ice and is the element considered above from an experimental standpoint. The second is the movement of the ice-mass as a whole relative to its bedrock

surroundings; this component of the total displacement is termed sliding or slip. There is a very long tradition of measuring glacier movement, particularly in the Swiss Alps, and as early as 1841 Agassiz had inserted stakes across the surface of the Unteraar Glacier and was able to demonstrate that the central zone moves faster than the margins. Since that time techniques have changed little in principle although improved equipment has enhanced the potential accuracy. The standard procedure is to bore stakes into the surface layer and to monitor both horizontal and vertical displacements by observation from a theodolite set up at convenient points on the valley walls. The tendency has been to concentrate measurement in the ablation rather

Fig. 11.7. Three examples of field measurements on glaciers in western North America: (A) variations in surface velocity, measurement being expressed as a proportion of the maximum velocity along the centre-line of the glacier (after Meier, 1960); (B) velocity changes with depth (after Paterson, 1981); (C) seasonal fluctuations in surface velocity and meltwater discharge. Both values increase in summer and decline in winter, but are significantly out of phase (after Hodge, 1974).

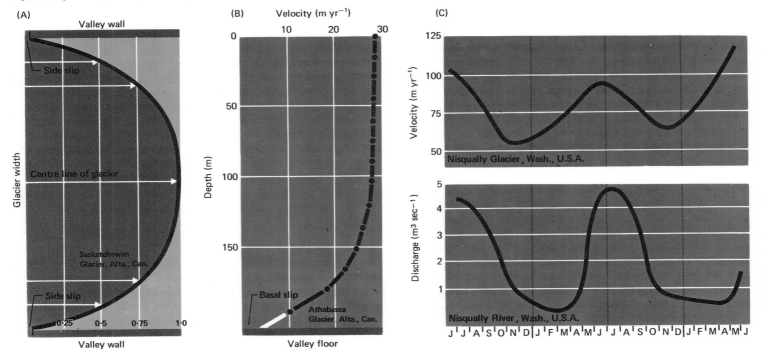

than the accumulation area, and only a few glaciers have been surveyed along their full length.

Measured rates vary greatly from one ice-mass to another. The interior of Antarctica with its lack of fixed reference points poses unusually severe problems of measurement but surface velocities almost certainly average less than 20 m yr^{-1}. In the peripheral areas values rise to over 100 m yr^{-1}, and along a series of ice-streams reach over 500 m yr^{-1}. These ice-streams are narrow radiating zones, sometimes delimited by prominent lines of crevasses, where the ice moves much faster than in the adjacent parts of the sheet; in many cases their positions appear to be related to depressions in the subglacial topography, and it is even possible that they are initiated where the base of very thick ice first attains pressure melting-point. Alpine valley glaciers have generally been recorded as moving at velocities of between 20 and 200 m yr^{-1}, but accelerating to over 1 km yr^{-1} down steeper slopes. The fastest rates are usually found on outlet glaciers descending from the Antarctic and Greenland ice-sheets; the very cold ice in the former area precludes exceptionally high values, but in Greenland velocities in excess of 7 km yr^{-1} have been recorded for such outlet glaciers as the Jakobshavn Isbrae.

The foregoing figures relate to surface layers near the centre-line of the glacier. Although velocities close to the rock margins are much reduced, they are of particular interest as an index of local slip or sliding (Fig. 11.8). For accurate assessment of side-slip, stakes must be inserted within 1 m or so of the ice edge; sometimes they have been installed horizontally with their ends virtually touching the valley wall. This slip velocity is often expressed as a percentage of the maximum velocity in the same cross-section, and values have been found to range from under 10 to over 65 per cent. More difficult to evaluate is basal sliding at the sole of the ice. Velocities at depth within an ice-mass can be ascertained but expensive procedures are involved. They normally entail drilling a hole to the bedrock interface and inserting some form of flexible casing. After surveying the position of the hole a specially designed inclinometer is lowered down the casing to measure the slope and alignment at closely spaced intervals. If the hole is then capped and the measurements repeated after the lapse of a year or more, a new profile of the casing can be drawn and the velocity variations with depth computed. This technique was first adopted in a study of the Aletsch glacier in Switzerland and has subsequently been repeated for a number of North American glaciers. All have conformed to the theoretical expectation that the surface layers move more rapidly than the lower, and that maximum differential movement is concentrated near the sole (Fig. 11.7). Another method of measuring basal slip has been to use tunnels excavated for commercial purposes beneath several glaciers; these have the advantage of permitting a relatively continuous monitoring of the sole of the ice and it has been shown, for instance, that substantial short-term fluctuations in velocity may occur. Expressed as a percentage of the maximum movement in the profile, basal sliding in warm-based ice-masses has been shown to vary from under 10 to over 80 per cent with a mean value of about 50 per cent. Even at adjacent points on the same glacier the figure can vary appreciably indicating highly complex stress relationships. Basal sliding is normally thought to be absent where cold-based ice is involved.

Weertman invoked two major mechanisms in explanation of basal sliding. The first is regelation where the very high pressure on the stoss sides of small bedrock protuberances induces local melting. The water then migrates to the lee sides where it refreezes under the conditions of diminished pressure. The latent heat released may be conducted through each protuberance to assist further melting on the onset side, although the poor thermal conductivity of rock means that small projections will be more effective than large. The second mechanism involves locally enhanced strain rates in the basal ice owing to greater-than-average stresses generated over bedrock bumps. The potentially uneven distribution of stress is well illustrated by the occurrence of subglacial cavities on the down-glacier side of such bumps where the ice temporarily separates from the rock surface. Theoretically, large bumps are likely to be more effective than small, so that there may be an intermediate size where neither regelation nor increased rate of deformation are particularly potent.

While bedrock form may thus explain some of the spatial vari-

Fig. 11.8. Two photographs illustrating conditions at the margin of a Swiss valley glacier. (Left) The irregular gap along the valley wall which rapidly channels any surface meltwater beneath the ice. (Right) Detail of the rock wall showing both the smooth face produced by abrasion, and the very rough surface resulting from freeze–thaw activity where there is no direct contact with the glacier ice. The scratch marks indicate the importance of sideslip as rock fragments such as that shown in the photograph are dragged along the valley wall.

ability in basal sliding, it does not account for an observed temporal variability. Laser measurements have shown that glacier movements conventionally measured over a period of weeks, months or even years are actually composed of innumerable discrete but exceedingly small displacements. This jerky motion is more intense in summer than in winter, and tends to peak during warm weather and after heavy rainfall. Since seasonal temperature fluctuations are confined to the surface layers of the ice, it seems likely that any acceleration is due to variations in basal water pressure which may produce temporary thickening of the aqueous film at the rock interface. Support for this view has come from investigations in subglacial tunnels where a direct correlation has been observed between diurnal pressure fluctuations and the sliding velocity. It has even been found that the surface of the Unteraar glacier in Switzerland rises vertically by as much as 0.6 m at the start of the summer melt season and gradually subsides again over the next 3 months; this uplift has been ascribed to a temporary increase in the volume of water stored in cavities at the glacier base. The resultant buoyancy may reduce both the effective weight of the ice on the rock and also the actual area of contact with the rock, thereby diminishing the friction and accelerating the sliding motion.

A final aspect of ice movement requiring mention is the longer-term velocity variation known as surging. Surges have now been recorded from several hundred glaciers, although detailed investigations remain sparse. In all cases a temporary acceleration is involved that may last from a few months to several years. During that time ice that has lain virtually stagnant for a decade or more is abruptly reactivated as a wave of fast-moving ice passes down the glacier. The Medvezhiy glacier in the Pamirs of central Asia was closely studied between two consecutive surges in 1963 and 1973. Both surges lasted about 3 months, during which time the terminus advanced by at least 1.5 km, temporarily achieving a velocity of 105 m in a day. Elevations of the ice surface in the up-glacier area declined by 100 m while the down-glacier region thickened by 150 m, implying a massive transfer of material from a so-called reservoir area to a receiving area. These last-named often do not coincide at all with the accumulation and ablation areas, and on the Medvezhiy glacier even the reservoir area was entirely within the ablation zone. The fundamental causes of surging have yet to be identified. It appears that certain ice-masses, but not all, have a cyclical tendency to become increasingly unstable until some unknown mechanism triggers accelerated sliding. Among possible mechanisms are progressive thermal changes at the sole of the ice, especially where part of the glacier is cold-based, the gradual accumulation of water at the rock interface which ultimately reduces the basal friction, and local steepening of the surface profile which eventually forces the basal shear stress beyond a critical value that is then propagated both up- and down-glacier. None of these suggested explanations is entirely satisfactory, and the underlying reasons for surging require further investigation.

Glacial meltwater

Supraglacial flow

Save in a dry snow zone and in the higher parts of the percolation zone, surface melting normally produces copious quantities of water that at first flow supraglacially but later disappear down crevasses and moulins before emerging once more at the ice-snout. Flow is particularly persistent in the ablation zone where summer discharge characteristically shows a regular diurnal cycle with peak values in the early evening and minimum values in the early morning. As on a normal water-eroded landscape both sheet and channelled flow can be found, but surface lowering takes place by melting rather than abrasion. In this way gullies are occasionally deepened to 10 m or more below the surrounding ice surface. Across steep ice-slopes near the margin and terminus of a glacier, high gradients and smooth channel walls combine to produce exceptionally fast flow.

The typically convex cross-profile of the ablation area on a valley glacier tends to shed meltstreams to a lateral position, although extensive marginal crevassing often ensures that a high proportion of the meltwater disappears subglacially before reaching the actual ice edge. Nevertheless, some surface melt, augmented by runoff from adjacent bedrock slopes, can sustain a truly marginal flow parallel to the valley walls. This is especially the case with cold-based ice, but with ice close to pressure melting-point an individual stream rarely flows any great distance before vanishing down a passage beneath the ice margin. Its place, however, is soon taken by a completely fresh stream so that, although marginal drainage may characterize the edge of a glacier, there need be no consistent down-valley increase in the volume of discharge. The channels of marginal streams may be excavated in either ice or bedrock, or may have one wall composed of rock and the other of ice; not uncommonly they pass from one material to the other and back again within the space of 100 m or less. A typically irregular long profile produces sharp contrasts in velocity. In places the meltwater may be ponded back into shallow lakes, while in others it plunges down precipitous slopes. An added characteristic of marginal stream courses is their general instability occasioned by down-valley movement of the glacier, by melting of the surface ice and by changes in the pattern of crevasses.

Englacial and subglacial flow

Once surface water has vanished beneath the ice, its precise route is difficult to chart. Attempts to use tracer dyes and injections of salt have had some success but much remains to be learnt. It seems clear that, in many instances, marginal drainage after disappearing from view does not immediately proceed to the valley floor but is channelled roughly parallel to the glacier edge in what is termed a submarginal position; it may then follow a zigzag route, alternating between short plunges downslope where the bedrock contact is not watertight and gentler reaches where the flow is forced laterally along the valley side. Two separate systems of passageways, one on each side of a glacier, are common, and the rates of flow calculated from straight-line distances normally fall in the range 0.2–1.5 m s^{-1}. Melting along the walls of ice-tunnels may add significantly to the volume of water to be discharged at the snout. Ejection at the snout can sometimes be quite violent, reflecting the very considerable hydrostatic pressures that may be generated by confinement within a network of tunnels and shafts. In a specially constructed laboratory beneath the Glacier d'Argentière in the French Alps, basal water pressures were monitored over a period of several years. During winter pressure proved to be relatively steady at 900–980 kN m^{-2}, but in summer to fluctuate daily between about 600 and 1 200 kN m^{-2}, presumably due to melting on the glacier surface. Detailed analysis of these and other measurements has been taken to imply the existence of a water-table within the glacier, with the meltwater locally ponded in subglacial and englacial cavities that are capable of acting as temporary storage chambers. Such water-filled cavities have been encountered during a number of tunnelling and drilling operations and their control on meltwater discharge has been

surmised for the Athabasca glacier where large intermittent floods recur without evident relationship to ambient temperatures; local blind passages and voids apparently fill with water and give rise to a sudden spate at the snout when temporarily integrated into the subglacial drainage system.

So far attention has been confined to the passage of meltwater originating at the ice surface, but a further obvious source of meltwater, at least where the ice is warm-based, is at the sole. Such water is extremely difficult to study directly and most work has been of a theoretical nature. The mean geothermal flux is capable of melting about 6 mm of ice per year; an equal thickness could be melted by the frictional heat released by basal sliding at a rate of 20 m yr^{-1}. Three possible drainage systems by which this liquid might be evacuated across an impermeable substrate have been considered. Some workers have argued that it will migrate towards the ice margin as a very thin film capable of little erosive or transporting activity. Others have maintained that it will progressively concentrate into a series of discrete channels reminiscent of an ordinary stream network. Two styles of channel have been hypothesized, each named after a distinguished glaciologist. Nye envisaged subglacial routeways incised into the local bedrock, while Röthlisberger argued in favour of tunnels through the base of the ice itself. In both cases the tendency towards closure by pressure from the overlying ice may be balanced by melting along the walls and roof of the conduit so that an equilibrium is established. The relative frequency of Nye- and Röthlisberger-channels has yet to be established, and it is possible that each characterizes different local circumstances. The whole position becomes even more complicated when the possibility of water migrating from a warm-based to a cold-based zone is envisaged, and much further research is clearly needed into subglacial plumbing systems.

Ice-dammed lakes

Impounded against the margins of modern glaciers are hundreds of small ice-dammed lakes. Most are situated at the lower ends of tributary valleys where the main valley is occupied by ice. Less frequently they occur where a major drainage artery is obstructed by a powerful glacier issuing from a tributary valley. A third common situation is around the frontal edge of a glacier in a hollow between the ice margin and an abandoned morainic ridge. Examples of ice-dammed lakes occur in virtually all regions of temperate glaciers. The most famous instance in the Alps is the Marjelensee ponded against the Aletsch glacier. This lake was 1 600 m long in 1878, but since that date it has gradually declined and owing to shrinkage of the impounding glacier is now almost extinct. The great majority of modern ice-dammed lakes are small, often less than 1 km² in extent. Among the larger examples are Lake George in Alaska which in 1965 covered some 120 km², and Graenalon in Iceland which is usually credited with an area of 18 km².

One difficulty in assigning sizes to particular lakes is the frequent fluctuations they exhibit. Although a few lakes spill across bedrock cols away from the ice edge, a situation tending to stabilize the water-level, most discharge intermittently through subglacial or englacial passages. Water-level may then decline so rapidly as to strand on the newly emergent hillsides icebergs that were previously floating near the shoreline. The concurrent discharge of water from the glacier snout will be equally erratic. A mere trickle may suddenly give way to an immense spate, often known by the Icelandic term *jokulhlaup*. The rampaging meltwater possesses great transporting powers, and besides moving exceptionally large boulders may also carry huge icebergs torn from the glacier front. In some *jokulhlaups* momentary discharges have been estimated to exceed 100 000 m³ s⁻¹.

Records are now available charting the changes in level of many ice-dammed lakes. Considerable diversity exists. Some lakes regularly discharge each summer while others continue to fill for several years and are only evacuated at infrequent intervals. Variations in height are often attested by minor strandline features closely analogous to those seen in flooded gravel pits after a fall in water-level. Much doubt still surrounds the precise mechanism by which discharge takes place. As long ago as 1954 Glen suggested that, owing to the differing densities of ice and water, stresses on an ice-cliff at the bottom of a lake might be adequate to force a breach through the dam; however, calculations indicate that for this mechanism to be really effective the lake would need to be at least 150 m deep, and this appears to place a severe limitation on the applicability of the hypothesis. At about the same time Thorarinsson argued that when a lake reaches a critical depth the ice-barrier will begin to float, thus allowing the water to escape beneath. It was pointed out by critics of this idea that if the lifted ice-margin had a tendency to sink back so as to reseal the outlet the result would merely be minor oscillations about the level required for flotation. However, Liestol and several more recent investigators have contended that, once water finds an initial escape route, the passage it uses will be so speedily enlarged by melting that, even if the ice falls back to its original position, drainage will still be able to continue. Attempts to test the flotation hypothesis, for instance by surveying the ice-dam for possible lift at the onset of discharge, have generally proved inconclusive, but the underlying concept has commanded wider support than most other ideas. In many cases the escape route may need to link with an existing tunnel system for drainage to be fully effective. Clarke described in some detail the outburst from Hazard Lake in Yukon Territory during the summer of 1978 when the water-level fell by 100 m in 2 days and over 19 × 10⁶ m³ of water was released to run for 13 km beneath the Steele glacier. Release of thermal energy from the lake water which had an input temperature of 6 °C was held to be the dominant factor contributing to tunnel enlargement.

References

Andreasen, J-O. (1983) 'Basal sliding at the margin of the Glacier Austre Okstindbre, Nordland, Norway', *Arct. Alp. Res.* **15**, 333–8.

Blatter, H. and Haeberli, W. (1984) 'Modelling temperature distribution in Alpine glaciers', *Ann. Glaciol.* **5**, 18–22.

Boulton, G. S. and Vivian, R. (1973) 'Underneath the glaciers', *Geogr. Mag.* **45**, 311–19.

Boulton, G. S. *et al.* (1977) 'A British ice-sheet model and patterns of glacial erosion and deposition in Britain', in *British Quaternary Studies* (ed. F. W. Shotton), OUP.

Burkimsher, M. (1983) 'Investigations of glacier hydrological systems using dye tracer techniques: observations at Pasterzengletscher, Austria', *J. Glaciol.* **29**, 403–16.

Clarke, G. K. C. (1982) 'Glacier outburst floods from "Hazard Lake", Yukon Territory, and the problem of flood prediction', *J. Glaciol.* **28**, 3–21.

Clement, P. (1984) 'The drainage of a marginal ice-dammed lake at Nordbogletscher, Johan Dahl Land, South Greenland', *Arct. Alp. Res.* **16**, 209–16.

Denton, G. and Hughes, T. J. (1981) *The Last Great Ice Sheets*, Wiley.

Dolgushin, L. D. and Osipova, G. B. (1973) 'The regime of a surging glacier', *Int. Ass. Hydr. Sci.* **107**, 1150–9.

Gilbert, R. (1971) 'Observations on ice-dammed Summit Lake, British Columbia, Canada', *J. Glaciol.* **10**, 351–6.

Glen, J. W. (1954) 'The stability of ice-dammed lakes and other water-filled holes in glaciers', *J. Glaciol.* **2**, 316–18.

Glen, J. W. (1955) 'The creep of polycrystalline ice', *Proc. R. Soc.* **A228**, 519–38.

Gow, A. J. *et al.* (1970) 'Antarctic ice sheet: preliminary results of first core hole to bedrock', *Science N.Y.* **161**, 1011–13.

Hallet, B. (1979) 'Subglacial regelation water film', *J. Glaciol.* **23**, 321–34.

Hansen, B. L. and Langway C. C. (1966) 'Deep core drilling in ice and core analysis at Camp Century, Greenland, 1961–1966', *Antarct. J. U.S.* **1**, 207–8.

Harrison, W. D. (1975) 'Temperature measurements in a temperate glacier', *J. Glaciol.* **14**, 23–30.

Hodge, S. (1974) 'Variations in the sliding of a temperate glacier', *J. Glaciol.* **13**, 349–69.

Iken, A. *et al.* (1983) 'The uplift of the Unteraargletscher at the beginning of the melt season – a consequence of water storage at the bed?', *J. Glaciol.* **29**, 28–47.

Jacobel, R. W. (1982) 'Short-term variations in the velocity of South Cascade glacier, Washington, U.S.A, *J. Glaciol.* **28**, 325–32.

Liestol, O. (1955) 'Glacier-dammed lakes in Norway', *Norsk. Geogr. Tidsskr.* **15**, 122–49.

Macheret, Y. Y. and Zhuralev, A. B. (1982) 'Radio echo-sounding of Svalbard glaciers', *J. Glaciol.* **13**, 295–314.

Mathews, W. H. (1974) 'Surface profiles of the Laurentide ice sheet in its marginal areas', *J. Glaciol.* **13**, 37–43.

Meier, M. F. (1960) 'Mode of flow of Saskatchewan glacier, Alberta, Canada', *U.S. Geol. Surv. Prof. Pap.* 351.

Nye, J. F. (1952) 'The mechanics of glacier flow', *J. Glaciol.* **2**, 82–93.

Nye, J. F. (1976) 'Water flow in glaciers: jokulhlaups, tunnels and veins', *J. Glaciol.* **17**, 181–207.

Oswald, G. K. A. and Robin, G. de Q. (1973) 'Lakes beneath the Antarctic ice sheet', *Nature, London* **245**, 251–4.

Paterson, W. S. B. (1981) *The Physics of Glaciers*, Pergamon

Post, A. (1960) 'The exceptional advances, Black Rapids and Susitna Glaciers', *J. Geophys. Res.* **65**, 3703–12.

Robin, G. de Q. (1955) 'Ice movement and temperature distribution in glaciers and ice sheets', *J. Glaciol.* **2**, 523–32.

Röthlisberger, H. (1972) 'Water pressure in intra- and subglacial channels', *J. Glaciol.* **11**, 177–203.

Savage, J. C. and Paterson, W. S. B. (1963) 'Borehole measurements in the Athabasca glacier', *J. Geophys. Res.* **68**, 4521–36.

Shreve, R. L. (1972) 'Movement of water in glaciers', *J. Glaciol.* **11**, 205–14.

Stenborg, T. (1969) 'Studies of the internal drainage of glaciers', *Geogr. Annlr.* **51A**, 13–41.

Sturm, M. and Benson, C. S. (1985) 'A history of jökulhlaups from Strandline Lake, Alaska, USA', *J. Glaciol.* **31**, 272–80.

Sugden, D. E. (1977) 'Reconstruction of the morphology, dynamics and thermal characteristics of the Laurentide ice sheet at its maximum', *J. Arct. Alp. Res.* **9**, 21–47.

Thorarinsson, S. (1953) 'Some new aspects of the Grimsvotn problem', *J. Glaciol.* **2**, 267–75.

Vivian, R. (1970) 'La nappe phreatique du glacier d'Argentière', *C. R. Hebd. Seanc. Acad. Sci. Paris* **270**, 604–6.

Vivian, R. (1980) 'The nature of the ice–rock interface: the results of investigation on 20 000 m² of the rock bed of temperate glaciers', *J. Glaciol.* **25**, 267–78.

Vivian, R. and Bocquet, G. (1973) 'Subglacial cavitation phenomena under the Glacier d'Argentière, Mont Blanc, France', *J. Glaciol.* **12**, 439–51.

Vivian, R. and Zumstein, J. (1974) 'Hydrologie sous-glaciaire au glacier d'Argentière (Mont Blanc, France)', *Int. Ass. Sci. Hydrol.* **95**, 53–64.

Weertman, J. (1979) 'The unsolved general glacier sliding problem', *J. Glaciol.* **23**, 97–111.

Weertman, J. and Birchfield, G. E. (1984) 'Stability of sheet water flow under a glacier', *J. Glaciol.* **29**, 374–82.

Selected bibliography

Almost all issues of the *Journal of Glaciology* contain papers adding significantly to our knowledge of modern ice-masses. The outstanding book covering the same field is W. S. B. Paterson, *The Physics of Glaciers* (2nd edn), Pergamon, 1981.

Chapter 12
Glacial erosion and deposition

The preceding chapter emphasized the contribution of glaciology to the understanding of glacial landforms. Yet it must be remembered that the erosional and depositional processes deep beneath an ice-sheet have never been directly observed. Admittedly, in recent years much information has come from artificial tunnels excavated beneath cirque and valley glaciers, but how representative these facilities are of conditions beneath thicker ice-masses remains difficult to ascertain. Laboratory experiments such as those that have tested the effect of sliding ice on prepared rock samples, and theoretical models of likely circumstances at the base of an ice-mass, have also contributed significantly to our present knowledge. Yet it is the study of landforms and deposits in formerly glaciated areas that has often provided the clearest guide to the way an ice-mass with its attendant meltwater can modify the landscape across which it passes. It is to the interpretation of these formerly glaciated regions that this and the next chapter are primarily directed.

Glacial erosion

The erosional processes

Glacial erosion is generally attributed to two separate processes, plucking and abrasion. Plucking refers to the quarrying effect when moving ice freezes on to bedrock and pulls out a block which it then carries away; the ice itself may assist in the initial rupturing by generating very different stresses on the up- and down-glacier side of an obstruction. Abrasion refers to the grinding effect when debris

transported in the sole of an ice-mass acts in the manner of a giant sandpaper. The existence of two distinct processes is manifested in the erosional form known as a *roche moutonnée* where the onset side is rounded and scored by abrasion while the lee side is angular and blocky due to the removal of joint-bounded fragments by plucking. There has been considerable discussion regarding the overall relative effectiveness of these two processes.

Many minor landforms bear testimony to the abrasive action of ice armed with rock debris. The most common is the striated bedrock pavement where fine lines have been scored parallel to the direction of ice movement; with diminishing size such striations may grade into polished surfaces on which the individual striae are only detectable with a hand lens. Larger striated surfaces often display a tendency towards fluting. Grooves 1–2 m deep and up to 100 m long have been described from Ohio, while in the Mackenzie valley in Canada giant examples attain depths of 30 m and can be traced for distances in excess of 10 km. Features of this dimension stand out conspicuously on aerial photographs and give certain areas of the Canadian Shield a prominent lineated appearance. The fashioning of such large-scale grooves obviously involves mechanisms additional to those responsible for fine striations, but both bear eloquent testimony to the importance of glacial abrasion. In theory the actual abrasion rate may be expected to depend upon the flux of rock particles embedded in the sole of the ice-mass, upon the relative hardness of the transported and bed materials, and upon the forces exerted by the mobile debris on the underlying surface. To test these ideas Boulton and Vivian bolted specially prepared basalt and marble slabs to the rock floor at the head of long subglacial tunnels in Iceland. After being overrun by the ice for just 3 months the slabs were found to have been abraded to mean depths of 1 and 3 mm respectively at locations where the basal slip in the same interval amounted to 9.5 m.

In another experiment on the bedrock floor beneath the Glacier d'Argentière transducers were installed that permitted both normal pressure and shear stress variations to be monitored as debris-laden ice moved across the site. The results demonstrated that both pressure and shear stress increased during the passage of an individual stone, but that over a longer time interval the relationship between the two values was much influenced by changes in the concentration of debris in the sole of the glacier. It always needs to be remembered that rock particles are being transported in a deformable medium so that, although abrasion theoretically increases with rising pressure and shear stress, a point may be reached where the friction between rock fragments and bedrock becomes so great that the ice simply moved past the debris; in effect the material has been deposited, at least temporarily, and abrasion ceases.

Many workers have contended that plucking is quantitatively a more effective means of erosion than abrasion. They cite in support the large blocks of unweathered rock incorporated in both glacial and glaciofluvial deposits. Moreover, rocks of similar mineralogical composition but with contrasting joint systems often respond very differently to glacial erosion, an outcome that is unlikely to be produced by abrasion but is readily explicable in terms of quarrying action. A striking example is provided by the Yosemite valley in California (Fig. 12.1) where massive granite has resisted glacial erosion but granite with closely spaced joints has been deeply quarried. Cliffs over 1 000 m high mark the junction between the two rock types but the most spectacular manifestation is Half Dome (Fig. 12.2). Here almost 50 per cent of an originally hemispherical mountain has been removed while the rest remains largely untouched. It is small wonder that Matthes, who made the outstanding scientific study of Yosemite valley, was an ardent advocate of the efficiency of glacial plucking.

Yet it is worth noting that, at first sight, there seem to be significant theoretical limitations to the plucking activities of both warm-based and cold-based ice. In the former case the ice may be separated from a smooth rock surface by a thin film of water, and in the latter case freezing on to the bedrock may itself inhibit the basal sliding required for effective quarrying. However, early workers realized that bed roughness at the sole of a warm-based ice-mass might generate sufficient localized pressure variations to induce sustained melting and refreezing around rock protuberances (see also p. 222). This idea was confirmed when several glacier tunnels afforded access to the ice–rock interface and it was seen that there was a distinctive layer of basal ice attributable to regelation processes; Kamb and LaChapelle, for instance, described such a layer up to 29 mm thick beneath the Blue glacier in Washington, and more recently Souchez and Lorrain demonstrated from the chemistry of even thicker basal layers beneath several Swiss glaciers that regelation is almost certainly operating.

Boulton, meanwhile, suggested another mechanism that may favour plucking beneath cold-based ice. He envisaged the existence of distinct thermal zones at the sole of many ice-masses (Fig. 12.3). In certain zones the heat generated at the base of the ice will exceed that which can be conducted upwards so that melting must occur. Surplus water produced in this fashion then migrates to other zones where it refreezes on to colder basal ice, raising the temperature of the ice and thereby ensuring continued basal slip. It is in these latter zones that plucking will be particularly effective, being facilitated not only by adhesion but also by the wedging effect of any water that penetrates into the rock joints. The concept of freezing-on of debris, as the process has been called, receives some support from the boreholes drilled through the Antarctic and Greenland ice-sheets and referred to on p. 216. Almost 5 m at the base of the core at Byrd Station in Antarctica was laden with stratified debris ranging from silt-sized particles to cobbles. Analysis of the enclosing ice showed that it had almost certainly been produced by refreezing, although how far the water may have migrated laterally is unknown. At Camp Century in Greenland the bottom 15 m of the core contained over 300 alternating bands of clean and debris-laden ice. Again the enclosing ice appeared to have frozen on to the sole, although in this particular instance, where the basal temperature is −13 °C, it is far from clear where the water could have originated and it may be necessary to postulate past changes in the thermal regime of the Greenland ice-sheet. A borehole that showed no evidence of freezing-on was sunk through the Devon Island ice-cap in northern Canada. Here, at a depth of 299 m and with a basal temperature of −18 °C, the lowest ice was free of rock debris and appeared to confirm the non-erosional character of an ice-mass frozen to its bed.

In summary, both abrasion and plucking can undoubtedly be very effective erosional processes. This does not imply that they will be uniformly active since so much depends upon conditions at the sole of an ice-mass. Much spatial and temporal variability must be anticipated as a glacier advances and retreats, or as an ice-sheet grows and decays.

Cirques

Being among the most distinctive of all landforms, these steep sided semicircular basins of glaciated highlands have long been accorded special descriptive names in a wide variety of languages. For

instance, the local name in Scotland is corrie, in Wales cwm, in Norway *botn* and in Germany *Kar*, but there is an increasing tendency to adopt the French term cirque in recognition of its use in a scientific context by de Charpentier as long ago as 1823. Many techniques may be employed in the study of cirques, but here attention will be confined to just two separate approaches. The first involves investigation of modern cirque glaciers, the second analysis of such properties as shape, aspect and elevation of the landform itself.

An early investigation destined to have a profound effect upon views about cirque development was made by Johnson at the turn of the present century when he was lowered down the bergschrund at the back of a cirque glacier on the northern face of Mount Lyell in California. He found that where the bergschrund intersected the headwall at a depth of some 40 m the rock surface displayed abundant evidence for active freeze–thaw processes. Huge angular boulders that had tipped forward to rest against the glacier were sheathed in clear ice. Similar material could be seen in the enlarged joint system of the headwall, amidst many partially dislodged blocks. Long icicles attested to migrating water that had later frozen. Johnson inferred that there must be numerous freeze–thaw cycles operating at the base of the bergschrund, possibly in a regular diurnal pattern during summer. Later workers have confirmed the substance of his observations but cast doubt upon his interpretation. Thermographs installed for lengthy periods have shown that air temperature variations in a bergschrund are generally of low amplitude and take place only slowly (Fig. 12.4). Even in summer there is no regular diurnal fluctuation, and intervals when the air is sufficiently warm to induce melting are infrequent. As a modification of the original Johnson hypothesis, Lewis suggested that most of the clear ice at the base of a bergschrund is water that has trickled down the back wall

Fig. 12.1. Yosemite valley in California, USA, a classic example of a glacial trough. The sheer face of El Capitan on the left is some 925 m high; the post-glacial fill on the valley floor at the foot of El Capitan is over 300 m thick (see Fig. 12.8). In the centre background is the summit of Half Dome.

Fig. 12.2. The distinctive form of Half Dome rising 1 480 m above Yosemite valley floor, Sierra Nevada, California, USA. The position of the 'missing' half coincides with a zone of well-jointed rock which succumbs much more readily to weathering and erosion than the rest. The massive structure of the rock in the vertical face contrasts with the superficial sheeting on the dome itself.

Fig. 12.3. Schematic representation of some of the different thermal conditions that may be encountered at the base of an ice-sheet (after Boulton 1972).

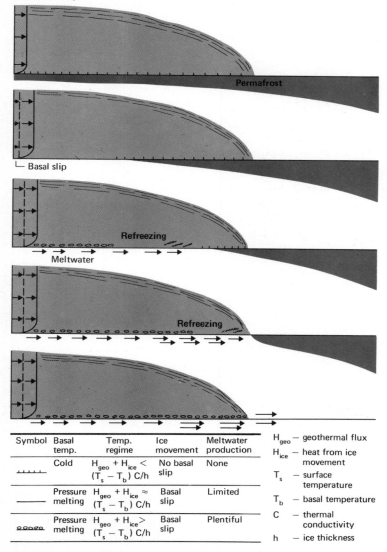

Symbol	Basal temp.	Temp. regime	Ice movement	Meltwater production
⊥⊥⊥⊥	Cold	$H_{geo} + H_{ice} <$ $(T_s - T_b)$ C/h	No basal slip	None
——	Pressure melting	$H_{geo} + H_{ice} \approx$ $(T_s - T_b)$ C/h	Basal slip	Limited
⊙⊙⊙⊙	Pressure melting	$H_{geo} + H_{ice} >$ $(T_s - T_b)$ C/h	Basal slip	Plentiful

H_{geo} — geothermal flux

H_{ice} — heat from ice movement

T_s — surface temperature

T_b — basal temperature

C — thermal conductivity

h — ice thickness

from the surface and frozen on entering regions of increasingly low temperature. However, by itself such a process would provide little basis for continued freeze–thaw activity since the headwall would presumably become coated with ice which would then function as a protective shield. Recently, therefore, less emphasis has been placed on the frost-sapping mechanism at the base of a bergschrund, and many workers have come to regard that process as restricted to a shallow zone near the uppermost edge of the glacier. In effect this means that frost-shattering, which is undoubtedly rife on the rock wall above most cirque glaciers, extends beneath the ice surface to a critical depth where temperatures are too stable for it to continue; the bergschrund often extends down below the critical depth.

Although headwall sapping around the upper margins of cirque glaciers is clearly an important process, it does not explain many of the salient characteristics of the cirque. In particular it does not account for the scouring of the rock basin which appears to be due primarily to the nature of the ice movement. This has been investigated on several occasions by the digging of tunnels, one of the most informative being that excavated in Vesl-Skautbreen in Norway under the leadership of Lewis in 1951. It had previously been postulated that the distinctive surface banding visible in the ablation zone of many cirque glaciers arose from annual accumulation layers being tilted so as to dip into the glacier. The rotational movement which this hypothesis implied received support from observation of the tunnel walls (Fig. 12.5). Stratification due to annual incremental layers was seen to dip inwards at 28° near the tunnel entrance, to become almost horizontal deep within the glacier, and then near the headwall to change rapidly until it dipped virtually parallel to the rock face. At the actual contact lay a narrow zone of bubble-free ice, almost certainly attributable to regelation and darkened by its high debris content; elsewhere the ice was remarkably clean with no sign of detritus migrating from the headwall into the body of the glacier. The disposition of the stratification, measured deformation of the tunnel wall and displacement of surface stakes, all served to confirm a degree of rotational movement. The motive force appears to be the excess of accumulation near the headwall, combined with maximum ablation near the toe. The potential steepening of the surface thus created demands slow rotation to maintain equilibrium. The internal deformation is relatively small, being restricted in the main to frictional retardation close to the headwall. Cirques are now widely held

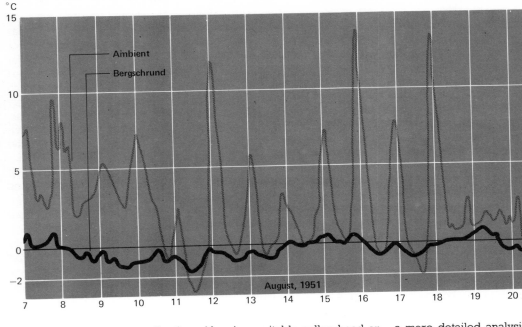

Ambient

Bergschrund

August, 1951

°C

Fig. 12.4. Temperatures recorded in Tverrabreen bergschrund, Norway, showing the pronounced damping effect when compared with the outside atmospheric temperatures (after Battle, 1960).

to begin by thick snow collecting either in a suitable valley head or in a depression enlarged by nivation beneath a snow patch. When the snow can accumulate to a thickness approaching 30 m the basal layers are transformed to true glacier ice. Thereafter abrasion and plucking at the base of the rotating ice, augmented by freeze–thaw processes at the top of the newly formed headwall, ensure development of the typical cirque form.

In considering cirque morphology, it must first be emphasized that there are great variations in size, ranging from the huge Walcott cirque in Antarctica reputed to be 16 km wide with a back wall over 2 000 m high, to more typical examples which are 1–2 km wide with back walls some 300 m high. The question immediately arises as to how far the shape varies with size and to what extent it is influenced by such factors as geological structure. Cirque form has been measured in various ways. Manley, for instance, examined the ratio between cirque length and height and found this normally lies in the range 2.8–3.2. He inferred that the nature of the local bedrock has relatively little effect on morphology, a conclusion reinforced during

a more detailed analysis of Scottish cirques by Haynes in 1968. She surveyed the long profiles of sixty-seven cirques and showed that these generally conform to a relatively simple logarithmic curve of the family $y = (1 - x)e^{-x}$. Aniya and Welch have subsequently shown that many cirque cross-profiles approximate a parabolic curve, a property they share with glacial troughs. Such consistency of form appears to justify the belief that process is the dominant control with rock structure playing only a subsidiary role. On the other hand, true cirques are confined to lithologies capable of sustaining precipitous headwalls, which in practice often means areas of igneous or metamorphic rock. The shales and mudstones of the Southern Uplands are noticeably lacking in cirques, whereas the form is well exemplified in the nearby English Lake District. Many observers have commented on the way rock structure influences morphological detail even if it does not determine overall form. Planes of weakness control the size and shape of blocks that are removed, and the slope at the foot of the headwall often comprises a staircase of treads and risers defined by bedding and joint systems. Equally conspicuous is the preferential widening of joints aligned parallel to the headwall; this suggests that rock dilatation may be a significant accessory process in cirque enlargement.

It has long been recognized that cirques display distinctive patterns of altitude and orientation. The former has been invoked as a guide to the elevation of Pleistocene snowlines since the initiation and development of cirques requires conditions normally found only near the snowline. However, one problem is knowing whether any particular group of cirques developed contemporaneously, and if so at which period, and the method is most satisfactorily employed where the former glaciers never developed beyond the cirque stage. Flint, for instance, used the technique in reconstructing the late

Pleistocene snowline in the Rocky Mountains and Cascade Range of the western United States. Yet even where a mountainous area has later been covered by an ice-sheet a regular pattern of cirque-floor elevations can often be discerned. In the Scottish Highlands, for example, cirque floors display a systematic rise from west to east (Fig. 12.6). In this and other cases the rough parallelism with modern isohyets has been held to indicate that the cirques probably lay close to a critical snowline at the onset of the last glaciation. Further support for this view comes from the simultaneous reoccupation of certain groups of cirques during climatic fluctuations at the close of that glaciation.

In the Northern Hemisphere cirques are preferentially orientated towards the north and east, and in the Southern Hemisphere towards the south and east. The obvious association is with contrasts in insolation and these have usually been regarded as the primary cause of the preferred alignments. During summer suitable valley heads would retain thick snow banks on shaded slopes while corresponding sites in full sun would be completely bare of snow. Nivation processes beneath the snow banks would slowly erode the rock floors until true cirque glaciers were able to develop. It might at first seem that similar conditions should exist at higher elevations on the sunny slopes, but this is not necessarily true; by localizing the preservation of snow banks the shadow effect can concentrate the attack of nivation processes in a way that is not normally possible on sunny slopes. A second but lesser influence stressed by certain writers is the effect of wind carrying snow on to the lee face of a mountain mass so as to leave the windward face relatively bare. Such blown snow will tend to collect in sheltered hollows and nourish the banks that are an essential preliminary to full cirque development.

Glacial troughs

As already emphasized, the degree of topographic modification achieved by valley glaciers and ice-sheets varies greatly from one locality to another. Some valleys that have undoubtedly carried large volumes of ice during a glaciation emerge almost unscathed, while others such as the Lauterbrunnen in Switzerland (Fig. 12.7) and the

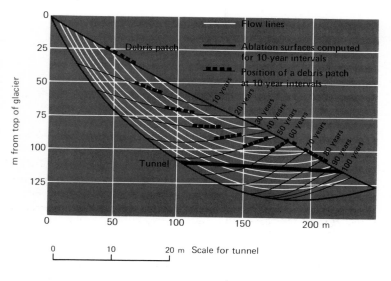

Fig. 12.5. Sections through Vesl-skautbreen showing flow lines and computed positions of ablation surfaces at 10-year intervals (based upon velocity measurements at glacier surface and along length of tunnel). Ablation surfaces as observed on the tunnel wall are shown below (after McCall, 1960).

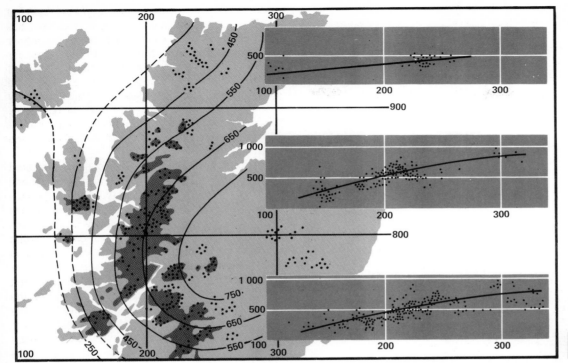

Fig. 12.6. Distribution of cirques in the Scottish Highlands. On the map the distribution is shown in relation to modern precipitation, with contours of cirque elevation interpolated from a diagram by Robinson *et al.*, 1971. The graphs on the right depict the elevation of the cirques between successive 100 km grid lines plotted in relation to their distance from the western edge of the map (after Linton, 1959).

Areas where annual precipitation exceeds 2290mm.

Yosemite in California are completely transformed. Yet the biggest changes are probably wrought where an ice-stream carves a completely new depression across a mountain mass, sometimes removing 1 000 m or more of solid rock in the process. In the following account attention will first be directed to the distinctive features of a valley reshaped by an Alpine glacier, and then to the forms produced by concentrated linear erosion beneath an ice-sheet.

Glacial erosion affects both the transverse and longitudinal profiles of a valley. The transverse profile is conventionally described as U-shaped although in relatively few cases do the valley walls approach the vertical. In Sweden Svensson found a transverse section through the Lapporten valley to conform quite closely to a parabola, but surprisingly few such morphometric analyses. have yet been attempted. In part this probably reflects continuing uncertainty about the degree of glacial modification suffered by many valleys. It is by no means unusual to find reaches of typical U-shaped profile separated by others in which a V-shaped profile dominates. In some instances the V-shaped reach merely represents an incision into a rock bar on the floor of a trough and can be explained by subglacial meltwater activity. In other cases, however, the variation in cross-profile affects the full depth of the valley and it is difficult to avoid the conclusion that ice has selectively widened certain reaches while leaving others almost untouched. Another difficult problem concerns the extent to which conversion into a U-shaped cross-profile has been accompanied by valley deepening. Rock basins testify to the ability of a glacier to deepen its bed, and hanging tributary valleys sometimes bear witness to the amount of vertical downcutting that has

Fig. 12.7. The glacial trough of Lauterbrunnen in Switzerland. Although excavated mainly in sedimentary rather than igneous rocks, it nevertheless displays a form similar in many respects to the Yosemite valley pictured in Fig. 12.1.

occurred (Fig. 12.8). However, the testimony of hanging valleys is by no means always clear. On the one hand, widening of the main valley may by itself produce a limited amount of hang, while on the other the tributary valleys could themselves have been subject to an unknown amount of glacial deepening. At one time it was argued that the gentle benches surmounting many Alpine troughs are remnants of the pre-glacial profile that might be reconstructed by extrapolating the remaining fragments towards the valley centre. Most Alpine workers strongly contest this view and are reluctant to admit that there has been wholesale deepening of the valley floors. Among the more plausible alternatives is that the Alps have long been subject to intermittent uplift and that the troughs already had a valley-within-valley form before being modified by glacial erosion.

The longitudinal profile of a glacially eroded valley is characteristically irregular. Besides frequent and sometimes dramatic changes in the down-valley gradient there are often reaches of reversed slope enclosing deep rock basins. Some basins are still occupied by lakes but many have been partially or totally infilled with sediment; for example seismic investigations on the floor of the Yosemite trough have revealed a detrital accumulation exceeding 600 m in thickness (Fig. 12.8). The factors determining the location of pronounced deepening of a valley floor have been the subject of many theories. It has been argued that an increase in erosive capacity leads to the development of a step immediately down-valley from the confluence of two glaciers. Since the valley is to a glacier as the channel is to a stream, it might be anticipated that as the channel enlarges below a stream confluence so, by analogy, the cross-sectional area of a valley will increase below a glacier confluence. It seems that part of this increase is often achieved by valley deepening. Many examples might be quoted to uphold this thesis, but the most consistent support comes from the excavation of trough headwalls just below the confluence of two or more cirque glaciers.

However, by no means all steps can be ascribed to greater ice discharge and alternative explanations must be sought. Many workers have endeavoured to relate them to changes in lithology, although several different arguments have been deployed. The simplest envisages selective bedrock erosion by the ice, often ascribing particular importance to the spacing of the joint systems as an aid to plucking. Where such an argument appears inappropriate, recourse may be had to the idea of preliminary deep weathering preparing the bedrock for removal; this might involve either decomposition during an interglacial phase or frost-riving during the cold period immediately preceding ice advance. Another hypothesis sometimes advanced is that where a valley is laterally constricted the glacier enlarges its cross-sectional area by eroding its floor more deeply. Despite this wide variety of possible explanations many specific rock steps still defy a full understanding, and a rather different approach to the problem was therefore adopted by Nye in 1952 when he emphasized the potential effect on ice movement of any initial profile irregularities. Employing the concept of extending and compressive motion outlined in Chapter 11, he suggested that each concave segment will tend to be subject to rapid erosion by compressive flow while any convexity will escape relatively lightly under extending flow. Irregularities will thus tend to be self-enhancing so that any factor capable of initiating them, including those discussed earlier in this paragraph, may ultimately be responsible for substantial bars, steps and basins.

Troughs fashioned by ice accumulating beyond the limits of the original watershed can be divided into several different types. The simplest comprises those shaped by outlet glaciers descending from the margins of an ice-cap or ice-sheet. These show few significant differences from the alpine type, although many outlet glaciers are unusually steep and fast-moving. A second type is associated with glacial diffluence in which an individual valley or outlet glacier divides into separate branches. This arises where a plentiful supply of ice fills the available valley to a level at which part of the glacier can escape across the surrounding watershed into an adjacent catchment. Numerous examples can be seen in present-day glaciated areas such as Alaska and Greenland. At first the diffluent lobe will carry a relatively small proportion of the discharge, but by vigorous erosion the route may be so deepened that eventually it is used by a much higher percentage of the ice.

A third type of trough is fashioned beneath an ice-cap or ice-sheet. If accumulation increases beyond a certain point valley glaciers ultimately merge to form a single continuous ice-mass. Although the domed form of the new ice-mass may demand a broadly

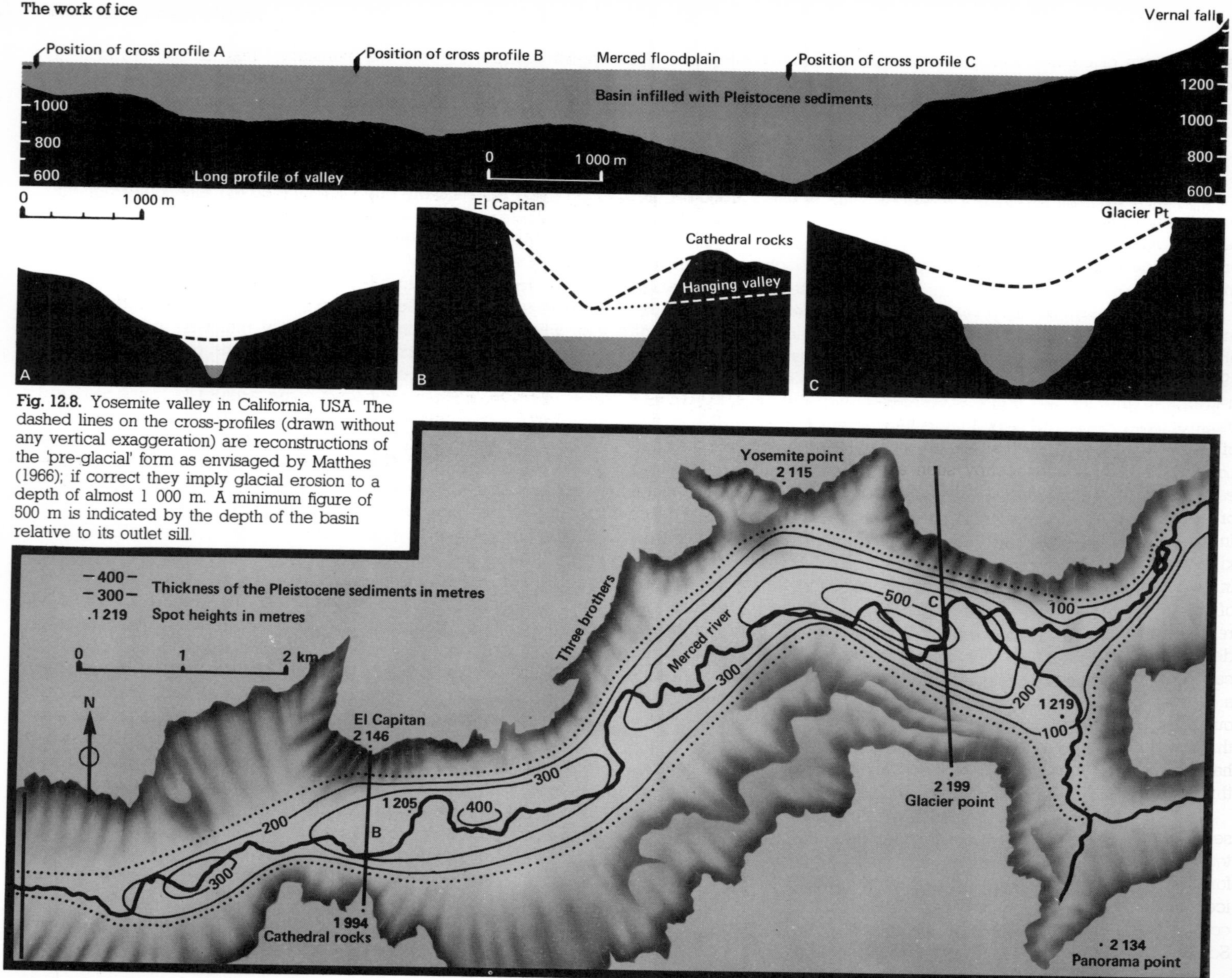

Fig. 12.8. Yosemite valley in California, USA. The dashed lines on the cross-profiles (drawn without any vertical exaggeration) are reconstructions of the 'pre-glacial' form as envisaged by Matthes (1966); if correct they imply glacial erosion to a depth of almost 1 000 m. A minimum figure of 500 m is indicated by the depth of the basin relative to its outlet sill.

radial pattern of outflow, this does not mean that the bedrock top-ography ceases to exert any influence. There is undoubtedly differ-ential streaming within many ice-masses, with maximum velocities concentrated along pre-existing valleys. However, instead of conforming completely to the original drainage pattern the ice may also gouge new routes across interfluves and impose on the under-lying relief a system of 'iceways' adapted to the transference of ice from the accumulation to the ablation area. Being almost entirely the result of glacial erosion, large troughs of this origin tend to be straight or gently curving with the classic U-shaped form uninterrupted by many tributary valleys.

Numerous breaches torn through old watersheds can eventually modify the pre-glacial relief almost beyond recognition (Figs 12.9 and 12.11), and by means of a simple statistical analysis Haynes in 1977 demonstrated an almost complete transformation of the valley network in the Scottish Highlands. Master troughs carved by the more active ice-streams may attain dimensions comparable to those of the original water-eroded valleys; in Scotland, for example, Loch Lomond is believed to occupy such a trough, scoured through an earlier interfluve when vigorously moving ice was diverted from its original outlet and locally lowered the rock surface by some 500 m. In the end the ice-shed on a large ice-sheet may be displaced many kilometres from the original watershed. This situation is found today in both Greenland and Antarctica, and its former existence in areas of Pleistocene glaciation is attested by erratics carried from source regions on one side of a mountain chain to be incorporated in glacial deposits on the other. In Scandinavia erratics on the Norwegian coast can be identified as coming from Sweden, while in northen Scotland a distinctive augen-gneiss in tills on the Atlantic coast is derived from outcrops on the opposite side of the North-west Highlands. Passes through both the Scandinavian and Scottish mountains thus appear to have been fashioned into typical glacial troughs by ice streaming through them towards the Atlantic coastlands.

Of course, an individual trough may have evolved under a varied set of conditions as ice-sheets waxed and waned during the Pleis-tocene period. However, several workers have argued that the main formative episodes are likely to have occurred during the maximum ice advances, and Sugden, for instance, contended in 1968 that the concept of ice streaming affords a far better explanation for many Scottish troughs than does the idea of diffluence to which they had

previously been attributed. He maintained that the outlines of many of the Cairngorm troughs could only have been shaped when the area was almost totally buried by ice. He envisaged a condition under which sluggish cold-based ice preserves the interfluves relatively intact, while streaming ice with its base warmed to pressure melting-point by locally generated frictional heat scours the intervening troughs; only in this way, it was suggested, can the frequently observed contrasts between upland and valley morphology be satis-factorily explained. In 1978 Sugden applied similar arguments in an analysis of the patterns of glacial erosion in North America. Here he endeavoured to reconstruct the thermal regime of the Laurentide ice-sheet and to show that there was a relationship between the nature of the erosional forms and temperature conditions at the base of the ice. He identified landscapes of areal scouring associated with central zones of basal melting. Areas with little or no sign of glacial erosion were found to coincide with zones calculated to have been covered with cold-based debris-free ice. Landscapes of selective linear erosion were noted as common on the uplands of Labrador where pre-existing valleys channelled the ice-flow and created a situ-ation in which there was warm-based ice over the valleys and protective cold-based ice over the intervening plateaux; some of the resultant valleys in Labrador are almost archetypal glacial troughs.

Fjords

In essence, fjords are troughs partially submerged by the sea. Just as troughs have a diverse origin, so therefore do fjords, and examples can be found associated with alpine valley glaciers, outlet glaciers, diffluent ice-lobes, and streaming within large ice-sheets. They often provide exceptionally fine instances of the morphological features characteristic of troughs. Thus many possess unusually steep sides and prominent rugged headwalls. However, their most distinctive attribute is undoubtedly the deep rock basin lying submerged near their seaward end. Separated from the open ocean by a shallow bar or threshold, the basins often descend many hundreds of metres below present sea-level. As the eustatic fall in sea-level during a glaciation did not greatly exceed 100 m (see p. 322), and in any case most fjord areas must have been isostatically depressed, it is difficult to avoid the conclusion that much of the glacial erosion took place below sea-level. Assuming a density for the ice of 900 kg m^{-3} a glacier 1 000 m thick would still be in contact with its bed when

A

B

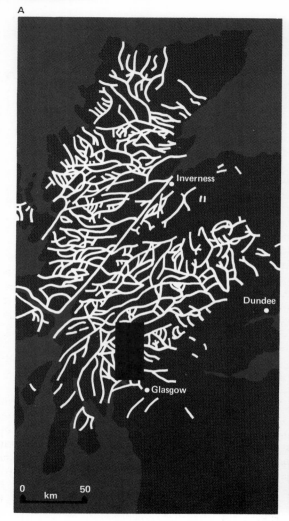

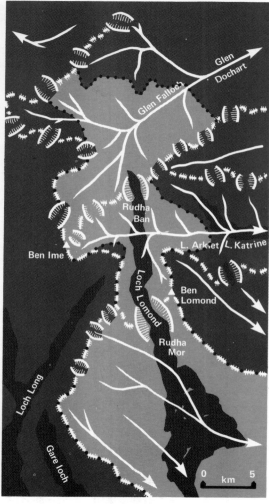

Fig. 12.9. Aspects of glacial erosion in the Scottish Highlands: (A) glacial troughs (after Linton, quoted by Clayton, 1974). Note in particular the anastomosing pattern in the most intensely glaciated areas; (B) scouring of the Loch Lomond trough by southward-moving ice from original eastward-trending river catchments. Glacial erosion at both Rudha Ban and Rudhá Mor is estimated to have removed 500 m of rock (after Linton and Moisley, 1960).

 Present Lomond catchment

 Reconstructed **pre-glacial** watershed

 Glacially scoured breaches in **pre-glacial** watershed

 Pre-glacial. drainage lines

in water almost 900 m deep, and there is no reason for believing it to be incapable of eroding at such a depth. Fjords may therefore be regarded as the product of thick, active ice-streams and glaciers that often descended steeply from high coastal plateaux and mountains. The threshold appears to signify a rapid decline in the erosive power of the ice, possibly resulting from accelerated basal melting at the point of contact with sea-water.

Erosional forms in lowland areas

Compared with the erosional forms of mountain areas those in lowland districts have often been rather neglected. In part this may be due to evidence that ice can move across such a region without disrupting the underlying surface at all. This is attested by till resting directly on such delicate materials as stratified sand and silt, laminated clay and, perhaps most diagnostic of all, fossil soil profiles. Although preliminary freezing may have rendered each of these more resistant than it appears today, it remains true that an ice-sheet operates neither uniformly nor universally as a scouring agent. On the other hand, there is abundant evidence that many lowland areas have been significantly degraded by the passage of an ice-sheet. The resultant forms vary according to the nature of the bedrock.

On the igneous and metamorphic basement rocks of Canada and Scandinavia there is occasional large-scale grooving, but more characteristic is seemingly haphazard relief with numerous lakes and marshy areas set amidst low, craggy hills of irregular outline. A rather similar 'knock and lochan' topography characterizes the coastal lowlands of Moine schist and Lewisian gneiss in north-western Scotland. Faults, shatter belts and other lines of weakness have been selectively eroded so that, except in such small-scale features as striated pavements and *roches moutonnées*, direct glacial lineation is not very obvious. Estimates of the amount of glacial erosion involved in the production of the present relief differ sharply. On the Canadian Shield regional scouring to a depth of 100 m or more has been claimed, while other workers, by contrast, have contended that features exhumed in pre-glacial times from beneath a cover of Ordovician rocks have escaped largely intact. Yet intense local degradation seems implicit in the depths of certain lakes; the floor of Great Slave Lake, for instance, lies 600 m or more below water-level. In Massachusetts the relationship between the present surface and older sheeting structures has been held to indicate a lowering

of 3–4 m by abrasion on the onset side of the average rock knob, and of up to 30 m by plucking on the lee side. In Scotland it is difficult to envisage parts of the knock and lochan relief being produced without at least 50 m of glacial erosion.

In lowland areas comprising isolated resistant outcrops set amidst much weaker strata, upstanding knolls are frequently fashioned into streamlined forms known as rock drumlins. Their onset sides are usually steep and craggy, their lee sides gentler and more smoothly rounded. Long curving depressions are often scored along both margins. Rock drumlins occur in considerable numbers throughout the Scottish Lowlands where intrusive igneous rocks provide the resistant cores. Where weaker rock is preserved on the lee side, the term 'crag and tail' may appropriately be used. It is worth emphasizing that the true crag and tail, such as that on which Edinburgh stands, can only be produced where there is significant glacial erosion.

In scarp-and-vale areas the nature of glacial modification seems to depend in part on the direction of ice movement relative to the strike of the rocks. The chalk cuesta of eastern England is believed to illustrate the effect of overrunning by obliquely moving ice (Fig. 12.10). South-west of Hitchin, beyond the generally accepted limit of glaciation, the Chalk forms the prominent escarpment of the Chiltern Hills with its crestline frequently exceeding 200 m in height. To the north-east, by contrast, the escarpment is much reduced in elevation and is progressively set back many kilometres from the western edge of the outcrop. The typical cuesta form is lost, being replaced by a low and relatively even-crested plateau locally plastered with drift. The view that the morphological contrasts between glaciated and unglaciated segments of the cuesta are due to ice scouring receives support from the very chalky nature of the till sheet in southern East Anglia. Even stronger confirmation of glacial erosion comes from quarries east of Hitchin. Here outcrops of apparent *in situ* bedrock are seen in reality to be composed of gigantic slabs of chalk, tens of metres in length and many metres thick, piled one on top of the other by the ice. The frequency of such mega-erratics is difficult to assess, but examples are known from other cuestas overrun by ice-sheets. Further north, for instance, Lincoln Edge has obviously suffered significant modification to judge from the widespread distribution of huge Lincolnshire limestone erratics. The pronounced straightness and regularity of the Edge has even been attributed to

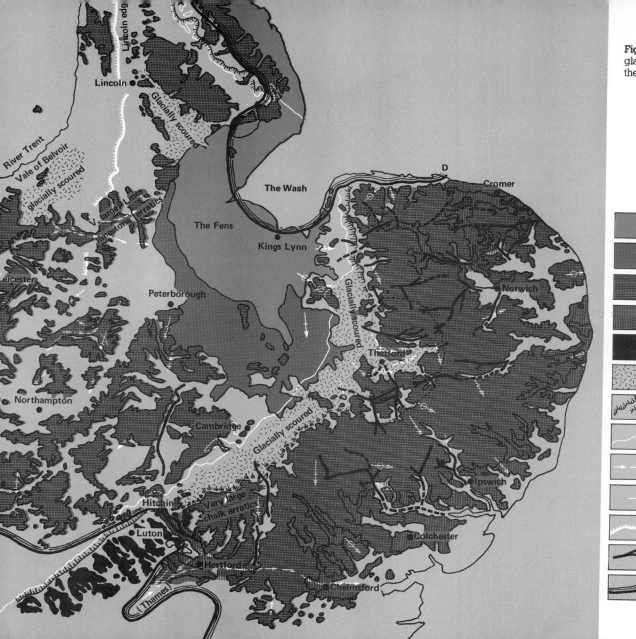

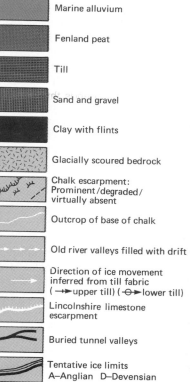

Fig. 12.10. Selected aspects of the glacial landforms in East Anglia and the East Midlands.

Marine alluvium

Fenland peat

Till

Sand and gravel

Clay with flints

Glacially scoured bedrock

Chalk escarpment:
Prominent/degraded/
virtually absent

Outcrop of base of chalk

Old river valleys filled with drift

Direction of ice movement
inferred from till fabric
(→upper till) (⊖►lower till)

Lincolnshire limestone
escarpment

Buried tunnel valleys

Tentative ice limits
A–Anglian D–Devensian

the trimming of former limestone promontories by ice moving parallel to the strike.

Where the ice advances at right angles to the trend of a cuesta, different erosional forms can be generated. These are well illustrated in southern Ontario where the Laurentide ice-sheet overwhelmed the Niagara limestone escarpment (Fig. 12.11). Instead of passing uniformly over the crest of the cuesta, the ice developed a series of more active streams, each of which scored a deep re-entrant in the scarp face. These effects may be compared with those on an even greater scale along the Allegheny scarp further east. Here the ice moving out of the Lake Ontario basin encountered a massive, partly dissected sandstone escarpment. Ice-streams developing along the axes of pre-glacial valleys scored the exceptionally deep rock basins in which the Finger Lakes now lie. Seneca Lake occupies a basin more than 300 m deep and is overlooked by hanging valleys that suggest an even greater amplitude of glacial erosion.

On relatively featureless lowlands erosion often seems to have been concentrated beneath lobes at the edge of the continental ice-sheet. In the Netherlands three or four deep rock basins have been identified, each apparently elongated parallel to the trend of ice movement and developed beneath a lobe of the last ice-sheet to affect the area (Fig. 12.12). Although the Netherlands today is known for its low relief, by the time the ice had arrived from Scandinavia a fall in sea-level may have induced sufficient downcutting by the rivers to promote ice-streaming and an associated lobate margin. Another possibility is to regard the basins as variants of the depressions found near the snouts of many Alpine glaciers and known by the German term *Zungenbecken*. The very deep lakes along the northern and southern edges of the Alps mark the positions of such basins and their consistent siting seems to reflect exceptionally active erosion where piedmont glaciers formerly spread out across lower ground. The modern Malaspina glacier is underlain by a similar basin and clearly there are processes inducing intense scouring where ice first fans out in a divergent pattern. Strongly compressive flow accompanied by radial shear planes passing upwards from the base of the ice appears the most likely explanation, and this might also be invoked where divergent movement occurs within the terminal lobe of a continental ice-sheet.

It seems worth concluding this review by stressing that glacial erosion in lowland regions of soft rock has often been underestimated in the past. Since the crucial evidence may be buried beneath later drift materials, reliance frequently has to be placed upon borehole logs which are notoriously difficult to interpret. Moreover, the surface morphology produced by areal scouring tends to lack readily diagnostic features so that degradation by an ice-sheet, even though it may be suspected, becomes extremely difficult to prove. Despite these problems, there is a steady accumulation of evidence indicating quite clearly that glacial erosion in lowland areas is more widespread than was formerly realized.

Glacial deposition

General considerations

According to its location, detritus being transported by an ice-mass may be divided into three broad categories: englacial, supraglacial and subglacial. The relative quantity of material in each category varies widely. Englacial detritus is normally very sparse, and deep cores in an ice-sheet or large valley glacier usually encounter great thicknesses of almost pure ice. The concentration of englacial debris only attains significant values where compressive flow is tending to carry material from the sole towards the surface. This is sometimes very apparent near a glacier snout where debris can be seen moving up shear planes towards the ablating surface above. Supraglacial material, other than aeolian dust, may be virtually absent on a large ice-sheet with few nunataks. On a valley glacier, by contrast, the ice surface is sometimes completely masked by avalanching from nearby rock faces. Often the supraglacial debris is concentrated into lateral moraines along the sides of a glacier, or into a medial moraine below the point where two glaciers merge. This material tends to become more widely distributed in the ablation zone owing to its insulating effect which results in the debris occupying sites which progressively stand proud of the surrounding clear ice. When this happens the fragments are prone to slip sideways and so be redistributed across the glacier surface. Subglacial detritus is more difficult to trace and measure. However, as already noted, most holes drilled to the sole of an ice-mass have encountered ice that is dirty in appearance and often studded with relatively large rock fragments. There seems little doubt that the subglacial load is extremely widespread, although rarely aggregating to any great thickness since concentrations by

Fig. 12.11. Glacial landforms around the southern and western margins of Lake Ontario. The inset shows a schematic section through the Finger Lakes, emphasizing the efficacy of ice-streaming and, in this limited area, the concurrent scouring of the interfluves. Further south and west interconnected troughs were gouged by the ice but the plateau surface itself was relatively little affected.

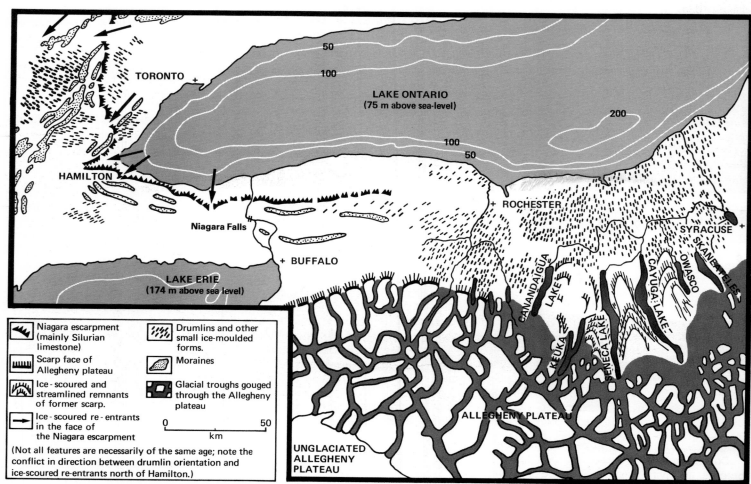

Niagara escarpment (mainly Silurian limestone)

Scarp face of Allegheny plateau

Ice - scoured and streamlined remnants of former scarp.

Ice - scoured re-entrants in the face of the Niagara escarpment

Drumlins and other small ice-moulded forms.

Moraines

Glacial troughs gouged through the Allegheny plateau

0   50
km

(Not all features are necessarily of the same age; note the conflict in direction between drumlin orientation and ice-scoured re-entrants north of Hamilton.)

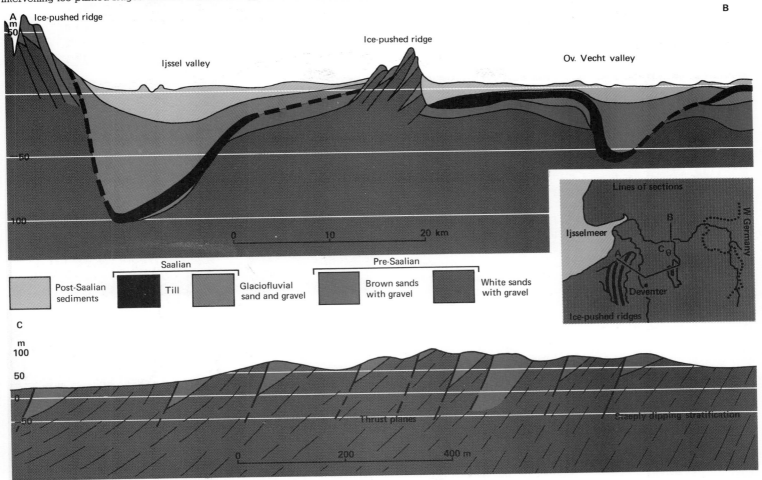

Fig. 12.12. Sections through the Pleistocene deposits of the eastern Netherlands (after de Jong). Section A–B portrays two deep ice-scoured basins with intervening ice-pushed ridges. Section C illustrates typical imbricate structures associated with the ice-pushed ridges.

weight are usually below 10 per cent and often below 1 per cent.

Deposition of the transported detritus is inevitably a complex process involving many different mechanisms. That fraction deposited directly from the ice is normally termed till. The most characteristic attribute of such glacigenic sediment is poor sorting, often accompanied by a lack of clear stratification. In addition the larger fragments or clasts, though rarely well rounded, may display edges and corners blunted by abrasion; occasionally a prominently striated face has been imposed by contact with the underlying bedrock. It is difficult to specify a typical mechanical composition since this varies widely according to the rock types over which the ice has moved. There are, for instance, great contrasts between the clast-dominated deposits of the Scottish Highlands and the matrix-dominated deposits of the clay lowlands in the English Midlands although both may be correctly identified as tills.

Traditionally two types of till were recognized, lodgement and ablation, the former being laid down subglacially when debris is released directly from the sole of the ice, and the latter accumulating initially in a supraglacial position but being later lowered to the ground surface by under-melting. However, as a result of detailed investigations along the glacier margins in Spitsbergen, Boulton in 1972 demonstrated that such a simple twofold genetic classification is totally inadequate. In particular it fails to take cognizance of the remarkable range of deposits and processes that characterize the surface of a wasting ice-edge (Fig. 12.13). Boulton distinguished two broad categories of supraglacial till. The first is flow till which consists of debris that has built up on the ice and after saturation with meltwater becomes so unstable that it flows or slumps into nearby hollows; the depressions in which it comes to rest may be either on the ice or else on earlier sediments. The second is melt-out till which is essentially englacial debris released by continued thawing beneath a detrital cover. Often the cover is a flow till or sheet of material laid down by meltwater, and the final result will be a succession in which the lower members have been released from the ice after the upper; under-melting of this type will frequently disrupt any earlier bedding. A possible variant of melt-out till is sublimation till in which disappearance of the enveloping ice takes place through direct evaporation rather than melting.

Similarly, in the subglacial environment lodgement is unlikely to be the only depositional process. Lodgement refers to progressive 'plastering on' of debris immobilized because friction with the underlying surface exceeds the drag induced by the moving ice; this is most likely to occur where basal melting loosens the grip of the transporting medium. On the other hand, where ice has stagnated and geothermal heat sustains a continued basal thawing, a form of subglacial melt-out till may be envisaged. If an ice-mass terminates

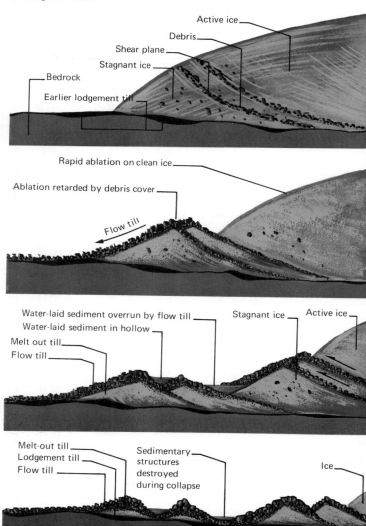

Fig. 12.13. Schematic diagram of sedimentation along the margin of a wasting ice-mass.

Active ice
Debris
Shear plane
Stagnant ice
Bedrock
Earlier lodgement till

Rapid ablation on clean ice
Ablation retarded by debris cover
Flow till

Water-laid sediment overrun by flow till   Stagnant ice   Active ice
Water-laid sediment in hollow
Melt out till
Flow till

Melt-out till   Sedimentary
Lodgement till   structures   Ice
Flow till   destroyed
                       during collapse

in water, whether it be the sea or a lake, sedimentation may occur beneath a floating ice-shelf. The resultant material will probably possess attributes of both an aqueous and a glacial deposit, but near the line where the shelf is grounded the dominant influence of the ice will often justify recognition of a distinctive 'waterlain till'. Gibbard, for example, has described from a number of locations clearly stratified tills that he believes accumulated in such a situation. Scandinavian workers recognize an additional category known as 'kalix till' which is also thought to have been deposited beneath a ponded ice margin but mainly by a slow-moving sheet of meltwater. These materials may be expected to grade laterally into a sediment where lacustrine or marine influences increase until the only sign of proximity to an ice-front lies in the presence of dropstones released from melting icebergs. Finally, examination of Pleistocene sections reveals that in some areas both drift deposits and adjacent bedrock have been so displaced and contorted by glacially generated folding and thrusting that a separate category of deformation till may need to be recognized.

One potential way by which glacigenic sediments of such diverse origin may be distinguished is on the basis of their internal structure or fabric. It was realized before 1900 that large erratics in a lodgement till are not randomly arranged but tend to have their long axes preferentially aligned. In the 1940s work on New England tills demonstrated a close relationship between clast shape and orientation, leading to the general conclusion that the long axes of most stones lie parallel to the presumed direction of ice movement. This property of lodgement till has been confirmed by many later studies, and has been shown to apply not only to the clasts but also to the finer particles in the matrix. It is the former, however, that have been the more widely studied. The orientation of an individual fragment is most conveniently specified in terms of the azimuth and dip of its long axis. These two measurements can be incorporated in a single diagram by plotting the points on an equal-area stereographic net. The array of points may be treated statistically to determine whether it deviates significantly from a random distribution; if it does, the mean vector is then normally calculated. In a lodgement till long axes are preferentially aligned parallel to the ice movement and with a gentle up-glacier dip. In many melt-out tills the particles seem to retain the disposition they had when being transported englacially. Boulton has shown that this can result in two contrasting fabrics. In recently deglaciated regions he found some sections displaying a preferred orientation parallel to the ice movement, and others at right angles. The difference is apparently related to the nature of ice-flow, with a strong compressional movement leading to a normal orientation and tensional movement to a parallel orientation. In flow tills clasts tend to be aligned parallel to the slope down which the material has slumped rather than to the original direction of ice movement. As might be anticipated, waterlain tills, unless disturbed by later subaqueous flowage, exhibit little preferential alignment.

As the above review suggests, combined fabric and compositional studies probably provide the best key to the origin of Pleistocene till sequences. However, it must be admitted that precise palaeo-environmental reconstructions often remain difficult to achieve.

Till plains

One of the most distinctive products of a continental ice-sheet is the relatively featureless plain underlain by a thick sheet of glacigenic debris. Most of this consists of lodgement till, but immediately after ice withdrawal there may also be a mantle of supraglacial detritus giving irregular but subdued topography. This superficial layer is the first to be removed by later erosion and so is absent from old till plains. Where well preserved, as in the Midwest of the United States, such plains can be remarkably smooth and cover many thousands of square kilometres. However, subaerial erosion will gradually dissect the original surface and eventually penetrate to the underlying bedrock. In this manner the once continuous cover is reduced to more or less isolated interfluve cappings such as can be seen today in the glaciated lowlands of midland and eastern England (Fig. 12.10).

The deposits underlying a till plain often reveal a complex history of accumulation. Beds of contrasting lithology and provenance may succeed each other with almost bewildering frequency. Moreover, since the surface relief is generally less than that of the buried topography the total thickness tends to be greatest over the former valleys and least over the interfluves; summits on the bedrock landscape may even protrude as local eminences rising above the till plain. Individual beds when traced from valley to interfluve commonly thin or disappear entirely and this can present formidable problems of correlation. It should also be emphasized that, although by definition the dominant surface material is till, the underlying deposits frequently include thick sequences of water-laid sediment, particularly along the lines of old valleys. It appears that the long history of sedimentation has usually involved a progressive reduction in relief and that the smooth till plain is merely a culmination of this trend.

Given the very small quantity of subglacial detritus beneath most modern ice-sheets, the requisite volume of lodgement till could only accumulate after prolonged deposition. The converse argument is equally valid; total stagnation of an ice-sheet would add a relatively insignificant layer of debris once melting was complete. Certain conditions must therefore have conduced to the release of the subglacial load from an actively moving ice-mass. As already indicated, the most important is likely to be substantial basal melting which will increase the frictional drag until it surpasses the tractive force of the ice. A minor variant may involve repeated stagnation of a debris-charged sole where enhanced friction with the floor leads to shear planes developing in the more easily deformed clear ice above; continued melting will later release the detritus from the immobilized basal layer. In either case, thermal conditions in the subglacial zone are obviously crucial. During the long history of an ice-sheet, conditions favouring deposition of lodgement till are likely to migrate over wide areas of the glaciated terrain; in general till plains are observed to be best developed on low ground some distance from the maximum limit of ice advance.

Difficult problems of interpretation are presented by the stratigraphy of till sheets. A single sheet is often divisible into layers, each characterized by a particular assemblage of erratics. There has sometimes been a tendency to assume that each layer represents a separate glaciation, but further supporting evidence is needed to justify such an assumption. A single glacial episode has often witnessed major shifts in the direction of ice movement with resultant changes in till composition. If sufficient sections are available it may be possible to demonstrate interdigitation of till types in a way that leaves no doubt about their broad contemporaneity. This is true for instance in the English Midlands where, during one glaciation, initial ice from the north-west was directly replaced by ice from the north-east. The earlier advance deposited a reddish till with a predominance of Triassic erratics preferentially aligned in a north-west–south-east direction. Occasional lenses within this red till contain abundant chalk and flint indicating that, even at an early stage, there was intermittent transport of erratics from the north-east.

Within a vertical interval of a few metres but with much interleaving of the deposits, the red till finally passes up into a greyish till bearing immense quantities of chalk and other erratics with a preferred north-east–south-west orientation.

Analogous variations in many other areas and in tills of many different ages can only be satisfactorily explained as the result of shifts in the direction of ice movement. Such shifts probably derive from changes in the pattern of ice accumulation, with different centres of dispersal achieving temporary dominance. This is most apparent around the margins of uplands lying outside the main centres of ice accumulation. At the onset of a glaciation, valley glaciers from these upland regions debouch freely on to adjacent lowlands. However, when the main ice-sheet arrives it constricts the locally generated ice. In consequence the till around the upland flanks often contains abundant local material at the base, replaced upwards by more far-travelled erratics marking the arrival of the main ice-sheet. The situation may conceivably be reversed during the declining phase of a glaciation with the production of an overlying till once more rich in local erratics.

With the potential addition of supraglacial tills and outwash during temporary melt phases, it is easy to see how an extremely complex succession can be built up in the course of only one major glaciation. Nevertheless, the thick sequences beneath certain till plains are undoubtedly the product of several independent glaciations. The finest examples are probably afforded by the till plains of the American Midwest. Over vast areas these are underlain by deposits averaging about 30 m in thickness, but with the infill along old valleys reaching over 150 m. The sediments belong to at least three separate glaciations. This is indicated by interglacial materials and fossil soil profiles at several horizons within the sequence. Such huge accumulations are fully capable of obliterating earlier drainage systems; boreholes supplemented by geophysical investigations have disclosed beneath Ohio, Indiana and Illinois a buried valley pattern totally different from that of today (Fig. 12.14). The major element in this ancient system was the Teays valley which ran westwards across the centre of the three states to the Mississippi. During an early ice advance its lower course was completely blocked and over 120 m of glacial deposits laid down. The valley was re-excavated during the ensuing warm period, but when the ice returned it was choked once more with glacial debris. The whole cycle was

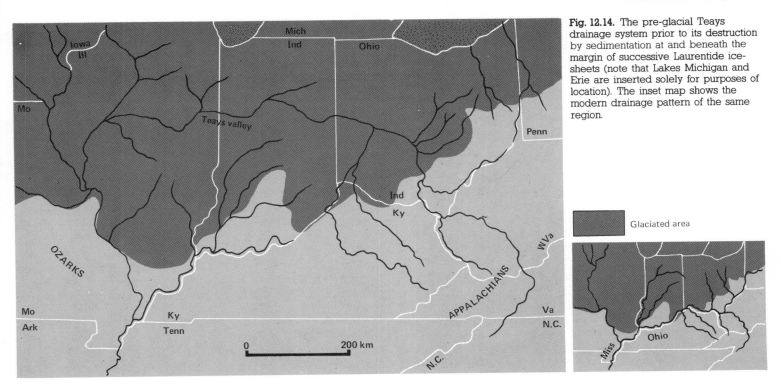

Fig. 12.14. The pre-glacial Teays drainage system prior to its destruction by sedimentation at and beneath the margin of successive Laurentide ice-sheets (note that Lakes Michigan and Erie are inserted solely for purposes of location). The inset map shows the modern drainage pattern of the same region.

Glaciated area

repeated a third time, but on this last occasion a new marginal drainage route was initiated along the present Ohio valley leaving the Teays system deeply buried beneath the till plain.

Drumlin fields

Drumlin fields resemble till plains in being composed primarily of glacially deposited materials, but differ from them in that the extremely subdued relief is replaced by distinctive patterns of low, elongated hills. An individual drumlin may vary in size from a swell only a few metres high to a hill that rises some 50 m above adjacent hollows. Its length and width are commonly measured in terms of a few hundred metres. A small isolated drumlin might prove difficult to identify as such, but the great majority occur in swarms and in this form constitute a prominent glacial landform. One of the greatest concentrations in the world lies in New York State south of Lake Ontario where some 10 000 drumlins occur within an area 250 × 60 km (Fig. 12.11). Other major swarms occupy parts of Wisconsin and New England, while in the British Isles rather smaller groups are found in Northern Ireland, the Glasgow area and the Vale of Eden.

Approximately ellipsoidal in plan, drumlins generally lie with their long axes parallel to the inferred direction of ice movement. Ratios of length to width usually lie in the range 2–3. A tendency towards a blunt onset end and a longer tapering tail has encouraged several workers to draw analogies with such streamlined forms as the cross-section of aircraft wings and to argue that drumlins will become more elongated where moulded by faster-moving ice. More recent analyses of shape have revealed that there may be hierarchies of drumlins with smaller examples superimposed on larger 'mega-drumlins', often over 1 km long, with the result that a single swarm may exhibit rather greater diversity of size, shape and orientation than was formerly supposed. In addition to morphometry, another technique that can profitably be employed in the study of drumlins is till fabric analysis. In Minnesota Wright found clasts to be aligned parallel to the orientation of the drumlin, but plunging up-glacier at the unusually large mean angle of 23°. He thought the high dip angle might be related to steep shear planes within exceptionally fast-moving ice, and suggested this as one possible explanation for the development of a drumlin field rather than a featureless till plain. Later work in England and Northern Ireland failed to detect a comparably steep dip, or any consistency in the direction of plunge either up-glacier or down-glacier. In most cases a parallelism has been found between clast and drumlin orientation, but in a few cases substantial divergences have also been noted. In some drumlins there are prominent large-scale fold and fault structures, indicative of localized high-stress conditions.

It has sometimes been maintained that drumlins develop where the basal ice is very unevenly laden with debris. Clean ice deforms more readily than that heavily encumbered with sediment. Dirty patches of the basal layer will therefore be the first to stabilize and be deposited, while adjacent clean ice continues to move past and mould the accumulation into a streamlined form. Whether full-sized drumlins could be adequately explained in this way is doubtful, and an interesting variant of the hypothesis attaches particular importance to the physical properties of the subglacial load which is viewed as a dilatant clay–water system. When dilated under high stress, this flows readily with the overlying ice, but if the local stress falls below a critical level, as may happen for instance in the lee of an obstruction, the material collapses to a more stable state and loses its mobility. It will then be bypassed by those sections of the load where local stresses have been either maintained or even increased. The effect of the enlarged obstruction will be cumulative so that a bigger streamlined form may evolve by accretion.

A totally different explanation for at least some drumlins has been suggested by Shaw. He noted from air photos a remarkable resemblance between drumlins in northern Saskatchewan and the casts of erosional marks left by water flow across cohesive sediments. This suggested that drumlins might conceivably originate from the injection of saturated subglacial detritus into the irregular underside of an ice-sheet scored by the passage of copious volumes of meltwater. This would be consistent with earlier observations that material must have moved upwards into the core of certain drumlins.

Finally it is worth noting that drumlins may conceivably be viewed as end members of a depositional series having, at its other extreme, delicate flutings less than 1 m high. These are common on recently deglaciated till surfaces where they trend parallel to the direction of ice movement, often in the lee of an obstacle such as a large boulder. They are, however, too fragile to persist for any lengthy period after the ice withdrawal. Intermediate forms include long narrow ridges of till which, in North Dakota, are recorded as attaining over 1 km in length but only 50 m in width and 5 m in height; they are highest near their onset ends and taper slowly in the lee direction. Drumlins would then represent subglacial conditions of increasing energy and sediment supply, with the largest megadrumlins evolving where these factors reach maximum values; drumlins superimposed upon mega-drumlins would develop under conditions of a diminishing energy regime.

End moraines

End moraines may be defined as ridges of glacigenic material accumulating along the terminal margins of active ice-masses. They vary in height from less than 1 m to well over 100 m, very large examples being associated with valley glaciers as well as ice-sheets. Across lowland regions end moraines may be traceable for hundreds of kilometres (see, e.g. Fig. 13.7), but their continuity is usually broken by gaps through which meltwater flowed either at the time of accumulation or during later ice withdrawal. Many different processes may be involved in their formation. Often stratified water-laid materials contribute substantially to the total volume of sediment, but if they are in the majority the feature is better termed a kame complex. From a detailed study of Icelandic moraines Okko concluded that two major depositing mechanisms might be invoked, namely dumping and pushing. Later investigations have often distinguished a third, known as squeezing. In many instances these are complementary, with large moraines the result of all three acting in concert.

The building of a large feature by dumping demands a stable ice-front and a plentiful supply of new debris. The exceptionally high moraines constructed by certain valley glaciers are probably to be explained by unusually large loads carried by the ice, which may in turn be linked with the erosion of *Zungenbecken* a short distance up-valley. Several workers have described one way in which dumping takes place. This involves upward movement of debris along shear planes in the marginal ice. Emergent particles mix with any supra-glacial detritus, and differential ablation then induces slumping so that all the material accumulates as a wedge at the foot of the slope. Melt-out till from any buried ice may later be added, but this is unlikely to amount to any large volume of sediment. In many instances a trough has been noted developing in the cleaner ice immediately up-glacier from the outcrop of the shear planes, thus depriving the ice-cored moraine, into which the stagnant glacier terminus has been converted, of any further debris supply (as e.g. in Fig. 12.13). To build up a substantial end moraine thus requires the whole process to be repeated many times, but this is not at all improbable in view of the constant minor fluctuations that characterize any active glacier snout.

Even at a virtually stable ice-front, short-term pulses of advance and retreat regularly disturb previously laid sediments. The action has been likened to that of a bulldozer, the ice pushing debris before it into a prominent ridge. The internal structure of moraines often reveals that such disturbance has taken place. Faulting, thrusting and imbrication may all be present, while in Iceland recent moraines produced by glacial pushing even contain contorted peat beds.

The terminal margins of many ice-masses are underlain by detritus saturated with meltwater. Even just beyond the ice edge the debris may be so waterlogged that anyone attempting to walk on it will immediately sink in above the knees. Such material deforms readily under stress and the overburden of ice may squeeze it upwards into crevasses and tunnels; an occasional reticulate pattern of till ridges left after melting of stagnant ice is believed to reflect this process. Some end moraines may be partly due to extrusion of similarly saturated debris from beneath the snouts of active ice-masses. In Iceland Price described individual ridges between 1 and 4 m high believed to have formed in this way. He analysed the orientation of clasts in the till and showed that their long axes are preferentially aligned at right angles to the ridge crests. On both sides of the crest the clasts tend to dip up-glacier, suggesting the ridges constitute a single dynamic unit and are not the product of slumping. In some localities the ridges appear to be annual features, possibly because summer meltwater reduces the shear strength of the subglacial debris well below its normal winter value. With a stable ice-front it is not difficult to envisage a relatively thick morainic accumulation building up by this mechanism.

While most end moraines have developed along the margins of ice-masses terminating on dry land, others have formed where the ice terminates in either ponded water or the sea. Termed cross-valley moraines, these features were the subject of detailed study by Andrews and Smithson on Baffin Island where a particular form of squeezing was invoked as the dominant process. On the gentler proximal slopes the stones in the till were preferentially aligned normal to the ridge crest and dipping up-glacier, while on the steeper distal side only a weak preferential alignment was recorded. This suggests the moraines may have evolved at the foot of a high ice-cliff where seepage of water was causing basal melting. Semi-liquid debris produced in this environment would be squeezed into the lake by the pressure of ice. On the proximal side where movement was closely controlled by the ice a marked orientation of the clasts would be imposed, whereas on the distal side greater freedom of movement would result in a significantly weaker orientation. Andrews and Smithson estimated that the cross-valley moraines they studied on Baffin Island, which were up to 20 m high, formed in water over 100 m deep.

Finally reference should be made to certain moraine-like features that appear to be due to ice-pushing on an exceptionally large scale. Sometimes termed ice-pushed ridges to distinguish them from conventional moraines, the material of which they are composed was never fully incorporated within the ice. Excellent examples are afforded by the hilly regions of the Netherlands, many of which are composed of sediments laid down by the Rhine in early Pleistocene times (Fig. 12.12). Deposited originally as fluvial or estuarine sands close to present-day sea-level, they are now found as steeply dipping, imbricated sheets building ridges that locally exceed 100 m in height. As the disruption of the former bedding is known to extend a further 100 m below sea-level the scale of the displacement is massive. Although the evidence for thrusting is inescapable, the precise mechanisms involved remain obscure. Several early Dutch workers invoked a bulldozing action as the ice forced its way south-wards into a valley system freshly rejuvenated by the glacio-eustatic fall in sea-level; a possible accessory factor was permafrost that concentrated any tendency towards shearing at the base of the frozen sands. However, severe doubt persists regarding the adequacy of this explanation, and later work has paid increasing attention to the potential role of porewater pressure in facilitating ice-pushing. As indicated in Chapter 6, water pressure is a major determinant of the shear strength of porous materials, and several factors might contribute to high subglacial pressures. Advance of the ice may block previous outlets for groundwater discharge and by reversing the direction of flow raise the pressure of interstitial water; permafrost beyond the ice-edge may also act as a barrier to the escape of groundwater with similar results. Where ice overrides a succession of poorly lithified materials, consolidation will tend to expel water from the fine-grained sediments and raise the pressure in the coarse-grained, more permeable strata. Yet the most potent factor of all, accentuating those already mentioned, is the production of meltwater at the base of the ice. Measurements at the sole of several glaciers have revealed pressures in excess of 10^6 N m^{-2} and these can certainly induce a very significant reduction in the shear strength of underlying permeable sediments. The most favourable conditions will exist where such strata rest on, and are confined laterally by, impermeable beds.

Moran has argued that large-scale block movement is probably a more important factor in the construction of ordinary end moraines than is commonly supposed. This, however, is difficult to demonstrate in the absence of detailed stratigraphic sections. Some thick accumulations of glacial debris undoubtedly include huge masses of translocated bedrock. At Cromer on the East Anglian coast blocks of chalk exposed in the cliffs are as much as 500 m in length. Other well-known occurrences are found in coastal sections in Denmark and southern New England. In each of these instances the bedrock masses are enveloped in normal till and the term moraines may appropriately be used for the topographic form. The fact that the best-authenticated cases have a coastal location almost certainly reflects the relatively complete sections provided by marine cliffs. On the basis of deep boreholes, several workers have suggested that inland moraines in both North America and Europe may also include large displaced slices of bedrock or even of earlier glacial deposits. Before leaving this topic, however, it should be emphasized that there is no proof in the foregoing examples that accumulation occurred at the actual ice-edge, and the possibility of formation some distance up-glacier from the ice-margin needs to be acknowledged. Strewn across the ice-pushed ridges of the Netherlands lie igneous and metamorphic erratics from Scandinavia, while in both Sweden and Canada workers have described till ridges with summits apparently grooved by actively moving ice. Since there is no evidence of later overrunning by a separate advance, it is inferred that some ice-pushed and till ridges probably develop subglacially. In Scandinavia these have been termed Rogen moraines, although precisely which mechanisms have been responsible for their construction often remains obscure.

Lateral moraines

Although lateral moraines are very prominent along the margins of many present-day Alpine glaciers, they rarely form ridges of comparable size in formerly glaciated areas. One factor is undoubtedly the ice-core that constitutes the bulk of so many modern lateral moraines. A second is the rapid erosion to which such a feature is prone once it has been deposited on a steep mountain slope. This is evident in the deep gullying that often characterizes moraines stranded a short distance above the current glacier surface. Most of the debris in a lateral moraine originates from rock fall rather than glacial erosion. Accumulating along the margin of the ice, it reduces the local ablation rate, and if the glacier is shrinking the clear ice near the centre will downwaste faster than that at the edges. Inversion of the relief will ensue, leaving an ice-cored ridge in place of the original outward-sloping glacier surface. When the ice-core finally melts, a line of debris will be left marking the earlier longitudinal edge of the glacier. In this way lateral moraines may be particularly helpful in the reconstruction of former glacier gradients. Small also used the volume of debris in lateral moraines on a Swiss glacier to calculate that rock faces above the ice-level must have been receding at least 0.75 mm yr^{-1} for the last 5 millennia.

References

Aario, R. (1977) 'Classification and terminology of morainic landforms in Finland', *Boreas* **6**, 87–100.

Andrews, J. T. and Smithson, B. B. (1966) 'Till fabrics of the cross-valley moraines of north-central Baffin Island, N.W.T., Canada,' *Geol. Soc. Am. Bull.* **77**, 271–90.

Andrews, J. T. and King, C. A. M. (1968) 'Comparative till fabrics and till fabric variability in a till sheet and a drumlin: a small scale study', *Proc. Yorks. Geol. Soc.* **36**, 435–61.

Aniya, M. and Welch, R. (1981) 'Morphometric analyses of Antarctic cirques from photogrammetric measurements', *Geogr. Annlr.* **63A**, 41–53.

Battle W. R. B. (1960) 'Temperature observations in bergschrunds and their relationship to frost shattering', in *Norwegian Cirque Glaciers* (ed. W. V. Lewis), R. Geogr. Soc. Res. Ser. 4.

Boulton, G. S. (1972) 'Modern Arctic glaciers as depositional models for former ice sheets', *J. Geol. Soc. Lond.* **128**, 361–93.

Boulton, G. S. (1974) 'Processes and patterns of glacial erosion' in *Glacial Geomorphology* (ed. D. R. Coates), State Univ. New York.

Boulton, G. S. (1975) 'Processes and patterns of sub-glacial sedimentation: a theoretical approach', in *Ice Ages: ancient and modern* (ed. A. E. Wright and F. Moseley), Seel House Press.

Boulton, G. S. and Vivian R. (1973) 'Underneath the glaciers', *Geogr. Mag.* **45**, 311–19.

Boulton, G. S. *et al.* (1979) Direct measurement of stress at the base of a glacier', *J. Glaciol.* **22**, 3–24.

Chorley, R. J. (1959) 'The shape of drumlins', *J. Glaciol.* **3**, 339–44.

Clayton, K. M. (1974) 'Zones of glacial erosion', in *Progress in Geomorphology*, Inst. Brit. Geogr. Spec. Pub. 7.

de Jong, J. D. (1967) 'The Quaternary of the Netherlands' in *The Quaternary*: 2 (ed. K. Rankama), Interscience.

Derbyshire, E. and Evans, I. S. (1976) 'The climatic factor in cirque variation', in *Geomorphology and Climate* (ed. E. Derbyshire), Wiley.

Evans, I. S. (1977) 'World-wide variations in the direction and concentration of cirque and glacier aspects', *Geogr. Annlr.* **59A**, 151–75.

Flint, R. F. (1971) *Glacial and Quaternary Geology*, Wiley.

Gibbard, P. (1980) 'The origin of stratified Catfish Creek till by basal melting', *Boreas* **9**, 71–85.

Goldthwait, R. P. (1979) 'Giant grooves made by concentrated basal ice streams', *J. Glaciol.* **23**, 297–307.

Gordon J. E. (1981) 'Ice-scoured topography and its relationships to bedrock structure and ice movement in parts of northern Scotland and west Greenland', *Geogr. Annlr.* **63A**, 55–65.

Gow, A. J.., Epstein, S. and Sheehy, W. (1979) 'On the origin of stratified debris in ice cores from the bottom of the Antarctic ice sheet', *J. Glaciol.* **23**, 185–92.

Gravenor, C. P., von Brunn, V. and Dreimanis, A. (1984) 'Nature and classification of waterlain glaciogenic sediments, exemplified by Pleistocene, Late Paleozoic and Late Precambrian deposits', *Earth Sci. Rev.* **20**, 105–66.

Gutenberg, B. *et al.* (1956) 'Seismic explorations on the floor of Yosemite valley, California', *Geol. Soc. Am. Bull.* **67**, 1051–78.

Haynes, V. M. (1968) 'The influence of glacial erosion and rock structure on corries in Scotland', *Geogr. Annlr.* **50A**, 221–34.

Haynes, V. M. (1977) 'The modification of valley patterns by ice-sheet activity', *Geogr. Annlr.* **59A**, 195–207.

Herron, S. and Langway, C. C. (1979) 'The debris laden ice at the bottom of the Greenland ice sheet', *J. Glaciol.* **23**, 193–207.

Hill, A. R. (1971) 'The internal composition and structure of drumlins in north Down and south Antrim, Northern Ireland', *Geogr. Annlr.* **53A**, 14–31.

Kamb, B. and LaChapelle, E. R. (1964) 'Direct observation of the mechanism of glacier sliding over bedrock', *J. Glaciol.* **5**, 159–72.

Koerner, R. M. and Fisher, D. A. (1979) 'Discontinuous flow, ice texture and dirt content in the basal layers of the Devon Island ice cap', *J. Glaciol.* **23**, 209–21.

Lewis, W. V. (ed.) (1960) *Norwegian Cirque Glaciers*, R. Geogr. Soc. Res. Ser. 4.

Linton, D. L. (1949) 'Watershed breaching by ice in Scotland', *Trans. Inst. Brit. Geogr.* **17**, 1–16.

Linton D. L. (1959) 'Morphological contrasts of eastern and western Scotland' in *Geographical essays in memory of A. G. Ogilvie* (ed. R. Miller and J. W. Watson), Nelson.

Linton, D. L. (1963) 'The forms of glacial erosion', *Trans. Inst. Brit. Geogr.* **33**, 1–28.

Linton, D. L. and Moisley, H. A. (1960) 'The origin of Loch Lomond', *Scot. Geogr. Mag.* **76**, 26–37.

Manley, G. (1959) 'The late-glacial climate of North-west England', *Lpool and Manchr Geol. J.* **2**, 188–215.

Markgren, M. and Lassila, M. (1980) 'Problems of moraine morphology: Rogen moraine and Blattnick moraine', *Boreas* **9**, 271–4.

Mathews, W. H. (1979) 'Simulated glacial abrasion', *J. Glaciol.* **23**, 51–5.

Matthes, F. E. (1966) *The Incomparable Valley*, Univ. Calif. Press.

McCall, J. G. (1960) 'The flow characteristics of a cirque glacier and their effect on glacial structure and cirque formation', in *Norwegian Cirque Glaciers*, R. Geogr. Soc. Res. Ser. 4.

Menzies, J. (1979) 'A review of the literature on the formation and location of drumlins', *Earth Sci. Rev.* **14**, 315–59.

Moran, S. R. (1971) 'Glaciotectonic structures in drift' in *Till: a symposium* (ed. R. P. Goldthwait), Ohio State Univ. Press.

Nye, J. F. (1952) 'The mechanics of glacier flow', *J. Glaciol.* **2**, 82–93.

Nye, J. F. (1965) 'The flow of a glacier in a channel of rectangular, elliptic or parabolic cross-section', *J. Glaciol.* **5**, 661–90.

Okko, V. (1955) 'Glacial drift in Iceland. Its origin and morphology', *Comm. Geol. de Finlande Bull.* **170**.

Price, R. J. (1970) 'Moraines at Fjallsjokull, Iceland', *J. Arct. Alp. Res.* **2**, 27–42.

Ray, L. L. (1974) 'Geomorphology and Quaternary geology of the glaciated Ohio river valley–a reconnaissance study', *U.S. Geol. Surv. Prof. Pap.* 826.

Rice, R. J. (1968) 'The Quaternary deposits of central Leicestershire', *Phil. Trans. R. Soc.* **A262**, 459–509.

References

Robinson G. *et al.* 1971 'Trend surface analysis of corrie altitudes in Scotland' Scot. Geogr. Mag. **87**, 142–6.

Rose J. and **Letzer, J. M.** (1977) 'Superimposed drumlins', *J. Glaciol.* **18**, 471–80.

Shaw, J. (1984) 'Drumlin formation related to inverted melt-water erosional marks', *J. Glaciol.* **29**, 461–79.

Small, R. J. (1983) 'Lateral moraines of glacier de Tsidjiore Nouve: form, development and implications', *J. Glaciol.* **29**, 250–9.

Smalley, I. J. and **Unwin, D. J.** (1968) 'The formation and shape of drumlins and their distribution and orientation in drumlin fields,' *J. Glaciol.* **7**, 377–90.

Souchez, R. A. and **Lorrain, R. D.** (1978) 'Origin of the basal ice layer from Alpine glaciers indicated by its chemistry', *J. Glaciol.* **20**, 319–28.

Stanford S. D. and **D. M Mickelson** (1985) 'Till fabric and deformational structures in drumlins near Waukesha, Wisconsin, USA', *J. Glaciol.* **31**, 220–28.

Straw, A. (1968) 'Late Pleistocene glacial erosion along the Niagara escarpment of southern Ontario', *Geol. Soc. Am. Bull.* **79**, 889–910.

Sugden, D. E. (1968) 'The selectivity of glacial erosion in the Cairngorm Mountains, Scotland', *Trans. Inst. Brit. Geogr.* **45**, 79–92.

Sugden, D. E. (1974) 'Landscapes of glacial erosion in Greenland and their relationships to ice, topographic and bedrock conditions', in *Progress in Geomorphology*, Inst. Brit. Geogr. Spec. Pub. 7.

Sugden, D. E. (1978) 'Glacial erosion by the Laurentide ice sheet', *J. Glaciol.* **20**, 367–91.

Svensson, H. (1959) 'Is the cross-section of a glacial valley a parabola?', *J. Glaciol.* **3**, 362–3.

Unwin, D. J. (1973) 'The distribution and orientation of corries in northern Snowdonia, Wales', *Trans. Inst. Brit. Geogr.* **58**, 85–98.

Vivian, R. (1980) 'The nature of the ice–rock interface; the results of investigation on 20 000 m^2 of the rock bed of temperate glaciers', *J. Glaciol.* **25**, 267–77.

Wright, H. E. (1957) 'Stone orientation in the Wadena drumlin field, Minnesota', *Geogr. Annlr.* **39**, 19–31.

Selected bibliography

Valuable general surveys of glacial geomorphology are provided by R. J. Price, *Glacial and Fluvioglacial Landforms*, Oliver and Boyd, 1973; C. Embleton and C. A. M. King, *Glacial Geomorphology*, Arnold, 1975 and D. E. Sugden and B. S. John, *Glaciers and Landscape*, Arnold, 1976. More uneven in its coverage, but reflecting a number of recent trends in the subject is N. Eyles (ed.) *Glacial Geology: an introduction for engineers and earth scientists*.

Much recent research on glacigenic sediments has been published in a whole series of symposium volumes, among which the following are worthy of mention: R. P. Goldthwait (ed.) *Till: a symposium*, Ohio State Univ., 1971; R. F. Legget (ed.) *Glacial Till: an interdisciplinary study*, Roy. Soc. Can. Spec. Pub. 12, 1976; W. Stankowski (ed.) *Till: its genesis and diagenesis*, Poznan, 1976; 'Till', *Boreas* **6**, 1977; C. Schlüchter (ed.) *Moraines and Varves*, Balkema, 1979; E. B. Evenson, C. Schlüchter and J. Rabassa, *Tills and Related Deposits*, Balkema, 1983.

Part 3 THE WORK OF ICE

Chapter 13
Meltwater erosion and deposition

The comments at the opening of Chapter 12 about the difficulty of observing glacial erosion and deposition in action apply with almost equal force to many meltwater processes. While drainage around the margins of modern ice-masses has proved a fruitful field of study, subglacial patterns of water movement are still very imperfectly known. In general, however, investigations of modern ice-sheets have tended to underline the potential importance of basal meltwater as an eroding and depositing agent, a conclusion reinforced by the analysis of forms and deposits in previously glaciated areas.

Glacio-fluvial erosion

In addition to meltwater in the strictest sense of the word, the flow here being considered may include ordinary stream drainage reaching the ice margin from surrounding valley sides. As the name 'glacio-fluvial' implies, the flow owes many of its characteristics to association with the ice. Confined within channels that may be either supraglacial, englacial, subglacial or marginal, its dominant feature is likely to be its changeability. This arises in the first instance from the mobility of the ice itself. As the ice moves, crevasse systems are modified and their relationship with the adjacent rock surface slowly altered. Original routeways over and through the ice are closed while new ones become available. A second cause of change is fluctuating discharge with large seasonal oscillations and, at least in summer, smaller diurnal variations.

One effect of altered pathways and contrasting discharge is very uneven flow velocities. At times the water is accelerated to abnor-mally high speeds and this may be the cause of the beautifully smoothed and rounded depressions, often several metres deep, which occur in considerable numbers in Scandinavia and and have also been described from Scotland. Known as plastically sculptured or p-forms, these have been attributed to cavitation erosion (p. 165) at points where water at the rock face attains exceptional velocities. Whatever the cause of p-forms, and there have been several conflicting views, there is little doubt regarding the general effectiveness of meltwater as an erosive agent. The rapidity with which marginal meltwater streams can incise their courses into solid rock is attested by numerous observations. The most famous are those of Tarr and von Engeln in their explorations of the Alaskan glaciers around the turn of the present century. They recorded how meltwater may carve gorges many metres deep within a period of a decade or less. Flint quotes an example of a small stream first superimposed on to resistant bedrock in 1935 which, by 1941, had excavated a gorge 15 m wide and more than 8 m deep. In favourable circumstances downcutting at a rate of 1 m yr^{-1} does not seem exceptional, and is adequate testimony to the power of fast-flowing meltstreams. There seems no reason why equally rapid meltwater erosion should not occur beneath the ice, although direct observational evidence is relatively limited. Artificial tunnels have occasionally offered opportunities for studying subglacial streams. Great seasonal variations in discharge have been recorded, with particularly violent flow characteristic of summer. Near Zermatt potholes 2.3 m wide and 2 m deep have been seen to form in ophiolites in a period of 3 years, and experimental quartzite blocks have been scored to a depth of 100 mm along the course of another stream in under 5 years.

In early studies of formerly glaciated areas most meltwater erosion was assumed to have taken place beyond the actual ice-edge. The ice was regarded as an impenetrable barrier with water either channelled along its margins or else impounded in proglacial lakes. The most influential early investigation was published by Kendall in 1902 on the North York Moors. In addition to simple 'in-and-out' channels excavated by marginal streams, he identified a series of channels believed to have been cut by overflows from ice-dammed lakes. Overspill was assumed to have occurred across the lowest available col. Often this would be at the edge of the ice barrier, in which case a 'marginal overflow' would result; occasionally it would be across an existing watershed in which case a 'direct overflow' would be initiated. This simple model formed the basis for innumerable further studies in the next 50 years or so. And yet even before Kendall wrote his paper, doubts had been expressed regarding the ability of ice to act as a totally watertight barrier. Prestwich, for instance, in discussing the parallel roads of Glen Roy, had pointed out that the Marjelensee in Switzerland regularly drains through the ice and not by surface overflow. Such reservations went unheeded as thousands of spillways from supposed ice-dammed lakes were identified. The diagnostic criteria most commonly quoted were steep sides, flat ill-drained floors and apparently underfit streams. However, by themselves such features do not demonstrate an overspill origin, and in 1949 Peel showed the morphology of two channels in Northumberland, earlier claimed as direct overflows, to be inconsistent with that interpretation. The anomalous aspect is the long profile which in both cases is conspicuously hump-backed, in one instance rising over 10 m above the supposed intake level (Fig. 13.1). When further

investigations disclosed many other putative overspills to have an up-and-down form it became apparent that an alternative explanation was needed.

Meanwhile in Scandinavia studies of meltwater erosion had proceeded along a rather different course. As early as 1945 Mannerfelt was relating channels in Sweden to the patterns of meltwater flow recorded on modern glaciers in Alaska and Spitsbergen, placing much greater emphasis on marginal and subglacial routeways than on ice-dammed lakes and overflow channels. One implication of this work was the inadequacy of the criteria then employed in Britain for identifying lakes and associated spillways. Recognizing this shortcoming Sissons instigated a radical reassessment of much of the British evidence. It is now generally acknowledged that many features earlier interpreted as overflow channels did not in practice originate as spillways from ice-dammed lakes. For present purposes, therefore, a basic fourfold division of meltwater channels will be employed: marginal and submarginal channels, subglacial channels,

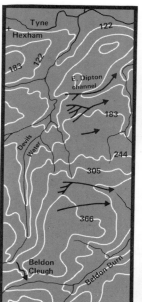

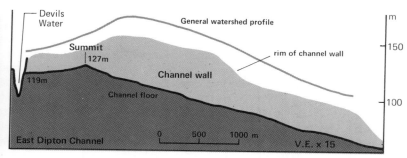

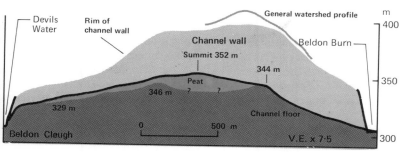

Fig. 13.1. The hump-backed long profiles of two meltwater channels originally interpreted as overflow channels. They constitute two of a series of channels transecting a northward-sloping ridge on the right bank of the Tyne in Northumberland (after Peel, 1949).

spillways and coulees.

Marginal and submarginal channels

While these two positions of meltwater flow are readily distinguishable along a modern glacier edge, it may be difficult to decide in which situation a relict channel was formed once the ice has melted. In theory a submarginal channel may contain debris which collapsed into it during final wastage of the ice, but such evidence is rarely very convincing. It has been argued that truly marginal channels are likely to have lower gradients than their submarginal counterparts, but the distinction is often difficult to draw and in any case the

reasoning is open to question. Another possible criterion is the presence of abrupt changes in direction. Unlikely in marginal streams, these may be very characteristic of submarginal streams in which flow is guided by crevasses penetrating to the base of the ice. Water will alternately be forced along the hillside and then allowed to plunge downslope, producing an oblique zigzag course. The probable submarginal origin of many meltwater features is now widely acknowledged, and swarms of shallow channels, often only a few metres deep, scored into the slopes of Scottish mountains are so interpreted (Fig. 13.2).

As already seen, small channels can be excavated extremely

Fig. 13.2. Part of a swarm of shallow meltwater channels mapped by Sissons in the central Grampians of Scotland. Most of the channels are only 1–4 m deep, but can often be traced for 1 km or more. They are believed to have been formed in a brief interval during the decay of a small ice-cap that existed in the period 10 800–10 300 years BP (after Sissons, 1974).

quickly. In some regions it has even been suggested that fresh channels may be eroded annually, each member of a series being the product of summer melt from a rapidly retreating ice-front. However, doubts remain about such an interpretation since it appears that several parallel channels can function simultaneously, each occupying a trough between ice-cored ridges that result from differential ablation of dirty ice bands. Moreover, from descriptions of modern glaciers there seems little doubt that submarginal streams occasionally form an anastomosing network; it consequently becomes very difficult to infer details of the former ice edge from a suite of abandoned channels.

A rather different type of marginal drainage occurs along the frontal edge of a glacier or ice-sheet. The simplest case is that of a valley glacier in which small streams emerging from tunnels near the valley wall flow along the ice edge to join the major meltwater stream issuing centrally from beneath the snout. Such watercourses sometimes cut shallow channels but generally leave little trace of their existence once the glacier has withdrawn. Around the lobate margin of an ice-sheet a similar pattern of drainage converging on a single master stream may be found. This is particularly true on lowlands where earlier arcuate end-moraines often guide the meltwater towards one outlet. The detail of the streamcourses is much influenced by the dominant wastage processes. For example, the water may be channelled between ice-cored moraines, frequently shifting its course as continued undermelting leads to collapse of individual ridges. Meltwater streams in such an environment are constantly abandoning old channels for new as lower routes become available. Fresh channels can be fashioned with remarkable rapidity, either through erosion of the unconsolidated sediments or through melting of residual ice, and in an area deglaciated for only a decade it may prove extremely difficult to plot all the various routes followed by meltwater in those few years. A further complication arises from the occasional meltwater stream that, after issuing from a tunnel and flowing along the ice margin, actually turns back under the ice-edge.

Much larger marginal channels are found where sizeable rivers, nourished by precipitation in extra-glacial areas, have been diverted along the edge of Pleistocene ice-sheets. Examples occur in both Europe and North America and have been termed *Urstromtäler* from their frequent occurrence on the north European plain. In some instances the routes initiated along the ice-margin are still occupied by large post-glacial rivers; the mouth of the Rhine was diverted westwards to its present position by the Scandinavian ice-sheet, while the modern course of the Ohio was initiated around the southern edge of the Laurentide ice-sheet (see Fig. 12.14).

Subglacial channels

Submarginal meltwater courses are obviously subglacial in the literal sense, but are commonly differentiated because they owe many of their characteristics to proximity to the ice-edge. When water penetrates deeper beneath the ice it may appropriately be termed subglacial and the linear depressions that it there excavates may be referred to as subglacial channels. In formerly glaciated areas steep ravines scored by meltwater can sometimes be seen running directly down hillslopes over a vertical distance of as much as 100 m. Known as subglacial chutes, these features are believed to mark sites where submarginal drainage plunged to the bottom of much thicker ice. They would presumably not form beneath cold-based ice, and might be especially favoured where thick, warm-based ice had stagnated. The latter conditions could also explain the tendency for certain chutes to pass downslope into eskers since the transporting capacity of the water might be expected to diminish across the lower ground.

Once marginal drainage has reached the base of thick mobile ice, the volume of water is still likely to be augmented by melting at the glacier sole and along the walls of tunnels. The precise nature of flow in this environment, where observations are so difficult to make, remains largely speculative. However, it seems increasingly clear that much of the available water tends to be concentrated into a network of passageways lying at the ice–rock interface. The monitoring of water-levels in several boreholes drilled to bedrock has revealed fluctuating water-pressure substantially less than the calculated overburden pressure. Such results appear consistent with Nye's advocacy of a thin sheet of water over most of the glacier sole, coexisting with, but remaining relatively independent of, discrete passageways that are responsible for draining most of the surplus meltwater. Shreve, moreover, has argued that large arteries will tend to grow at the expense of smaller ones so that eventually there may be a dendritic network not unlike that seen under subaerial conditions. In most modelling of subglacial flow patterns it is argued that the tendency towards tunnel closure through deformation of the ice is balanced by melting along the tunnel walls so that an equilibrium is established. This does not necessarily mean that, once established, a flow pattern remains unchanging since the few direct observations currently available have revealed several instances of tunnel abandonment. This was found, for instance, when shafts associated with a Norwegian hydroelectric scheme were driven under the Bondhusbreen outlet glacier near Bergen, and was also discovered beneath the Glacier d'Argentière when a subglacial routeway that had been in regular use for several years was suddenly abandoned and a new one began to function some 340 m from the original.

In aggregate, the available evidence suggests that subglacial passages are subject to dramatic changes in water-pressure and therefore in flow velocity, and that individual passages may fall into disuse in an erratic fashion. One source of particularly large fluctuations in discharge may be the rapid subglacial evacuation of ice-dammed lakes. In Scotland Sissons argued in 1979 that the final drainage of the 260-m lake in the Glen Roy area (see Fig. 13.5) involved 5 km^3 of water escaping beneath the ice within a single week, with a peak flow estimated at 22 500 m^3 s^{-1}. Giant potholes up to 15 m deep along the Spean gorge were ascribed to this catastrophic event.

The hump-backed channels in formerly glaciated areas that were once attributed to overspill from lakes are now generally agreed to be subglacial in origin. The primary reason for this interpretation is the inability of water to cut such forms except under considerable hydrostatic pressure. The amplitude of the hump precludes an explanation in terms of normal stream processes and the conclusion must be that the water was capable of flowing uphill. Yet it is not essential for a channel to have an up-and-down form before being regarded as fashioned beneath the ice. The disproportionate size of some channels, relative to their potential catchment as marginal features, argues for their having originated subglacially. Similarly, the complex anastomosing pattern of some channel systems defies explanation in terms of shifting ice-margins, but may be readily accounted for by large volumes of water flowing subglacially. The relationship to depositional forms that develop beneath the ice can also be significant; in a number of cases channels have eskers running along part of their length.

Despite all this evidence for meltwater erosion at the sole of an ice-mass, there has still been controversy regarding the relative importance of tunnel enlargement and channel incision in accommodating the discharge (see p. 225). Individual meltwater channels are often only 1–2 km in length, and therefore other parts of the original drainage network must have disappeared virtually without trace. At one time the discontinuity of the features was ascribed to their erosion by englacial streams, segments of which might have been progressively superimposed on to bedrock ridges during a stagnation phase. It was argued that if the elevation of the routeway declined more rapidly through the ice than across the rock spurs, the water would only be able to maintain its course by flowing uphill under hydrostatic pressure; this it might continue to do until further melting offered a completely new, lower passageway. While this hypothesis explains many of the observed characteristics of meltwater channels, it is now thought improbable that englacial streams exist on the scale envisaged and greater emphasis is placed upon potential velocity variations along subglacial passageways. These may be controlled by the local pressure field attributable to the ice cover, with accelerated flow across upstanding bedrock ridges where almost all the erosion is concentrated. It can scarcely be denied that more supporting field evidence is required, but meltwater erosion beneath active as well as stagnating ice would certainly accord with the limited observational data that are currently available.

Analogous in many respects to the subglacial channels just discussed are the remarkable slot-like gorges that sometimes transect rock bars on the floors of glacial troughs. The most famous example is the Aareschlucht in Switzerland where the River Aar cuts through a resistant limestone barrier in a narrow, vertical-walled gorge over 200 m deep. The rock bar itself appears to be the result of differential ice erosion, but there is no indication that it ever constituted an unbreached barrier impounding a lake on its upstream side. It seems much more likely that subglacial meltwater initiated the gorge while the whole valley floor lay beneath the ice.

Of contrasting shape are the must larger features known as *tunneldale* in Denmark and as *Rinnentäler* in northern Germany. Excavated in a sheet of glacial drift but locally penetrating the subjacent bedrock, these tunnel valleys are commonly 1–2 km wide, up to 100 m deep and in some cases traceable over a distance of 75 km (Fig. 13.3). They tend to have steep sides and fairly flat floors, but in long profile are irregular with enclosed lake-filled hollows. Their subglacial origin seems beyond doubt. Cut in tills of

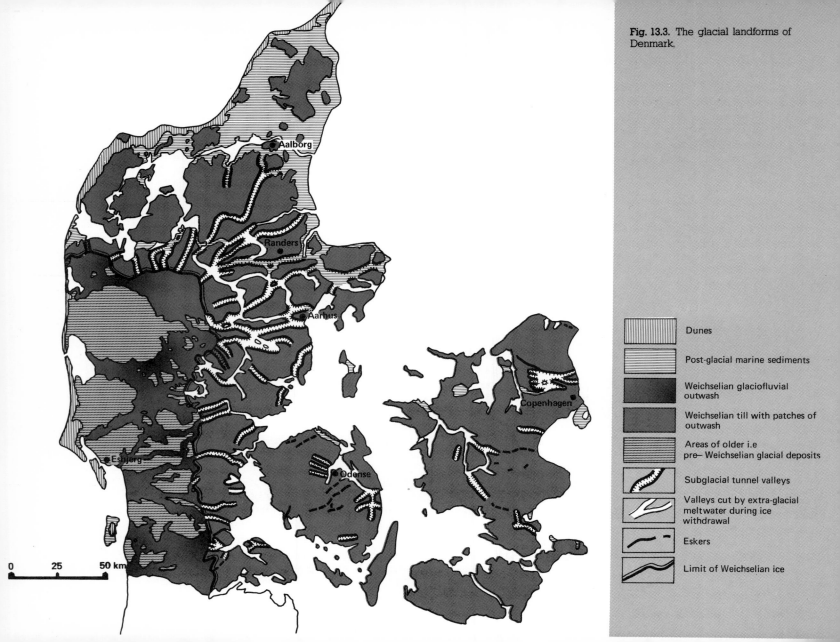

Fig. 13.3. The glacial landforms of Denmark.

Dunes

Post-glacial marine sediments

Weichselian glaciofluvial outwash

Weichselian till with patches of outwash

Areas of older i.e pre–Weichselian glacial deposits

Subglacial tunnel valleys

Valleys cut by extra-glacial meltwater during ice withdrawal

Eskers

Limit of Weichselian ice

Aalborg

Randers

Aarhus

Esbjerg

Odense

Copenhagen

0   25   50 km

the last glaciation they sometimes have well-formed eskers on their floors. They end very abruptly, being replaced beyond the ice margin by low outwash fans deposited where the emergent meltwater lost its transporting capacity. The nature of meltwater streams able to erode such broad channels is a puzzle. It is tempting to envisage vast streams completely filling the valleys, yet this would require an ice roof of most unlikely span and one which has no known counterpart today. Most workers have preferred the idea of a smaller tunnel migrating laterally, and this certainly appears more consonant with the late-stage development of eskers. It is strange that such distinctive feature are prolific in Denmark and north Germany, but have been reported with much less frequency elsewhere. Certain valleys in eastern Scotland do, however, display affinities with tunnel valleys, as do several deep elongated pits on the floor of the North Sea.

Another type of subglacial channel is that which Woodland termed a buried tunnel valley. This is a deep linear depression filled with glacial deposits until it has little or no surface expression. Owing to its burial such a feature is difficult to distinguish from a normal stream-eroded valley subsequently infilled with glacial drift. Nevertheless, as more borehole evidence becomes available it is increasingly clear that closed depressions quite unlike normal valleys are of widespread occurrence. Woodland documented sixteen examples in East Anglia where boreholes are believed to have penetrated the fill of subglacially eroded channels (see Fig. 12.10). The channels characteristically have very steep sides, even approaching the vertical in places, and extend to depths which preclude normal outfall to the sea. For example, just north of Hitchin far inland from the coast the glacial fill is over 125 m thick and its base lies more than 75 m below sea-level. The conclusion is inescapable that the boreholes have penetrated a very deep enclosed depression. Further evidence shows the hollow proved in this way to be a particularly deep segment of an undulating channel traceable for at least 20 km.

Although Woodland stressed the morphological similarities between the buried channels in East Anglia and the tunnel valleys in Denmark, two striking contrasts also need emphasis. The first is the relative narrowness of the East Anglian channels, possibly due to their incision into Chalk rather than drift and poorly consolidated Cenozoic rocks. The second and more important is the infill of the English examples (Fig. 13.4). This consists predominantly of water-laid sediments having obvious lithological affinities with the tills on adjacent interfluves. In general, sand and gravel occur at the distal end of each channel, loams, silts and clays at the proximal end; in the middle reaches the finer material tends to overlie the coarser. Woodland envisaged an initial erosional phase during which the channel is scoured and any excavated debris is deposited as outwash beyond the ice margin. As the ice front melts back, subglacial erosion continues but the distal segment of the channel is gradually filled with gravel. With further melting and downwasting, the subglacial water loses its hydrostatic head and becomes so slow-moving that the proximal end of the channel is filled with silts and clays. No satisfactory explanation has yet been offered for an accumulation phase affecting tunnel valleys in East Anglia but not in Denmark, and until it has it would perhaps be wise not to stress the apparent analogies too strongly. Although East Anglia provides the greatest known concentration of buried tunnel valleys in Britain, they have been found in many other glaciated regions. In addition they have been recorded on the floor of the North Sea.

Spillways

Although most modern proglacial lakes empty beneath the impounding ice-dam, others conform to the Kendall model described earlier and spill either as marginal streams along the ice-edge or as direct overflows across bedrock cols. The overflowing water is commonly assumed capable of modifying any col it occupies and thereby producing a steep-sided flat-floored channel very like those excavated in marginal and subglacial situations. This raises the problem of the criteria by which channels originating as lake overspills are to be distinguished. In practice such terms as spillway and overflow channel are probably best restricted to those cases where there is independent evidence for the existence of a lake. Such evidence might consist, for example, of either lacustrine sediments or strandline features. Of course, lacustrine sediments can accumulate in a lake that drains subglacially so that by themselves they do not prove that a nearby channel is a true spillway. More significance possibly attaches to strandline features. For a durable assemblage of shoreline forms to develop, the lake surface must remain at the same elevation for a relatively long period, and this seems most likely to occur where the discharge is across a bedrock col. Moreover, an old shoreline permits a relatively precise estimate

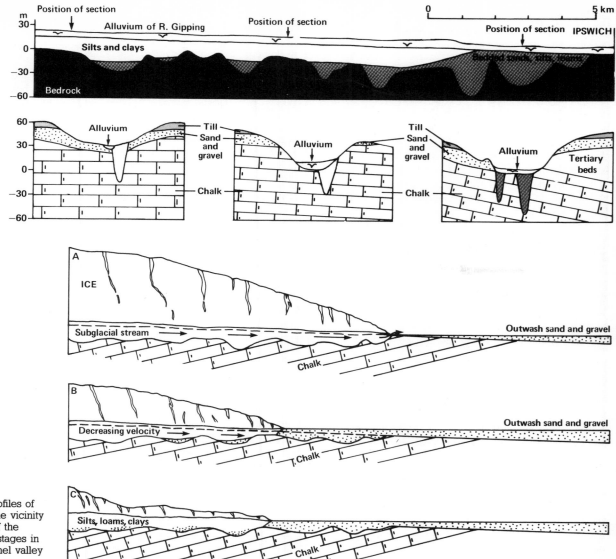

Fig. 13.4. Long- and cross-profiles of the buried tunnel valley in the vicinity of Ipswich. The lower part of the diagram shows hypothetical stages in the evolution of a buried tunnel valley (after Woodland, 1970).

Fig. 13.5. The parallel roads of Glen Roy. The diagrammatic sketches below show the three stages in the evolution of the lake system as the edge of the main ice-mass withdrew towards the south-west.

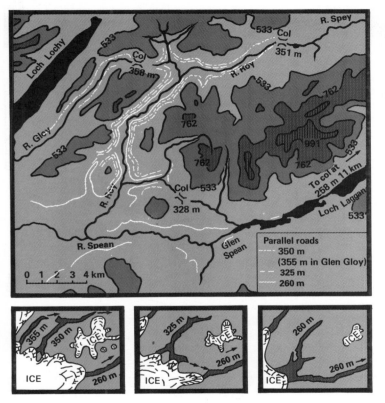

of the glen allowing the lake-level to fall to 325 m. Further wastage permitted drainage via Glen Spean with the water-level stabilized at 260 m. Not only do the strandline and spillway heights approximately correspond, but the shore features disappear in those areas presumed to have been occupied by ice. Similarly, a shoreline along the south-eastern edge of glacial Lake Harrison in the English Midlands can be traced for 55 km before terminating at the presumed position of the ice-front (Fig. 13.6). Its elevation of 125 m is virtually identical to that of the Fenny Compton gap through which the lake water drained, first into the Cherwell valley and thence into the Thames. A striking aspect of both the Glen Roy and Lake Harrison overflow routes is their lack of what were once regarded as the diagnostic morphological criteria for spillways. This is especially obvious in the case of the Fenny Compton gap which is broad and gentle-sided and quite unlike the traditional picture of an overflow channel. It is ironic that so many overflow channels were wrongly identified by early workers on the basis of their steep sides and flat, ill-drained floor, while some of the best-authenticated spillways lack the supposed distinctive shape.

The frequent assumption that any overflow route suffers pronounced lowering needs careful qualification. Several factors will influence the erosion of a spillway. Among them are the nature of the material composing the outlet, the volume of water discharging through it, the length of time involved and the original form of the col. Taking the Fenny Compton gap as an example, the present floor is composed mainly of clay, and although buttressed by thin limestone bands, would certainly not be classified as particularly resistant. The volume of water that flowed through it is unknown; although there is no necessary correlation between lake size and overspill volume, extensive lakes fringed by long ice-cliffs seem likely to generate most water and on this basis Lake Harrison would have discharged prolifically through the gap. There is at present no reliable way of assessing how long the overflow functioned, but it cannot have been a particularly short episode since appreciable time is required for the cutting of the associated lake bench. Finally, the original col can at most have been 7 m higher than at present since an alternative outlet exists at that elevation. This means that overflow from the lake must have entered a gently sloping valley which it failed to modify to anything like the degree commonly associated with other types of meltwater channel.

of the former water-level, capable of being compared with the height and morphology of the putative spillway.

The classic illustration in the British Isles is provided by the parallel roads of Glen Roy (Fig. 13.5). It was Jamieson in 1863 who initially interpreted the 'roads' as strandlines of a lake controlled in elevation by three separate overflow routes. During the first stage the lowest available outlet was a direct overflow at 350 m near the head of Glen Roy. Ice withdrawal then uncovered a col to the east

Fig. 13.6 Lake Harrison close to its maximum extent as visualized by Shotton (1953). At the time of the overflow through the Fenny Compton gap the lake may have been very much reduced in size; the minimum size is presumably indicated by the extent of the lake bench. The inset shows the form and geology of the Fenny Compton gap with a notable absence of any trenching on the floor of the col.

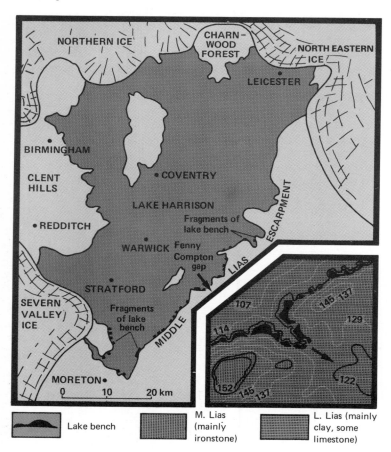

It is instructive to consider an example from a recently deglaciated area where the evolution of an ice-dammed lake is recorded in a succession of strandlines. One such example is provided by the southern end of the Lake Michigan basin in North America (Fig. 13.7). Here Lake Chicago, dammed against the Laurentide ice-sheet, discharged via the valley of the Des Plaines River. Successsive lake-levels are marked by four shorelines at heights of 18, 12, 7 and 3 m above Lake Michigan. Bretz in 1966 suggested that the highest level was stabilized when a boulder pavement at the point of outflow protected the underlying weak drift from erosion. The pavement was destroyed by an abrupt increase in discharge when waters from the Lake Erie basin suddenly began escaping into Lake Chicago. The outlet was lowered some 6 m before further ice withdrawal led to temporary abandonment. Following a minor re-advance, ponded water again escaped down the Des Plaines River and it was at this stage that the 12 m shoreline received its final shaping. Once more a sudden accession of Erie meltwater lowered the outlet until a bedrock sill stabilized the lake-level at the height of the 7 m shore-line. Melting later permitted temporary discharges eastwards along the edge of the main Laurentide ice-sheet, and on two separate occasions water-level fell below that of the present Lake Michigan. After total deglaciation, however, isostatic rebound elevated the more northerly and easterly outlets from the Great Lakes region and the Des Plaines valley briefly functioned as an overflow for the last time. The catchment was a vast lake incorporating both the Superior and Michigan basins of the present day. The sill, however, was lowered very little during this final phase; other outlets were functioning simultaneously and it was these rather than the Chicago outlet that apparently controlled the lake-level from 3 000 years BP onwards.

The complete sequence of events at the Chicago outlet dates from the period between 14 000 and 2 000 years ago, but for about half this time there was no actual overflow. During the spells of active discharge the spillway was lowered some 11 m in two distinct phases of incision. For much longer periods the sill remained stable, permitting strandlines to be etched on the slopes around the southern shores of Lake Michigan.

Coulees

In the preceding pages attention has already been drawn to two common ways in which glacially impounded lakes discharge meltwater, namely through subglacial tunnels or by direct overflow across subaerial cols. A third possibility, although much rarer, is by sudden destruction of the ice barrier with virtually instantaneous

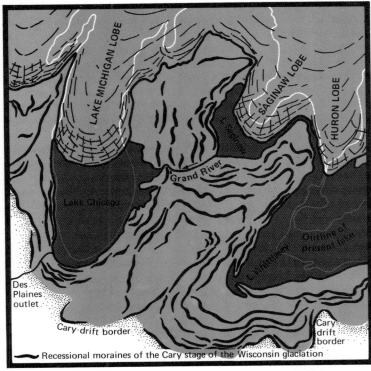

Fig. 13.7. The pro-glacial drainage system that formerly drained through the Des Plaines spillway. The stage depicted is that responsible for a sudden increase in discharge as water escaped from the Erie basin via the Grand River outlet (c. 13 000 years BP). Below is a plot of changes in water-level around Chicago during the last 14 000 years.

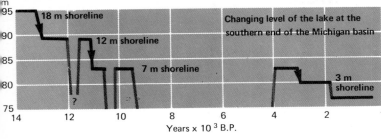

release of the complete contents of a large ice-dammed lake. Although flow through ice-tunnels can produce violent *jokulhlaups*, their scale is small compared with the cataclysmic events that can accompany a total collapse of the retaining dam. The term 'coulee' is appropriately used to describe the channels scoured by the escaping lake waters, since this is the local name for such features in the north-western United States where the most startling results of a dam burst have been described. Here the Pend Oreille lobe of the cordilleran ice-sheet pushed southwards into Washington and impounded a major lake in the Clark Fork valley (Fig. 13.8). Known as Lake Missoula, this has left prominent strandlines which indicate that, at its maximum extent, it had a superficial area of 7 500 km² and a total volume of 200 km³. Beside the dam the water was at least 700 m deep, resting against ice that cannot have been less than 1 070 m thick. So much can be inferred from the distribution of glacial and lacustrine sediments, but the immediate cause of the ice barrier collapsing is unknown. Within the area of the lake several remarkable features indicate an abrupt fall in water-level. Cols that were originally submerged are scored by deep, beautifully stream-lined depressions. It is believed that as the dam broke, a steep surface gradient developed across the lake, and water started rushing towards the outlet. At each submerged gap the constriction acted like the throat of a gigantic flume with a fall in pressure sufficient to induce cavitation (see p. 165). By this process hollows up to 30 m deep were rapidly gouged from resistant bedrock. On the slopes below some of the gaps are giant ripples, 10 m high and with 75 m between successive crests.

However, it is the area downstream from the ice-dam that carries the most spectacular results. Shortly below the dam the floodwaters were over 200 m deep and beyond Spokane they fanned out across the Columbia plateau basalts. Velocities are estimated to have attained 30 m s⁻¹ in places. So great was the spate speeding down the valleys to the west that it spilt across divides and surged up minor tributary valleys, bearing huge boulders with it. Stripping earlier loess from the plateau surface, the floodwaters quarried deep channels into the underlying basalt; steep-walled and irregularly floored, these are the coulees. They owe many details of form to the strong vertical jointing of the basalt, large blocks of which were ripped away as huge vortices developed in the raging torrents. Dry waterfalls with deep plunge pools characterize many of the coulees. Although the

Fig. 13.8. Lake Missoula and the 'channeled scablands' of the north-western United States.

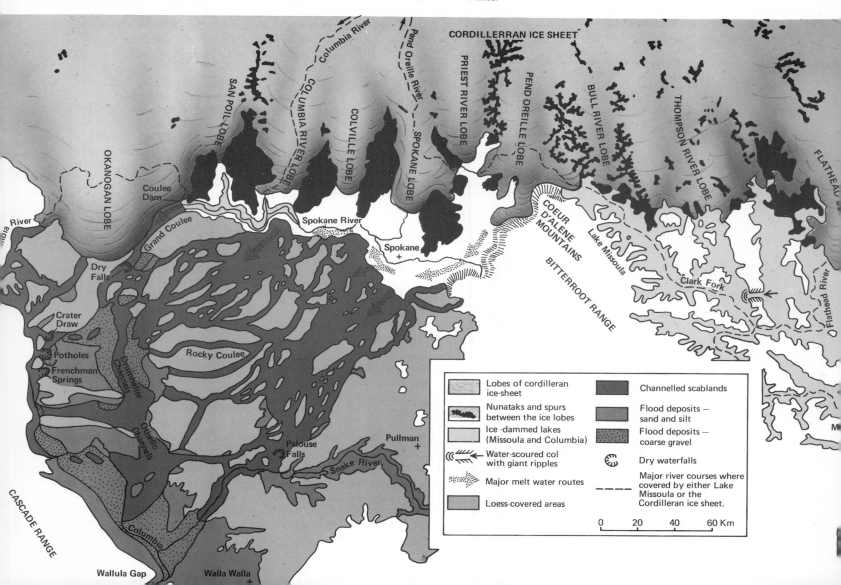

CORDILLERRAN ICE SHEET

Columbia River

Pend Oreille River

SAN POIL-LOBE

COLUMBIA RIVER LOBE

COLVILLE LOBE

PRIEST RIVER LOBE

SPOKANE LOBE

PEND OREILLE LOBE

BULL RIVER LOBE

THOMPSON RIVER LOBE

FLATHEAD

OKANOGAN LOBE

Coulee Dam

Spokane River

Spokane

COEUR D'ALENE MOUNTAINS

Lake Missoula

Clark Fork

Flathead River

Grand Coulee

BITTERROOT RANGE

Dry Falls

River

Crater Draw

Rocky Coulee

Drumheller Channels

Potholes

Frenchman Springs

Othello Channels

Pullman

Palouse Falls

Snake River

CASCADE RANGE

Columbia

Wallula Gap

Walla Walla

Legend	
Lobes of cordilleran ice-sheet	Channelled scablands
Nunataks and spurs between the ice lobes	Flood deposits — sand and silt
Ice-dammed lakes (Missoula and Columbia)	Flood deposits — coarse gravel
Water-scoured col with giant ripples	Dry waterfalls
Major melt water routes	Major river courses where covered by either Lake Missoula or the Cordilleran ice sheet.
Loess-covered areas	

0   20   40   60 Km

channels are sometimes over 100 m deep, no one by itself could accommodate the whole spate and therefore anastomosing networks developed; together these constitute the area known as the 'channeled scablands'. Meanwhile the extremely coarse sediments swept along by the flood were fashioned into giant bars, ripples and even pool-and-riffle sequences.

There now seems little doubt that most of the landforms in the 'channeled scablands' result from large-scale meltwater flooding. Yet when Bretz first advanced the idea in 1923 its acceptance was far from unanimous. In the intervening years evidence has accumulated to suggest that there was not a single inundation but a succession of such events as, on several separate occasions, the water built up in Lake Missoula. In the most recent major flood, probably around 15 000 years ago, peak discharge has been estimated at 21.3 × 10^6 m^3 s^{-1} or 100 times the mean flow of the Amazon. The whole lake seems to have been evacuated in a few weeks so that the fashioning of the coulees and scablands was a remarkably brief episode. This serves as a salutary reminder of the possible catastrophic origin of certain landforms. There is occasionally a tendency to assume an attitude of extreme uniformitarianism in which the potential significance of the cataclysmic event is denied. There is no intrinsic reason why similar devastating floods should not have occurred in other glaciated regions, although the necessary supporting evidence has rarely been identified.

Glacio-fluvial deposition

Meltwater streams on the surface of an ice-mass vary greatly in the amount of sediment they transport. On an ice-sheet or near the centre of a valley glacier they can be quite clear, but where the ice abuts against a rock wall they often carry substantial quantities of debris picked up from the lateral moraine. Similarly, near the frontal edge of an ice-mass the ablation moraine may supply sediment to the streams before they disappear down moulins or crevasses. The streams that gush forth from beneath the ice-edge are even more heavily laden. Not only do they appear very turbid but it is sometimes possible to hear large cobbles clashing together in the water. Most of this material is believed to be englacial or subglacial debris picked up along the tunnel perimeter. A further potential source is the bedrock floor if the meltwater is actively deepening submarginal or subglacial channels. The range of particle size being transported is often very wide. The coarsest tends to be deposited quite close to the ice-front, but the finer silt and clay is carried long distances and gives many meltwater streams a characteristic milky appearance.

The diversity of meltwater paths provides a great variety of environments in which glacio-fluvial deposition can take place. Theoretically a fundamental distinction may be drawn between ice-contact materials laid down on, within or beneath the ice, and pro-glacial materials deposited beyond the contemporary ice limit. The two types grade into each other near the ice-edge, while the proglacial sediments tend to lose their distinctive glacio-fluvial character the further their point of accumulation from the ice-margin. In practice, additional complexity is almost invariably introduced by shifts in the position of the ice-front.

Glacio-fluvial deposits naturally display many of the features found in other water-laid sediments. They tend to be reasonably well sorted, at least rudely stratified and frequently to exhibit dune and ripple structures. They also display certain attributes of their own. Part of the constituent material commonly originates from sources outside the present catchment, but can be matched in local tills. A variable percentage of the pebbles retain traces of glacial faceting and striation. The water-laid sediments are often interleaved with lenses of till. Numerous abrupt changes in particle size reflect the frequent fluctuations in meltwater discharge. Perhaps most diagnostic of all are various collapse structures resulting from decay of the ice over or against which many of the materials accumulated.

The terminology that has evolved for the classification of ice-contact glacio-fluvial deposits is extremely confused. In part this arises from the use of terms taken from different languages and inadequately defined before being adopted for scientific work. More fundamental is the confusion that results from mixing criteria based upon form, constituent material and presumed origin. The problem is well illustrated by the term kame. This was derived from the Scottish word 'kaim' which refers to any steep-sided ridge irrespective of origin: for instance, many sharp-crested ribs of solid rock in Scotland still retain the name. Yet when adopted to refer to a glacio-fluvial landform, its use was often restricted to circular mounds with ridges specifically excluded. For elongated shapes the term esker, derived from the Irish word *eiscir*, was commonly substituted. To make matters even more confused, certain workers later broadened the term esker to include any hillocky glacio-fluvial form and introduced the Scandinavian word *ose* to refer to the elongated shapes. At present it is impossible to formulate universally agreed definitions, so that it is proposed to adopt here those usages that seem most widely employed among English-speaking workers. The term kame will be used to designate hillocky glacio-fluvial deposits that owe their current form primarily to removal of the ice against or over which they originally accumulated. Esker will be used to connote an elongated ridge which is essentially the cast of a meltwater channel on, within, or beneath the ice. The Icelandic term *sandur* will be employed to indicate the relatively smooth surface built up by meltwater streams beyond the ice-margin.

Kames

Kames may exist either as isolated steep-sided mounds or as an assemblage of such features forming what has appropriately been called sag-and-swell topography. They range from low hillocks only a few metres high to conspicuous conical hills as much as 50 m high. They characteristically consist of sand and gravel which may be relatively well bedded in the core of the mounds but often shows evidence of slumping to its angle of repose around the margins. Striated boulders and till lenses are quite common within the deposit, and some kames carry a veneer of water-washed glacigenic sediments.

Kames almost certainly originate in a variety of ways. One potential mechanism is the building along an ice-edge of either coalescent fans or more gently sloping deltas, especially where heavily laden meltwater streams issue into standing water. The proximal side will initially be supported by the ice but will later collapse. Such a feature will tend to have an asymmetric outline with a smooth distal slope but an irregular ice-contact face. This interpretation for certain kames is supported by internal bedding that consistently dips towards the outer slope. Where a prominent flat upper surface is preserved, some authors have wanted to distinguish a category of delta-kame.

A more common origin for kames is believed to be accumulation in an area of stagnant, downwasting ice. The term 'glacier karst' has been employed to describe the complex of sink holes and tunnels that may then develop. Clayton described the essential characteristics from the Martin River glacier in Alaska. A cover of ablation till is required to prevent rapid melting and to allow evolution of a tunnel

system below the surface. Over the outer edge of the Martin River glacier the till is up to 3 m thick and has been accumulating sufficiently long to support 100-year-old spruce and hemlock trees. The surface of the glacier is pock-marked with funnel-shaped sinks that are typically 100 m across and locally occur at a density of 30 km^{-2}. Many are occupied by lakes and may be as much as 90 m deep. The dominant process is melting by water, either around the lake margins or along the passages through the ice. Coalescence of sinks creates compound features over 1 km^2 in extent, while tunnel collapse initiates new depressions. Sediment from the surface will either fall or be carried by meltwater into the growing cavities. After the final melting the irregularly distributed sediments will produce a chaotic assemblage of materials and forms, closely resembling those found in many Pleistocene kame complexes.

Many kames have undoubtedly originated under conditions of ice stagnation. Till lenses enclosed within deposits may be explained by intermittent flowage off the surrounding ice surfaces, while the capping of ablation till seen on many kames is probably due to superficial debris being released and lowered during the final melt phases. Several authors have envisaged a wasting ice-mass so riddled with caves and passages that an englacial water-table develops; above this, erosion is able to continue, but below it the dominant process is deposition. As early as 1958 Sissons showed that seemingly irregular kame areas sometimes have approximately accordant summits which could mark the level of an englacial water-table. In certain cases a succession of accordant levels may be discerned, each apparently related to an intermittently falling water-table whose elevation was controlled by the heights of successive cols unblocked by ice wastage. Intrinsic to this argument is the view that most of the kames accumulated directly on the subglacial floor, since lowering from an englacial or supraglacial position would presumably destroy the accordance. The same reasoning applies with even greater force to the occasional flat-topped form situated in the midst of a kame complex, or even standing as an isolated hill. Termed a crevasse filling, this is usually much larger than that name seems to imply. The remarkably even summit may extend over several square kilometres until it comes to resemble a fragment of sandur. Yet the margins are clearly of ice-contact origin, consisting of glacio-fluvial sediments that have slipped to their angle of repose above a series of fringing kettle holes. It appears that, penetrating the complete thickness of a stagnant ice-slab, there must have been an opening into which meltwater streams poured immense quantities of debris.

A third way in which kames may form is by collapse of an originally smooth sandur laid down over buried ice. As the ice melts the surface becomes pitted with irregular depressions. In some instances the process goes no further than to produce a few kettle holes dimpling the surface of an otherwise even plain. Yet in other cases it is capable of totally destroying the original surface and leaving an irregular, hummocky topography virtually indistinguishable from that formed within a stagnant ice-mass. Successive aerial photographs of recently formed sandar have shown conclusively that smooth plains can be converted into a jumbled assemblage of steep-sided mounds and kettles.

Under the general heading of kames, mention must also be made of kame-terraces and kame-moraines. Kame-terraces are the result of marginal glacio-fluvial deposition where the ice abuts against an earlier sloping surface. Aggradation takes place partly on the surface and partly on the ice. When the ice melts, the outer edge of the terrace collapses but the inner part retains its form almost intact. Kame-moraine is a rather unfortunate term occasionally used to denote a kame complex believed to have accumulated along a stable ice margin. The form may therefore have a similar chronological significance to an end-moraine but a totally different origin. Linear complexes are quite common, often seeming to consist of coalescent marginal fans or deltas; many display deltaic structures and appear to be truly subaqueous. A good example is afforded by the ridge of glacio-fluvial sediments at Galtrim in southern Ireland (Fig. 13.9). Rising 15 m above the adjacent lowland, this ridge is bounded on one side by a steep ice-contact face, on the other by a more gentle delta front. The channels through which the glacio-fluvial sediments were fed to the ice margin are marked by a series of eskers. A thin capping of unsorted debris on the ridge crest probably represents englacial material that slumped to the base of a former ice-cliff. On a much larger scale are the Salpausselka ridges in Finland. Bounded by irregular ice-contact slopes, these locally exceed 100 m in height and 2 km in width. Composed almost entirely of stratified deposits, they have been regarded by some workers as deltaic accumulations laid down in a shallow sea when Scandinavia was still glacio-isostatically depressed. Virkkala, however, claimed that ice-contact faces occur on both proximal and distal sides of the ridges which, he contended, were laid down in deep fracture zones running parallel to, but some distance from, the actual ice margin.

Eskers

Eskers are among the most distinctive of all ice-contact forms. They retain their characteristically elongated, steep-sided and narrow-crested form over an immense size range. At one extreme are ridges that locally rise as much as 100 m above their surroundings and are traceable, apart from short gaps, for distances of several hundred kilometres; at the other are ridges less than 2 m high and traceable over distances of only a few hundred metres. In plan eskers are usually sinuous and occasionally even meandering. They may occur singly, form a confluent pattern resembling a normal drainage net, or constitute part of an anastomosing system. Their crestlines are smooth or gently undulating while the bounding slopes may approach the angle of repose of the constituent material. This can range from large cobbles to fine silt and even laminated clay, but the most common component is sand and gravel with cross-bedding indicative of the direction of water flow. The strata sometimes display an arched arrangement, although there may then be some doubt whether this is a depositional structure or the result of later slumping. Many studies have been made of the provenance of pebbles in an esker. This is particularly instructive where a long esker crosses a series of distinctive rock types. Most materials seem to have travelled only a few kilometres from their point of origin, although resistant rocks are sometimes carried very much further. In Canada esker deposits have been used as a guide to mineralization of the bedrock, and a study of the 400 km long Munro esker showed that the maximum concentration of a mineral occurs 3–15 km downstream from the point of outcrop.

There are remarkably few records of really large eskers being observed in the course of formation at modern ice margins. Many workers have recorded ridges of gravel, some of them as high as 16 m, appearing from beneath the receding snouts of glaciers, but with only a few exceptions they have been ice-cored and once melting was complete were reduced to insignificant proportions. In view of the thorough studies made during recent years along current ice margins, it might be inferred either that present conditions are not conducive to esker development and preservation, or that they

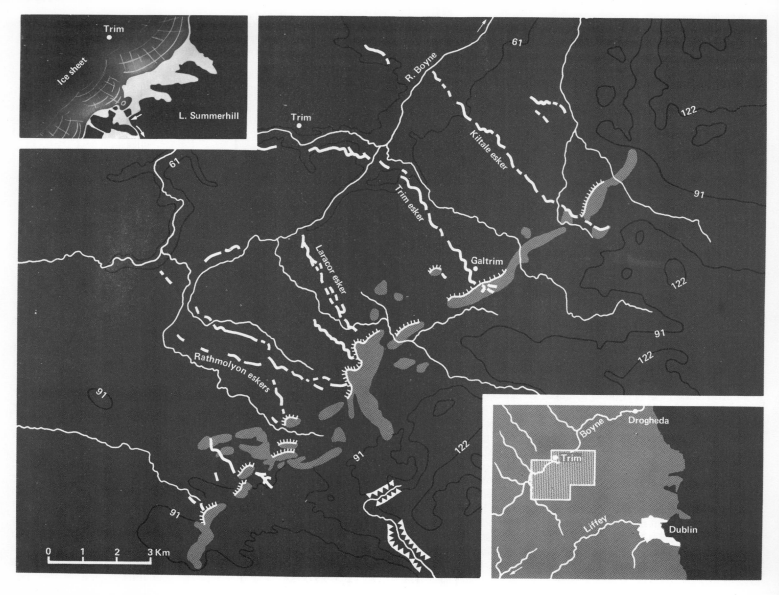

Trim

Ice sheet

L. Summerhill

Trim

R. Boyne

61

122

91

Kiltale esker

Trim esker

Laracor esker

Galtrim

122

91

Rathmolyon eskers

122

91

91

91

0 1 2 3 Km

Boyne

Drogheda

Trim

Liffey

Dublin

Fig. 13.9. The linear kame-complex at Galtrim in southern Ireland with a series of feeding eskers along its north-western edge. Conditions at the time of accumulation are reconstructed in the inset at the top left (based on Synge, 1950).

 Sand and gravel deposits of kame complex

 Prominent ice-contact faces.

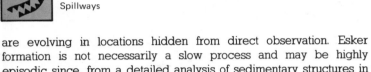 Eskers feeding into kame complex

Spillways

are evolving in locations hidden from direct observation. Esker formation is not necessarily a slow process and may be highly episodic since, from a detailed analysis of sedimentary structures in a Swedish esker, it has been argued that the deposits in a ridge over 20 m high could even have accumulated in a single melt season.

Eskers are generally agreed to originate as some form of fill along a meltwater routeway. As early as 1897 De Geer suggested they develop as deltaic accumulations where subglacial tunnels discharge into ponded water or the sea. He noted that certain Swedish eskers are divisible into segments, each of which grades downstream from coarse gravel to sand. The gravel reaches tend to be steep and narrow, the sand reaches broader and shaped like deltas. De Geer inferred that each segment represents an annual increment at the edge of a rapidly retreating ice-mass. Most of the deposition takes place in summer, the coarse debris being laid down in the last few hundred metres of a meltwater tunnel and the finer sediments accumulating in a delta protruding just beyond the ice margin. There seems little doubt that many eskers have originated in this way, although they may still constitute a small fraction of the total number.

A currently more favoured hypothesis envisages development entirely within an ice-tunnel; this is more consistent with the narrow, steep-sided form of many eskers which lack any sign of deltaic 'beading'. There are two possible locations for the tunnels, subglacial and englacial. The ice-core of many small esker-like ridges emerging from modern glaciers points quite clearly to their englacial origin. Deposition has first taken place along a channel through the ice, and retreat of the glacier margin has progressively exposed the line of debris. The steep-crested form of the ridge in this instance is not an original structure but is due to slumping and differential ablation. However, englacial tunnels of the dimensions needed to accommodate the sediments of a full-sized esker are almost unknown in modern glaciers; moreover, slump structures seem to be much less widespread than the englacial hypothesis would entail. For these reasons it is usually argued that the tunnels must be subglacial and that an observed tendency for certain eskers to run uphill for short distances must be due to water flowing under hydrostatic pressure. Accumulation of so much sediment seems to imply that the final fill takes place under conditions of diminishing discharge, possibly due to competition from the opening of alternative routeways. In some eskers the meltwater may have been finally debouching into a lake or the sea, but this does not seem to be an essential requirement. The fragmentation of many eskers into short segments separated by gaps is probably due to erosion by meltwater streams after exposure of the ridge at the ice-margin.

A suggestion that some eskers may represent the fill of supra-glacial channels receives little support from the study of modern ice-masses where suitable accumulations of water-laid sediment are extremely rare. Another possibility advocated by several Finnish workers is that the materials of certain eskers were laid down in open channels bounded by very high walls of stagnant ice; this is believed to apply in parts of Lapland where the Scandinavian ice-sheet terminated on dry land rather than in the sea or ponded water. One other small group of eskers appears to have been laid down submarginally. Termed subglacially engorged eskers, these originate at points on valley sides where meltwater streams, having passed beneath the glacier edge, deposit sufficient material to build glacio-fluvial ridges. Trending obliquely down the hillslope, such features are usually small in comparison with those formed near the ice-front and do not conform to the general rule that eskers trend parallel to the direction of ice movement.

Sandar

Beyond the immediate ice margin heavily laden meltwater streams may aggrade their beds to build up extensive accumulations of glacio-fluvial outwash. The lateral extent of the depositional zone is largely controlled by the local relief; below the snout of an Alpine glacier a relatively narrow valley sandur is normal, but where an ice margin and associated outwash are laterally unrestrained a much broader plain sandur may develop. Both types are characterized by braided channels, by abundant coarse debris and by marked variations in seasonal discharge.

The valley sandur below the Emmons glacier in Washington was the subject of a detailed study by Fahnestock. Like most similar features it has a steep long profile falling generally at over 1 in 100 and locally at over 1 in 8. The slope is very closely related to particle size and discharge. Along an individual channel with no significant change in discharge, the median particle diameter declines at a rate of about 45 mm km^{-1}. Presumably attributable to sorting, this down-valley change in debris size is associated with a pronounced concavity in the long profile. Great contrasts exist between winter and summer conditions. During the cold season meltwater discharge is low and a single meandering channel tends to develop; when ablation increases in high summer the much greater volume of water occupies a complex of braided channels. Frequent shifts of water-course preclude establishment of a continuous vegetation cover and the whole environment is one of almost constant change. It is uncertain how fast a valley-sandur surface can be aggraded; in one reach Fahnestock recorded a net gain of 0.36 m in 2 years, but it is most unlikely that this rate can be sustained over a long interval; in Alaska a mean value of 16 mm yr^{-1} has been calculated for a period covering several millennia. Much probably depends upon movements of the ice-front. During a retreat phase the meltwater streams are likely to degrade the steep surface previously built up close to the ice edge. During an advance, on the other hand, as the glacier overrides earlier outwash the meltwater will rapidly construct a new frontal apron at a relatively steep angle. Further downstream the changes are less conspicuous and there often seems to be a slow but sustained aggradation. This may still be much faster than that achieved by unglaciated tributaries, and the latter in consequence can suffer damming by the fill across their mouths. The greater the distance from the ice-front, the higher the proportion of the discharge that is normal stream flow; in the end, the specifically glacio-fluvial characteristics are almost totally lost.

Many formerly glaciated areas display fossil valley sandur, but rarely is one preserved in its original state. The steep long profile encourages rapid dissection by river erosion with the result that the earlier surface remains only as an elevated terrace. Incision often takes place in several stages so that the former accumulation is fashioned into a flight of terraces; unpaired meander terraces are frequent where the downcutting river encounters buried bedrock spurs.

Plain sandar are particularly extensive in front of the outlet glaciers descending from Vatnajokull in Iceland (see Fig. 9.13). They exhibit many similar properties to those already described for valley sandar. The constituent materials are very coarse close to the ice-edge but diminish in size as the intervening distance increases. Gradients commonly lie between 1 in 25 and 1 in 200, with an appreciable concavity to the overall profile. The braided channels carrying the meltwater cover only part of the sandur surface at any one time, and it may be several decades before they migrate backwards and forwards across the full width of the plain. The inactive areas may become partially vegetated but still lack the capacity to stabilize the watercourses. Recent glacier recession in Iceland has offered an opportunity to study the contact zone between a plain sandur and the ice margin. In several cases, prior to retreat, faces some 20–30 m high had been constructed against the ice-front. At one point, for example, a narrow proglacial lake impounded against the retreating ice edge and overflowing across the crest of the old sandur was found to be 23 m deep; the lake-floor elevation is only reached again on the sandur surface at a distance of 3 km. The implied minimum aggradation of 23 m at the crest of the sandur seems to demand that meltwater streams can issue from the ice-front at progressively higher levels. Low eskers feeding into the proximal side of many sandar also indicate that meltwater is forced upwards, presumably under hydrostatic pressure, before debouching over the surface.

Fossil plain sandar are found around the lowland margins of several former ice-sheets. A particularly fine example extends westwards from the limit of the last glaciation in Jutland (Fig. 13.3). The outwash buries most of an older landscape composed of earlier till, although the highest parts remain uncovered and protrude as low hills. Coarse gravel in the east grades away from the ice edge into fine gravel, sand and silt. The tunnel valleys of eastern Jutland end abruptly at the ice margin and the meltwater flowing through them must have ascended steeply before spreading out over the sandur. Less well preserved but of great importance in the history of Pleistocene studies are sandar along the northern fringe of the German

Alps. Here in 1909 Penck and Brückner identified four separate glaciations, which they termed Günz, Mindel, Riss and Würm, each attested by a sandur observed to pass southwards into a till. The two earlier sandar were once gravel plains extending at different levels across the Alpine foreland, but now so dissected that only small remnants survive. According to Penck and Brückner, fluvial incision after the Mindel glaciation was sufficient to confine the later Riss and Würm outwash within newly formed lowlands as valley sandar.

Glacio-lacustrine deposition

Characteristics of the sediments

Much of the coarse debris discharged into lakes by meltwater streams is normally trapped in marginal deltas. Owing to their large sediment loads these streams can construct big deltas with remarkable rapidity. In Canada the Assiniboine delta, built out into glacial Lake Agassiz, covers an area of 6 400 km² and is composed of materials that locally reach over 70 m in thickness; yet the period of formation seems to have been restricted to a single millennium between about 13 000 and 12 000 years BP. The finest sediments, however, are normally carried to the central parts of a lake to accumulate as an extensive sheet of clay and silt. This may also contain a varied assortment of pebbles and even boulders rafted into the deeper parts of the lake by icebergs; such 'dropstones' may be clearly attested in suitable sections by the distorted bedding their impact causes.

It is, however, the lamination of the finer sediments in many ice-marginal lakes that has attracted most attention. The tendency to divide into laminae is due to easy parting along silt horizons while the intervening cohesive clays resist splitting. This is readily seen in hand specimens where the light silt and dark clay layers also give the material a characteristic banded appearance. Each pair of silt and clay layers is termed a couplet, and may vary in thickness from less than 1 mm to over 100 mm. The deposit as a whole is best termed a rhythmite, a name to be preferred to the more frequently used 'varved clay' since the word 'varve' strictly connotes an annual increment which is still unproven for many deposits. The idea that each couplet represents an annual accumulation is a very old one,

but it was only with the detailed investigations of De Geer in Sweden starting in the late nineteenth century that its validity was widely established. De Geer contended that the silt and clay layers generally represent summer and winter deposits respectively. During summer-time meltwater introduces abundant fresh sediment into the ice-free lake. The coarser fraction of the new input settles rapidly to the floor, while the relatively unstable water conditions also trigger turbidity currents from the marginal slopes. During winter-time, with meltwater flow curtailed and the lake frozen, sedimentation is restricted to the very fine particles settling from suspension. Onset of the next spring will see a relatively abrupt reversion to accumulation of coarser material. Within both the summer and winter horizons there is often a tendency towards particle-size grading. It is the coarser fraction that is most frequently graded, sometimes passing up progressively into the succeeding clay. Such couplets are termed diatactic by Scandinavian workers to distinguish them from the symmict in which the grading is much less complete. Diatactic couplets are believed to accumulate in fresh water, whereas the symmict type are associated with salt or brackish water in which the presence of electrolytes flocculates the clay particles and causes them to settle rapidly.

The belief that the majority of couplets in glacial rhythmites are annual increments may be justified on several grounds. Where climatic conditions are not too severe, the pollen at successive horizons within an individual couplet has been shown to exhibit a seasonal cycle corresponding with the period of plant flowering. The abundance of diatoms may also vary systematically within a couplet, being much greater per unit volume of sediment in the silt than in the clay layers. For certain lakes the aggregate quantity of material in each recent couplet has been shown to equal the current annual input by meltwater streams. Finally, after the advent of radio-carbon dating it became possible to compare assumed ages with those indicated by C^{14} assays; in general, agreement has been close and it has even been suggested that ages calculated from rhythmite studies should be used as a correction factor for C^{14} dates.

Chronological calculations

Working on the assumption of annual couplets in both glacio-lacustrine and glacio-marine sediments, De Geer attempted to establish a chronology for the deglaciation of Sweden. At many hundreds of localities he measured varve thicknesses, plotting the data for each site as a curve of relative thickness against time. Curves from adjacent localities, generally less than 1 km apart, were then placed side by side and moved up and down until a visual match was obtained. The theoretical basis for this method is that climate acts as an overall control of varve thickness, primarily by determining the amount of summer ablation. Varves thicker than usual for their respective sites will form during a warm summer with exceptional meltwater flow; conversely, unusually thin varves will develop over a wide area during a cool summer. On this basis correlation may be established from one location to another. Assuming that the northern limit of an individual varve marks the contemporary ice edge, De Geer was able to trace the withdrawal of the Scandinavian ice-sheet over a distance of 1 000 km covering a period of about 12 500 years (Fig. 13.10). A critical point in De Geer's time-scale was the sudden drainage of a large ice-dammed lake in central Jamtland. He believed this took place when the Scandinavian ice-sheet finally split into two smaller remnants. Known as the bipartition, this event produced a particularly thick varve which De Geer employed as his zero varve in working backwards in time across central and southern Sweden. Liden subsequently claimed that the zero varve corresponds to 6839 BC, although research has now revealed bipartition as a more complicated process than formerly supposed. The so-called zero varve actually seems to consist of two varves separated by an interval of as much as 80 years. Despite unresolved problems of this type, De Geer's chronology has stood up remarkably well to critical examination by later workers. There seems no doubt regarding the annual duration of most of the couplets he employed, nor the existence of a single overriding determinant of varve thickness, presumably the climate.

It is in Denmark that most severe criticism of De Geer's time-scale has been voiced. Several workers have contended that couplets which he used to extend his chronology across the Danish islands of Zealand and Fünen are not annual layers but represent shorter-term fluctuations in meltwater discharge. These could be due either to warm and cold spells of weather or possibly to periods of storminess; it has even been inferred that some couplets may be no more than daily layers. It is also worth recording that in Canada Shaw and Archer have interpreted certain sand layers within glacio-lacustrine sequences as due to turbidity currents set off by temporary winter-

Fig. 13.10. Dating by varve chronology of the retreat of the ice-front across Sweden. The dashed lines show presumed ice-fronts but are not firmly tied to the lines of the main time-scale; the latter are at 100-year intervals and extend from 10 500 to 6800 BC (after Lundqvist, 1965).

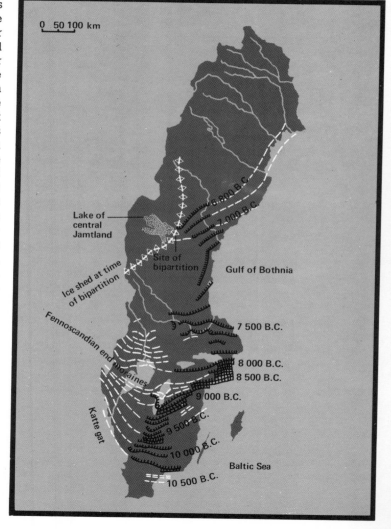

time falls in water-level. Despite these potential limitations, it is still true that varve analysis provides one of the most valuable guides not only to the age of the materials themselves but also to the speed with which various geomorphological changes take place. For example, in Sweden it is calculated that during the 400 years immediately prior to bipartition the ice-front receded a distance of 130 km, a mean rate of 325 m yr^{-1}; in Finland for over a millennium after deposition of the Salpausselka ridges the rate of retreat averaged about 260 m yr^{-1}.

References

Allen, J. R. L. (1971) 'A theoretical and experimental study of climbing-ripple cross-lamination, with a field application to the Uppsala esker', *Geogr. Annlr.* **53A**, 157–87.

Atwater, B. F. (1984) 'Periodic floods from glacial Lake Missoula in the Sanpoil arm of glacial Lake Columbia, northeastern Washington', *Geology* 12, 464–7.

Baker, V. R. (1973) 'Paleohydrology and sedimentology of Lake Missoula flooding in eastern Washington', *Geol. Soc. Am. Spec. Pap.* 144.

Baker, V. R. and **Nummedal, D.** (1978) *The Channeled Scabland*, NASA, Washington.

Bretz, J. H. (1966) 'Correlation of glacial lake stages in the Huron, Erie and Michigan basins', *J. Geol.* **74**, 78–9.

Bretz, J. H. (1969) 'The Lake Missoula floods and the channeled scabland', *J. Geol.* **77**, 504–43.

Clayton, L. (1964) 'Karst topography on stagnant glaciers', *J. Glaciol.* **5**, 107–12.

Dahl, R. (1965) 'Plastically sculptured detailed forms on rock surfaces in northern Nordland, Norway', *Geogr. Annlr.* **47**, 83–140.

Dury, G. H. (1951) 'A 400-foot bench in southeastern Warwickshire', *Proc. Geol. Assoc.* **62**, 167–73.

Fahnestock, R. K. (1963) 'Morphology and hydrology of a glacial stream – White River, Mount Rainier, Washington', *U.S. Geol. Surv. Prof. Pap.* 422-A.

Flint, R. F. (1971) *Glacial and Quaternary Geology*, Wiley.

Fromm, E. (1970) 'An estimation of errors in the Swedish varve chronology', in *Radiocarbon Variations and Absolute Chronology* (ed. I. U. Olsson), Almqvist and Wiksell.

Goldthwait, R. P. (1974) 'Rates of formation of glacial features in Glacier Bay Alaska', in *Glacial Geomorphology* (ed. D. R. Coates), State Univ New York.

Gray, J. M. (1981) 'p-forms from the Isle of Mull', *Scot. J. Geol.* **17**, 39–47.

Hagen, J. O. (1983) 'Subglacial processes at Bondhusbreen, Norway: preliminary results', *Ann. Glaciol.* **4**, 91–8.

Hantz, D. and **Lliboutry, L.** (1983) 'Waterways, ice permeability at depth, and water pressures at glacier d'Argentière, French Alps', *J. Glaciol.* **29**, 227–39.

Huddart, D. and **Lister, H.** (1981) 'The origin of ice-marginal terraces and contact ridges of East Kangerdluarssuk Glacier, SW Greenland', *Geogr. Annlr.* **63A**, 31–9.

Lee, H. A. (1965) 'Investigations of eskers for mineral exploration', *Geol. Surv. Can. Pap.* 65–14, 1–17.

Lundqvist, J. (1965) 'The Quaternary of Sweden', in *The Geologic Systems: The Quaternary* (ed. K. Rankama), Wiley.

Lundqvist, J. (1980) 'The deglaciation of Sweden after 10 000 B.P.', *Boreas* **9**, 229–38.

Mannerfelt, C. M. (1945) 'Some glaciomorphological forms and their evidence as to the downwasting of the inland ice in Swedish and Norwegian mountain terrain', *Geogr. Annlr.* **27**, 1–239.

Nye, J. F. (1976) 'Water flow in glaciers: jokulhlaups, tunnels and veins', *J. Glaciol.* **17**, 181–207.

Peel, R. F. (1949) 'A study of two Northumbrian spillways', *Trans. Inst. Brit. Geogr.* **15**, 75–89.

Shaw, J. and **Archer, J.** (1978) 'Winter turbidity deposits in Late Pleistocene glaciolacustrine varves, Okanagan valley, British Columbia, Canada', *Boreas* **7**, 123–30.

Shotton, F. W. (1953) 'The Pleistocene deposits of the area between Coventry, Rugby and Leamington and their bearing upon the topographic development of the Midlands', *Phil. Trans. R. Soc.* **B237**, 209–60.

Shreve, R. L. (1972) 'Movement of water in glaciers', *J. Glaciol.* **11**, 205–14.

Shreve, R. L. (1985) 'Esker characteristics in terms of glacier physics, Katahdin esker system, Maine' *Bull. Geol. Soc. Am.* **96**, 639–46.

Sissons, J. B. (1958) Supposed ice-dammed lakes in Britain, with particular reference to the Eddleston valley, southern Scotland', *Geogr. Annlr.* **40**, 159–87.

Sissons, J. B. (1974) 'A Late-glacial ice-cap in the central Grampians, Scotland', *Trans. Inst. Brit. Geogr.* **62**, 95–114.

Sissons, J. B. (1978) 'The parallel roads of Glen Roy and adjacent glens, Scotland', *Boreas* **7**, 229–44.

Sissons, J. B. (1979) Catastrophic lake drainage in Glen Spean and the Great Glen, Scotland', *J. Geol. Soc. Lond.* **136**, 215–24.

Synge, F. M. (1950) 'The glacial deposits around Trim, Co. Meath', *Proc. R. Irish Acad.* **53**, 99–110.

Teller, J. T. and **Clayton L.** (ed.) (1983) *Glacial Lake Agassiz*, Geol. Assoc. Can. Spec. Pap. 26.

Virkkala, K. (1963) 'On ice-marginal features in south-western Finland', *Comm, Geol. de Finlande, Bull.* 210.

Weertman, J. and **Birchfield, G. E.** (1984) 'Stability of sheet water flow under a glacier,' *J. Glaciol.* **29**, 374–82.

Willman, H. B. (1971) 'Summary of the geology of the Chicago area', *Ill. State Geol. Surv. Circ.* 460.

Wold, B. and **Østrem, G.** (1979) 'Subglacial constructions and investigations at Bondhusbreen, Norway', *J. Glaciol.* **23**, 363–78.

Woodland, A. W. (1970) 'The buried tunnel valleys of East Anglia', *Proc. Yorks. Geol. Soc.* **37**, 521–78.

Selected bibliography

The texts listed in the selected bibliography at the end of Chapter 12 all include valuable sections on glacio-fluvial as well as purely glacial topics. One publication dealing specifically with the characteristics of meltwater sediments is A V Jopling and B C McDonald (eds), *Glaciofluvial and Glaciolacustrine Sedimentation*, Soc. Econ. Palaeont. and Mineralog. Spec. Pub. 23, 1975.

Chapter 14
Periglacial forms and processes

The term periglacial was introduced by Lozinski in 1909 to refer to the area bordering an ice-sheet and characterized by the severity of its climate. There is no necessary identity between the two ideas contained in this statement, and usage gradually changed until the word came no longer to demand an ice-marginal location but assumed only an intensely cold climatic regime. It was thus acknowledged that the periglacial zone is distinguished from the cool temperate regions primarily by the importance of frost action and prolonged surface freezing; indeed, periglacial regions might be defined as non-glacial areas in which geomorphic processes are still dominated by the presence of water in its solid phase. However, frost action and surface freezing are related but not exactly the same concept, and in attempts at further refinement two schools of thought emerged. On the one hand were workers who restricted the term to areas underlain by permafrost or perennially frozen ground, and on the other were those who wished to include, in addition, regions experiencing a large number of freeze–thaw cycles. It is the latter interpretation which is today the more generally accepted and which will be used in the following pages. However, for purposes of presentation permafrost will first be discussed as a separate topic and only then will attention be turned to some of the more distinctive surface sediments and structures arising, in part at least, from repeated freeze–thaw action.

Despite accelerating research, much remains to be discovered about periglacial environments. An early impetus for intensive study came from strategic considerations during the Second World War when it was realized that correct engineering practice in such areas demands a proper understanding of periglacial processes. More recent stimulus has come from the desire of commercial firms to exploit the mineral resources of these sparsely populated regions. An outstanding example is the extraction of oil on the north slope of Alaska where the drilling teams had to cope with some of the harshest conditions on the North American continent. The success of British Petroleum in proving large reserves highlighted another argument for intensified research. This is the conservationist argument. To move the oil to the consuming areas it must be transported by specially designed pipeline across Alaska to the ice-free Pacific coast. Yet little is known of the potential long-term effects of the pipeline on what is generally agreed to be a fragile environment that recovers extremely slowly from any disturbance. It seems likely, therefore, that the pace of research will continue to quicken.

One further aspect of research in modern periglacial areas is its potential for specifying the environmental controls over the development of features now found in fossil form in more temperate regions. Such information may ultimately be useful in defining climatic zones as they existed during critical periods of the Pleistocene epoch, thereby supplementing the biological evidence that may be available for the same purpose. It must be stressed, however, that climate is by no means the sole determinant of many periglacial processes; the texture and moisture content of the soil, the character of the vegetation, the duration of the snow cover and the aspect of the slope are among a range of other important influences.

Permafrost

Distribution

Defined as a layer of earth materials whose temperature has remained continuously below 0 °C for several years in succession, permafrost has been estimated to underlie more than 15 per cent of the world's contemporary non-glacial land surface (Fig. 14.1). The more extreme climatic zones are underlain by continuous permafrost where the only unfrozen areas occur beneath the largest lakes and rivers. Such conditions exist in Canada north of a line from Hudson Bay to the Mackenzie delta, in Alaska along the north slope, and in Siberia throughout a broad belt extending 1 000 km southwards from the Arctic Ocean. Within this area the thickness of the permafrost

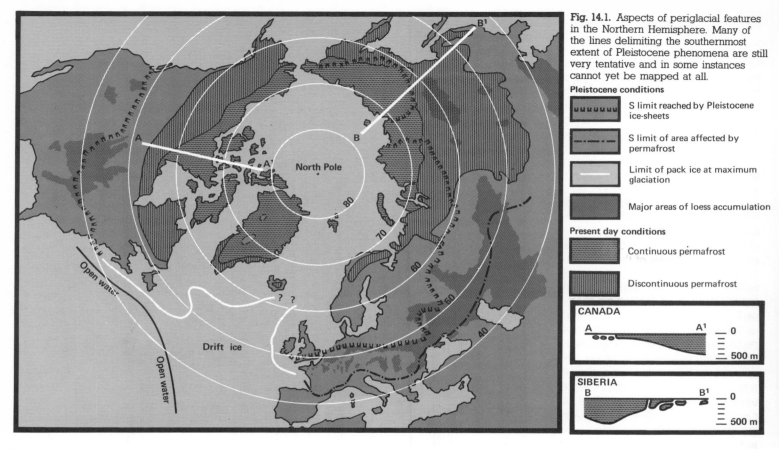

Fig. 14.1. Aspects of periglacial features in the Northern Hemisphere. Many of the lines delimiting the southernmost extent of Pleistocene phenomena are still very tentative and in some instances cannot yet be mapped at all.

Pleistocene conditions

- S limit reached by Pleistocene ice-sheets
- S limit of area affected by permafrost
- Limit of pack ice at maximum glaciation
- Major areas of loess accumulation

Present day conditions

- Continuous permafrost
- Discontinuous permafrost

varies considerably. In general it is greatest in the most northerly latitudes, locally exceeding 600 m near the arctic coast of Siberia. In northern Canada and Alaska thicknesses of over 500 m have been recorded in a number of boreholes, while on Spitsbergen colliery workings are said to encounter frozen ground at depths of 320 m. Near the southern margin of the continuous permafrost zone a more typical figure would be about 50 m. Beyond this margin the permafrost becomes discontinuous. Local geological, topographical and botanical factors combine to limit the development of frozen ground so that its distribution becomes patchy. There is a gradation from regions where all but a small minority of the area is underlain by permafrost to those where perennially frozen layers occur as isolated pockets. In Canada the zone of discontinuous permafrost extends from the southern end of James Bay eastwards to Labrador and westwards to northern Alberta. It reaches virtually to the Pacific coast of Alaska and in central Asia has been recorded as far south as 45 °N. Sporadic outliers are found in such elevated areas as the Scandinavian mountains, the Urals and the Rockies of British

Columbia.

Relationship to climate

The primary control of permafrost is obviously climate. An equilibrium may be established when the perennially frozen layer attains a depth determined by the atmospheric temperature and the geothermal flux. Since, on average, the flow of internal heat is minute compared with that from the sun, there is a strong tendency towards seasonal equalization of the atmospheric and surface soil temperatures (Fig. 14.2). Below the surface the amplitude of the seasonal fluctuations gradually diminishes until, at a depth of about 15 m, the ground temperature remains stable throughout the year. This is the level of 'zero annual amplitude' and the temperature at this horizon can be anywhere between just below freezing-point and about −16 °C; in practice it tends to approximate very roughly to the mean annual air temperature. At greater depths the temperature steadily rises, normally at a rate of 1 °C per 40–50 m, until freezing-point is reached at the level of the permafrost base.

Many attempts have been made to examine the relationship between permafrost distribution and climate. In theory, perennial freezing may occur wherever the mean annual temperature is below zero; in practice, however, the mean temperature is usually a degree or so below freezing-point before permafrost actively develops. Studies in Canada have shown that perennially frozen layers are rarely encountered south of the −1 °C isotherm, and that they are largely restricted to peatlands between the −1 and −4 °C isotherms. It is only north of the −4°C line that permafrost becomes widespread, while it is virtually continuous beyond the −6.5 °C line. These figures are surprisingly at variance with those recorded in Asia. In the eastern USSR permafrost is widespread south of the −2°C isotherm and is by no means rare where the mean annual temperature is at freezing-point. The cause of this discrepancy is believed to lie in the fact that much Siberian permafrost is a relic of late Pleistocene times when the cover of insulating ice-sheets was far less extensive than in Canada. Russian workers think that the extent of permafrost is currently decreasing along the southern margin of the discontinuous zone, although it may still be deepening in parts of northern Siberia. Implicit in these interpretations is the idea of a long time-lag between a change in climate and restoration of equilibrium conditions.

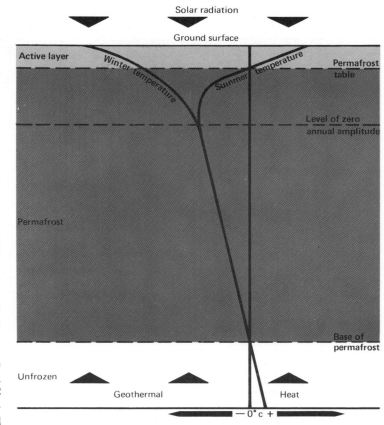

Fig. 14.2. Idealized underground temperature distribution in an area of permafrost.

As the climate of a temperate area cools at the onset of a periglacial phase, the top few metres of the ground will first be subject to annual freeze – thaw cycles. With further refrigeration the winter freezing will penetrate deeper than the summer thawing so that a layer at depth will be left perennially frozen. Each year this layer will thicken as new permafrost is added at the base. The temperature of the perennially frozen ground will vary seasonally until the level of zero annual amplitude is reached. Thereafter, as the temperature at the level of zero annual amplitude falls, more permafrost will be added until eventually it may attain a thickness of several hundred metres. With a climatic amelioration the process is reversed, with the temperature of the frozen layers slowly rising and the depth of permafrost declining. Complications of this simple model may be readily visualized, particularly with rapid climatic oscillations, and actual field relationships tend to be very intricate. For example, many boreholes have encountered unfrozen layers, known as taliks, within otherwise continuous permafrost; conversely, temperatures below the level of zero annual amplitude sometimes seem excessively cold, possibly due to the recent climatic warming of many permafrost areas.

The active layer and associated freezing processes

A period of perennial freezing has little lasting effect on deep-seated rocks. It is the shallow surface layer subject to summer thawing that suffers the most severe disturbance. Designated the active layer, its thickness can vary from a few centimetres to more than 3 m. The base of the active layer is known as the permafrost table; while in general parallel to the ground, it tends to show many local irregularities related to detailed surface conditions. In spring and summer the active layer thaws from the surface downwards so that the resulting water cannot percolate freely to greater depths. With the possibility of additional liquid from snowmelt it is scarcely surprising that waterlogging is one of the most characteristic features of permafrost areas in summer. In autumn when the temperatures begin to fall it is the surface horizons that freeze first, trapping a layer of saturated sediments immediately above the permafrost table. As was shown experimentally by Pissart in 1973, before the whole active layer is refrozen in winter the final remnants of the trapped water may be placed under considerable 'cryostatic' pressure.

The freezing process in a porous medium like the soil involves much more than simple conversion of the interstitial water to ice at 0 °C. It is now several decades since Taber demonstrated that the freezing of saturated sediments can lead to the segregation of large pockets of clear ice. The results are seen in frost-heaved soils underlain by layered bodies of ice, and in boreholes within the permafrost that have encountered patches of pure ice well over 1 m in thickness. It is obvious that these were never cavernous voids

filled with liquid, but that interstitial water must have migrated towards the growing crystalline body. The processes involved are highly complex and will not be pursued here in detail. There are really two related problems. The first concerns the reason for water moving through the porous medium to feed the expanding ice crystals. Like all crystals, ice grows by virtue of the electrochemical attraction exerted by one stable ion for another in the surrounding medium. On account of its structure, the solid phase exerts a far greater attraction than that linking the liquid molecules, with the result that water is drawn very strongly towards the growing crystal. Schenk has argued that the suction with which water molecules are attracted to an ice-crystal must equal the force that the crystal itself can exert. With a value of over 200 MN m^{-2} this exceeds the force by which most water films are held to sedimentary particles, including even the hygroscopic water adsorbed on to clay minerals.

The second problem concerns the reason for part of the water remaining in a liquid state while the rest is frozen solid. The progressive freezing of water in a porous substance is affected by several variables. Numerous experiments on sediment–ice–water mixes have demonstrated the coexistence of ice and water at temperatures much below 0 °C, and it is now firmly established that in fine-grained materials over 10 per cent of the water may remain unfrozen at temperatures below −2 °C. This is primarily because such water is held by very high tensions on the smaller soil particles which in turn depresses its freezing-point. Burt and Williams have demonstrated in the laboratory the ability of such water films to migrate through sediments cooled well below 0°C. It is worth stressing that most frost heaving is not simply the result of expansion on conversion of water into ice, but involves additionally the migration of supercooled water. A related effect is the desiccation of those zones that are supplying the water.

Taber showed by a series of experiments that the tendency towards segregation is much greater in some sediments than in others. In general ice segregates most readily in silts. This has long been recognized by engineers who need to identify what they term 'frost-susceptible' materials prone to particularly severe upheaval as a result of freezing; as a guideline they usually reckon that debris containing more than a few per cent silt-sized particles will be especially frost susceptible. In coarser sediments little supercooling occurs and the whole mass freezes as a body when the temperature falls to 0 °C. In finer sediments segregation may be inhibited by the intrinsic impermeability of the materials. These variations, combined with more obvious contrasts in the distribution of interstitial water, can lead to extremely erratic development of ground ice. For example, in areas of discontinuous permafrost patches of peat are often elevated a few metres by their locally high ice content to produce mounds known as palsas. If climatic warming or some other agency initiates degradation of permafrost, the uneven distribution of ice can produce a very fitful melting and subsidence. This frequently involves the development of lakes, termed alases, and a very poorly integrated drainage system; the topography originating in this way is known as thermokarst. In southern Siberia new thermokarst is currently being created by the clearance of natural vegetation for purposes of cultivation; this permits the deeper penetration of summer heat and irregular thawing at the permafrost table. When the process is complete, any stratification in the surface layers may be thoroughly disturbed producing superficial structures resembling the involutions to be described on p. 291.

Characteristic weathering and transport processes

Coarse slope deposits

Mechanical fracture of rock material is one of the most distinctive attributes of the periglacial environment. It results from the pressure exerted by ice crystallization, and is particularly effective in permeable rocks saturated with water (p. 96). Debris produced in this way ranges from large boulders to fine angular chips depending on the fissility of the parent material. On flat surfaces the detritus may accumulate as a bouldery or rubbly mantle, but more often it will proceed downslope as sheet of mobile debris.

A wide variety of materials and forms develops as a direct result of frost shattering. Where the debris is exceptionally coarse, blockfields and rock glaciers may be produced. Blockfields, also known by the German term *Felsenmeer*, consist of angular and poorly rounded boulders, many of them 1 m or more across. Large voids promote the washing down of any finer debris so that the impression of an uneven boulder pavement is created. The arrangement of the materials is seemingly haphazard and there may be no evidence of

individual fragments having moved any great distance. On progressively increasing gradients, blockfields give way to sheets of rubble in which the component fragments tend to lie with long axes aligned downslope. In modern periglacial areas blockfields are frequently seen on flat summit surfaces; they are known in fossil form from such regions as Scandinavia and the Appalachians where the lack of fines retards plant colonization so that they may appear as small, poorly vegetated patches in the midst of dense forests. Rock glaciers consist of lobate, slow-moving bodies of frost-riven debris. They tend to originate at the foot of precipices from which fresh material is constantly being wedged free. A common source is the headwall of an old cirque from which the rock glacier moves down-valley in a form very reminiscent of a true glacier. The debris is often around 30 m thick, with a rough grading from coarse boulders at the surface to a much thicker layer of mixed boulders, sand and silt at the base. The margins tend to be very steep but the upper slopes generally lie at angles between 5 and 20°. Both active and fossil forms are known. Current movement may be indicated by concentric wrinkling of the surface, and superficial debris has been measured travelling at about 1.5 m yr^{-1} in Switzerland and 0.5–0.7 m yr^{-1} in Alaska. There is no doubt that in the latter region the lower layers of several active rock glaciers are perennially frozen, and it has been inferred that the current motion is almost entirely due to the presence of interstitial ice. When the climate ameliorates a rock glacier is therefore likely to be stabilized and slowly colonized by vegetation. Such fossil forms are known from many parts of the world, although other possible origins for tongues of coarse angular debris need consideration; it has been suggested that some spreads are merely an exceptionally thick ablation moraine let down from a stagnant glacier.

Another prominent feature consequent upon active freeze–thaw processes is the talus or scree accumulating at the foot of a steep rock face. The scree surface in a periglacial environment is rarely stabilized. Frost-riving continues to comminute the debris, while relatively rapid downslope movement is sustained by such factors as the growth of interstitial ice and the impact of occasional snow avalanches. Sections in talus sometimes reveal a pronounced stratification composed of alternating layers of fine and coarse particles. Known by the French term *grèzes litées*, these are commonest where the source rock is highly fissile and the average size of fragment correspondingly small. The stratification presumably reflects some form of environmental fluctuation, but its nature remains uncertain. Experimental research in France has indicated that comparable results could be produced by seasonal meltwater from snowdrifts depositing a slurry of mixed debris from which the finer material is later eluviated, but some workers have evisaged much longer-term changes in factors like the surface runoff and the duration of snow cover. A further form that is undoubtedly affected by the snow cover is the protalus rampart. Material wedged by ice from the face of a precipice can skip and slide across a basal snowbank to build a ridge of coarse debris that may locally reach as much as 10 m high. If the climate becomes warmer and the snowbank melts, the ridge or rampart appears to be isolated from the debris source and can easily be mistaken for a small cirque moraine. However, its composition and fabric will be different and the materials will show no sign of glacial transport.

Solifluction sheets.

The combination in many periglacial regions of deep freezing and varied weathering processes tends to generate hillslope mantles of ill-sorted water-saturated debris with constituent particles ranging in size from clay to large boulders. In 1906 Andersson proposed the term 'solifluction' for the movement of such material which he observed to be taking place very actively in the Falkland Islands. Indeed, he regarded the removal of waste in that environment as occurring at a rate unsurpassed in other parts of the world. It is important to recognize that in solifluction, as the word is currently employed, two processes normally operate simultaneously. The first is flowage which can be induced by any tendency towards saturation and is therefore not a uniquely periglacial phenomenon. For this reason some workers have wished to introduce the term 'gelifluction' for flowage specifically across a frozen substrate. Gelifluction occurs when the surface layers melt in the spring but the underlying horizons remain frozen, either perennially or at least until much later in the summer. Eventually the moisture content may be raised to the Atterberg liquid limit, drastically reducing the strength of the thawed material. High porewater pressures may be generated with similar consequences. It has also been suggested that during thaw the soil colloids are flocculated and lose some of their cohesion. The reason may lie in a concentration near the surface of free ions which, during

the melt phase, coagulate the clay minerals to produce a crumb structure which is mechanically very weak. Moreover, the material becomes thixotropic so that any disturbance tends to accentuate the mobility of the debris. All these factors combine to promote rapid gelifluction down quite gentle slopes.

The second solifluction process is creep induced by freeze–thaw cycles. In the heaving phase debris is lifted normal to the slope, but in the settling phase it tends to fall back vertically under the pull of gravity. Surface layers will generally be displaced more than the lower, while further differential movement may result from complex and irregular patterns of freezing. A particularly effective form of freezing that can affect the top few centimetres of the regolith is the development of pipkrake or needle ice. As the air temperature drops, acicular ice crystals develop beneath stones and patches of soil that are relatively good conductors of heat. In one major cycle such material can be raised as much as 0.1 m from its original position before sinking back vertically when melting takes place. In New Zealand rock debris on an 11° slope was found to be moving over 0.6 m yr^{-1} by this process.

An active solifluction sheet tends to generate a number of distinctive surface forms. Uneven movement frequently produces conspicuous lobes of rather thicker debris. On steeper slopes terrace-like features often evolve, sometimes extending hundreds of metres parallel to the contours, but more commonly dividing and uniting to create a series of irregular crescent-shaped steps known as garlands. They can be divided into two main types, differentiated by the appearance of the riser. In some instances this is composed of large stones, in others it is covered by a binding turf layer. The precise mechanism responsible for such stone-banked and turf-banked terraces remains uncertain. They are of the same general dimensions with the risers commonly 1–5 m high and the treads 5–25 m across. However, current research suggests that rather different processes operate in their formation. The turf probably acts as a restraint to fast downhill movement, with more mobile debris collecting on the upslope side as a terrace. This is indicated by the occasional rupture of the turf allowing detritus to spill across the next lower tread. Yet the riser itself is not usually stationary but moving slowly downhill. This is evident where trenches dug through terraces have revealed old, overridden soil profiles. The boulders in stone-banked terraces may also act as barriers to downhill movement with finer debris piling up behind them. However, several workers have suggested that a sorting mechanism within a solifluction sheet must first drive the largest fragments towards the frontal margin, and only when concentrated in this way do they lose their capacity for accelerated motion. The continued downslope migration of stone-banked terraces is also attested by the overrunning of old humic layers. From radio-carbon dates the rate of migration in the Colorado Rockies has been estimated to range from under 1.5 to over 20 mm yr^{-1}.

As the weathered regolith in a temperate environment normally moves much more slowly than the corresponding solifluction sheet, refrigeration of the climate tends to increase the rate at which material is conveyed to the valley floor. The transporting capacity of the river may not be proportionately increased since, for long periods in winter, there will be no discharge at all. In consequence, soliflucted sediments can accumulate to considerable thicknesses and in suitable sections individual debris sheets may be seen piled one on top of another. A further attribute of solifluction is its ability to move sizeable boulders down quite gentle slopes. Incorporated as part of the solifluction sheet these will be carried towards the valley floor and, if it is beyond the competence of the stream to move them, the largest may be left as an exceptionally coarse residual deposit.

When the climate warms at the close of a periglacial episode, the immobilized solifluction sheets and the thick valley-floor accumulations may remain as witness to the operation of a distinctive set of slope processes. On gentle clay hillsides in eastern England slickensided shear planes at a depth of about 1 m below the surface have been widely identified. These planes are now stable, and permafrost inhibiting downward percolation of water was almost certainly involved in their initiation. As long ago as 1839 de la Beche drew attention to poorly sorted bouldery deposits at the foot of valley sides in south-western England. He called the material 'head', a term still widely employed to designate the produce of former periglacial slope processes. The poor sorting of head can make it difficult to distinguish from till. One of the more obvious diagnostic features of a solifluction sheet is its purely local derivation, but even this criterion may be hard to apply where there is potential contamination from earlier glacial deposits. It is generally claimed that soliflucted debris is more angular, its texture less compact and its fabric composed of particles aligned with long axes pointing downslope. Nevertheless, there are many cases where a distinction between till and head has been disputed. In Wales, for example, detrital sediments enveloping the cliffs and hills of the Cardigan Bay coast have been interpreted by most workers as till, but the Watsons argued they could equally well be periglacial deposits and drew attention to similarities with deposits in north-western France beyond the normally accepted limit of ice advance.

Fluvial sediments

Fluvial sediments accumulating in modern periglacial areas often contrast sharply with those in more temperate regions. One reason is that arctic river regimes differ substantially from those of lower latitudes. In winter all but the largest rivers are totally frozen, and for extensive areas stream flow is confined to less half the year. When the thaw arrives it briefly releases huge volumes of water causing widespread inundation. However, in most cold climates the precipitation totals are so low that river-levels subside well before the onset of the next winter freeze. In several instances it has been shown that over 50 per cent of the total annual discharge passes down a stream channel in 2 or 3 weeks during the spring. Conditions are particularly conducive to the braiding of streamcourses. Violent fluctuations in discharge, much coarse debris resulting from accelerated movement on hillslopes, weak soil development and poor binding by the vegetation cover, all combine to promote braiding and frequent shifts of streamcourse. Lenses of silt and clay may accumulate where abandoned distributaries are being infilled by floodwaters. The latter deposits in their fossil form frequently enclose organic horizons which, from their faunal and floral remains, serve to confim the extreme climatic regime under which the aggradation took place. One extra factor peculiar to periglacial regions and conducive to braiding is worthy of note. This is the ability of the river to undercut its banks by thawing the frozen sediments at and below the water-level. Recession of a niche produced in this way has sometimes been observed to create an overhang 10 m wide before final collapse occurs. As yet relatively few reliable records of sediment transport by periglacial streams have been compiled. However, those that are available imply the proportion of total sediment moved as bedload may be exceptionally high, with values in excess of 75 per cent being suggested in some cases.

Where large rivers flow polewards into areas of more severe climate, unique conditions may be encountered. Since the upper courses often thaw weeks before the lower, masses of floating ice dislodged from the headwaters can become jammed against the still frozen downstream reaches. Water is ponded back and forced to spill across the floodplain while large ice-floes continue to grind against the channel banks. When the blockage is finally released a huge surge of water proceeds down the original channel, capable of carrying with it exceptionally large boulders. The overall effect is to build up an alluvial cover of generally coarse grade, poorly sorted and with occasional boulder beds.

Wind-blown materials

Wind-blown sediments are widely distributed in modern periglacial environments. Several factors favour the aeolian transport of silt- and sand-sized particles under such a climatic regime. The prevalence of strong winds is a first requirement that is fully met in Arctic areas, with the sparse vegetation of many high-latitude regions ensuring that the wind can operate to full effect. Secondly, an abundant supply of suitable fine detritus is furnished by broad alluvial flats on which the discharge at many times throughout the year is so low as to allow the surface to dry out completely. Moreover, away from the valley floors the growth of ground ice tends to desiccate the uppermost layers of the regolith so that during the cold season a dusty surface is exposed to aeolian deflation. Winter snowfall can act as a protective blanket, but often the snow itself is blown about in blizzards and carries much loose sediment with it.

Periglacial wind-blown materials vary considerably in grain size, but for convenience may be divided into two major classes, loessic dust and aeolian sand. Loessic dust consists in the main of silt-size particles having diameters in the range 0.015–0.05 mm. As explained in Chapter 10, such dust can be carried fully suspended in the atmosphere and so may be transported long distances without settling. Although it has been shown that material of this type is currently being deflated during summer from the dried-up distributaries of braided rivers in Alaska, it remains true that most of our knowledge of wind-blown dust in periglacial environments comes from the study of Pleistocene loess accumulations in modern temperate areas.

Pleistocene periglacial loess normally comprises an unstratified, highly porous deposit that, in some localities, attains a thickness of over 50 m although a figure of under 5 m is more typical. Its distribution (Fig. 14.1) points very strongly to glacial outwash as a primary source of the sediment. It tends to be thickest immediately east of those rivers which during Pleistocene times carried prolific meltwater, either from a major continental ice-sheet or to a lesser extent from large valley glaciers. Examples are afforded by the Bug, Don and Volga in the USSR, by the Rhine and Danube in central Europe and by the Mississippi in the United States. However, not all periglacial loess comes from such sources. Much wind-blown material in northern France, for instance, has a composition that varies according to the calcareous or non-calcareous nature of the underlying bedrock, and the boundaries are too sharp to be consistent with the wind simply deflating alluvial deposits which would be much more mixed in their general composition. One region where loess tends to be relatively sparse is along the cool temperate oceanic margins. In part this may be due to the humid climate restricting sediment availability, firstly by impeding the desiccation of the surface, and secondly by favouring rapid colonization of new outwash by plant communities. Yet even more significant may be the mobility of moist regolith incorporating any freshly fallen silt into the topmost soil horizons so that loess does not accumulate as a separate unit. In Great Britain, for example, recognizable loess deposits have a very patchy distribution. In south-eastern England the material known as brickearth probably originated as wash from local hillslopes mantled with wind-blown silt, but the most widespread identification of fine aeolian sediments has come from careful analysis of the soils. Such research has disclosed that a very wide range of soil types include an unexpectedly high proportion of silt falling within the normal limits of loessic grain size. Even more diagnostic have been mineralogical studies that have revealed important components of the silt that could not have been derived from weathering of the local bedrock but must have been introduced from outside. The consistency of the findings leaves little doubt about the aeolian origin of the material.

From research in the Mississippi valley and elsewhere a number of inferences may be drawn about the range of conditions under which loess accumulation occurs. The wind-blown silt often contains immense numbers of molluscan shells. They are dominantly land snails, and being much too fragile to have been transported long distances must represent part of the local fauna that was interred as deposition progressed. Occasionally so many woodland species are found that it is difficult to avoid the conclusion that some loess accumulated beneath a forest cover. Where radio-carbon dating is available it seems to show that sedimentation was normally very slow, the build-up of each metre taking many centuries. Thick loess sequences often contain several fossil soil horizons, implying that conditions favouring accretion were repeated on many separate occasions. Most of the palaeosols have been attributed to interglacial or interstadial episodes when loess deposition virtually ceased; they constitute an important line of evidence in the formulation of Pleistocene chronologies for both central Europe and the Midwest of the United States (Fig. 2.5, p. 29).

Aeolian sand in periglacial regions may assume two forms. In some areas it is blown into dunes, in others spread out as a relatively smooth plain. Constructional relief implies an abundant supply of sand, mostly in the size range 0.1–0.25 mm, which in the majority of cases has been provided by heavily laden meltwater steams. In northern Germany, for example, prominent dune groups fringe the *Urstromtäler*, while in parts of Alaska dunes of various ages and degrees of stability are conspicuous elements of the landscape. Poorly cemented sandstones may also be deflated under periglacial conditions; the Sand Hills of Nebraska, covering some 35 000 km² were apparently activated during the last glaciation, although how far destruction of the binding turf was caused by aridity rather than low temperatures is an open question. Featureless spreads of blown sand are particularly characteristic of the Low Countries in north-western Europe. Known as coversands, these deposits are usually only a few metres thick. The materials are less well sorted than either dune sand or loess and show a crude stratification. They are believed to have been deposited during blizzards when both snow and fine sediments were blown by the wind. When the snow melted the particulate matter was left behind and slow accretion took place. The resulting deposits are sometimes termed niveo-aeolian sands.

Superficial structures due to frost action

As mentioned in the introduction to this chapter, one interest of relict periglacial features in current temperate areas is the contribution they can make to palaeoenvironmental reconstruction. However, before such guidance can be obtained it is necessary to know the range of conditions under which a particular feature may evolve. It is obviously desirable that this should encompass not only a perceived spatial correlation between feature and climate, but also an understanding of the formative processes. This is required, if for no other reason, because there were almost certainly Pleistocene periglacial environments that have no precise analogue today; as an illustration, lowland periglacial conditions in the United States extended to within 38° of the equator so that the elevation of the sun and duration of daylight would have been without exact counterpart today.

In the following account, attention will be directed, firstly, to those phenomena believed to be associated with permafrost, and then to those that demand active freeze–thaw cycles but do not necessarily require permanently frozen ground.

Ice wedges

In many parts of the current periglacial zone thermal contraction of frozen ground engenders a network of frost cracks. The coefficent of linear expansion of ice is relatively high, about 50×10^{-6} per °C, so that a fall in temperature of 20 °C can theoretically generate fissures some 25 mm across around a polygon 25 m in diameter; in practice contraction is likely to be substantially less than this because the medium is a sediment–ice mix rather than pure ice. Although cracks of this origin are common in intensely cold regions, they need to develop much further before they attain the dimensions of a typical ice wedge. This consists of a tapering mass of relatively pure ice 1 m or more across at the surface and extending to a depth that may locally reach 10 m (Fig. 14.3). The evolution of a wedge of this size requires gradual widening of the original fissure. During winter hoarfrost accumulates in the crack, while during spring water trickles down from the active layer to refreeze below the permafrost table. As the temperature of the permafrost rises in summer, expansion may contort the bedding adjacent to the enlarged wedge. Renewed thermal contraction during the next winter reopens the crack and the whole process is repeated, each additional ice film endowing the ice wedge with a characteristic vertical foliation. In 1974 Mackay published the results of monitoring the changes in ice wedges in the Canadian Arctic over a 6-year period and demonstrated that most were growing at less than 1 mm yr⁻¹, although a few were achieving about 2 mm yr⁻¹. Similar observations in the Antarctic appear to confirm the belief that a thick ice wedge usually takes a minumum

Fig. 14.3. Diagrammatic sketches of three types of superficial structure that arise owing to repeated frost action.

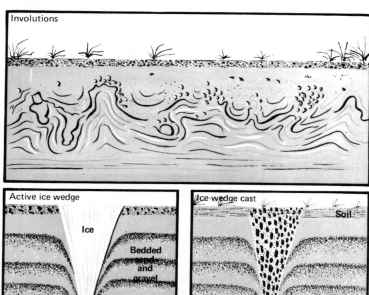

Involutions

Active ice wedge

Ice

Bedded sands and gravel

Ice-wedge cast

Soil

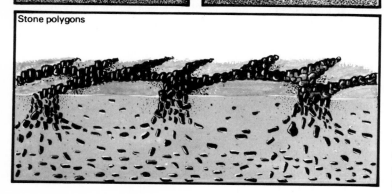

Stone polygons

of several centuries to form.

Ice-wedge networks are particularly conspicuous on river terraces and floodplains (Fig. 14.4). The patterns they form vary from one location to another, but near a river there is often a major set of wedges parallel to the water's edge and a second set at right angles producing what has been called an oriented orthogonal system. Elsewhere the commonest pattern appears to be a random orthogonal system in which the junctions in the network are at approximately 90°, but otherwise there is no regular geometrical arrangement. Where wedges have developed on alluvial sediments a distinction may be drawn between epigenetic and syngenetic types. The former are those that have evolved where there is no longer significant fluvial aggradation, whereas the latter grow concurrently with sedimentation and extend upwards as the level of the ground is raised. In general, syngenetic wedges are likely to be narrower and deeper than their epigenetic counterparts.

It is generally agreed that a large and rapid fall in temperature at the onset of winter favours surface cracking and therefore the development of ice wedges. In 1966 Péwé argued that such conditions are only likely where the mean annual temperature is −6 °C or less, and this accords with the view that permafrost is essential to the creation of ice-wedge networks. One problem in examining the relationship between climate and wedge distribution is that the melting process may be so slow that, even in decline, the ice persists for long periods without actually vanishing. A possible clue may be provided by the observation that active wedges are accompanied by raised borders to each polygon, declining wedges by raised centres.

Fossil ice wedges, or ice-wedge casts as they may appropriately be termed, are often seen in old terrace deposits. They typically take the form of narrow V-shaped fillings cutting across the normal horizontal bedding. The filling comprises debris that has either slumped from the walls of the wedge or been introduced from the surface. During climatic warming, an ice wedge melts mainly from the surface downwards. As the permafrost table is lowered, the active layer collapses into the vacant space and a shallow surface trench develops. Water channelled into the trench percolates downwards around the dwindling ice wedge, carrying fine sediment which accumulates in the interstices of the collapsed debris and gives the final infill a different texture from the surrounding beds.

Fig. 14.4. Two examples of patterned ground produced by ice wedges. (A) Active ice-wedge polygons in northern Alaska, some 50 km south-east of Barrow. Note in particular the oriented orthogonal system close to the river. (B) Crop-mark patterns attesting to a system of fossil ice wedges beneath the fields of the English Midlands. The ice wedges are believed to have developed during the last (Devensian) glacial period.

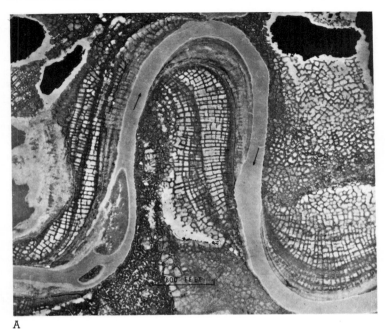

A

B

Aerial photography has disclosed widespread fossil networks in the modern temperate zones of Europe and North America. Soil contrasts produce crop marks which delineate very clearly the pattern of ancient wedges. Excellent examples have been described from the Midlands and East Anglia in southern Britain (Fig. 14.4), while similar features can be traced across a broad zone of the north European plain into the USSR. Ice-wedge casts can often be dated by reference to the age of the deposits in which they occur. For example, in southern Britain they are particularly common in terrace sediments dated by radio-carbon to the middle part of the last glaciation; they imply very widespread permafrost since that time. Stratigraphical studies in the same area have disclosed a number of earlier periods during which wedges developed. Dates in these cases are usually inferred from the age of any overlying sediments that remain unaffected. Thus, in East Anglia, ice-wedge casts have been observed beneath the Cromer Forest Bed. This middle Pleistocene interglacial horizon is older than any unequivocal till in Britain. The underlying wedge must obviously have formed at an even earlier date and is evidence of intense cold despite the lack of identifiable glacigenic deposits of similar antiquity.

Before leaving the subject of ice wedges it is worth noting that not all wedge-shaped forms originate in this way. Several workers in Antarctica have described sand wedges that are developing by the continual infilling of thermal contraction cracks with wind-blown sand. Each year more aeolian sediment is added to the growing structure until eventually it may reach a width of over 1 m. It is conceivable that some features which have been identified as ice-wedge casts are actually sand wedges, although the latter are unlikely to be at all common since they appear to require an exceptionally dry as well as an extremely cold climatic regime.

Pingos

The term 'pingo' is used by the Eskimo of the Mackenzie delta in Canada to denote scattered dome-shaped hills rising abruptly above the alluvial plain. Within the area of the delta there are some 1 500 such hills, ranging from minor hillocks only 2 m high to large examples over 50 m high. Diameters vary from 10 m to over 200 m, but the sides are almost always steep and often exceed 20° in angle. The smaller examples may be perfectly dome-shaped, but many of the larger are breached by crater-like depressions 5 m or more in depth. The cores range from huge lenses of pure ice to beds of silt and fine sand interstratified with ice layers that all dip outwards from the centre. Each pingo on the Mackenzie delta is surrounded by either a shallow lake or dried-out lake bed, and in 1962 Mackay suggested that the insulating effect of a lake plays a critical role in the evolution of such hills (Fig. 14.5). At first the surface water is assumed to be adequate to prevent local permafrost, but if the climate deteriorates, or the basin fills with sediment to the point where its insulating effect is lost, the lake floor may become perennially frozen. Water trapped in the underlying sands is placed under increasing pressure as permafrost advances from all sides, and in response it may be expelled and injected towards the surface, pushing up the overlying layers and finally freezing as a large mass of ice in the core of the dome-shaped hill that has been created. Pingos originating in this way are described as being of the 'closed-system' type and in the 1970s Mackay measured the growth of a number of such examples that were seen to be forming along the Canadian arctic coast after 1950. He concluded that initial upward growth may be at a rate of about 1.5 m yr^{-1}, but soon this figure diminishes and later development is much slower so that a large pingo can take over a millennium to mature. He also inferred that, in the early stages of development, ice-segregation mechanisms may be more important than the injection of pressurized interstitial water; it is possibly the late injection processes that cause the rupturing and cratering of many older pingos.

After early work on pingos had concentrated on the Mackenzie delta, it was then found that similar features occur in other environmental situations where different causal processes would need to be invoked. This led to formulation of the 'open-system' explanation, whereby many pingos are believed to be supplied by underground water migrating through materials affected by only thin or discontinuous permafrost. In essence they become analogous to the springs of more temperate regions, although under cold climates the pathways followed by the water, and the growth of any artesian pressure, are determined in part by the distribution of permafrost. Open-system pingos tend to be concentrated in certain topographic situations, most notably along the valley sides where the hydrostatic head impels water towards the surface so that it freezes just below the ground in the form of a large, lens-shaped mass of ice. Again it is probably the steady supply of water rather than its forcible ejection that is responsible for the continued upward growth of the pingo. It is far from certain that all pingos belong to the closed- and open-system types so far described, and any situation capable of generating a steady supply of underground water to a perennial ice lens may theoretically be capable of creating an ice-cored hill. When attention was initially confined to the pingos of the Mackenzie delta, it was held that such features could only evolve under periglacial conditions of extreme severity. However, with the discovery of open-system types in Alaska developing under subarctic forests with a mean annual temperature no lower than −2 °C, ideas about the requisite climate had to be relaxed although the association of all pingos with permafrost still seems firmly established.

Once a pingo has been isolated from its supply of underground water, it will cease to grow although it may persist for long periods with little outward change. Any climatic warming, however, will lead to progressive melting until eventually all that remains is a shallow pool surrounded by a ridge comprising sediment that either slumped or was pushed to the margin of the ice-core. The pool may later be infilled by the accumulation of organic materials, and fossil pingos composed of a circular rampart around a peaty hollow have now been identified across a broad belt of Europe extending eastwards from southern Ireland into central Germany. In the Ardennes, for example, the remnants of over 100 ice-cored hills have been identified, while other major groupings occur in central Wales and in eastern England near the Fenland margin. In all these instances the features are believed to have been of the open-system type and so to delineate a former zone of discontinuous permafrost; radio-carbon dates from the base of the associated organic materials imply the ice-

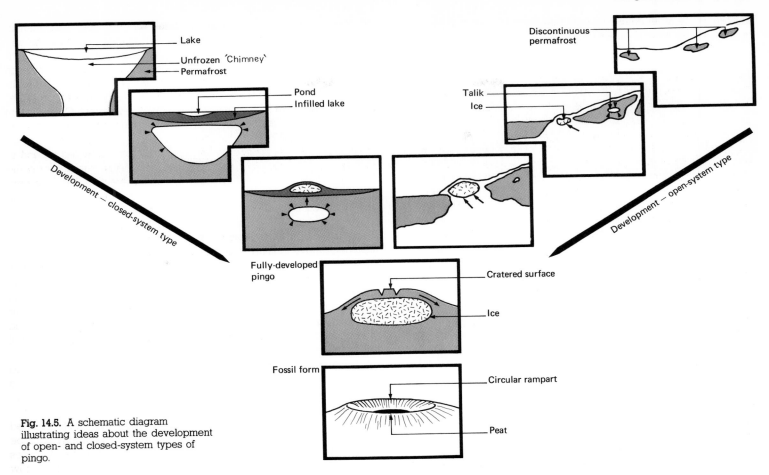

Fig. 14.5. A schematic diagram illustrating ideas about the development of open- and closed-system types of pingo.

cores finally melted a little over 10 000 years ago. In addition to these relatively youthful forms, possible much older examples have been identified beneath the river gravels under central London.

Cambering, gulling and valley bulging

In areas of gently dipping sedimentary rocks where thick clays alternate with thinner, more competent beds, one consequence of a periglacial climatic regime appears to be the development of large-scale superficial structures that profoundly disturb the original disposition of the strata. This is particularly clearly seen in the English Midlands where ironstones comprising part of the Jurassic succession were at one time viewed as a commercially workable reserve and so were mapped in great detail. It was early recognized that, despite a simple regional dip of less than 1° towards the south-

Fig. 14.6. Diagrammatic section through a hillslope that has been subject to cambering.

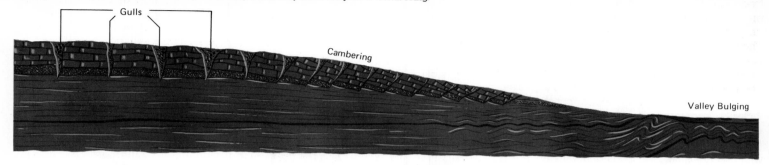

east, the ironstone beds are draped over the underlying clays in such a fashion that they consistently dip towards the valley floors (Fig. 14.6). The beds conform so closely to topography that the structures are clearly not tectonic in origin but must result from superficial movements that have taken place in the course of denudation. The term cambering aptly describes the dip from each interfluve towards the adjacent valleys, with the beds in places cambered down more than 30 m below their regional level. Such beds also show clear signs of extension in the form of widened joints trending parallel to the contours. These enlarged joints may be several metres across and are known by the quarryman's term of gull. They were often exposed in ironstone pits near the feather edge of the outcrop where they could be seen to be filled with either rubbly debris or till. In areas of cambering and gulling a third type of disturbance known as valley bulging, is frequently present. This involves disruption of the original stratification at and below the level of the valley floors. Structures in this situation are rarely well exposed, but where available for study they generally show severely contorted clays elevated far above their regional level; thus the crest of an anticlinorium at the dam impounding Rutland Water in Leicestershire contained one horizon that had been displaced upwards by some 25 m. As with cambering and gulling, there is a clear association with topography since the disturbances are often traceable for many kilometres along winding valley floors.

There is still doubt concerning the origin of the structures. It is generally agreed that excess stress along the margins of each flat-topped interfluve has depressed the edges of the competent beds and extruded clay along the line of the valley. The problem is the cause of the stress beyond the limit which the clays were capable of sustaining. An early suggestion was that rapid stream incision by itself might have induced the structures, but there are several reasons for rejecting this idea. A second hypothesis invoked burial beneath a thick ice-sheet as the source of the deforming stress, but this would not explain known cases of cambering and valley bulging situated beyond the generally accepted limit of Pleistocene glaciation. The third and most widely supported suggestion is that the crucial factor in initiating the structures was not so much an increase in stress as a temporary weakening of the clay. This could have been occasioned by an interval of permafrost during which segregated ice lenses developed along the valley floor. On melting, high porewater pressures would have been generated with a critical reduction in the shear strength of the clay, possibly to the point where is was unable to sustain the load of cap-rock that it had previously borne. Two additional observations lend some support to this interpretation. The first is that in a number of cases the superficial structures can be shown to have evolved in a relatively brief interval spanning the extremely cold periods on either side of a major glacial advance. The second is an unusual brecciation of certain of the disturbed clays which has been independently attributed to perennial freezing.

Involutions

At this stage in the review of superficial periglacial structures, consideration turns to phenomena that arise from repeated freeze–thaw activity without the necessity for permafrost being present. The term involution refers to a variety of features that may develop in both bedrock (Fig. 14.7) and unconsolidated sediments. They range from relatively regular festoons, in which originally horizontal strata are deformed into cup-shaped structures, to highly irregular contortions in which individual beds may be twisted and deformed beyond recognition (Fig. 14.3). The diversity of forms probably reflects different causes, but these have never been satisfactorily separated and defined. Some involutions are almost certainly connected with the development of earth hummocks or 'thurfurs'. In many modern periglacial areas the surface is covered with numerous small mounds generally about 0.5 m high and 1–2 m across. Sections show cores of sediment that have clearly been forced upwards, but the precise mechanism remains obscure. Most workers have invoked differential freezing, but have clashed over the way this operates. Some have argued that frost first penetrates along incipient waterlogged depressions, and that the resultant pressure eventually squeezes the intervening material into mounds. Others have contended that the elevation of the hummock arises from frost heave repeatedly affecting the mounds before the encircling depressions. If permafrost is present, yet another mechanism might be responsible. As the active layer cools during autumn, cryostatic pressures in saturated sediments trapped above the permafrost table will sometimes force them to the surface where they freeze into individual mounds. With climatic warming, all earth hummocks will tend to subside but the surface horizons may remain contorted to a depth of 1–2 m.

Other involutions probably have little direct surface expression. They may result from stresses that develop when layers of different sedimentary composition are subjected to alternate freezing and thawing. As already indicated, freezing in such circumstances is an irregular process. Frozen zones expand and exert cryostatic pressures on surrounding material. Segregation leads to lenses and patches of pure ice which, on melting, collapse in disorderly fashion. In most cases, moreover, the materials are so saturated that they will deform readily under small shear stresses. Horizons of silt and clay are particularly prone to disturbance, and some of the most spectacular results are found where terrace gravels rest upon a clayey bedrock. The basement material may then be thrust up several metres through the overlying fluvial sediments, wrapping round individual pebbles; in such cases it may simply be differential loading above a highly fluidized basal stratum that causes the intrusion.

The effectiveness of cyclic freezing and thawing in producing involutions is well attested both by laboratory experiments and by field observations. However, at present, few climatic inferences can be drawn from the very widespread occurrence of fossil involutions since they are obviously capable of forming in several different ways, and in individual cases it is often impossible to determine the precise mechanism that was responsible.

Patterned ground

The ground in many periglacial areas is covered by geometrical patterns that may assume a multitude of different forms. In the preceding pages reference has already been made to several examples, such as the network of ice wedges or the myriads of earth hummocks, both of which typify patterns that are classified as unsorted because they do not involve segregation of material of different sizes. Unsorted patterns can vary in dimensions from polygons that are only a few centimetres across to others that may be tens of metres across. The great majority are attributable to contraction cracking that arises from either intense freezing or desiccation; later an intimate relationship often evolves between the pattern of cracks and the vegetation cover so that cause and effect become difficult to disentangle. On sloping ground non-sorted stripes sometimes develop, with the pattern in that case due to parallel bands of vegetated and bare debris orientated down the steepest gradient.

It is, however, patterns generated by sorting processes that produce the most distinctive ornamentation of the ground in periglacial areas. Among the many patterns that can result from debris being segregated into different grain sizes, two major groupings are discernible, namely polygons and stripes. Stone polygons occur both singly and in groups. Their diameter is commonly between 0.5 and 3 m, with the large fragments concentrated around the periphery and the fine debris in the centre. They may be observed in many high-latitude areas at the present day, and are by no means confined to regions underlain by permafrost. They develop best on flat or gently sloping surfaces. On steeper gradients the shapes become deformed, elongating downslope until ultimately, on angles of more than about 5°, they pass into sorted stripes. These latter consist of

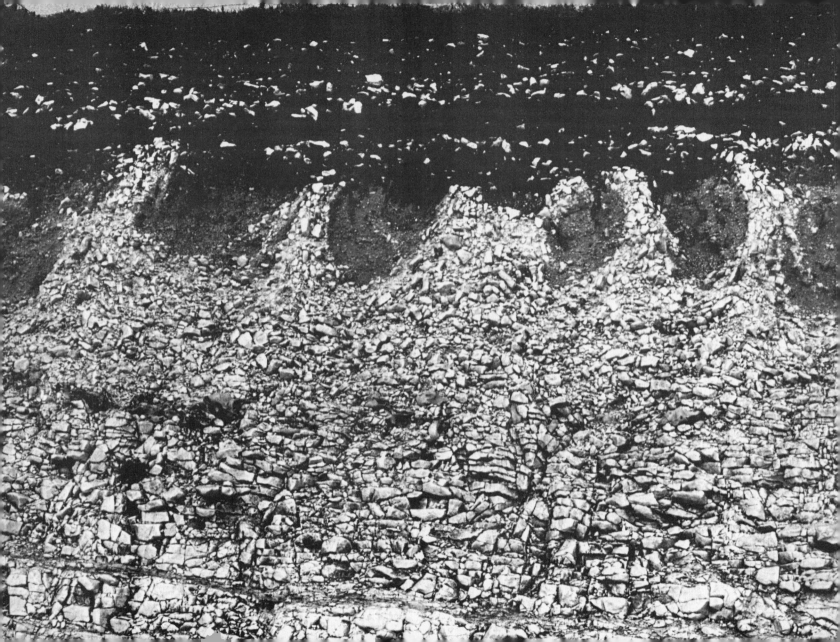

Fig. 14.7. Regular periglacial involutions, sometimes known as frost boils, exposed at the top of a chalk cliff on the English Channel coast near Brighton. The structures affect a thickness of about 1.5 m of the Chalk.

parallel ribbons of debris, divided according to fragment size and normally running directly downhill. The distance between coarse stripes varies widely, but in the majority of cases is under 2 m. In well-developed examples the stripes can be traced continuously down-slope for distances of 100 m or more.

Much uncertainty still surrounds the processes involved in the development of sorted polygons and stripes. It is tempting to assume that the same basic mechanism is responsible for both, but few of the hypotheses advanced in explanation of one seem entirely appropriate for the other. At the present time it appears likely that several different processes, often acting in concert, are capable of producing polygons, but that only a limited number of these can also be responsible for stripes.

Both field and laboratory investigations have been pursued in efforts to determine the cause of the sorting. One of the foremost workers, Corte, has stressed the need for examining not only the surface configuration but also the relationships as seen in section; to this end he regarded a bulldozer as an essential part of the geomorphologist's equipment. He found sorted polygons to be underlain by amorphous masses of ground ice composed of large crystals where the debris was coarse and of small crystals where the debris was fine; sorting appeared to be restricted to areas where the superficial layers contained an appreciable proportion of fine sediment. Under experimental conditions in 1963 he showed that repeated freeze–thaw cycles can induce differential movement within mixed materials. Two situations may be envisaged. In the first the freezing front moves vertically downwards. Under such circumstances, the larger fragments tend to be elevated towards the surface while the finer particles are displaced downwards. Two possible mechanisms have been proposed. The first, known as frost-pull, envisages the fine sediments, as they freeze, gripping the upper half of a stone and lifting it vertically before the freezing front has penetrated to its base. The underlying void may then be filled by the slumping of adjacent loose material so that the stone is unable to return to its former position. The alternative view, known as frost-push, is that freezing commences beneath the larger fragments on account of their greater thermal conductivity. This promotes differential heaving and possibly ice segregation during the freezing phase, with the new relative positions maintained by uneven collapse during thawing. A sorting process by either or both of these mechanisms appears well established, but of course it does not explain the surface pattern. It is here that the second situation may need to be envisaged. This involves the freezing front migrating laterally through the soil. Corte showed experimentally that, under such circumstances, the fine material moves in the same direction as the cooling front and leaves behind a zone enriched in coarse debris. This mechanism might be expected to operate if the initial frost penetration were extremely irregular, but by itself it scarcely affords a full explanation for sorted polygonal networks. Many supplementary processes have therefore been invoked.

The peripheral migration of the large debris has been ascribed to upward bulging of the centres when subjected to intense freezing; the frost susceptibility of the core material is indicated by measurements showing the centres as relatively elevated in winter but depressed in summer. A variant of this hypothesis visualizes needle ice developing centrally as being responsible for the outward displacement of large stones. A third theory that has attracted considerable support envisages freezing starting in the coarse debris and progressing towards the core, which is then bulged upwards under cryostatic pressure; concurrent migration of small particles ahead of the freezing front may accentuate and enlarge the initial core areas. One shortcoming of all these ideas is that they seem appropriate to the maintenance of sorted polygons but do not adequately explain their initiation. There is observational evidence that desiccation cracking may act as a trigger in certain cases, but the more general problem has been tackled by Ray and a group of American workers who have argued that, in a saturated superficial layer, weak convective cells may develop in response to unstable density stratification where the surface water is at 4 °C and the basal water near any buried ice–liquid interface is at 0 °C. The significance of any circulatory system is not in its ability to transport sediment but in its downward transfer of heat that will lead to a regular undulatory pattern in the top of the frozen layer (Fig. 14.8). Theoretical calculations suggest that the convective cells should generate frozen peaks at distances equal to 3.8 times the thickness of the aqueous layer. Measurement of the sorted cores in stone

Fig. 14.8. A diagram illustrating the idea that three-dimensional convective cells may develop in an unfrozen porous horizon owing to temperature and density variations. (A) The basic driving mechanism of the cells; (B) the calculated dimensional relationships that would arise from such a driving mechanism.

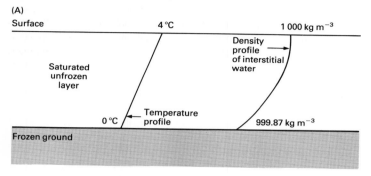

(A)

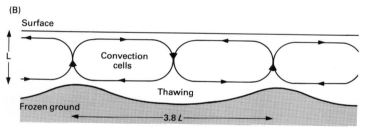

(B)

polygons has revealed width-to-depth ratios close to this figure, suggesting that thermal convection may influence the spacing of polygons even if it is not directly responsible for their sorting. As yet, though, no complete model accounting for all aspects of the features has been propounded.

The size of the consituent fragments in stone stripes varies widely and appears to influence the dimensions of the pattern. Where the fragments are exceptionally large with many of the stones over 0.5 m in diameter, the distance between successive coarse stripes may be as much as 10 m; where the material is appreciably finer with few particles over 0.1 m in diameter, the distance is reduced to 1 m or less. These delicate forms are the more common, and in many of them a proportion of the stones are conspicuously standing on edge.

Measurements show the fine stripes to be moving downhill much faster than the coarse, with differential velocities as large as 15 cm yr^{-1} being recorded on some Antarctic islands. The surface sorting does not extend to any great depth, normally dying out between a few centimetres and 1 m below ground-level. Although the coarser material has occasionally been reported to stand higher than the fine, the reverse is the case in the majority of actively developing examples. This accords with the belief that the chief mechanism in maintaining stripes is differential frost heaving. In the English Lake District Caine showed in 1963 that, during prolonged freezing, the surface of the fine debris is raised some 30 mm above that of the adjacent coarse debris. This is a significant superelevation where the coarse stripes are only 0.25 m apart. It is due to the growth in the fine material of thin ice layers, individually only 2 mm thick but in aggregate attaining several centimetres. Of eighty marked stones placed on the fine debris Caine found that, in one winter, ten had migrated into the neighbouring coarse stripes. Movement was mainly by sliding although there was also evidence of some debris having overturned. Needle ice has sometimes been invoked as the agent causing displacement, but none was observed in the Lake District. Although frost-heaving may explain the maintenance of the stripes, by itself it appears inadequate to account for their genesis.

Ray and his associates have argued that the idea of convective water movements in the surface layer, as originally applied to polygons, may also apply to stripes. However, the system of hexagonal cells characteristic of flat ground is modified on slopes and replaced by what are called two-dimensional roll cells. These are believed to be capable of melting the buried frozen surface into a regular pattern of parallel ridges and troughs orientated downslope. This undulatory pattern may then significantly affect the nature of freeze–thaw processes in the overlying mobile layer as it migrates down the hillside.

Patterned ground due to sorting can develop under many different periglacial regimes. It is currently found in Greenland underlain by thick permafrost, but is also evolving on recent outwash in Iceland under much less rigorous conditions. Active stone stripes have been recorded in arctic Canada, but also on hillslopes in Wales and Scotland. Moveover, within even quite small areas active and fossils forms can be found side by side. In upland Britain, for instance, delicate stripes destroyed to a depth of 0.3 m have been shown to re-form within a few years; yet much larger fossil stripes can also be seen to pass beneath a recent peat cover. Great care is obviously needed in attempting to use fossil stone polygons and stripes for environmental reconstruction.

Other aspects of landscape development

As will be discussed again in Chapter 19, a number of early writers envisaged a sequence of landform changes under a periglacial climate that they believed sufficiently distinctive to justify recognition of a 'periglacial cycle of erosion'. In general such a concept has proved of only limited value in the study of periglacial landscapes, although there remain a number of features characteristic of arctic areas that merit further examination before this chapter is concluded.

Cryoplanation terraces and tors

In many arctic areas hillslope profiles are observed to comprise a series of broad, gently inclined steps separated by rocky bluffs. The treads of the stairway may be up to 1 km or more in width, with intervening risers from 5 to 25 m high. Often the hilltop summits appear to be merely the highest of the steps, with small frost-riven bedrock residuals rising above them. The term cryoplanation terrace has been applied to these benches cut into solid rock. There is sometimes a cover of coarse debris over both the tread and the riser so that the bedrock surface is partially concealed, but even then it may well possess a stepped form. Most workers have ascribed cryoplanation terraces to nivation processes where snowbanks persist in the concavity at the back of each tread. Solifluction and snowmelt remove the frost-shattered debris as fast as it accumulates so that there is a steady recession of the steep face into the hillside, leaving behind a gentle surface sloping outwards at around 7°. Initiation of the terraces is generally attributed to original slope irregularities, such as structural benches, which favoured the accumulation of snowbanks in their lee. Examples of actively evolving cryoplanation terraces have been described from Alaska, the Canadian Arctic and Siberia, and similar features in fossil state almost certainly occur widely in former periglacial regions, although positive identification is often difficult owing to the problem of eliminating alternative explanations.

The backwearing of free faces by nivation has a further potential consequence in the fashioning of residual rock-masses into tors. The origin of tors has been a matter of heated controversy (p. 105), but there now seems little doubt that, in periglacial areas, many owe their form to the stripping of earlier regolith and active frost-riving of the exposed bedrock. Two situations may be visualized. In the first the initial regolith is deep enough to conceal upstanding conical masses of little-altered rock so that removal of the cover by solifluction exhumes tors that have, in their main essentials, already been fashioned. On the other hand, if the base of the regolith is originally smooth, the craggy outline of the tor must be produced either during or after the stripping. Cryoplanation terraces demonstrate the capacity of nivation processes to generate steep, frost-riven bluffs. In such circumstances varying susceptibility to freeze–thaw activity is likely to be a significant influence on the rate of bluff recession. Frost-wedging and solifluction will preferentially remove well-jointed rock but leave zones of more massive rock relatively untouched. Both these sequences thus lead to tors sited where the rock is poorly jointed. Particular uncertainty has attached to the contribution of freeze–thaw action in the shaping of tors now found in temperate areas. In some localities the amount of coarse debris incorporated in a solifluction apron leaves no doubt that substantial modification must have taken place under periglacial conditions. It is difficult to obtain accurate measurements, but around some quite small tors the total volume of boulders has been estimated at over 250 000 m³

Dry valleys

Many modern temperate regions that formerly experienced a periglacial climate exhibit valley systems in which part of the network is today permanently dry. There is a clear implication that the drainage net was at one time more extensive, and one plausible explanation is that a perennially or seasonally frozen subsoil rendered large areas less permeable than at present. The greatest change in drainage density might be expected where the bedrock is naturally permeable, and attribution of an extended drainage net to former periglacial conditions has most commonly been made in areas of clastic and calcareous sedimentary strata. Polish workers have applied this explanation to numerous short valleys on Quaternary sands and gravels in the area south-west of Warsaw. They believe that in this region snowbanks provided abundant run-off in springtime when the ground was still deeply frozen; it was only when the climate warmed at the close of the glaciation that infiltration became possible and surface flow ceased in the network of small valley heads.

In applying this interpretation to dry valleys, due attention must be paid to alternative explanations and it is often difficult to decide which is the most plausible. Shrinkage of drainage nets has been ascribed to lowering of the water-table through, among other processes, scarp recession, stream incision and reduced precipitation. The last-named has also been invoked directly as the cause of diminished surface flow. In addition, care must be exercised to ensure proper regard for the exceptional storm under the present climatic regime, thus avoiding the need to invoke any environmental change at all. The dating of valley abandonment, though rarely easy to achieve, can help to discriminate between the competing hypotheses. In south-eastern England, for example, a striking valley in the escarpment of the Chalk in Kent has been shown to have evolved in a relatively brief period of about 500 years at the close of the last glaciation. In this case the nature of the sediments evacuated from the valley, together with the known environmental changes that characterized the end of the last glacial period, has allowed a clear identification of the dry valley with a periglacial episode. It is unusual for the evidence to be so strong, but the general concept of network shrinkage at the end of a periglacial period seems well established.

Asymmetrical valleys

In a periglacial environment where freeze–thaw processes are so dominant, aspect is obviously a potential factor of great importance. There have been numerous descriptions of asymmetrical valleys believed to owe their form to processes operating with different efficiencies on the sunny and shaded slopes. However, there is still remarkably little agreement about the precise nature of the controls. Insolation is not the only climatic factor that needs to be considered since the prevailing wind direction, particularly as it affects snow-drifting, could also be crucial.

French has tabulated over a dozen reports of valley asymmetry from modern periglacial regions in the Northern Hemisphere, about two-thirds of which record north-facing slopes as steeper, and under one-third south-facing slopes as steeper. The argument most frequently deployed to explain the asymmetry involves greater solifluction on south-facing slopes producing a mobile sheet that accumulates at the foot of the slope until it pushes the stream across the valley floor to undercut and steepen the opposite hillside. This was the situation described from north-west Alaska by Currey who found that, of over 200 cross-sections, approximately one-third display no asymmetry, in one-half the north-facing slope is steeper, and in only one-eighth is the south-facing slope steeper. In arctic Canada French described a more unusual situation where the steeper slopes face west and south-west. He argued that this is partly due to snow, blown by dominant westerly winds, accumulating on east-facing slopes and promoting particularly active solifluction and nivation processes as it melts during the summer. A significant feature of this investigation was its concern with an area at 74°N since it needs to be remembered that insolation contrasts due to aspect will actually diminish as the pole is approached.

The records from current arctic areas provide a puzzling contrast with those from modern temperate regions where valley asymmetry is held to be a periglacial relic. In Europe the vast majority of workers identify south or south-west-facing slopes as steeper than north-facing. Of course, given that the valley sides at the onset of a periglacial episode are of equal declivity, asymmetry can be produced in a variety of ways. One side may be steepened more than the other, one side may be flattened more than the other or one side may be steepened while the other is flattened. Greater freeze–thaw activity on south-west-facing slopes is most frequently invoked in Europe, although precisely how this leads to steepening of the profile, especially since it does not appear to displace the drainage lines to undercut the opposite hillside, has never been entirely explained. Clearly, a much greater understanding of the relationship between microclimate and slope processes is required before periglacial valley asymmetry can be used in palaeoenvironmental reconstruction.

References

Ball, D. F. and **Goodier, R.** (1970) 'Morphology and distribution of features resulting from frost action in Snowdonia', *Field Studies* 3, 193–217.

Ballantyne, C. K. and **Matthews, J. A.** (1983) 'Desiccation cracking and sorted polygon development, Jotunheim, Norway', *Arct. Alp. Res.* 15 339–49.

Benedict, J. B. (1970) 'Downslope soil movement in a Colorado alpine region: rates, processes and climatic significance', *Arct. Alp. Res.* 2, 165–226

Burt, T. P. (1981) 'Factors influencing the growth of miniature ice lenses' *Earth Surf. Processes Landf.* 6, 179–82.

Burt, T. P. and **Williams, P. J.** (1976) 'Hydraulic conductivity in frozen soils' *Earth Surf. Processes* 1, 349–60.

Caine, T. N. (1963) 'The origin of sorted stripes in the Lake District, northern England', *Geogr. Annlr.* 45, 172–9.

Caine, T. N. (1972) 'The distribution of sorted patterned ground in the English Lake District', *Rev. Geomorph. Dyn.* 21, 49–56.

Catt, J. A. (1977) 'Loess and coversands', in F. W. Shotton (ed.) *British Quaternary Studies* OUP.

Catt, J. A. (1979) 'Distribution of loess in Britain', *Proc. Geol. Assoc.* 90, 93–5.

Chambers, M. J. G. (1967) 'Investigations of patterned ground at Signy Island, South Orkney Islands', *Brit. Antarct. Surv. Bull.* 12, 1–22.

Chandler, R. J. (1970) 'Solifluction on low-angled slopes in Northamptonshire', *Q. J. Eng. Geol.* 3, 65–9.

Corte, A. E. (1962) 'Vertical migration of particles in front of a moving freezing plane', *J. Geophys. Res.* 67, 1085–90.

Corte, A. E. (1963) 'Particle sorting by repeated freezing and thawing', *Science N. Y.* 142, 499–501.

Currey, D. R. (1964) 'A preliminary study of valley asymmetry in the Ogotoruk Creek area, northwestern Alaska', *Arctic* 17, 84–98.

Eden, D. N. (1980) 'The loess of north-east Essex, England', *Boreas* 9, 165–77.

French, H. M. (1971) 'Slope asymmetry of the Beaufort Plain, northwest Banks Island, N. W. T., Canada', *Can. J. Earth Sci.* 8, 717–31.

Giardino, J. R. (1983) 'Movement of ice-cemented rock glaciers by hydrostatic pressure: an example from Mount Mestas, Colorado', *Z. Geomorph.* 27, 297–310.

Hall, K (1983) 'Sorted stripes on sub-Antarctic Kerguelen Island', *Earth Surf. Processes Landf.* 8, 115–24.

Hamilton, T. D. and **Obi, C. M.** (1982) 'Pingos in the Brooks Range, northern Alaska, U.S.A.', *Arct. Alp. Res.* 14, 13–20.

Horswill, P. and **Horton A.** (1976) 'Cambering and valley bulging in the Gwash valley at Empingham, Rutland', *Phil. Trans. R. Soc.* A283, 427–62.

Hutchinson, J. N. (1980) 'Possible late-Quaternary pingo remnants in central London', *Nature, London* 284, 253–5.

Kellaway, G. A. and **Taylor, J. H.** (1953) 'Early stages in the physiographic evolution of a portion of the East Midlands', *Q. J. Geol. Soc. Lond.* 108, 343–75.

Kerney, M. J., Brown, E. H. and **Chandler, T. J.** (1964) 'The late-glacial and post-glacial history of the Chalk escarpment near Brook, Kent', *Phil. Trans. R. Soc.* B248, 135–204.

Klatkowa, H. (1965) 'Vallons en berceau et vallées séches aux environs de Lodz', *Acta Geogr. Lodziensa* 19, 124–42.

Lill, G. O. and **Smalley, I. J.** 1978 'Distribution of loess in Britain', *Proc. Geol. Assoc.* 89, 57–65.

Mackay, J. R. (1962) 'Pingos of the Pleistocene Mackenzie River delta area', *Geogr. Bull.* 18, 21–63.

Mackay, J. R. (1972) 'The world of underground ice', *Ann. Ass. Am. Geogr.* 62, 1–22.

Mackay, J. R. (1973) 'The growth of pingos, western Arctic coast, Canada', *Can. J. Earth Sci.* 10, 979–1004.

Mackay, J. R. (1974) 'Ice wedge cracks, Garry Island, N.W.T.', *Can. J. Earth Sci.* 11, 1366–83.

Meentemeyer, V. and **Zippin, J.** (1981) 'Soil moisture and texture controls of selected parameters of needle ice growth', *Earth Surf. Processes Landf.* 6, 113–25.

Miller R. *et al.* (1954) 'Stone stripes and other surface features of Tinto Hill', *Geogr. J.* 120, 216–19.

Perrrin, R. M. S., Davies, H. and **Fysh, M. D.** (1974) 'Distribution of late Pleistocene aeolian deposits in eastern and southern England', *Nature, London* 248, 320–4.

Péwé, T. L. (1959) 'Sand-wedge polygons in the McMurdo Sound region, Antarctica', *Am. J. Sci.* 257, 545–52.

Péwé, T. L. (1966) 'Paleoclimatic significance of fossil ice wedges', *Biul. Peryglacjalny* 15, 65–73.

Péwé, T. L. *et al.* (1969) 'Origin and paleoclimatic significance of large-scale patterned ground in the Donnelly Dome area, Alaska', *Geol. Soc. Am. Spec. Pap.* 103.

Péwé, T. L. (1983) 'Alpine permafrost in the contiguous United States: a review', *Arct. Alp. Res.* 15, 145–56.

Piggott, C. D. (1965) 'The structure of limestone surfaces in Derbyshire', *Geogr. J.* 131, 41–4.

Pissart, A. (1963) 'Les traces de "pingos" du Pays de Galles (Grande Bretagne) et du plateau des Hautes Fagnes (Belgique); *Z. Geomorph.* 2, 147–65.

Pissart A. (1973) 'Resultats d'experiences sur l'action du gel dans les sol', *Biuletyn Peryglacjalny*, 23 103–113.

Pissart A. (1983) 'Remnants of periglacial mounds in the Hautes Fagnes, Belgium: structure and age of ramparts', *Geol. en Mijnbouw* 62, 551–5.

Psilovikos, A. and **Van Houten, F. B.** (1982) 'Ringing Rocks barren block field, east central Pennsylvania', *Sed. Geol.* 32, 233–43.

Ray, R. J. *et al* (1983) 'A model for sorted patterned-ground regularity', *J. Glaciol.* 29, 317–37.

Schenk, E. (1955) 'Die periglazialen Strukturbodenbildung als Folgen der Hydratationsvorgänge im Boden', *Eiszeitalter und Gegenwart* 6, 170–84.

Soons, J. M. (1971) 'Factors involved in soil erosion in the Southern Alps, New Zealand', *Z. Geomorph.* 15, 460–70.

Soons, J.M. and **Greeenland D. E.** (1970) 'Observations on the growth of needle ice', *Water Resources Res.* 6, 579–93.

Sparks, B. W. *et al.* (1972) 'Presumed ground ice depressions in East Anglia', *Proc. R. Soc.* A327, 329–443.

Wahrhaftig C. and **Cox, A** (1959) 'Rock glaciers in the Alaska Range', *Geol. Soc. Am. Bull.* 70, 383–436.

References

Washburn, A. L. (1979/80) 'Permafrost features as evidence of climatic change', *Earth Sci. Rev.* **15** 327–402.

Watson, E. (1972) 'Pingos of Cardiganshire and the latest ice limit', *Nature London.* **236**, 343–4.

Watson, E. and **Watson, S.** (1972) 'The coastal periglacial slope deposits of the Cotentin peninsula', *Trans. Inst. Brit. Geogr.* **49**, 125–44.

Williams, P. J. (1979) *Pipelines and Permafrost, Physical Geography and Development in the Circumpolar North,* Longman.

Selected bibliography

Admirable surveys of periglacial features, together with extensive bibliographies, are provided by the following three texts: C. Embleton and C. A. M. King, *Periglacial Geomorphology*, Arnold 1975; H. M. French, *The Periglacial Environment*, Longman, 1976; and A. L. Washburn, *Geocryology: a survey of periglacial processes and environments*, Arnold, 1979.

Part 4
THE ROLE OF THE SEA

Chapter 15
Coastal processes

In certain respects the coast provides the geomorphologist with exceptionally fine opportunities for studying contemporary processes. A beach, for example, is one of the most changeable of all landforms, and even cliffs are subject to more rapid modification than most. Yet the coast also presents particular difficulties for process studies. Many of the changes take place in the zone of wave action where direct observation is by no means easy. Moreover, there is an increasing awareness of the intricate exchange of sediment between the foreshore and offshore zones. At one time coastal geomorphology tended to concentrate almost exclusively on two basic subjects, the erosion of cliffs and wave-cut platforms and the redistribution of sediment by longshore drifting. It is now recognized, however, that the narrow ribbon between high- and low-water marks cannot be treated in isolation but must be regarded as part of a much broader zone of erosion and accretion. Study of sediment movement across the full width of that zone poses obvious problems, although a long tradition of investigation for engineering purposes has fortunately evolved a sophisticated instrumentation.

The study of water movement is clearly fundamental to an understanding of coastal processes. For a long time tides were regarded as the dominant influence in shaping the edge of the land, and virtually all marine features have, on one occasion or another, been attributed to their action. With inadequate observational data to form a sound judgement, a reaction ensued by which almost every coastal form was ascribed entirely to wave activity. Recent investigations indicate that the pendulum swung too far and that in certain circumstances tides can play a significant role in the fashioning of the coast.

Their role is twofold; they raise and lower the level of wave attack, and by mass transfer of water generate currents capable of transporting large quantities of fine sediment. The way in which water movement arising from both tides and waves is translated into processes of erosion and sediment transport forms the major theme of this chapter.

Tides

Tide-generating forces

Tides differ from other types of energy input in geomorphology by being attributable to neither solar nor terrestrial heat. Admittedly their amplitude may be affected by temporary meteorological conditions, but basically they are the outcome of gravitational attraction between the earth, moon and sun. The whole topic of tide generation is one of immense complexity, but for present purposes is most easily approached by considering first a highly idealized situation in which the earth is entirely covered with water and subject

only to attraction by the moon moving in a circular 28-day orbit. For the whole system to remain in equilibrium the earth and moon must move round a common centre of gravity since the gravitational and centrifugal forces must be exactly balanced. The centre is found to lie within the earth at a depth of about 1 700 km. The simplest type of motion about this point capable of maintaining the equilibrium is revolution of the earth (Fig. 15.1A). The centrifugal force in this case is everywhere the same. The lunar gravitational attraction, on the other hand, being inversely proportional to the square of the distance from the centre of the moon, must vary from place to place on the globe. It has a mean value equal to the centrifugal force at the centre of the earth, rises to a maximum at the sublunar point closest to the moon and falls to a minimum at the antipodal position. At any location, therefore, the tide-generating force is essentially equal to the deviation of lunar gravity from its mean value, resulting in a maximum residual force acting towards the moon on one side and away from it on the other (Fig. 15.1B).

Strictly speaking, the tide-generating force needs to be resolved into a horizontal tractive force to explain the actual movement of the

Fig. 15.1. Schematic portrayal of the tide-generating forces.

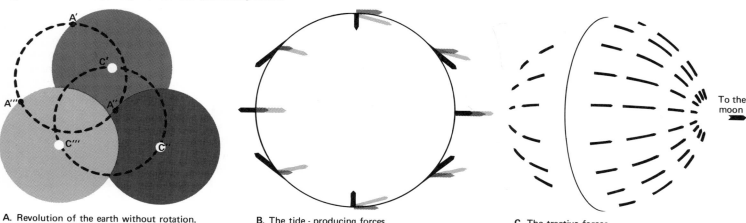

A. Revolution of the earth without rotation.

Note that all points on and within the sphere describe a similar circle (e.g. C', C'', C''' and A', A'', A'''), indicating that under these conditions the centripetal force is everywhere the same.

B. The tide - producing forces.

➡ Vectors of uniform centripetal force
➡ Vectors of variable lunar gravitational attraction
➡ By vector subtraction, resultant tide-producing forces

C. The tractive forces.

These reach a maximum at 45° from the line to the moon and decline to zero at 0° and 90°.

To the moon

water. Tractive values are zero at the sublunar point and its antipodes, and also along the great circle at an angular distance of 90°; they reach a maximum at an angular distance of 45°. The tractive forces are all directed towards the locations closest to the moon or furthest therefrom (Fig. 15.1C), and water being drawn thither transforms the spherical shape of the globe into a prolate spheroid with its major axis aligned on the moon. The actual distortion from sphericity will be minute since the extreme deviation of lunar gravity from its mean figure is equivalent to no more than one-nine-millionth part of the earth's own gravity. In the hypothetical circumstances being visualized, each antipodal tidal bulge will proceed round the globe in 28 days.

It is now possible to relax some of the artificial constraints so far imposed. The moon moves in an elliptical rather than circular orbit so that its tide-generating force varies with a periodicity of approximately 28 days. The tidal range will consequently show an oscillatory cycle depending on the distance between the moon and earth. An even greater fluctuation is introduced by the sun. The solar tide-generating force is also capable of deforming the ocean surface into a prolate spheroid, although the potential deviation from the sphere in this case is just under half that due to the moon. In practice, of course, the tidal effects of sun and moon interact. When all three bodies lie in a straight line the lunar and solar tides reinforce each other to produce the greatest tidal ranges. Conversely, when the bodies lie in quadrature the gravitational pulls counteract each other and the tidal ranges are at a minimum. This means in effect that the amplitude of the tides tends to be greatest during periods of full and new moon, and least at times of the first and third quarter. The former are known as spring tides and the latter as neap tides. A minor complication is introduced by the lunar orbit lying at an angle of 5° to the plane of the ecliptic; one consequence is that spring and neap tides themselves show systematic oscillatory patterns.

Observed oceanic tides

The preceding section confined attention to the tide-generating forces that arise from relative movements of the earth, moon and sun. The tide produced under these idealized conditions is known as the equilibrium tide, and is obviously very different from the changes in sea-level observed at coastal gauging stations. Three major factors explain the difference between the equilibrium and observed tides:

the earth's rotation, the configuration of the ocean basins and temporary meteorological conditions. The first two are systematic and can be allowed for in tidal predictions; the third is random and capable of causing significant deviations from the predicted levels.

Rotation of the earth has been ignored so far since it is not a tide-generating force. The globe is simply spinning within the gravitational fields of the sun and moon. With respect to the sun each full rotation takes 24 hours. However, owing to passage of the moon along its orbital path a full rotation with respect to that body takes 24 hours 50 minutes. Because lunar gravity is the dominant influence high tides are commonly observed at intervals of 12 hours 25 minutes; these are known as semidiurnal tides. If the moon's orbit and the equator lay in the same plane successive high tides would reach approximately the same level. However, the rotational axis of the earth is tilted with respect to the lunar orbit. This means that only twice in one orbit does the sublunar point move along the equator. At other times it moves along parallels of latitude as far north and south as the two tropics. When it moves along the Tropic of Cancer the antipodal point will be moving along the Tropic of Capricorn and vice versa. At any location close to the tropics this will produce a marked inequality between successive semidiurnal tides. Another important consequence of rotation is the Coriolis effect which becomes particularly significant when division of the earth's surface into separate ocean basins is considered.

The configuration of any water-filled basin is of fundamental importance in determining its response to the tide-generating forces. It is simplest to consider first a hypothetical basin of great length but negligible width. When the water in such a unit is disturbed, it moves backwards and forwards heaping up first at one end and then at the other. The form is known as a standing oscillation, with a central nodal line at which the water elevation remains constant. The period of oscillation depends upon the rate of travel of a normal progressive wave along the trough. Since a wave travels at a speed equal to $\sqrt{gD}$, where g is gravitational acceleration and D is depth, the period may be defined as $2L/\sqrt{gD}$, where L is the length of the basin. In other words it is the time taken by a wave to move from one end of the basin to the other and back again. If it is approximately equal to the interval between successive high tides a resonance effect will increase the amplitude of the tides; if the period is either much shorter or much longer, the oscillations will be out of phase and the tidal range will be correspondingly smaller. In this way the natural period of oscillation has a profound effect upon local tides, often exaggerating constituents of the tide-generating cycle to the point where the simple semidiurnal pattern is significantly distorted.

Of course in a normal ocean basin the water does not merely move to and fro as in a narrow trough but is capable of lateral movement. In such circumstances the Coriolis effect deflects the water to the right in the Northern Hemisphere and to the left in the Southern. The result is a high tide that sweeps round the margins of a basin in an anticlockwise direction north of the equator and in a clockwise direction south of the equator. Instead of a nodal line there is a central point of zero tidal range known as an amphidromic point. Most of the ocean and sea basins act as one or more large amphidromic systems. The exact configuration of some of these systems is still imperfectly known since tidal changes away from the coast are difficult to measure. However, in the North Sea basin three separate systems have been identified, giving the complex pattern of tidal ranges and co-tidal lines illustrated in Fig. 15.2

Tidal streams

Tidal rise and fall obviously demands large-scale horizontal transfer of water. In the open ocean this motion may assume a slow rotatory pattern, but for the geomorphologist much more interest attaches to tidal streams engendered in shallow water closer to the shore. Here the rotatory movement tends to be replaced by an alternating one in which the water moves for just over 6 hours in one direction and then approximately reverses that direction for the next 6 hours. Two aspects of such tidal streams are of vital importance: the maximum velocity at the sea bed which will determine the size of any sediment that can be entrained, and the circulatory pattern which will control the net transport direction. An exact reversal of tidal streams every 6 hours or so must certainly not be assumed since field investigations nearly always reveal a more complex circulatory system.

Several techniques have been devised for measuring water movements on the sea floor. The simplest involves the use of sea-bed drifters. These consist of plastic discs carefully weighted to travel close to the sea floor and capable of moving under the influence of low-velocity currents. Released at predetermined points, they drift under the impetus of tidal streams and other currents until washed up on the coast. They often have an attached card which the finder is asked to return to the investigator with details of time and place of recovery. Their value lies in the guidance they provide regarding overall directions of bottom movement; unfortunately the precise route followed between the points of release and recovery remains unknown, although it may often be inferred from other evidence. To obtain fuller information more elaborate methods must be employed. One technique involves submerging to known depths small bottles partly filled with hot liquid jelly and enclosing a compass card. The alignment of each bottle is recorded by the setting of the jelly, and from its tilt and orientation the current velocity can be assessed. In water 7 m deep off the East Anglian coast tidal streams have been measured by this method at velocities up to 5 km hr^{-1}. In the Bay of Biscay a tidal stream moving at 1 km hr^{-1} has been recorded at a depth of over 150 m. It is not practicable to attempt continuous monitoring with jelly bottles, and for this much more complex instrumentation is required. In areas of commercial exploitation like the North Sea, this work may be combined with other engineering investigations and by installing arrays of specially designed current meters a continuous record of water flow over a complete tidal cycle can be obtained. When plotted in the form of a vector diagram, the resulting data will often disclose significant residual net movements (Fig. 15.3).

The most elaborate studies have been undertaken in estuaries where patterns of tidal flow tend to be especially intricate. Movement may there be concentrated in relatively narrow threads of faster-flowing water, with the positions of maximum-velocity threads on the flood tide differing from those on the ebb. In this way flood and ebb streams may be spatially differentiated, as is well exemplified in Fig. 15.9. Many factors impinge upon details of the circulatory system. The shape and depth of the estuary will have an important bearing on the velocity attained by the inflowing and outflowing water. In all cases there must be a net outflow as a result of fresh-water discharge into the head of the estuary; however, the ratio of river discharge to the volume of the tidal prism entering and leaving the estuary varies greatly from one example to another. The differing densities of river- and sea-water can have a profound effect upon the circulation. In some sharply stratified estuaries the fresh water passes seawards over a saline bottom layer with little mixing; owing to the Coriolis effect the interface, as viewed from the land, often slopes down to the right in the Northern Hemisphere. In other cases

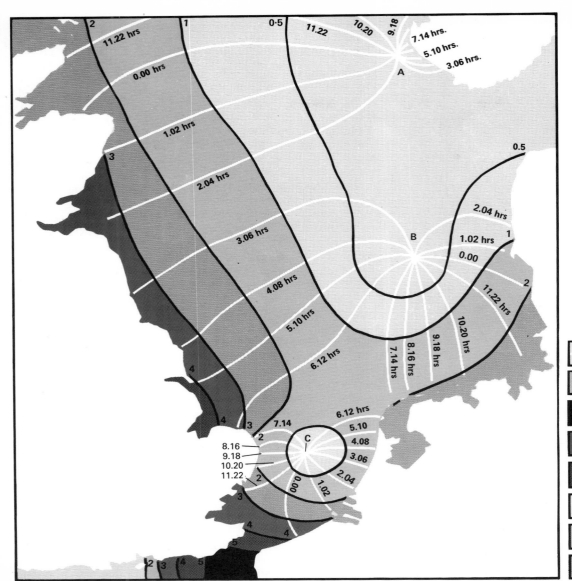

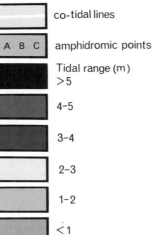

Fig. 15.2. The tidal system in the North Sea basin.

co-tidal lines

A B C   amphidromic points

Tidal range (m)
>5

4-5

3-4

2-3

1-2

<1

Fig. 15.3. An example of a tidal cycle plotted as a vector diagram from measurements made at a single point in the Elbe estuary. The plot is of water movement; sediment movement is likely to be significantly different since only at certain periods will the water reach the critical velocity required to entrain bed material.

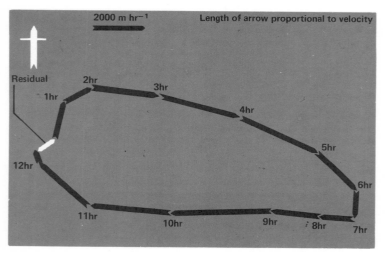

mixing is so complete that the estuary waters can be regarded as virtually homogeneous. A final factor worthy of mention is the meteorological situation which, among other things, influences the volume of the tidal prism entering and leaving the estuary. It is evident that many of the influences on the circulatory system vary from time to time so that, for a full understanding, observations need to be made over a wide range of conditions.

Many of the ideas derived from estuarine studies are also applicable to the open coast and, to a lesser extent, to the continental shelf as a whole. The most fundamental finding as far as the geomorphologist is concerned is that tidal movement comprises not a uniform, low-velocity drift regularly reversing in direction, but a concentration into distinct tidal streams which can have a profound effect on both sediment transport and sea-bed morphology.

Waves

Wind-generated waves in deep water

The vast majority of sea waves are generated by wind blowing over the water. They represent a transference to the hydrosphere of the kinetic energy of the atmosphere, part of which is ultimately expended where the waves break along the coast. It is important to distinguish between waves in deep water which are oscillatory in character and those that break along the shoreline which are translatory in character. In the former the individual water particles pursue an approximately circular path so that they return very nearly to their original position after the passage of a wave. In a wave of translation, on the other hand, the water particles move forward at approximately the same speed as the wave form.

A standard terminology is applied in the description of oscillatory waves (Fig. 15.4). It is found that the profile of a wave with a very low steepness ratio approximates to a sine curve, but that steeper waves are more closely trochoidal in form, A trochoid is the curve traced out by a point within a circle when that circle is rolled along a straight line; it has a flatter trough and sharper crest than a sine curve. By common observation, the orbits which the water describes during the passage of a wave are not exactly circular but involve a slow forward movement. This gradual creep in the direction of wave advance is known as the mass transport velocity. It increases rapidly with wave steepness and can be extremely important near the coast where it leads to the piling up of water and the raising of the tide above its predicted level.

The precise mechanism by which waves are generated remains uncertain. When the wind blows across a smooth water surface, pressure differences due to turbulence will cause random stress variations. These will be adequate to generate irregularities of different shape and size; it is the way in which certain of these irregularities grow into fully-fledged waves while others decline that is not fully understood. Phillips suggested that there may be oscillatory pressure fluctuations of such a frequency as to cause resonance with a selection of the incipient waves which then develop preferentially. A second important factor is the increasing effect of the disturbed water surface upon the pattern of air flow. Many workers have argued that there must be significant pressure differences between

Fig. 15.4. Diagrammatic sketches to illustrate aspects of wave motion.

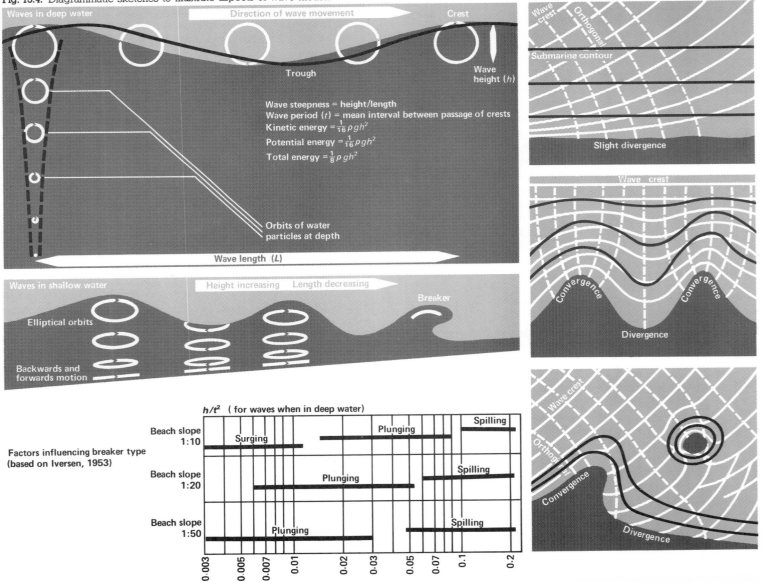

Waves in deep water

Direction of wave movement

Crest

Trough

Wave height (*h*)

Wave steepness = height/length
Wave period (*t*) = mean interval between passage of crests
Kinetic energy = $\frac{1}{16}\rho g h^2$
Potential energy = $\frac{1}{16}\rho g h^2$
Total energy = $\frac{1}{8}\rho g h^2$

Orbits of water particles at depth

Wave length (*L*)

Waves in shallow water

Height increasing   Length decreasing

Breaker

Elliptical orbits

Backwards and forwards motion

Wave crest
Orthogonal
Submarine contour
Slight divergence

Wave crest
Convergence   Convergence
Divergence

Wave crest
Orthogonal
Convergence
Divergence

Factors influencing breaker type
(based on Iversen, 1953)

h/*t*² (for waves when in deep water)

Beach slope 1:10   Surging   Plunging   Spilling

Beach slope 1:20   Plunging   Spilling

Beach slope 1:50   Plunging   Spilling

0.003 0.005 0.007 0.01 0.02 0.03 0.05 0.07 0.1 0.2

the windward and leeward sides of a developing wave, complemented by a greater drag stress on the windward side. A third factor emphasized by some writers is the possible transfer of momentum from short to long waves; while the former are constantly being regenerated, their momentum is being swept up by the larger waves which are thus enabled to develop further.

In the absence of an entirely satisfactory theory of wave generation, much attention has been devoted to the collection and analysis of empirical data. It is generally agreed that there are three basic controls of oscillatory deep-sea waves: wind velocity, wind duration and fetch. Any one of these can act as a limiting factor. For example, fetch constitutes an important limiting control on wave development in both the North Sea and the Irish Sea, whereas in the Atlantic and Pacific Oceans it is wind duration and velocity that impose the ultimate restriction. For purposes of wave prediction, tables have been prepared relating the three variables to observed wave height. For example, the fetch of an easterly wind blowing across the North Sea to the English coast is about 600 km. With a wind blowing at 50 km hr^{-1} for a period of 30 hours, waves with a maximum height of just over 4 m might be expected. In the Irish Sea, with a fetch restricted to 200 m, the identical wind would generate waves under 3 m high; in the Atlantic the same velocity could produce waves up to 6 m high, but it would need to be sustained for substantially more than 30 hours.

In practice such figures need to be treated with caution. Other factors can have an influence, and storms produce a complex spectrum of waves with different periods and lengths. Interacting waves can result in locally exaggerated highs and lows where two crests or two troughs temporarily coincide. The waves may then steepen dramatically until the crests overturn and 'whitecaps' are formed. As they move away from the storm centres, where they compose true 'sea', the waves gradually separate into trains of more uniform character known as 'swell'. The waves of longest period travel the fastest and thus arrive first at points distant from the original storm. Many individual wave trains have now been traced thousands of kilometres from their source regions, and the changes that they undergo have been analysed. Wave height gradually declines but the velocity remains fairly constant. There is also a small increase in the wave period indicative of growing wave length. Storm waves initially 10 m high diminish to swell little more than 2 m high after travelling 2 000 km, while the typical period may extend from under 10 seconds to over 12 seconds. These changes are important for the geomorphologist since they gradually alter the characteristics of the ocean waves approaching the coastal zone.

Wind-generated waves in shallow water

In the open sea the orbital motion during passage of a wave diminishes rapidly from the surface downwards. At a depth equal to one wave length the orbital movement is almost imperceptible, theoretically being about one-five-hundredth of that at the surface; at a depth equal to one-half of the wave length it approximates to one-twentieth of that at the surface. Not surprisingly, therefore, waves only begin to 'feel bottom' and suffer rapid modification of form when they enter water of a depth equivalent to under one-half of the wave length. Although the period remains constant, nearly all other properties change. Velocity diminishes until, in very shallow water, it conforms to the equation for other wave forms of $V = \sqrt{gD}$. Since the period is unchanging, there must be a proportionate decrease in length and the waves appear to be piling up along the shoreline. Wave height increases only slowly at first but with increasing rapidity as very shallow water is entered. The opposed changes in height and length mean that the steepness increases dramatically before the wave actually breaks. An associated change affects the orbital paths of the water. In the open sea these are approximately circular, but in shallow water they become elliptical with the long axis horizontal. Close to the sea floor the orbits are flattened until there is a simple to-and-fro motion of the water. During the passage of a sharpening wave crest the water moves rapidly landwards, while during the ensuing broad trough it recedes slowly seawards.

The primary reason for waves breaking on entering very shallow water is that the foregoing changes in form are accompanied by an acceleration in orbital velocity. When the increased velocity at the wave crest surpasses the diminishing speed of the wave form, the water surges forward unsupported. A secondary factor in the process is the decreasing volume of water within each wave despite the sustained orbital movement. Ultimately a point may be reached when the water is inadequate to complete the orbit and a cavity forms on the landward side of the wave; thus the over-steepened crest lacks support and so plunges towards the beach. It is a matter of observation that breaking normally occurs when the depth of water is about

1.3 times the wave height.

Three different types of breaker may be distinguished. The first is the spilling breaker in which the water never falls freely but constantly spills down the front of the wave as a prominent foaming crest. The wave form is preserved but gradually declines in height as the water advances up the beach. Several spilling breakers can often be seen advancing simultaneously one behind the other. The second type is the plunging breaker. In this case the water at the crest falls vertically and traps a pocket of air within the wave. Following the plunge the wave form disintegrates and the water rushes forward as a turbulent foaming mass. The third type is the surging breaker, in which it is the base of the wave that rushes up the beach to leave behind a crest that rapidly collapses. Although waves can often be assigned to one or other type with confidence, it is also possible to encounter intermediate forms. For instance, some waves first break by spilling but later plunge when they pass into much shallower water. The factors determining whether a wave spills, plunges or surges are still incompletely understood. A plunging breaker has been held to occur where the velocity of water in the wave crest significantly exceeds that of the wave body, and a spilling breaker where the two velocities are approximately equal. Empirically it has been shown that the two dominant factors are wave steepness and beach gradient. Spilling breakers occur with steep waves, plunging breakers with intermediate waves and surging breakers with gentle waves. Beaches of low gradient are more commonly associated with spilling and plunging than with surging breakers.

The shoaling of water also has an important effect on the angle at which the waves strike a coast. There is a pronounced tendency for wave crests, irrespective of their direction of approach, to swing round until more nearly parallel to the shoreline. Known as refraction, this process begins as soon as the waves feel bottom and are thereby retarded. It is seen in its simplest form where a train of waves approaches a straight and uniformly sloping coastline at an oblique angle. The wave crests curve round until they approach the shore almost at right angles. In practice refraction is often incomplete so that slight obliquity remains. In the case of an irregular, indented coastline the situation is more complex. It must be remembered that the basic control is water depth so that the pattern of refraction depends ultimately upon the submarine contours. However, in most instances the waves first feel bottom opposite the headlands, while the advance into the bays continues relatively unimpeded. A vital consequence of refraction in such circumstances is the concentration of wave energy on the headlands. The breakers here will be appreciably larger than those in the bays.

A final aspect of waves in shallow water is their capacity for inducing nearshore circulatory current systems. These consist of two major components, longshore currents parallel to the coast and rip currents normal to the coast. It has been shown that, landward of the breaker zone, there exists a rise in mean water-level due to wave activity known as the wave set-up. This provides the head to drive the circulatory currents, but the pattern that they form and in particular the factors governing the position of the rip currents has been the subject of some dispute. The amplitude of the set-up varies with wave height, and longshore currents might therefore be expected to flow from positions of greatest breaker height towards rip currents sited where the breakers are relatively small. Such a pattern is commonly seen on indented coastlines where longshore currents move from the headlands into the bays and rip currents pass seawards from the bay-heads. On straight uniform coasts, where wave refraction cannot be invoked as the primary cause of differences in breaker height, the regular spacing of rip currents poses more of a problem. Longshore currents flowing at velocities of 3 km hr^{-1} have been recorded in such circumstances, with rip currents up to twice that speed occurring at intervals between 100 and 1 000 m. It was early noted that, on the same beach, large waves generate a few strong rips while smaller waves generate numerous weaker rips. Later measurements have shown the presence of what are termed edge waves. These are long standing waves with crests at right angles to the shoreline (Fig. 15.5); they are believed to originate from resonance between the waves incident on the shore and those reflected from it. At the nodes there is no observable up-and-down motion, whereas at the antinodes the full amplitude of the edge waves is experienced. Because the edge waves often have the same period as the incoming swell, they interact to give a consistent spacing of above- and below-average wave heights parallel to the coast. Again it is where the breakers are lower that rip currents form. Some workers have doubted whether regular edge waves are of sufficiently common occurrence to explain widespread rip currents, and there remains the possibility of other phenomena being

Fig. 15.5. The idea of a standing edge wave and its interaction with incident swell. (A) The basic spatial relationships; (B) the effect upon wave-height parallel to the coast; (C) one possible consequence in the regular spacing of rip currents. (Note that the interaction between the swell and the standing edge wave is crucially dependent on the periods t_i and t_e; only where these are the same will the regular pattern depicted above be likely to occur.)

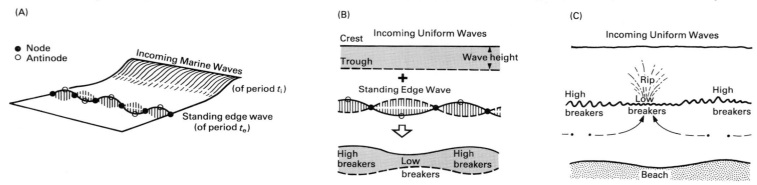

involved in the circulatory systems.

Other coastal waves

Of all coastal waves, infinitely the most common are the normal wind-generated type described above. However, brief reference must also be made to occasional waves of other origins since some of these are capable of modifying coastal landforms out of all proportion to their frequency. Two types are of particular importance, tsunamis and storm surges. As wave forms, both are characterized by their immense length and extended period.

A tsunami is generated by some form of submarine disturbance, most commonly a displacement of the sea floor during seismic activity. The resulting oscillatory waves can travel thousands of kilometres with little loss of energy. The wave length often extends to several hundred kilometres and the period to over an hour. Velocities inferred from trans-oceanic travel times have yielded speeds of about 700 km hr^{-1}. Yet tsunamis in the open ocean have a very small vertical amplitude, often less than 1 m, so that their passage at sea is usually obscured by the much higher wind-generated waves. On their approach to the coast, dramatic changes occur. The shoaling causes marked deceleration and a rapid growth in height, sometimes to ten times its original value. These transformed and very formidable waves do not normally break in the ordinary fashion, but owing to

their great concentration of energy can wreak immense damage on low-lying coasts. They have been responsible for a number of major disasters with great loss of life, and this has encouraged international cooperation in their study. A thorough investigation of the waves generated by the Alaskan earthquake of 1964 showed those spreading across the Pacific to have a period of about 108 minutes, although as they entered shallower water they often gave rise to secondary waves of higher frequency. The study underlined the role played by the configuration of the continental shelf in focusing the waves on certain segments of coast rather than others, and thereby influencing their liability to attack by tsunamis.

A storm surge may be regarded as a form of long wave initiated by exceptional weather conditions. It normally requires a cyclonic centre of unusually low pressure together with strong winds capable of inducing mass transport on a very large scale. By themselves these two factors can raise local sea-level by about 2.5 m, but in the largest storm surges a further contributory factor is some degree of resonance. This occurs, for instance, where the induced wave moves at the same velocity as the storm centre itself. Many tropical hurricanes produce surges of exceptional magnitude where they pass over wide, shallow continental shelves. Under such circumstances sea-level may be elevated by at least 5 m and vast areas of low-lying coast inundated. Surges of this type are well exemplified by those afflicting the Ganges delta when a tropical hurricane develops over the Bay of Bengal. In November 1970 an estimated 600 000 people were killed in what has been described as the greatest natural disaster ever experienced. Storm surges in certain other areas of the world owe their severity primarily to coastal configuration. The North Sea falls into this category. Because it open northwards into the Atlantic Ocean periods of extreme northerly winds drive huge volumes of water into the basin. During a great storm surge in 1953, for instance, it is estimated that 42.5×10^{10} m^3 of water, enough to raise the mean sea-level of the North Sea by over 0.6 m, entered the basin in a 15-hour period. The passage of this water could be traced as a gigantic wave moving southwards along the east coast of England and gradually rising in height. By the time it reached the Rhine estuary it had attained a height of about 3 m and in flooding across 1 800 km^2 of reclaimed land was responsible for 1 800 deaths. It was against the recurrence of such an event that the improved sea defences of the Delta Plan in the Netherlands were conceived, and in the 1980s the Thames tidal barrier at Greenwich in eastern London was completed.

Bedrock erosion by marine processes

Mechanical erosion

Waves approaching the coast possess kinetic energy by virtue of the orbital motion of the water, together with potential energy arising from the elevation of the wave crest above mean water-level. When the waves break the potential energy is converted into kinetic energy, greatly enhancing the capacity of the water to produce such features as cliffs and shore platforms. At least two mechanical processes are commonly involved, hydraulic action and corrasion.

Hydraulic action on cliffs actually encompasses two separate but related mechanisms. The first is simply the impact of the moving water. Anyone who has walked along a coastal promenade subject to battering by large storm waves can testify to the veritable hammer-blows that it receives. Dynamometers installed in a number of coastal protection structures have recorded water pressures temporarily exceeding 10^5 N m^{-2}. The second mechanism is the shock pressure exerted by the abrupt compression and decompression of any pocket of air trapped between a rock face and a breaking wave. Shock pressures of this origin have been recorded as attaining values at least an order of magnitude greater than the direct wave pressure. The peak values are sustained for extremely brief periods, generally less than 0.01 second, and owing to random fluctuations are generated by only a minute proportion of the incident waves during any one period of investigation. Even though the measured pressures are much below the crushing strength of nearly all rocks, repeated battering of this ferocity is demonstrably capable of enlarging incipient fracture patterns and helping to divide apparently massive rock into lesser joint-bounded blocks. These blocks need not be small before they can be dislodged, since the momentum of the waves is capable of displacing fragments weighing several tonnes. The process is sometimes described as quarrying, and the significance of rock structure is often clearly visible in the detailed form of the cliff face. Major joint and fault planes are preferentially attacked, and deep narrow inlets excavated along lines of structural weakness. It is here that maximum shock pressures are likely to be created until eventually, in extreme cases, blowholes develop far to the landward of the cliff and forcibly expel large volumes of air during periods of storm waves.

The efficiency of hydraulic action, examined in the laboratory by Sunamara in the 1970s, depends very much upon the precise location of the wave break-point. If a wave breaks more than a few metres in front of a cliff face much of its energy is dissipated in the turbulent motion of the water. If, on the other hand, it fails to break before reaching the cliff, its form will simply be reflected seawards with minimal expenditure of energy on the rock face; the interference of the advancing and reflected waves produces the form known as clapotis. It is when a breaker makes its initial plunge directly against a cliff face that the greatest pressures are generated. Since it is shoaling that causes a wave to break, vertical cliffs descending directly into deep water are must less prone to hydraulic action than those rising from a moderately wide basal platform.

Corrasion is the erosive action of rock fragments regularly set in motion by the sea. Although it undoubtedly contributes to cliff recession where waves break a short distance in front of a rock face, and especially where the junction of cliff and beach is just above still-water level, it is more commonly associated with downwearing of a shore platform. The mobile particles may range in size from sand grains to large cobbles, and the effect of abrasion is readily apparent in the prominent rounding of marine shingle. In part this is doubtless due to attrition as the fragments collide with each other, but it is also the result of impact with exposed rock surfaces. Smoothed and polished faces bear witness to this action, while frequent circular potholes also attest to the effectiveness of swirling water armed with sand and small pebbles.

Both hydraulic action and corrasion are most effective under conditions of high-energy storm waves. At the global scale this suggests that they will be a dominant influence in cool temperate latitudes where the westerly wind system is characterized by a high frequency of gales; in lower latitudes storm waves are less common, while along Arctic coasts the formation of sea-ice will preclude full wave action during at least part of the year. At the regional scale, exposure is one of the most important factors determining the effectiveness of mechanical erosion. A long fetch favours the development of the most powerful storm waves, while at the approach to the coast refraction will often concentrate the attack on local headlands. At an even more detailed scale, such influences as the growth of seaweed can profoundly modify the impact of the waves by severely restricting the velocity of the water in contact with the rock.

Chemical erosion

The chemical alteration of rocks in contact with sea-water is generally referred to as corrosion. It is not restricted to the immediate vicinity of the breakers, but may occur where spray drifts high above the surf zone. Investigations have tended to concentrate on limestone coasts, although even here much remains to be learnt about the solutional processes involved. The surface waters of the oceans are generally saturated or even supersaturated with calcium carbonate so that they do not appear particularly aggressive towards limestone. Yet minor solutional forms are very common in the intertidal zone, and their formation is inferred to depend either upon precipitation or upon organic processes that increase the acidity of rock pools during low tide. Biochemical activity is certainly significant on many limestone coasts. Both plants and animals are capable of boring into calcareous rocks. Blue-green algae are especially important in this respect since they also form the food source of many marine browsers which tear away the weakened rock surface during feeding. Other borers include various species of sponge, mollusc and echinoid, all dissolving limestone and at the same time reducing the strength of the remaining rock.

There is as yet little detailed evidence regarding the relative effectiveness of hydraulic action, corrasion and corrosion along most coasts, and many other simultaneous processes certainly contribute to the overall erosion rate. For instance, rocks between high and low tide are subject to very regular wetting and drying. This can directly promote a wide range of chemical processes, can induce salt crystallization where the waters are highly saline, and in cold climates may favour disintegration of the foreshore by freeze–thaw action. Some of the fastest rates of marine erosion are found where cliffs are composed either of clay or of till. In such cases wetting reduces the cohesion of the material to the point where it constantly slides or slumps to the cliff base, is disaggregated by the waves and finally transported to the seaward side of the surf zone. In this and all other instances the ability of the sea to remove the products of earlier erosion is an important factor in the continued effectiveness of the marine processes.

Transport of sediment

Movement in suspension

Waves and currents are responsible for transporting vast quantities of sediment, in part as a suspension load and in part as a mobile bed layer. In estuaries such as the Wash, on coasts backed by clay cliffs undergoing rapid erosion and at the mouths of large rivers, huge amounts of suspended mud are redistributed by the sea. There is, moreover, an almost constant rain of fine terrestrial sediment over the ocean floors, although in these areas far from the continents the sediment concentration by weight in the sea-water is often no more than 2.5×10^{-8}. The primary concern here, however, is with coarser material in the silt and sand grain sizes. In the earlier part of this chapter three types of water movement potentially capable of lifting sea-floor sediment into suspension were identified: namely, currents, oscillatory waves and translatory waves. The motions due to currents and waves can interact and this has greatly complicated our understanding of the transportation processes to which they give rise. Near the coast the sediment concentration in breaking waves can be measured by using special suction equipment to collect water samples from different depths below the surface. The results have disclosed considerable diversity, and much further research is required to establish all the controls. Representative mean concentrations by weight lie in the range $1–5 \times 10^{-3}$, but in individual waves close to the break-point the vertical distribution has sometimes been found to vary by only a small amount, whereas in other cases a rapid exponential increase with depth has been detected. In the swash advancing up the beach high turbulence may well keep the sand evenly distributed, but in the backwash there is a marked segregation close to the bottom. Some monitoring has located exceptionally high sediment concentrations where the backwash down the beach collides with the incoming swash. The capacity of breakers to raise sand into suspension is a crucial element in the longshore transport of sediment. Although longshore currents resulting from wave set-up may not, by themselves, be capable of generating a suspension load, they will nevertheless carry material that the passage of waves is constantly lifting off the bottom. Ultimately such material may be moved seawards by a rip current, or else stabilized as part of the beach, but in the meantime it constitutes a major component of longshore littoral transport.

Seawards of the breaker zone it was at one time believed that nearly all transport occurred through the sliding, rolling and skipping of a mobile bed-layer. However, it is now clear that both tidal streams and the to-and-fro motion associated with the passage of oscillatory waves are capable of raising fine sand and silt into suspension. As an illustration, a large tripod lowered 90 m to the Pacific sea floor off the coast of Washington was furnished with, among other equipment, a camera and a device for measuring water turbidity. Both these instruments detected fine sediment being lifted into suspension during severe storms. Off the East Anglian coast a frame with sampling nozzles set at six heights between 0.1 and 1.75 m above the sea floor was placed in water 7 m deep. In this case a tidal stream is believed responsible for disturbing the local sand, and a concentration by weight of 1×10^{-3} was recorded from the nozzle situated 0.25 m above the bottom. Calculating actual transportation rates with instrumentation of this type is still far from easy, and much effort is at present being devoted to checking whether such concepts as critical erosion velocity can be satisfactorily transferred from fluvial to marine studies.

Movement as bedload

The relative importance of suspension and traction as processes of sediment transport in the marine environment remains a matter of dispute. Komar, for example, has argued from gross disparities in the estimated total sediment flux and the suspension-load flux that at least three-quarters of longshore littoral transport must take place by traction. On the other hand, the investigations off the East Anglian coast discussed in the preceding paragraph were part of a larger project that appeared to show transport by suspension exceeding that by traction. These apparently divergent findings are both concerned solely with sand transport, and part of the difficulty in resolving any conflict comes from the problems associated with attempts at direct bed-load measurement. These are analogous to those encountered with rivers, and although special traps have been deployed in a number of studies their reliability has been called into question.

There is obviously no such conflict over the nature of shingle movement in the offshore zone, but most of the data relevant to transport rates has come from tracer experiments, discussed in the next section, rather than from direct measurement. The greatest mobility of coarse debris occurs in the swash zone where particles may be swept up the beach under the influence of oblique waves and return directly down the beach under the influence of gravity. This longshore swash transport theoretically involves a zigzag route along the face of the beach. In deeper water increasing use is being made of underwater TV cameras to monitor shingle movement, a good example being afforded by investigations in the west Solent near the Isle of Wight. Here a camera was employed in conjunction with an array of current meters and also an acoustic device that could pick up the impact of moving pebbles. It proved possible to relate the velocities of the tidal stream to initial displacement of gravel ranging upwards in size to 5 cm.

Tracer experiments

The use of tracers for studying the movement of marine sediments has a long history. Early experimenters were content to use a wide variety of materials despite the obvious desirability of matching the physical properties of the tracer to natural sediments. Later work was carried out by marking natural shingle with marine paint, or colouring the local sand with a distinctive dye. As early as 1951 King used the latter technique to elucidate the maximum depth to which sand is disturbed within the breaker zone. At low tide a column of sand was removed from the beach, dyed and replaced. After one tidal cycle the site was re-excavated and the depth of disturbance ascertained. The results revealed a close relationship with wave height. In few instances did the depth exceed 0.04 m and King concluded that, even under very high waves, sand would be unlikely to be disturbed to depths in excess of 0.2 m. It appears that a relatively thin sheet of sediment, far from acting as an abrasive agent, can sometimes form a protective cover for a rock platform. It should be noted, however, that the observations apply only where the beach profile is approximately the same at the beginning and end of the tidal cycle; under storm conditions many beaches suffer very substantial accretion or depletion, as the repeated profile surveys to be discussed below so clearly demonstrate.

Today the most widely used tracers are of two types, radioactive and fluorescent. The advantages and limitations of both these have already been outlined in the discussion of sediment movement by rivers (p. 171) and need not be repeated here. However, the marine environment poses special problems that do merit further comment. In early work with both fluorescent and radioactive sand the material was normally placed on the beach between high- and low-tide marks and its subsequent dispersal examined at frequent intervals, in the case of fluorescent dyes with an ultraviolet lamp after dark, and in the case of radioactive tracers with a scintillation counter. Such simple procedures yielded little quantitative data other than the maximum observed distance of travel. Various more elaborate techniques were subsequently developed. Sampling was improved by the use of adhesive cards which, when pressed on to the surface of the beach, retain a large number of sand grains that can be examined at leisure to determine the proportion of marked material. This proved particularly valuable in work with fluorescent tracers since, in addition to providing much greater flexibility, it also permitted the construction of isopleth diagrams indicating the changing concentration of tracer away from the point of initial insertion (Fig. 15.6). Even more important was the opportunity it afforded to extend observations below low-tide level. This facility was already available with radioactive sand where a scintillation counter could be dragged across the sea floor on a sled (see, e.g. Fig. 15.9). Its application to fluorescent tracers was particularly important at a time when it was becoming increasingly clear that considerable transfer of material between beach and offshore zone might be taking place. So long as observations were confined to the inter-tidal zone, only longshore movement could be traced and the potential loss of marked material

seawards was an unknown factor. Moreover, with a little ingenuity the method could be adapted to study sediment movement in much deeper water. Various techniques for injecting the tracer have been devised. Scuba-divers were originally employed to release the sand, but containers that automatically split open on reaching the sea floor, or others that gradually dissolve, have now been designed. Large volumes of fluorescent tracer are always required, and sometimes these have been pumped down a pipe lashed to the side of a ship and ending just above the sea bed. For sampling the dispersal pattern, divers have occasionally used greased cards, but more often special adhesive weights have been allowed to sink to the sea floor. Alternatively, grab samples may be recovered from a ship, and in some offshore experiments significant quantities of fluorescent tracer have still been obtainable by this method 9 months after the original injection.

Shingle movement demands such high water velocities that pebble tracers have mainly been used to determine patterns of transport within and close to the surf zone. Considerable attention has been focused upon the rates at which materials of different size move. For instance, on beaches at Deal and Winchelsea in southeastern England, Jolliffe inserted selected natural pebbles marked with marine paint, together with artificial pebbles made of concrete chips and a fluorescent constituent. Observation lasted for periods of 4–17 days. The most mobile material travelled over 150 m and proved to be, not the smallest of the particles, but those of intermediate size. This important result demonstrated that longshore swash transport is not simply a winnowing action by which fine debris is sorted from the coarse but involves much more complex relationships. Work in slightly deeper water is well exemplified by investigations on the East Anglian coast close to the shingle spit of Orfordness. Here 600 radioactive pebbles were carefully located in water where the depth varied between 6 and 8.5 m according to the state of the tide. Their position was surveyed at intervals over a period of almost 6 weeks but no movement was detected. For comparison 2 000 pebbles were placed in much shallower water 11 m seaward of the low-tide mark. Some of these travelled northwards more than 1.6 km during a prolonged spell of light southerly winds, but when the winds reversed an even more rapid southward migration occurred. This study clearly demonstrated the need for maintaining observation over a sufficient period of time to cover a wide range of weather conditions.

Fig. 15.6. An example of seaward transport by a rip current, identified by the use of fluorescent sand tracers. The isopleths represent conditions 40 minutes after release of the tracer and refer to the number of fluorescent grains per 6.45 cm^2 (after Ingle, 1966).

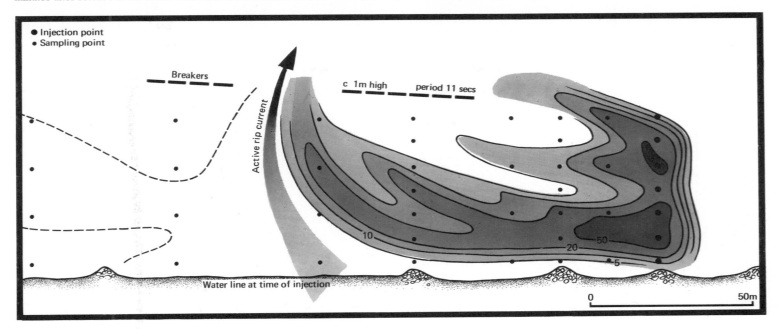

Short-term changes in coastal morphology

The effects of sediment movement are also seen in short-term changes of such features as beach profiles, the distal ends of spits and the bottom morphology of estuaries. These are all coastal forms that can be profitably studied either by the techniques of repeated surveys or by recourse to the evidence of old charts and maps.

Beach profiles constitute one of the most mobile of all coastal features. They often display systematic patterns of erosion and accretion related to such controls as tidal cycles and weather sequences. Foreshore profiles normal to the water-line are most conveniently surveyed during low tide. It is extending the profiles into the offshore zone that presents the greatest difficulty. On a gently shelving beach normal survey methods can continue to be used with the staff-holder wading into shallow water. On steeper beaches it may be necessary to employ sounding procedures with

a consequent loss in precision. Once the selected profiles have been surveyed several times, the results for each can be superimposed so as to allow visual inspection of any changes that have occurred (Fig. 15.7). The area between the highest and lowest lines is known as the sweep zone and is a measure of the maximum changes that have taken place within the period of study. Volumetric computations can be made by assuming that the individual profiles are representative of contiguous strips across the beach.

Rather longer intervals are usually treated when considering such constructional forms as spits. Nevertheless, significant changes can often be detected within a period of a few years and repeated surveys may provide valuable evidence regarding the processes involved. Yet care needs to be exercised in the interpretation of the results since it has been shown in a number of cases that the sequence of change is cyclical; in such circumstances the continu-

Fig. 15.7. An example of superimposed beach profiles and the 'sweep zone' that they define (after King, 'Beaches and Coasts' 1972).

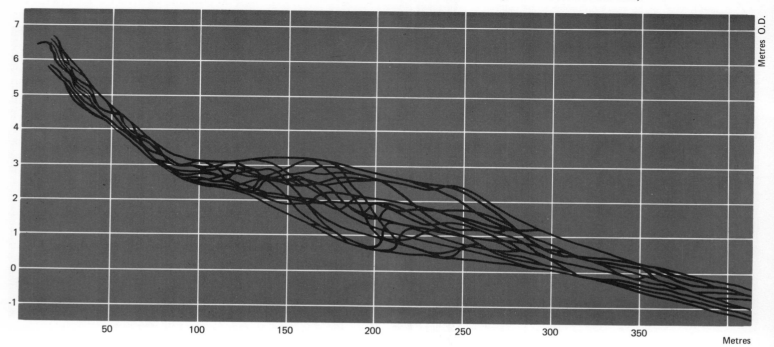

ation of the observed trends cannot be assumed. The longer the time interval to be considered, the greater the reliance that needs to be placed upon historical records. Old charts and documents can prove extremely informative if interpreted with due caution, remembering in general the older they are the greater the doubt regarding their reliability. The value of documentary evidence is well illustrated by the historical study of Spurn Point by de Boer. He showed that this spit at the mouth of the Humber estuary undergoes a phase of elongation followed by narrowing of its neck as the coast to which it is attached recedes under the influence of storm waves (Fig. 15.8). Eventually the neck is breached, leaving the distal end of the spit as an island. This persists briefly but is ultimately destroyed by further wave action. A new spit is then initiated and grows rapidly southwards. Each cycle is believed to last about 250 years, with new spits

starting to develop around the years 600, 850, 1100, 1360 and 1610. In the mid-nineteenth century the present spit was breached by storm waves and a gap of almost 500 m cut across the neck. This was artificially closed, and since that date the form has been carefully stabilized as a matter of policy to protect the Humber estuary.

In the hands of an experienced hydrographer charts of different date can provide valuable insights into the processes modifying the sea-floor topography. Unchanging features may first be identified and used as reference points; investigations generally show these to be composed of materials that local currents are incapable of moving. In contrast are the topographic features that vary in either form or position from one chart to another. These often display a distinctive morphology permitting their origin to be inferred with considerable confidence. Among the most common are sand waves

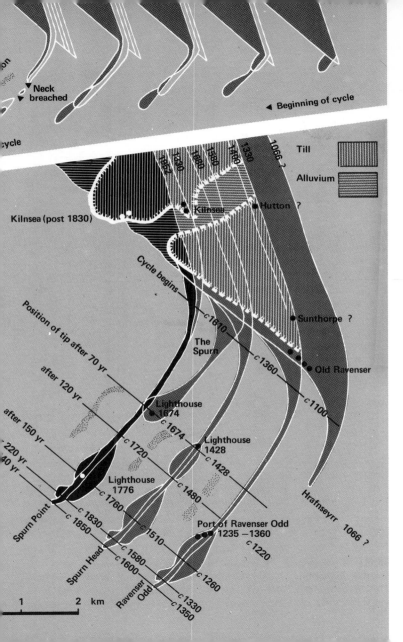

and systems of ebb and flood channels. Examples of both have been intensively studied to elucidate the general principles of formation. Sand waves are long undulations that corrugate the surface of many sandbanks. They vary widely in size. In large examples individual crests may be over 1 km apart, rise more than 10 m above the intervening troughs and be traceable laterally for long distances. Smaller waves have lengths of less than 10 m and heights of under 1 m. The characteristic common to all mobile sand waves is asymmetry, with the steeper face pointing in the direction of movement. Ascribed to tidal streams, they are believed to be a manifestation of large-scale sediment transport. It has been estimated that 40×10^6 m³ of sand move northwards along the coast of the Netherlands each year, with a compensating return movement southwards past the East Anglian coast.

Ebb and flood channels are a prominent feature of many hydrographic charts close to the coast. They are particularly conspicuous in estuaries where abundant mobile sediment is available for shaping by tidal streams. The ebb and flood components interdigitate, with the mobile sand between the channels moulded into elongated banks. Both channels and banks shift position with time, but it is often possible to discern cyclical patterns with the original bottom configuration restored after a matter of a few years or decades. The direction of dominant water movement in the channels can usually be deduced from their morphology. At their intake end they are relatively deep and narrow but flare in a downstream direction. Velocity measurements show that water frequently continues to flow inwards along flood channels for a short time after the tide has begun to recede. Ebb channels are similarly occupied by outflowing water after the tide has turned. There is a net transport of sediment landwards along flood channels and seawards along ebb channels (Fig. 15.9). Individual grains may pursue a gyratory course, the residual movements first carrying them from the flood into the adjacent ebb channel and later returning them to the seaward end of the original flood channel. Changes in channel pattern are of obvious economic importance in estuaries widely used by shipping,

Fig. 15.8. Evolution of Spurn Head since AD 1100 as envisaged by de Boer. Above, schematic diagram of a single cycle of growth and destruction. Below, application of the cycle concept to the present Spurn Point and its immediate predecessors, Spurn Head and Ravenser Odd (after de Boer, 1964).

and may also be highly significant where waste material is discharged into the sea.

Sediment budgets

The preceding pages have included reference to numerous techniques for measuring sediment movement in a variety of coastal environments including both the nearshore and offshore zones. Methods have also been outlined for evaluating the temporary storage of sediment through analysis of changing beach and sea-floor profiles. It is clear that if all these approaches were to be applied to a suitable section of coast a budget could be compiled in which sediment inputs would be balanced against sediment outputs, with any difference being reflected in net changes in beach and sea-floor morphology. It would first be necessary to define suitable discrete cells to which such a budgetary approach might be applied. Possible input sources to such a cell might include sediment discharge from local streams, cliff erosion, onshore marine transport, longshore littoral drift, biogenic production and aeolian activity; a corresponding list of output sinks might include offshore marine transport, longshore littoral drift and aeolian deflation. The objective would be to quantify each of the above items as precisely as possible. If all inputs and outputs have been correctly identified and measured, the balance should then correspond with net accretion or depletion over the time interval under consideration.

In practical terms, few coasts of the world have yet been adequately studied to permit accurate sediment budgets to be compiled. Nevertheless, the concept is an important one which is increasingly influencing the design of research programmes. It not only affords a framework for a better understanding of coastal processes in general, but may also assist in pinpointing areas of investigation where our current knowledge is particularly deficient. As extra information is gathered, and as techniques of measurement improve, there are likely to be many more attempts at compiling coastal sediment budgets.

Use of scale models

Two types of physical model are in common use for the study of coastal processes. The first consists of a large tank equipped with a means of producing artificial waves. Through its use investigations are made into the relationships between various wave properties and the movement of beach and sea-floor sediment. The second type is a scale reproduction of a specific section of coast. This is particularly valuable where engineering works are proposed since it allows the likely changes in coastal morphology to be studied. Considerable difficulties arise in the process of scaling. Certain factors, such as gravity, obviously defy direct scaling. Nevertheless, as far as possible the geometrical, kinematic and dynamic properties of the prototype must be retained. This can theoretically be achieved by adherence to what is known as Froude's law. This states that time and velocity scaling should be equal to the square root of the linear scaling. This means, for example, that in a model constructed at a scale of 1 : 1 000 the tidal cycle should be reduced to just under one-thirtieth of its natural duration; for a normal semidiurnal tide this will be about 25 minutes. Geometrical scaling is therefore accompanied by an acceleration of coastal change which is an obvious advantage to the user. However, even simple geometrical scaling poses problems. At a scale of 1: 1 000 water that is 1 m deep would be represented by a layer 1 mm thick, but the nature of flow in that layer would be entirely different. To overcome this difficulty most coastal models are built with a vertical exaggeration and the motion of the water is regulated by numerous small pins having no natural counterparts. A further problem is occasioned by the fact that sediment properties do not vary as a simple function of grain size. It is no use employing clay particles to simulate sand grains since their physical response is so different; they require higher velocities to set them in motion, but once in suspension have slower settling speeds. To surmount this difficulty special particles lacking the cohesion of clays and of different density from sand are often employed.

These points illustrate a few of the many problems involved in model building. Yet the practical utility of models can scarcely be doubted in view of the time and money spent both on their construction and on the large buildings required to house them. Well-known models include those of the Mersey, Humber and Thames estuaries in Britain, the Rhine delta in the Netherlands, and San Francisco Bay, Puget Sound and the Delaware estuary in the United States.

Fig. 15.9. An illustration of the use of radioactive tracers for studying sediment movement in an ebb and flood channel system in the Firth of Forth. Both diagrams represent conditions one day after release of the tracer; measurements at the points indicated were made by lowering a scintillation head to the sea floor (after Smith et al., 1965).

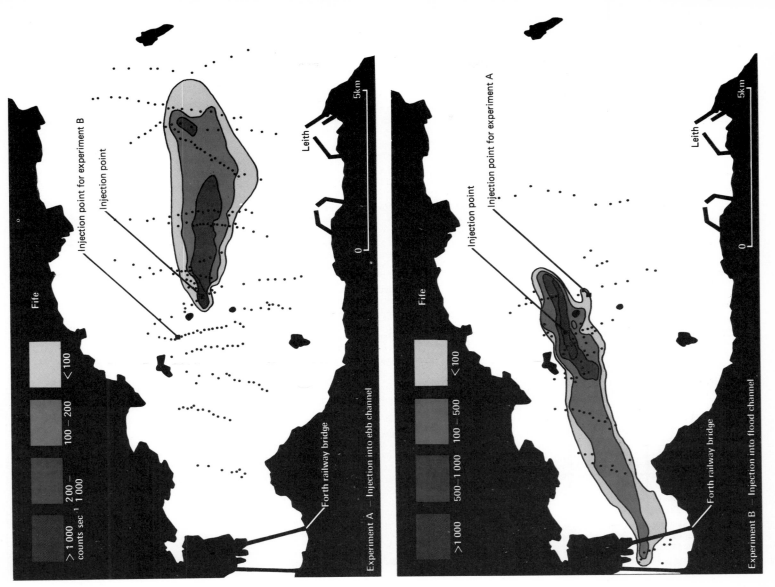

Fife

Leith

□ <100

□ 100 — 200

■ 200 — 1 000

■ >1 000
counts sec⁻¹

Injection point for experiment B

Injection point

Forth railway bridge

Experiment A — Injection into ebb channel

5km

0

Fife

Leith

□ <100

□ 100 — 500

■ 500 — 1 000

■ >1 000

Injection point for experiment A

Injection point

Forth railway bridge

Experiment B — Injection into flood channel

5km

0

References

Carr, A. P. (1965) 'Shingle spit and river mouth: short term dynamics', *Trans. Inst. Brit. Geogr.* **36**, 117–29.

Carr, A. P. (1980) 'The significance of cartographic sources in determining coastal change ', in *Timescales in Geomorphology* (ed. R. A. Cullingford, D. A. Davidson and J. Lewin), Wiley.

Carruthers, J. N. (1962) 'The easy measurement of bottom currents at modest depths', *Civ. Eng.* **57**, 486–8.

Carruthers, J. N. (1963) 'History, sand waves and nearbed currents of La Chapelle Bank', *Nature, London* **197**, 942–7.

De Boer, G. (1964) 'Spurn Head: its history and evolution', *Trans. Inst. Brit. Geogr.* **34**, 71–89.

Hammond F. D. C., Heathershaw, A. D. and Langhorne D. N. (1984) 'A comparison between Shields' threshold criterion and the movement of loosely packed gravel in a tidal channel', *Sedimentology* **31**, 51–62.

Hansom, J. D. (1983) 'Shore platform development in the South Shetland Islands, Antarctica', *Mar. Geol.* **53**, 211–29.

Ingle, J. C. (1966) 'The movement of beach sand', *Developments in Sedimentology* **5**.

Iversen, H. W. (1953) 'Waves and breakers in shoaling water', *Proc. 3rd Conf. Coast. Eng. (Council on Wave Research)*, 1–12.

Jolliffe, I. P. (1964) 'An experiment designed to compare the relative rates of movement of different sizes of beach pebbles', *Proc. Geol. Ass.* **75**, 67–86.

Kidson, C. and Carr, A. P. (1962) 'Marking beach materials for tracing experiments', *J. Hydraulics Div. Proc. Am. Soc. Civ. Eng.* **3189**, HY4, 43–60.

King, C. A. M. (1951) 'Depth of disturbance of sand on beaches', *J. Sed. Pet.* **21**, 131–40.

Larsen, L. H. *et al.* (1981) 'Field investigations of the threshold of grain motion by ocean waves and currents', *Mar. Geol.* **42**, 105–32.

Lees, B. J. (1979) 'A new technique for injecting fluorescent sand tracer in sediment transport experiments in a shallow marine environment', *Mar. Geol.* **33**, M95–8.

Lees, B. J. (1983) 'The relationship of sediment transport rates and paths to sandbanks in a tidally dominated area off the coast of East Anglia, U.K.', *Sedimentology* **30**, 461–83.

Mackenzie P. (1958) 'Rip current systems ', *J. Geol.* **66**, 103–13.

McCave, I. N. (1971) 'Sand waves in the North Sea off the coast of Holland', *Mar. Geol.* **10**, 199–225.

Phillips, O. M. (1957) 'On the generation of waves by turbulent wind', *J. Fluid Mech.* **2**, 417–45.

Robinson, A. H. W. (1960) 'Ebb–flood channel systems in sandy bays and estuaries', *Geography* **45**, 183–99.

Rossiter, J. R. (1954) 'The North Sea storm surge of 31 Jan and 1 Feb, 1953', *Phil. Trans. R. Soc.* **A246**, 371–99.

Smith, D. B *et al.* (1965) 'An investigation using radioactive tracers into the silt movement in an ebb channel, Firth of Forth', *UKAEA Research Group Rep.*, HMSO.

Sternberg, R. W. and Marsden, M. A. H. (1979) 'Dynamics, sediment transport, and morphology in a tide-dominated embayment', *Earth Surf. Processes* **4**, 117–39.

Sunamara, T. (1975) 'A laboratory study of wave-cut platform formation', *J. Geol.* **83**, 389–97.

Sunamara, T. (1977) 'A relationship between wave induced cliff erosion and erosive force of waves', *J. Geol.* **85**, 613–18.

Sunamara, T. (1982) 'A wave-tank experiment on the erosional mechanism at a cliff base', *Earth Surf. Processes Landf.* **7**, 333–43.

Thorne, P. D., Heathershaw, A. D. and Troiano, L. (1984) 'Acoustic detection of seabed gravel movement in turbulent tidal currents', *Mar. Geol.* **54**, M43–8.

Trenhaile, A. S. (1983) 'The width of shore platforms: a theoretical approach', *Geogr. Annlr.* **65A**, 147–58.

Selected bibliography

An admirable account of coastal processes is provided by J. Pethick, *An Introduction to Coastal Geomorphology*, Arnold, 1984. More limited in scope but providing an excellent review of selected topics is P. D. Komar, *Beach Processes and Sedimentation*, Prentice-Hall, 1976.

Chapter 16
Sea-level changes

The processes reviewed in the preceding chapter are virtually time-independent. The longest time interval to which reference was made amounted to no more than a few centuries. Yet many coastal land-forms are clearly the result of a much longer history, and a greatly extended period must be taken into account when considering their evolution. As soon as that period exceeds a few millennia it is essential to have regard to sea-level changes which will normally constitute a major factor in their development.

Relative changes of land and sea-level may arise from a number of different causes. Eustatic changes, in theory affecting all coasts equally, may be occasioned by variations in the volume of oceanic water, or by fluctuations in the capacity of the ocean basins. Local changes having little effect beyond their immediate area of occurrence can arise either from tectonic movement (Fig. 16.1) or from crustal flexuring due to loading by such phenomena as ice-sheets. Given these circumstances the ideal procedure would seem to involve construction of a eustatic sea-level curve against which local variations might then be compared; major deviations between global and local levels would presumably be ascribable to either tectonic or isostatic causes. Yet even this apparently simple objective has never been fulfilled. In part this may be due to the problem of identifying a section of coastline that has been stable sufficiently long to provide a eustatic yardstick. There is always a grave risk of circular arguments being employed. A coast may first be classified as stable on the basis of apparently undeformed shorelines, and then used as a model to which shoreline sequences in other parts of the world are referred. Thereafter similarities tend to be stressed, and it is very easy for these to be interpreted as strengthening the claims of the original area to provide a global reference standard. However, such is the complexity of the record that superficial similarities must be treated with the utmost caution and, as will be seen below, the very concept underlying eustasy has come under increasing challenge from some workers. The whole topic of long-term changes in sea-level remains fraught with uncertainty.

The nature of the evidence

Marine features located above present sea-level

Elevated marine features fall into two major categories, those due to erosion and those due to deposition. The former include abandoned cliffs, shore platforms, fossil stacks and marine caves. Where all are present identification is relatively easy, but often the features occur individually and in such degraded form that there may be real doubts concerning their authenticity. For example, gently sloping benches backed by steeper bluffs on the flanks of Exmoor in south-western England have been interpreted by some workers as marine and by others as subaerial in origin. Of possible criteria by which raised shore platforms might be differentiated, the most widely employed is consistency in the elevation of the concavity at the foot of the presumed cliff. Yet even after positive identification of features as due to marine erosion, it may still be difficult to infer the exact height of the contemporaneous sea. Two factors militate against a precise determination. The first springs from likely original variations in altitude relative to any such datum as the high-water mark of ordinary spring tides; lithology, exposure and fetch can all influence the height at which a shore platform abuts against a cliff. The second is the progressive degradation that initially clear forms suffer. Even recently abandoned cliffs tend to have their bases obscured by slumping and rockfall. With older cliffs the destruction may be so great that the modern break of slope affords very poor guidance regarding either original form or even exact position.

Marine deposits sited above the reach of present-day storm waves afford some of the most convincing and informative evidence on sea-level changes. They range from old shingle ridges to extensive flats of estuarine clay and silt. Distinguishing unfossiliferous marine shingle from other coarse gravels can prove troublesome, although the fabric

Fig. 16.1. One of a suite of raised marine shorelines that characterize the Pacific coast of California, USA, between San Francisco and Los Angeles. In this tectonically active region the elevations of the shorelines provide little direct guide to eustatic changes in sea-level.

and surface textures of the material may aid in identification. The task is much easier where fossils are present, and such biological evidence can also shed light on the contemporaneous water temperature. This can be particularly valuable in indicating whether the high sea-level occurred during a glacial or interglacial phase. It may also exclude as improbable correlations which altitude alone suggests as quite reasonable. As with erosional features, it is not always easy to determine the exact height of sea-level at the time deposition was taking place. Large waves build storm ridges to a significantly greater height on exposed than on sheltered sections of coast. Fine sediments are subject to much compaction and it would be unwise to assume that the surface of raised estuarine silt coincides with, say, old mean tide-level. Occasionally isolated pockets of sediment are found that contain the remains of organisms known to live within clearly defined depth limits. It is tempting to use such information to reconstruct the former sea-level even though no contemporaneous beach features have survived. While sound in theory, in practice this procedure often gives rise to dispute, particularly over such questions as whether the shells are still in their original growth positions. It is often safest to adopt the conservative argument that the upper limit of marine sediments indicates the minimum level attained by the sea.

Emphasis above has deliberately been placed on some of the difficulties of reconstructing former high sea-levels, since in the past the problematic nature of much of the evidence has been neglected. Nevertheless, this approach should not be allowed to detract from the prolific evidence in many parts of the world that the sea formerly stood at higher levels relative to the land than at the present day. Both erosional and depositional forms attest quite clearly to this fact.

Terrestrial and shoreline features below present sea-level

The increasing pace of research on the continental shelves has yielded abundant indications that parts of the present sea floor were once dry land. Both physical and biological evidence attests to one or more phases of low sea-level. Among the most prominent physical signs is the seaward prolongation of many continental valleys in the

form of sediment-filled channels. These channels are not restricted to glaciated areas in which some form of subglacial scour might be invoked but also occur in regions far beyond the limits reached by the ice-sheets. They commonly descend to depths of about −90 m, although there is no real uniformity. Besides these submerged valleys, other identifiable features include old soil profiles, deltas, cliffs and even former spits. In some northerly latitudes forms due to glacial deposition can be traced from the land surface across the adjacent sea floor without observable change; it must be presumed that at the time of accumulation the sea floor was dry land.

Biological materials recovered from below sea-level tell the same story. Submerged 'forests', generally consisting of tree stumps in position of growth on the foreshore, have long provided an indication of a rise in sea-level. During the nineteenth century when large docks were being excavated along the coasts of both Europe and North America, many beds of freshwater peat were encountered below modern sea-level. As investigations have extended into areas of deeper water signs of terrestrial life have continued to be recovered. Peat has been dredged from the floor of the North Sea, bones of such animals as mammoth, bison, elk and moose have been found off the Atlantic seaboard of North America, and shell banks comprising marine species restricted to water under 10 m deep have been located at depths of over 100 m in the western Pacific. In most cases there is good reason to regard such material as *in situ* and undisturbed by recent submarine transport. Further important sources of evidence are the great deltas of the world where boreholes have frequently penetrated organic deposits indicative of fresh- or brackish-water conditions at depths well below modern sea-level. Tectonic subsidence and sediment compaction complicate these environments, but the magnitude of these factors can often be computed and an appropriate allowance made. Minimum sea-levels of −100 m or more are often indicated by such procedures.

The evidence for lower sea-levels is usually inadequate to fix the precise height of the sea. Peat and animal remains merely show an area was once dry land without any indication of elevation above the contemporaneous shoreline. More significance attaches to the remains of organisms requiring a habitat in either brackish or very shallow salt water. However, such finds constitute only a small proportion of those that are made. Correlation from one locality to another is extremely difficult when dealing with submarine features, and often all that can be done is to record the minimum depression of the sea consistent with a particular find.

Models of sea-level change

The amplitude of glacio-eustatic fluctuations

It was recognized soon after the idea of an ice age was propounded in the nineteenth century that the growth of large ice-sheets would have a profound effect upon the volume of ocean water and therefore upon global sea-level. Indeed, the effect upon sea-level was suggested as one means by which the whole concept of an ice age could be tested. When evidence of multiple glaciation was later adduced, the explanation for several large-scale eustatic fluctuations was recognized to be at hand. Yet it has subsequently proved very difficult to obtain an unequivocal basis for estimating the amplitude of such glacio-eustatic fluctuations.

There are two obvious approaches to the subject. The first is to see whether the evidence for sea-level changes sheds light on oscillations that occurred simultaneously with major glacial advances. The difficulty with this approach has proved to be gross uncertainty over dating and correlation; while most observations are consistent with large glacio-eustatic fluctuations, in the absence of satisfactory dating they can rarely be held to sustain one thesis rather than another. The second approach involves calculating the water volumes locked up in major continental ice-sheets during a specific glaciation. Once these volumes are known the figures can be converted into the change in sea-level they would represent. Many workers have attempted the necessary computations. Flint, for instance, employed the concept of a model glacial age in which ice is assumed to cover simultaneously all the territory glaciated at any time during the Pleistocene epoch. In such circumstances the ice volume, over and above that in present-day ice-sheets, is estimated at about 50×10^6 km³, which corresponds to a water volume of 47×10^6 m³ and a gross fall in sea-level of 132 m. A correction is then required to convert this theoretical maximum to the figure appropriate for any particular glaciation; for the last glaciation the correction is believed to be a reduction of about 10 per cent. Flint stressed at least three sources of potential error in his calculations. The first is that global ice advances may not be exactly synchronous. The second is the prob-lematic thicknesses of the Pleistocence ice-sheets; most three-dimensional reconstructions rely upon analogy with existing counterparts. The third is the isostatic consequences of abstracting large volumes of water from the ocean basins. The ocean floors almost certainly sink and rise in sympathy with the alternate loading and unloading that they experience, and the magnitude of changes in water-level might conceivably be reduced by as much as one-third by this so-called hydro-isostatic factor. It is also worth noting that this analysis assumes current conditions to be typical of an interglacial period. Were the Antarctic and Greenland ice-sheets to melt, it is estimated that a water layer some 65 m thick would be returned to the oceans, and with due allowance for isostatic effects sea-level would rise by just over 40 m. However, it now seems clear that neither the Antarctic nor the Greenland ice-sheets disappeared during recent interglacial periods, and in all probability neither deviated greatly from its present dimensions.

Since the 1970s an alternative method of evaluating glacio-eustatic fluctuations has presented itself. As mentioned on p. 23, measurement of the O^{18}/O^{16} ratio in the shells of benthic marine organisms has permitted changes in the isotopic composition of oceanic water to be calculated. Since these changes are believed to derive from alterations in the quantity of O^{16}-rich water locked up on the land masses, their amplitude should in the main reflect variations in the aggregate volume of continental ice. Computations based upon this argument have yielded gross figures for sea-level fluctuations at least as large as those derived by Flint, with some estimates reaching as much as 160 m. Given all the possible sources of error, it seems reasonable to conclude that a major glacio-eustatic oscillation has a net amplitude of about 100 m with an uncertainty of ±20 m.

The concept of secular decline in Pleistocene sea-levels

If glacio-eustatic oscillations were the sole control of sea-level changes, then after each glaciation the sea should return to its original elevation. In such circumstances any raised shoreline would imply either tectonic or glacio-isostatic disturbance. Yet early investigations revealed so many elevated strandlines beyond the limits of glaciation and without obvious signs of deformation that it was widely agreed some other control must be operating. The most influential pioneer studies were those of French workers around the western

Mediterranean. De Lamothe in Algeria, Dépéret in southen France and Gignoux in southern Italy identified a succession of strandlines to which the following names and heights were conventionally accorded: Calabrian 150–180 m; Sicilian 80–100 m; Milazzian 55–60 m; Tyrrhenian 30–35 m; and Monastirian 15–20 m. Each of these beaches was held to be undeformed so that the sequence implied a slow but sustained eustatic fall. This then constituted the basis for a simple model of sea-level change that long dominated thinking on the subject. The model envisaged glacio-eustatic oscillations superimposed on a secular decline (Fig. 16.2). Even general testing of this concept proved remarkably difficult. Profiles across many coasts of the world from, say, 200 m above to 200 m below modern sea-level undoubtedly disclose signs of periods when the sea stood both higher and lower than at present. Yet arranging the various levels in a chronological framework poses immense problems. The fragmented nature of the evidence often precludes use of simple stratigraphic principles. Any organic remains are usually too young to permit application of the normal precepts of biological evolution, while ecological arguments, because of the repeated

climatic cycles of the Pleistocene epoch, have limited value even for relative dating. Except in recently deglaciated regions, elevated shorelines are generally too old for radio-carbon dating and although the use of uranium-series and amino-acid techniques is locally very important, much of the evidence is not amenable to these approaches.

Despite these problems over validation, theory continued to demand glacio-eustatic fluctuations and observations to imply a secular decline. Yet certain postulates of the model were not always substantiated by the field evidence. This was as true in the Mediterranean region as in most others, and a vast literature accumulated on a number of controversial issues. For instance, some preliminary studies on the relationship between raised strandlines and Alpine glacial advances suggested a synchroneity that would have undermined one of the basic tenets of the model. Although many of these early studies were later discredited, even today it remains difficult to establish a clear link between glacial advances and sea-level changes. A second approach to the climatic associations of the Mediterranean beaches lay through the faunal assemblages they

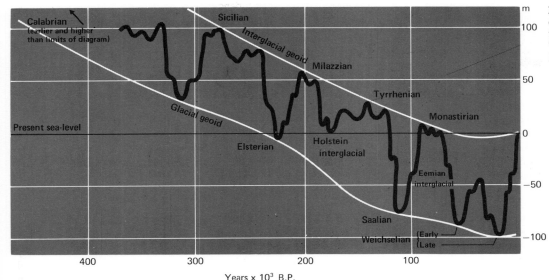

Fig. 16.2. An example of a Pleistocene eustatic curve based upon the traditional Mediterranean sequence. This particular interpretation is that of Fairbridge, 1961.

Calabrian (earlier and higher than limits of diagram)

Sicilian

Interglacial geoid

Milazzian

Glacial geoid

Present sea-level

Tyrrhenian

Monastirian

Elsterian

Holstein interglacial

Eemian interglacial

Saalian

Weichselian (Early / Late)

m
100
50
0
−50
−100

400   300   200   100

Years x 10³ B.P.

contain. This also produced results not entirely accordant with predictions of the model. Whereas the Milazzian, Tyrrhenian and Monastirian beaches yielded assemblages with subtropical affinities and therefore acceptable as of inter-glacial origin, the Calabrian and Sicilian beaches contain a number of molluscan species with northern affinities implying water temperatures lower than today. This problem is still unresolved, possibly because the western Mediterranean presents a singularly difficult area in which to make close faunal comparisons. Water temperature and salinity are much influenced by the pattern of currents through the Straits of Gibraltar. At the time of the Calabrian and Sicilian beaches the straits were both wider and deeper than now, allowing a more ready exchange of water with the Atlantic. Precise evaluation of this effect is difficult, but it does underline the need for caution in any comparison with the present day. A further indication of the complexities of the Mediterranean basin comes from southern Italy where studies to fix the stratigraphic position of the Plio-Pleistocene boundary have identified a sudden influx of cold species but without any obvious fall in sea-level.

The long chapter of controversies surrounding the Mediterranean beaches undoubtedly eroded their claim to be regarded as a prototype for Pleistocene eustatic fluctuations. Nevertheless, in the decades following their first scientific description they were widely used as a reference standard and many apparently sound correlations were established. The classical sequence implied an early Pleistocene sea-level of between 150 and 200 m, and this value was accepted by many workers throughout north-western Europe. For example a Calabrian age was frequently assigned to a prominent bench around the uplands of south-western England and Wales at about 180–200 m (see, e.g. Fig. 4.11 p. 68). Although marine sediments are generally lacking the dating was often held to be strengthened by analogy with south-eastern England where shingle at the same elevation yielded a fauna consistent with an early Pleistocene age. Numerous other examples could be quoted, and although there is a very real risk of circularity of argument as was outlined earlier in this chapter, it is difficult to ignore the success that attended what might be termed the Mediterranean model of sea-level change.

Yet as research proceeded in other parts of the world, this Mediterranean model came under increasingly critical scrutiny. The Atlantic seaboard of the United States, for instance, has been claimed as more suited to establishing eustatic changes than the Mediterranean region owing to its greater tectonic stability, and when the findings from these two areas are compared considerable conflict emerges. In the coastal zone from Virginia to Florida Pleistocene marine features, in the view of many workers, extend no higher than about 30 m. This is in sharp contrast to the classic sequence of the Old World, although it should be added that available dating in North America indicates a younger age for the highest shorelines than would normally be accorded the Calabrian, or even the Sicilian, beaches in Europe. If it be contended that the earlier and higher episodes are unrepresented along the American seaboard, the question obviously arises why that should be so if the Mediterranean levels are truly eustatic. The enigma is all the greater when it is recalled that the Atlantic coastal plain bears the imprint of several mid-Cenozoic transgressions; it requires special pleading to argue that the early Pleistocene examples have been selectively destroyed. Another area claimed as particularly suited to the study of eustatic changes owing to its presumed tectonic stability is Australia, and the height ranges recorded here are generally closer to those of North America than those of the Mediterranean. Such findings encouraged several workers to embrace the concept of a secular decline in sea-level, but to argue that the Mediterranean model greatly exaggerated its magnitude.

Alternative concepts of sea-level change

One weakness of all models invoking long-term eustatic decline is undoubtedly the lack of any agreed mechanism. Moreover, the altitude of the Calabrian shoreline in the Mediterranean model approaches the theoretical maximum value for any simple eustatic change (p. 90) and is therefore inherently unlikely. The oxygen isotope record from marine cores lends no support to the concept of slow secular change, and increasing numbers of workers appear to subscribe to the view that interglacial sea-levels have differed relatively little from each other during much of the Pleistocene. This idea gains credence from the growing evidence relating to the last 250 000 years when correlation by means of uranium-series dating and amino-acid investigations is feasible. In Bermuda, for instance, where an unusually complete record of late Pleistocene marine fluctuations has been compiled, it has been concluded that only twice

Fig. 16.3. Two examples of inferred curves depicting sea-level changes during the last 250 000 years. In each case the horizontal scale is based upon either uranium-series disequilibrium dating or amino-acid dating. The upper curve derives from a tectonically stable area where the falls in sea-level are particularly difficult to evaluate. The lower curve derives from an area of tectonic uplift where direct evidence for depressed sea-levels is potentially more accessible, but an allowance must be applied to offset the effects of earth movements.

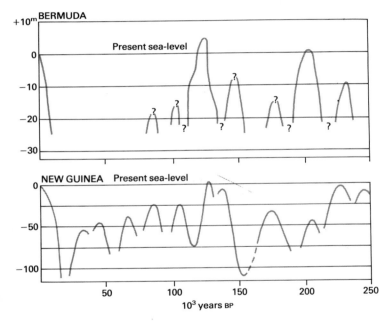

during the past 250 000 years has the sea risen higher than at present: approximately 200 000 years ago it stood at about +2 m and 125 000 years ago it temporarily reached +5 m (Fig. 16.3). A comparison of these values with those portrayed in Fig. 16.2 illustrates the radical change that is involved.

Adoption of any model envisaging relatively constant interglacial sea-levels provides no immediate explanation for the remarkable number of coasts where elevated strandlines have been recorded. In some cases, and this may be true of the western Mediterranean, uplift along a plate margin can be invoked, but this does not have universal applicability and two other independent factors have been conjectured as responsible. The first is a tendency for hinge-line tilting along the edge of many land masses which induces simultaneous depression of the continental shelf and upheaval of the adjacent coast. The precise forcing agency in any such movement has not yet been identified, but isostatic elevation consequent upon subaerial denudation is one idea worthy of much further evaluation (see also p. 389). The second factor, invoked particularly by the Swedish worker Mörner, implicates change in the form of the geoid. As pointed out on p. 3 the geoid is not a regular ellipsoid but has a surface relief amounting to as much as 180 m. If bulges of this amplitude migrate across the surface of the earth, large local variations in sea-level will be induced without any alteration in either the volume of marine water or the capacity of the ocean basins; in effect, the whole concept of global eustasy is undermined since one area will be experiencing a fall in sea-level while at the same moment another will be experiencing a rise. At present the fundamental causes of geoidal irregularity are poorly understood, and it is not known how rapidly the configuration can change. Moreover, without more precise dating techniques than are currently available, the whole idea of geoidal alterations is proving extremely difficult to test.

The conceptual framework for sea-level studies underwent a revolution in the 1970s. The idea of relatively constant water volumes and ocean-basin capacities during successive interglacials appears to offer a more reliable model than that provided by the early Mediterranean studies. Yet this revised model must certainly be combined with such factors as hydro-isostasy, plate-margin tectonics, hinge-line tilting at the coast and shifts in geoidal relief if local changes in sea-level, even in non-glaciated regions, are to be properly understood.

Eustatic movements during the last 35 000 years

General pattern

An analysis of sea-level changes associated with the last glaciation is crucial to an understanding of the way many coastal landforms have evolved. The selection of the 35 000-year period for special discussion reflects the approximate interval spanned by radio-carbon dating since this technique has contributed immeasurably to our knowledge of recent sea-level changes. As indicated earlier in this chapter, elevated beaches outside the glaciated regions are normally too old to be dated by radio-carbon techniques, and in such areas it is material recovered from below sea-level that has proved amenable to the method. This is not surprising inasmuch as we live just after the close of a major continental glaciation and might expect the return of water to the oceans to have produced a sharp rise in sea-level.

A common procedure adopted by workers in many different parts of the world has been to plot the age of terrestrial organic remains against their depth of recovery, sometimes with the added refinement of distinguishing those finds that are particularly sensitive indicators of sea-level. From a wide variety of coastal environments the results have been reasonably similar (Figs 16.4 and 16.5). They nearly all imply an extremely rapid rise in sea-level during the period between 15 000 and 7 500 years ago. Most estimates of sea-level for 15 000 years ago lie within the range from −100 m to −60 m, and those for 7 500 years ago from −15 m to −10 m. This gives a sustained rise averaging about 1 m per century. Rapid flooding of the upper parts of the continental shelves took place. Where these are gently sloping very extensive areas of dry land were drowned, as is well exemplified by the recent evolution of the North Sea basin (Fig. 16.6).

Matters of continuing dispute

Differences of opinion about sea-level changes during the last 35 000 years have centred on three major issues: movements prior to 15 000 years ago, the steadiness of the rise during the main transgression and minor fluctuations within the last 7 500 years. The ice-sheets of the last glaciation are generally acknowledged to have reached their greatest extent in both Europe and North America between 18 000 and 20 000 years ago. Since the lowest sea-level would be expected to coincide with that maximum advance, many workers have inferred that a slow rise was probably taking place for several millennia prior to 15 000 years ago. Estimates for the elevation of the lowest sea-level have ranged from −150 to −80 m, but as Fig. 16.4 suggests an average assessment would lie close to −100 m. On the subject of changes prior to 20 000 years ago two schools of thought emerge. The first contends that sea-level remained substantially below that of the present day, while the other argues that, at least for a brief

Fig. 16.4. A diagram illustrating some of the views that have been expressed regarding eustatic sea-level changes during the last 35 000 years.

interval, it actually rose a little higher than its modern elevation. Although several interstadials have been identified during the last glaciation, the prevailing temperatures seem to have remained sufficiently low to ensure that ice-sheets would have persisted in adequate volume to maintain a depressed sea-level. Yet a few radio-carbon assays on marine materials at or just above current sea-level have yielded ages of around 30 000 years: it is in unresolved conflicts of this type that Mörner sees confirmation of his view that the basic geoidal form has varied significantly over relatively short time-scales.

A number of workers have claimed to identify minor regressive phases during the main rise in sea-level between 15 000 and 7 500 years ago. Fairbridge, for instance, suggested that four separate periods when the sea temporarily receded could be recognized. However, past lower sea-levels can rarely be assessed with any great precision. There is a range of uncertainty attaching to both dates and depths. For radio-carbon assays this may be expressed statistically in the form of confidence limits, but it is difficult to apply such an objective scheme to depths. When the data come from widely separated coasts that may well have suffered slight but varying degrees of tectonic displacement, the problem of computing an exact sea-level curve becomes almost insurmountable. For this reason many workers have greeted attempts to identify minor cyclical patterns with considerable scepticism; yet, as subsequent research on the last few millennia has shown (see below), a conflict based upon comparative regional studies could also spring from an over-simplified view of glacial eustasy.

The rate of sea-level change was clearly beginning to decline quite sharply about 7 000 years ago. During the preceding 2 millennia the sea had risen over 20 m, but in the next two that figure was to be virtually halved. Thereafter, any fluctuations lay within relatively narrow limits. Dissension early centred on whether the sea has ever risen above its current level within the last 5 000 years. Fairbridge collated observations from many regions that seemed to demonstrate rises to 1 m or more above modern datum around 5 000 and 3 500 years ago. By contrast, painstaking investigations in the Netherlands were believed to show that global sea-level was below

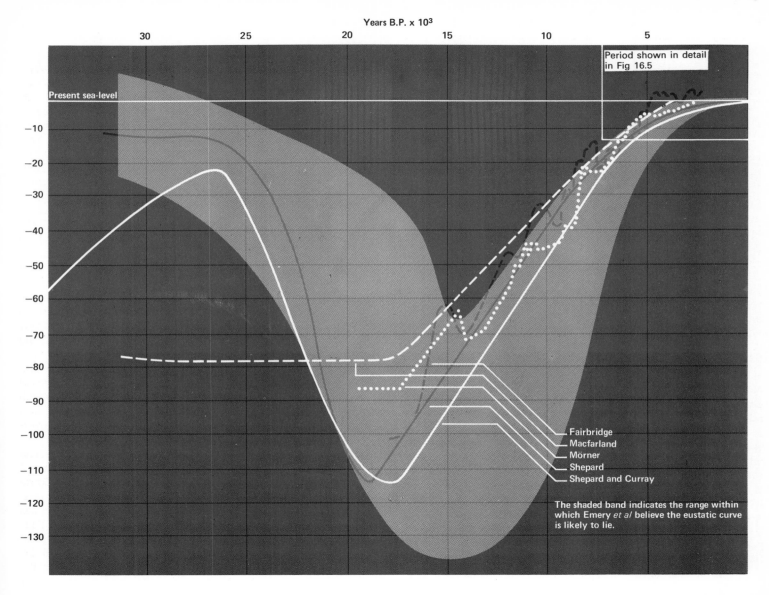

Years B.P. x 10³

Period shown in detail
in Fig 16.5

Present sea-level

Fairbridge
Macfarland
Mörner
Shepard
Shepard and Curray

The shaded band indicates the range within
which Emery *et al* believe the eustatic curve
is likely to lie.

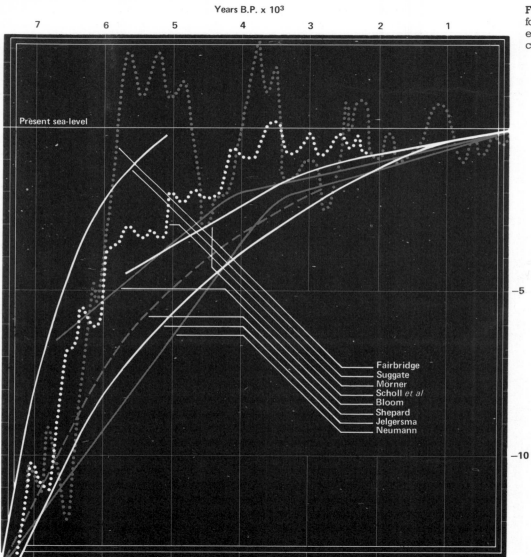

Years B.P. x 10³

Present sea-level

−5

−10

Fairbridge
Suggate
Mörner
Scholl *et al*
Bloom
Shepard
Jelgersma
Neumann

Fig. 16.5. Examples of eustatic curves for the last 7 500 years constructed by eight different authorities (based on a compilation by Mörner, 1971).

its present height 5 000 years ago and has since exhibited a slow but consistent rise. A similar conclusion was drawn from intensive studies on the Everglades coast of Florida where the final transgression was across a broad limestone shelf claimed as the best site in the United States for measuring recent sea-level changes. Here the sea 5 000 years ago was some 4 m lower than at present, and the current level represents the highest point reached at any subsequent date. Again resolution of this disagreement could well lie in simple global eustasy proving to be a conceptual mirage. The case is especially interesting because, in relative terms, so much more data are available. This has encouraged several workers to look for regional rather than global patterns.

One group of Americans, for example, has made an analysis of 768 radio-carbon dates and shown that different patterns characterize different latitudinal groupings. Between latitudes 30 and 45 °N the curve of sea-level change resembles that of Bloom in Fig. 16.5, but nearer the equator the curve corresponds more closely to that of Fairbridge. Clark, meanwhile, has endeavoured to construct a numerical model of likely global sea-level changes after an episode of deglaciation. The model attempts to take into account deformation of the ocean floor and distortion of the ocean surface that will result from the redistribution of ice and water on the earth's surface and the associated migration of rock material in the earth's interior. It predicts that at least six different zones of sea-level changes may be recognizable, and that in four of these a phase of higher Holocene sea-level will be present even though the ocean water volume is assumed to have been constant for the last 5 000 years. Most coastlines of large land masses well away from the limits of glaciation will show such an emergence because the return of water from the ice-sheets will cause a delayed flow of the asthenosphere from beneath the ocean basins towards the continents, thereby causing the latter to rise. Oceanic islands may be expected to behave differently, since they will move in sympathy with the changing elevation of the sea floor to which they are attached.

It can be seen that in many cases of inter-regional sea-level comparisons, a false antithesis was first drawn because an unrealistic global uniformity was anticipated. It now seems that regional diversity should be regarded as the norm, and more data should be collected to establish the details and causes of this spatial variability. It need hardly be added that such work must be undertaken with great thoroughness since the changes of sea-level are often so small that their effects could conceivably be simulated by other processes, such as the growth of protective coastal barriers. However, within an individual region a broad parallelism of sea-level trends should still be discernible, even if extrapolation to a global scale now seems inappropriate. That there is still scope for legitimate conflict is illustrated by work within southern Britain where some workers still argue for a consistent rise in sea-level and others for transgressive and regressive phases.

Sea-level changes in areas of glacio-isostatic uplift

General principles

It is over a century since the likely depression of the earth's crust by continental ice-sheets was first mooted. Support for this idea came from the association of recent raised beaches with severely glaciated terrain in Canada and Scandinavia. Not only did the preservation of delicate marine forms testify to the post-glacial age of the beaches, but the limits of the areas over which they could be found also ran roughly parallel to the limits of glaciation. It was soon recognized that the beaches are tilted and that lines drawn through places of equal elevation on the same strandline tend to encircle the areas of major ice accumulation. Gravity measurements disclosed negative anomalies over both Scandinavia and North America, pointing to a disequilibrium such as might be expected from the sudden removal of a thick ice load. These and other observations left little doubt about the reality of glacio-isostasy. It was concluded that the growth of any large ice-sheet leads to a basin-like depression of the crust, in general form resembling that beneath Antarctica today. As an ice-sheet grows in size the downwarping must increase in amplitude at the centre and slowly extend outwards. With melting the sequence is reversed. As the ice-sheet decays the depression becomes shallower and less extensive. However, there is a delay between the melting of the ice and the recovery of the land surface. It is this relaxation time that explains the current gravity anomalies over both Scandinavia and north-eastern Canada.

In many instances depression of the land surface was sufficient to permit the sea to flood across the terrain vacated by the ice. This

Fig. 16.6. Flooding of the North Sea basin in the millennium between 9 300 and 8 300 years ago (after Jelgersma, 1961).

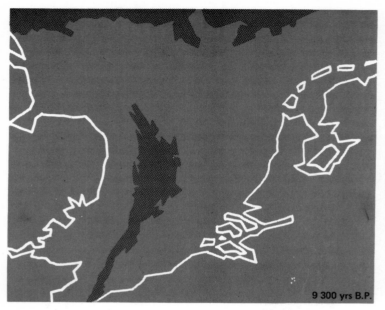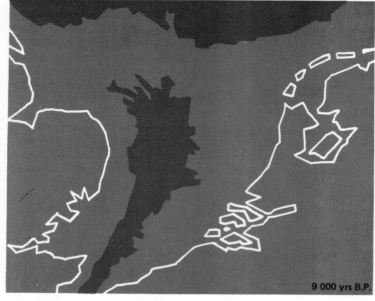

9 300 yrs B.P.

9 000 yrs B.P.

occurred despite the contemporaneous lower sea-level, and the invading water fashioned shoreline features destined to be warped during later uplift. So long as the rate of uplift exceeded that of sea-level rise, successively lower strandlines could be formed. If the rates were temporarily reversed a transgression took place. By careful dating and mapping of strandlines it should be possible to reconstruct in some detail the history of isostatic recovery.

Of the many diagrammatic techniques that have been employed for recording the history of uplift three will be quoted to illustrate the range of possibilities. The first involves the plotting of lines known as 'isobases' (see, e.g. Fig. 16.8). The term isobase has been used in slightly different ways by different authors. In its simplest form it refers to an isopleth depicting the varying elevation of an individual strandline above sea-level. By extension it has also been used for an isopleth that indicates the amount of isostatic uplift suffered by an area within a specified time interval; the calculations then require that an

appropriate allowance be made for eustatic influences in the sense of oceanwide average movements of mean sea-level. The second common diagram is the profile drawn parallel to the presumed direction of tilt (see, e.g. Fig. 16.7). Although very convenient for comparing the heights and gradients of a group of beaches, this method does not illustrate directly the amount of uplift that has occurred since it takes no account of eustatic fluctuations. Great care must also be taken in selecting the orientation of the profile. To show true gradients it must be aligned in the direction of maximum tilt; should the direction of tilt have varied from one beach to another, its value for comparative purposes would be severely curtailed. A third method of depicting sea-level changes within a small area is to plot the altitudes of local beaches against their ages. An 'emergence curve' may then be drawn by linking the plotted points; in effect this portrays the variations that would have been noted by an observer situated at the locality during the post-glacial period. It can be

Fig. 16.6 (continued)

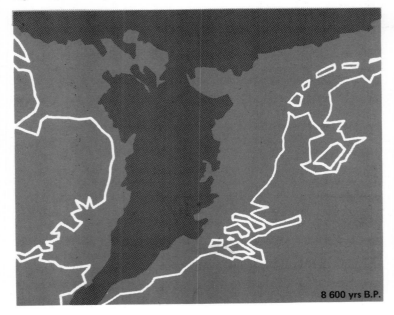

8 600 yrs B.P.

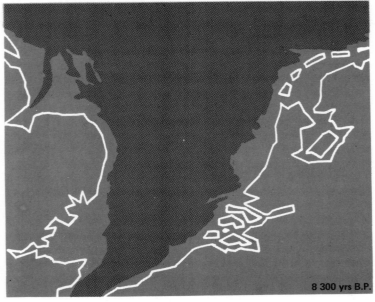

8 300 yrs B.P.

converted into an 'uplift curve' by making the necessary corrections for eustatic changes. From such corrected curves for a series of different localities it is possible to compare rates and times of maximum uplift.

The last-mentioned procedure has demonstrated one very important aspect of isostatic movement; the uplift rate is normally at a maximum as a site is deglaciated and thereafter declines almost exponentially. At present any uplift that occurs prior to ice withdrawal cannot be reliably measured, but it seems almost certain that a significant proportion of the total recovery precedes deglaciation. For example, if the Scandinavian ice-sheet were originally 3 500 m thick, as seems likely, the aggregate depression of the crust could well have approached 1 000 m. Yet the maximum uplift registered by marine strandlines is some 520 m, and even when allowance is made for perhaps 100–150 m of uplift still to come, a large component of the overall recovery remains unrecorded. In Canada,

Andrews inferred that the complete uplift curve is likely to be S-shaped, with maximum velocity at the moment of deglaciation; many of his calculations suggested that here about half the total recovery preceded deglaciation.

One final aspect of glacio-isostasy that is still far from fully understood is the development of a peripheral forebulge around an ice-sheet. Most workers have assumed that viscous flow in the asthenosphere away from the downwarped basin must generate an annular zone of uplift, but the shape and dimensions of this feature have yet to be defined. In 1980 Mörner estimated that from beneath the Scandinavian ice-sheet there was transfer of a mass volume of about $0.7 \times 10^6 \ km^3$, and he believed this generated a forebulge up to 170 m high sited between 1 000 and 1 500 km from the centre of ice dispersal. Other workers have argued that the volume of material may be distributed over a wider area and therefore produce lower elevations. The problem with all these proposals is that, while sound in principle, they have proved very difficult to test by field investigation; forebulge collapse during deglaciation has yet to be adequately separated from other factors influencing post-glacial sea-level changes.

Following this review of some of the basic concepts of glacio-isostasy, attention may now be directed to two specific areas of rather different dimensions that have undergone recent glacial unloading.

Scotland

Although the idea of glacio-isostasy was enthusiastically endorsed by many early workers in Scotland, it is ironic that a rigid mode of thought evolved that was reluctant to acknowledge the tilt affecting Scottish raised beaches. In part this was due to an unfortunate terminology, apparently developing almost by accident, in which the beaches were classified by their approximate height above sea-level. It was only in the 1960s, largely at the instigation of Sissons, that serious attempts were made to set them in a more rational framework. The recent age of the beaches can only be reconciled with eustatic changes if they are tilted. It follows that it is totally inappropriate to designate them by their elevation. This led to abandonment of the classic terminology of '100-foot', '50-foot' and '25-foot' raised beaches, and adoption instead of a simple division, based upon ages, into Late-Glacial and Post-Glacial beaches.

Early investigators in western Scotland had noted that certain high shorelines, clearly visible at the seaward end of some of the glens, terminated abruptly when traced inland. As the points where the shorelines disappeared were often marked by moraines or thick glacio-fluvial accumulations, it was inferred that the heads of the glens were still occupied by ice. A similar conclusion was drawn by Sissons and his collaborators from their study of the Forth valley in south-eastern Scotland. They distinguished ten separate high-level beaches, each rising gently west-north-westwards and terminating at its highest point in a mass of outwash believed to mark the position of the contemporaneous ice-front (Fig. 16.7). This whole assemblage of Late-Glacial forms was interpreted as recording the spasmodic recession of an ice-sheet, with the area vacated by the ice immediately submerged by ingress of the sea. The earlier beaches slope more steeply than the later, confirming that uplift and ice retreat were taking place simultaneously; the oldest of the ten beaches

has a gradient of 1.63 m km^{-1}, the most recent a gradient of 0.55 m km^{-1}. In order to measure the total isostatic recovery represented by these shorelines it is obviously necessary to know their ages. This remains uncertain, but the most recent probably dates from about 13 000 years ago. At this time global sea-level was still 60 m below its present height. If this figure is added to the 37.5 m by which the shoreline rises above current sea-level near Stirling, isostatic recovery at that point approaches 100 m in the last 13 000 years.

Around 13 000 years ago the land must have been rising very rapidly. This can be deduced from the fact that although sea-level was rising at a mean rate of about 1 m per century it was still being outstripped by isostatic recovery. After the last of the high-level beaches had been formed in the Forth valley, relative sea-level swiftly fell to the elevation of thé Main Late-Glacial shoreline. The next event of note was a minor rise and accumulation of the features which Sissons termed buried raised beaches. These are a group of three beaches that span the boundary between Late-Glacial and Post-Glacial times. The term 'buried' is used because they were all submerged during a later transgression in which a cover of fine estuarine sediments was laid down. The highest is almost contemporaneous with the 10 300-year-old Menteith moraine. The intermediate or main beach, dated to about 9 500 years ago, is approximately 1 m lower and can be traced from the Menteith moraine downstream to beyond Stirling. Sea-level near Stirling was then 8 m above its present height, and since the contemporaneous eustatic level was around −30 m the area must have been elevated isostatically some 38 m during the last 9 500 years. When this is compared with the 100 m of uplift since the higher suite of beaches was formed, it is evident that in the preceding 3 500 years isostatic displacement totalled 62 m at a mean rate of almost 1.8 m per century. The lowest buried beach indicates a relative fall in sea-level of about 2 m and is dated to some 8 800 years ago.

The next formative episode in south-eastern Scotland was a major transgression. For a brief interlude during the Post-Glacial period the continuing eustatic rise outstripped the decelerating uplift of the land. The sea flooded across areas that had already become dry land. Without the plentiful coarse debris previously supplied by nearby glaciers, most of the accumulating sediment was fine estuarine silt, known locally as carse clay. The sea reached its culminating height

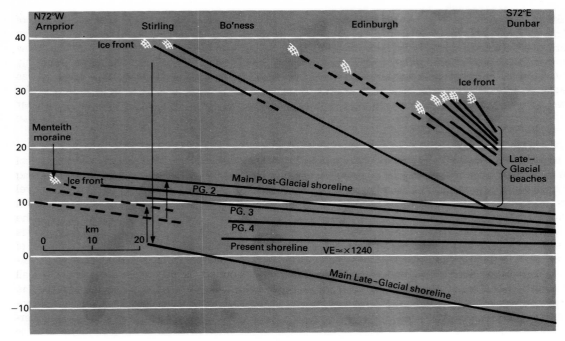

Fig. 16.7. The raised beaches of the Firth of Forth (after Sissons *et al*, 1966). The beaches fall into four major categories: (a) the elevated Late-Glacial beaches formed during wastage of the main Devensian ice-sheet; (b) the Main Late-Glacial shoreline formed after rapid isostatic uplift; (c) the slightly younger buried raised beaches (shown by dashed lines), the earliest of which corresponds to the Zone III Loch Lomond Advance; (d) the Post-Glacial beaches which exhibit the lowest angle of tilt of the whole sequence. Note the immense vertical exaggeration required to portray the slope of the beaches satisfactorily.

5 500 years ago. The main Post-Glacial shoreline lies at an elevation of 16 m near the Menteith moraine, but descends eastwards to under 6 m near Dunbar. There has clearly been both overall uplift and differential tilting during the last 5 millennia. Relative sea-level has fallen spasmodically so that at least three further shorelines have been fashioned, each now sloping more gently than its predecessor.

Attention has deliberately been focused on the pioneering work of Sissons and his associates in south-eastern Scotland, since it laid the foundations for most of the later research on raised beaches elsewhere in Scotland. With appropriate modification of detail, the basic sequence identified in the Forth valley has also been found applicable to coasts around the Scottish Highlands. These further studies permitted definition of an area of domed uplift centred a little to the north of Loch Lomond. Nevertheless, in a review published in the early 1980s, Jardine still concluded that there is a real need for the collection, from a large number of sites in Scotland, of new data by which former sea-level elevations may be determined and sea-level curves constructed.

Arctic Canada

Uplift associated with decay of the Laurentide ice-sheet was first studied by means of the deformed strandlines that encircle the Great Lakes. However, since about 1950 increasing attention has been devoted to the coasts of north-eastern Canada and Hudson Bay where analysis of marine shorelines has helped clarify the later history of isostatic recoil. The earliest major event was a transgression along the St Lawrence valley known as the Champlain Sea. This is dated to about 12 000 years ago when global sea-level was some 50 m lower than at present. The Champlain strandlines locally attain heights of over 200 m, giving a total uplift since deglaciation of more than 250 m. Melting back very rapidly across southern Canada the ice-sheet finally split into a number of separate remnants around 8 000 years ago. A prime factor in this fragmentation was penetration by sea-water into the Hudson Bay depression. For a long time the low eustatic sea-level and the barrier of high ground along the Atlantic seaboard had prevented such penetration. However, by about 10 000 years ago the rapidly rising sea-level promoted vigorous calving in the vicinity of the Hudson Strait, eventually permitting the water to leak into the Hudson Bay area and lift the thinning residual ice into a floating shelf. Continuing glacial recession created a large marine embayment known as the Tyrrell Sea by about 8 000 years ago, and this finally divided the former ice-sheet into three separate masses, each sited on higher ground beyond the reach of glacier calving. The mass of Baffin Island has persisted to the present day, but those on the Ungava peninsula and in the Keewatin district both disappeared completely between 6 000 and 5 000 years ago.

Early studies of glacio-isostasy in arctic Canada tended to concentrate on the marine limit, that is, the maximum elevation attained by the sea in a particular locality since deglaciation. At first few attempts were made to map individual strandlines as was done in south-eastern Scotland, and under the leadership of Andrews a rather different approach was adopted. Andrews constructed twenty-one curves depicting isostatic uplift at sites where dates were available not only for the marine limit but also for several lower strandlines. Identifying a common mathematical shape to all the curves, he inferred that they approximate to an exponential decay form and indicate a uniform response by the earth's crust to deglaciation. As a consequence, realistic uplift curves could be constructed for a large number of sites where relevant information was restricted to the age and elevation of the marine limit, and Andrews used this technique to construct isobase maps for the whole of arctic Canada (Fig. 16.8). On certain further assumptions, he was also able to assess the amount of isostatic recovery that is still to come, concluding that along the eastern shores of Hudson Bay further uplift is likely to surpass 140 m.

As Andrews emphasized in his pioneering investigation, his predictions needed checking by additional field studies, and since 1970 much further research has been undertaken. This has not radically altered the preliminary findings, but has contributed a lot of extra detail. For example, a striking sequence of raised beaches along the

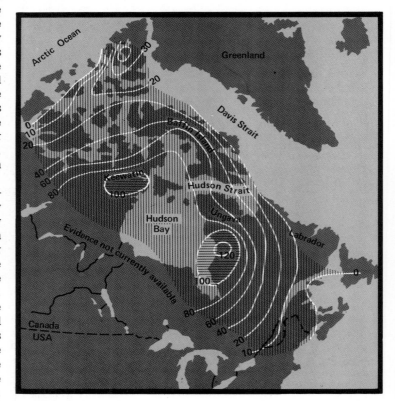

Fig. 16.8. Isosbases depicting emergence (i.e. discounting eustatic sea-level changes) in north-eastern Canada during the last 6 000 years (after Andrews, 1970).

eastern coast of Hudson Bay was analysed by Hillaire-Marcel in 1980. He here recognized no less than 185 separate strandlines formed in the period since 8 300 years ago. The highest strandline lies at an elevation of 270 m, and according to Hillaire-Marcel the region displays one of the fastest known rates of Holocene isostatic uplift; with a displacement of at least 300 m in 8 000 years, the rate of elevation averaged 3.75 m per century.

References

Andrews, J. T. (1970) *Post-glacial Uplift in Arctic Canada*, Inst. Brit. Geogr. Spec. Pub. 2.

Andrews, J. T. and Peltier, W. R. (1976) 'Collapse of the Hudson Bay ice center and glacioisostatic rebound', *Geology* 4, 73–5.

Bloom, A. L. (1971) 'Glacial eustatic and isostatic controls of sea level since the last glaciation', in *Late-Cenozoic Glacial Ages* (ed. K. K. Turekian), Yale.

Chappell, J. (1983) 'Aspects of sea levels, tectonics, and isostasy since the Cretaceous', in *Mega-geomorphology* (ed. R. Gardner and H. Scoging), OUP.

Clark, J. A. (1980) 'A numerical model of worldwide sea level changes on a viscoelastic earth', in *Earth Rheology, Isostasy and Eustasy* (ed. N.-A. Mörner), Wiley.

Cronin, T. M. et al. (1981) 'Quaternary climates and sea-levels of the U.S. Atlantic coastal plain', *Science N.Y.* 211, 233–40.

Emery, K. O. et al. (1971) 'Post-Pleistocene levels of the East China Sea', in *Late-Cenozoic Glacial Ages* (ed. K. K. Turekian), Yale.

Fairbridge, R. W. (1961) 'Eustatic changes in sea level', in *Physics and Chemistry of the Earth*, Vol. 4, 99–185.

Flint, R. F. (1971) *Glacial and Quaternary Geology*, Wiley,

Harmon, R. S. et al. (1983) 'U-Series and amino-acid racemization geochronology of Bermuda: implication for eustatic sea-level fluctuation over the past 250,000 years', *Paleogeogr., Paleoclim., Paleoecol.* 44, 41–70.

Hillaire-Marcel, C. (1980) 'Multiple component postglacial emergence, eastern Hudson Bay, Canada', in *Earth Rheology, Isostasy and Eustasy* (ed. N.-A. Möner), Wiley.

Hillaire-Marcel, C. and Fairbridge, R. W. (1978) 'Isostasy and eustasy of Hudson Bay', *Geology* 6, 117–22.

Jardine, W. G. (1982) 'Sea-level changes in Scotland during the last 18,000 years', *Proc. Geol. Assoc,.* 93, 25–41.

Jelgersma, S. (1961) 'Holocene sea-level changes in the Netherlands', *Med. Geol. Sticht.* Ser. C, 6.

Moore, W. S. (1982) 'Late Pleistocene sea level history', in *Uranium-series Disequilibrium: applications to environmental problems* (ed. M. Ivanovich and R. S. Harmon), OUP.

Mörner, N.-A. (1969) 'Eustatic and climatic changes during the last 15,000 years', *Geol. en Mijn.* 48, 389–99.

Mörner, N.-A. (1971) 'The Holocene eustatic sea-level problem', *Geol. en Mijn.*, 50, 699–702.

Mörner N.-A. (1976) 'Eustasy and geoid changes', *J. Geol.* 84, 123–51.

Mörner, N.-A. (1980) 'The Fennoscandian uplift: geological data and their geodynamical implication', in *Earth Rheology, Isostasy and Eustasy* (ed. N.-A. Mörner), Wiley.

Mörner, N.-A. (1983) 'Sea levels', in *Mega-geomorphology* (ed. R. Gardner and H. Scoging), OUP.

Nasu, N. et al. (1983) 'Remnants of an ancient forest on the continental shelf of northwest Japan', *Boreas* 12, 13–16.

Newman, W. S. et al. (1980) 'Eustasy and deformation of the geoid: 1000–6000 radiocarbon years B. P.', in *Earth Rheology, Isostasy and Eustasy* (ed. N.-A. Mörner), Wiley.

Nunn, P. D. (1984) 'Occurrence and ages of low-level platforms and associated deposits on South Atlantic coasts: appraisal of evidence for regional Holocene high sea level', *Prog. Phys. Geogr.* 8, 32–60.

Sissons, J B. (1976) *Scotland*, Methuen.

Sissons J. B. et al. (1966) 'Late-glacial and postglacial shorelines in southeast Scotland', *Trans. Inst. Brit. Geogr.* 39, 9–18.

Thom, B. G. (1973) 'The dilemma of high interstadial sea levels during the last glaciation', *Prog. Geogr.* 5, 167–246.

Tooley, M. J. (1978) *Sea-level Changes*, OUP.

Tooley, M. J. (1982) 'Sea-level changes in northern England', *Proc. Geol. Assoc.* 93, 43–51.

Ward, W. T. et al. (1971) 'Interglacial high sea levels – an absolute chronology derived from shoreline elevations', *Paleogeogr., Paleoclim., Paleoecol.* 9, 77–99.

Selected bibliography

Because of the ambiguous nature of so much of the evidence, there is no single authoritative source on Pleistocene sea-level changes. In addition to the works listed among the references, useful reviews can be found in C. Kidson, 'Sea level changes in the Holocene', *Quat. Sci. Rev.* 1 (1982), 121–51, and K. W. Butzer, 'Global sea level stratigraphy: an appraisal', *Quat. Sci. Rev.* 2 (1983), 1–15.

Collections of papers dealing with glacioisostatic recoil are offered by N.-A. Mörner (ed.), *Earth Rheology, Isostasy and Eustasy*, Wiley, 1980, and D. E. Smith and A. G. Dawson (ed.), *Shorelines and Isostasy*, Inst. Brit. Geogr. Spec. Pub. 16, Academic Press, 1983. Sea-level changes around the British Isles during the last 15 000 years form the subject of a special issue of the *Proceedings of the Geologists' Association* 93, pt 1 (1982), 3–125.

Chapter 17
The evolution of coastal landforms

The sea-level changes discussed in the preceding chapter have two important implications for the study of coastal landforms. In the first place, on a stable coast the sea only reached its present level about 5 000 years ago, so that neither erosional nor depositional features can properly be attributed to marine processes acting unchanged over a protracted period. Secondly, all models of eustatic change require that the sea has stood at its current level at a number of periods in the past. This is clearly the case if there has been no secular fall in sea-level since the water surface should return to approximately the same position during each interglacial. If there has been a long-term decline, each glacio-eustatic oscillation should have carried sea-level below its present height once the interglacial geoid had fallen to about 100 m. One potentially important distinction between these two models is that, without secular decline, sea-level should have stabilized close to its present elevation for the duration of each interglacial, whereas otherwise it should have coincided with it only on the rising and falling leg of each glacio-eustatic fluctuation, and there then seems no intrinsic reason why in aggregate the sea should have stood longer at that height than at many others.

The notion of sea-level change has formed the basis for many attempts at coastal classification. Since these often incorporate important ideas about coastal evolution, they will be briefly reviewed before attention is turned to more specific aspects of landform history.

Schemes of coastal classification

For many years ideas of coastal classification were dominated by the work of the American geomorphologist, Johnson. In his classic work, *Shore Processes and Shoreline Development*, published in 1919, he placed immense stress on the dual concepts of submergence and emergence, although admitting the need to recognize two further classes which he termed neutral and compound coasts. Neutral coasts according to Johnson are those where the form is due, not to sea-level change, but to essentially non-marine processes such as faulting and vulcanicity. Compound coasts are those which involve some combination of the other three. In many ways the most unfortunate aspect of Johnson's classification proved to be its emphasis on sea-level change. He proposed two major models of coastal evolution, one to represent the sequence of changes after a rise in sea-level, the other after a fall. The former involves the drowning of earlier river valleys, the concentration of wave attack upon headlands and the filling of initial inlets with bay-head beaches. Gradually the promontories are cut back as a series of cliffs until the plan of the coast is eventually converted from crenulate to straight. Emergence, on the other hand, was held to lead typically to the uncovering of a low coastal plain and creation of an offshore bar. The resultant tidal lagoon is slowly silted up with both marine and fluvial sediments. Concurrent wave attack pushes the bar landwards until it ultimately reaches the original coastline. Any further erosion induces cliffing, but this only happens after a very protracted period of stable sea-level.

There is no doubt that Johnson intended these models of coastal development as illustrative of two commonly occurring situations. He also realized that submergence can sometimes produce features resembling those described as typical of emergence. He wrote: 'A shoreline of submergence (emergence) is one in which the dominant, not necessarily the latest, features reflect submergence (emergence) – or in which the features resemble those which would be produced by submergence (emergence).' Although this sentence might be construed as implying a morphological rather than a genetic classification, the very terms emergence and submergence clearly refer to coastal evolution, and many of Johnson's disciples treated them as indicating specific sea-level changes. The results were often totally paradoxical. The Atlantic seaboard of the United States was quoted as the prototype of an emerged coast, whereas all the available evidence identifies it as a classic example of a gently shelving area drowned by the post-glacial marine transgression. Fjords are conventionally described as features of submergence, but often occur on coasts that have experienced substantial isostatic rebound and a net fall in sea-level since deglaciation.

It is clear that if emergence and submergence are to be retained as classificatory criteria, the time interval under consideration must be more rigorously defined. In the last 10 000 years nearly all coasts have experienced a net inundation, the only common exceptions being those where isostatic uplift has outstripped the Holocene rise in sea-level; in the longer term, on the other hand, many coasts must have experienced relative uplift judging by the widespread occurrence of raised shorelines. In this limited respect the Atlantic seaboard of the United States could still be regarded as an emerged coast, although the value of such long-term changes as a basis for coastal classification seems dubious.

Although the ideas of Johnson long exercised a profound influence, many alternative classificatory schemes have been suggested. Shepard proposed a division into coasts fashioned mainly by terrestrial agencies and those shaped mainly by marine processes. The former are subdivided into land erosion coasts, subaerial deposition coasts, volcanic coasts and coasts shaped by tectonic movements; land erosion coasts include both rias and fjords, while subaerial deposition equally embraces both fluvial and glacial activity. Coasts shaped by marine processes are divided into three major classes, those due to erosion, deposition and the activity of organisms. Shepard's classification has much to commend it; its major weakness in application is the subjectivity in deciding when a coast has been sufficiently modified to warrant a designation as shaped mainly by marine agencies. A second interesting variation on the traditional approach to coastal classification was proposed by Valentin. He suggested that the primary division should be between coasts that are advancing and those that are retreating. The changes on the former may be due to either emergence or deposition, on the latter to either submergence or erosion. Various combinations can be readily visualized, and in certain instances a balance may be struck between the forces leading to advance and retreat so that the position of the coastline remains stable.

Inman and Nordstrom have proposed a classificatory scheme based upon the ideas of plate tectonics (Fig. 17.1). They recognize three primary classes: collision coasts at convergent plate margins, trailing-edge coasts where continental blocks are moving away from spreading centres, and marginal-sea coasts. Trailing-edge coasts are divided into three types depending on whether continental separation is geologically recent (neo-type), both continental margins are trailing-edge (afro-type), or one margin is trailing edge and the other collision (amero-type).

Finally, mention should be made of the approach to coastal classification advocated by Davies. Owing to the significance of wave action in shaping coastal forms he has maintained that it is vital to distinguish contrasting wave environments. Four major types are recognized (Fig. 17.1). The first is the storm-wave environment in which a significant proportion of the waves are generated by local gale-force winds. These winds are most common in the temperate storm belts so that such an environment occurs with greatest frequency in middle-to-high latitudes where waves of high energy produce prominent cliffs and shore platforms. Constructional features are also abundant and often composed of shingle. The second environment is that of west-coast swell. This occurs primarily within the tropics where gale-force winds are rare and most of the large waves are produced by swell originating from temperate storms. The mean wave energy is moderately high near the tropics but decreases towards the equator. Cliffing is less active and constructional forms more commonly consist of sand than of shingle. Complexity is introduced by some coasts being subject to tropical cyclones and others experiencing monsoon-generated waves. The third wave environment is that of east-coast swell. Owing to the strong westerly component in most temperate storms the swell reaching east coasts is on average lower than that on west coasts. Mean energy levels may be characterized as low to moderate, although tropical cyclones afflicting certain of these coasts can have an effect out of all proportion to their frequency and duration. The final wave environment is that of the protected sea. Here either coastal configuration prevents the full penetration of oceanic swell, or an ice cover damps the energy of waves approaching the coast. Typical features of such a low-energy environment are intricate constructional forms related to the complex patterns of fetch and wind. It is worth noting that while wave action may be damped by an ice cover, shoreline frost weathering way be enhanced so that cliff recession is less inhibited than might otherwise be the case.

No single classificatory scheme commands the attention once given to that of Johnson, so much depending upon the purpose to which the classification is to be put. Morphology, evolution and contemporary processes have all been suggested as primary criteria, sometimes with unfounded assumptions about the relationships

Fig. 17.1. Two different methods of classifying the coasts of the world; (A) according to tectonic type; and (B) according to the wave environment.

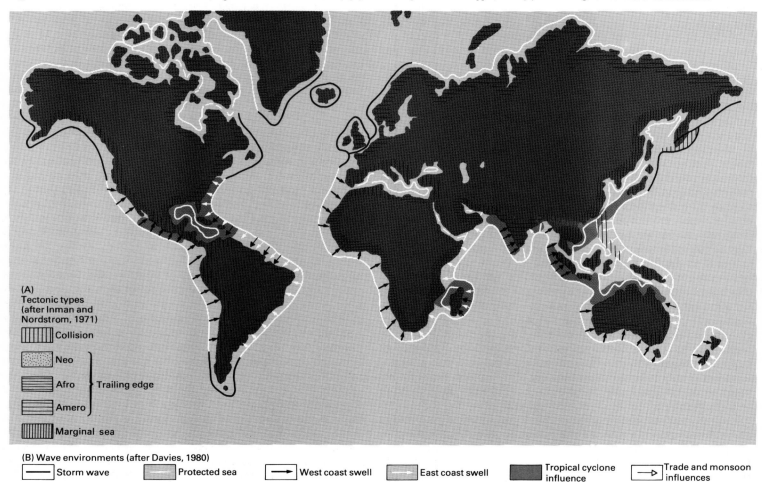

(A)
Tectonic types
(after Inman and
Nordstrom, 1971)

|||||| Collision

Neo

Afro   } Trailing edge

Amero

|||||| Marginal sea

(B) Wave environments (after Davies, 1980)

Storm wave   Protected sea   West coast swell   East coast swell   Tropical cyclone influence   Trade and monsoon influences between them. This led Russell, one of the foremost students of coastal landforms, to argue that attempts at coastal classification are premature until we possess a much better factual knowledge.

Forms due to marine erosion

Cliffs

Although coastal cliffs can be among the most spectacular of all landforms, their origin and development has attracted surprisingly little detailed investigation. On all active cliffs two groups of processes operate simultaneously. At the base the sea performs the dual functions of erosion and transportation. Higher on the cliff face a variety of subaerial processes induces collapse and supply of debris to the waves below. The nature and balance of these activities together determine the form of the cliff. In many respects the evolution of marine cliffs can be regarded as a particular case of slope development. The full range of subaerial processes as outlined in Chapter 8 can be observed to operate on different sections of coast. The main distinguishing feature is the constant backwearing at the base. As long as sea-level remains stable this can only act horizontally, and it is easy to visualize a simple equilibrium form being conserved during continuous recession. A slope is first steepened by wave attack until the subaerial processes are able to keep pace with the basal retreat. On some materials this situation may be reached when the cliff angle is only 20° or so, but on massive bedrock the profile may be steepened to the vertical before the whole face begins to retreat uniformly. A particular form can only be maintained so long as the balance between marine and subaerial processes remains unchanged; in theory this is unlikely to persist over a protracted period since wave action will gradually be modified by the increasing width of the shore platform.

One way of classifying cliffs is by reference to the dominant subaerial processes acting on them. In weakly coherent materials movement may take place by means of flowage, gullying, or sliding. Often all three processes can be observed in operation within a short distance of each other. Flowage is particularly characteristic of cliffs formed in clay where interstitial water substantially reduces the mechanical strength of the material; the toe of each earthflow may be regularly trimmed back by wave action at high spring tides. On

certain cliffs weakly coherent deposits are rapidly degraded by surface runoff. This is especially true of poorly cemented sands and silts which are sometimes deeply gullied to form veritable badland areas. Detrital fans built out on the beach after heavy rain are obliterated when waves next wash the cliff base. It is, however, sliding that constitutes the dominant mechanism on many coastal cliffs, particularly those incorporating thick clay sequences. Shallow slab slides are extremely common along cliffs cut in till, while deep-seated arcuate slides tend to develop most fully on well-bedded sedimentary rocks. Especially favourable circumstances occur where massive strata overlie an impermeable clay. By attacking the clay the waves steepen the cliff until the basal material is unable to sustain the weight of overlying rock. The strength of the clay may be reduced by percolating water after heavy rain, while a more gradual reduction often arises from long-term chemical alterations. The precise shape of the slip plane is much influenced by structural details, but in many instances rotational movement can be demonstrated with back-tilting of the uppermost strata and elevation at the toe of the slide. The latter occasionally produces a prominent ridge on the foreshore which is subject to later destruction by wave action. However, the main slumped mass itself is rarely removed in total before the next segment of cliff is dislodged; the result is an under-cliff of highly irregular relief and jumbled structure. Amid the displaced slices of rock, flows of saturated clay often help to carry part of the debris further seawards.

Well-known examples of rotational slipping occur at several localities on the coast of the English Channel. At the Warren, east of Folkestone, Chalk and Upper Greensand rest upon impermeable Gault clay. Water percolating through the Chalk is heavily charged with calcium carbonate, and on reaching the glauconitic Gault clay the calcium ions replace the original potassium ions. This reduces the strength of the clay and is a contributory factor in the frequent slides that characterize this section of coast. Ten major slips were recorded in the period between 1765 and 1915, since when there has been rather greater stability. Further west near Lyme Regis the cliffs also consist of Chalk, Upper Greensand and a rather sandy facies of the Gault clay, the whole succession here resting unconformably on Liassic clays, marls and limestones. Deep-seated rotational movement is very common. The largest recorded displacement involves a mass known as Goat Island, some 6 hectares in extent and over 8 million tonnes in weight. The movement took place in 1839, leaving a chasm at the foot of the main cliff 60 m deep and 100 m wide. In this particular instance there remains some uncertainty about the precise form of the slip plane, but many smaller slides undoubtedly occur along arcuate surfaces. An interesting aspect of many of these is their accompaniment by earthflows indicative of the clay being temporarily converted to a liquid state in which it undergoes relatively rapid flowage. A few kilometres east of Lyme Regis, Brunsden in the 1970s monitored the movement of a large multiple slide and showed that on certain arcuate shear-planes sporadic displacements may total as much as 80 m in a year.

On less deformable materials the primary mechanism of cliff recession is rockfall. This is the process associated with the traditional concept of marine cliffs as high rock faces, often approaching the vertical and sometimes overhanging due to the undercutting action of the waves. Structure and lithology are the dominant controls of form. Where massive, horizontally bedded strata are involved, collapse of huge vertical slabs and pinnacles is probably the most significant means of recession. Isolated by erosion along prominent joints, they are eventually rendered unstable by continual undercutting at the base. Sea-stacks are a characteristic feature of cliffs retreating in this way. Where similar massive strata are tilted, the cliff profile depends on the disposition of the planes of potential movement (see Fig. 8.3, p. 141). Where the joints are more closely spaced, numerous minor collapses tend to bring variously sized blocks crashing down. Many limestone and chalk cliffs recede in this way, an excellent example being afforded by the high chalk cliffs of the Seven Sisters in Sussex.

As on all slopes, it needs to be remembered that many different processes operate simultaneously. Although the fundamental cause of rockfall in marine cliffs is instability due to wave action, the actual triggering will often be a period of exceptionally heavy rain or intense freeze–thaw activity. In cold climates rockfall may be so rapid that it prevents the cliffs attaining the angle of slope they would reach in other climatic zones. Decomposition and disaggregation of rocks has been little studied on the face of marine cliffs, but there seems little doubt that in such an exposed position individual mineral grains are particularly prone to loosening. One possible factor worth recalling is the great volume of salt spray carried high up on the cliff face during a severe storm.

Structure and lithology affect not only the profiles of cliffs but also their development in plan. This can be seen at several different scales. On the regional scale lithology is particularly influential. Where the geological strike runs normal to the trend of the coast, rapid erosion of weaker rocks produces indented bays between promontories of more resistant material; where the strike runs parallel to the coast, alternating weak and resistant outcrops can engender a highly irregular outline. Both situations are beautifully exemplified on the coast of the English Channel in Dorset (Fig. 17.2). At the more detailed scale structure often appears to be the prime influence. The crenulate cliffs of north Cornwall illustrate very well the way in which bedding planes, joints and faults are all subject to preferential erosion (Fig. 17.3). It is not always clear why one set of fractures has been more successfully eroded than another. Other things being equal, fault planes may suffer especially rapid enlargement owing to the shattered nature of the adjacent bedrock. However, it is the orientation of the fractures relative to the direction of wave approach that is probably most significant. On the coast of north Cornwall features aligned parallel to the approach of the largest storm waves are said to be the most rapidly attacked. On the coast of northern Oregon, on the other hand, the fractures suffering most rapid erosion are those facing north-west, despite the approach of the most powerful waves from the south-west; the explanation appears to lie in the prevalence or greater frequency of the marginally smaller storm waves generated by the north-westerlies.

So far it has been assumed that cliff forms are to be explained purely in terms of processes seen to be operating at the present day. Yet this assumption needs to be treated with caution since it is now clear that many cliff lines fashioned from resistant rocks have had a much more complex evolution than was formerly supposed. On a stable coast wave erosion at the cliff base can have been occurring for 5 000 years at most. Prior to that there must have been a minimum of 25 000 years during which the sea was too low to have had any effect. Two specific lines of evidence point to the sea having recently reoccupied cliffed coastlines in Britain that were abandoned earlier in the Pleistocene epoch. Firstly, in a number of localities old cliffs buried by till have been identified. One of the most striking instances occurs along the Welsh coast in Cardigan Bay. Here the apparently modern cliffs pass laterally behind a plug of drift with very little change in form. It is evident that a major function of wave action

Fig. 17.2. The relationship between geology and coastal configuration in south Dorset. The lower diagram depicts the area around Lulworth cove in greater detail.

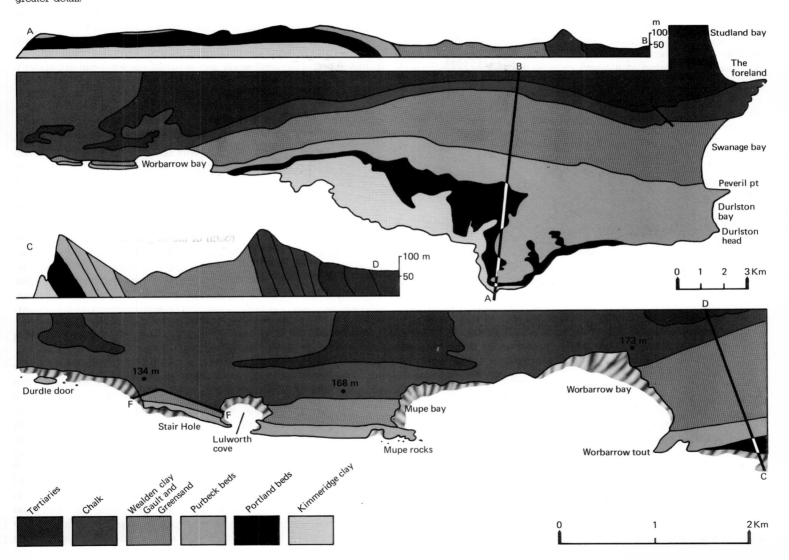

during the last 5 000 years has been removal of a drift cover plastered against the ancient cliff line. The current cycle of erosion may have trimmed and freshened the forms but has certainly not shaped them *ab initio*. Secondly, south of the limit attained by Devensian ice it is not uncommon to find an old beach lying 1–2 m above present sea-level and backed by ancient cliffs beyond the reach of contemporary wave action. In many instances this beach can be shown to predate the last glaciation. In these localities the sea has clearly returned to a position near, but not exactly coincident with, one occupied earlier in the Pleistocene epoch. Where recent marine erosion has destroyed the old beach, it may still be reasonable to infer that the sea is reoccupying an ancient cliff line.

To a certain extent recognition of this two-stage evolution removes a predicament for the geomorphologist, since it is often difficult to conceive that major cliff lines have been produced within the last 5 000 years. Yet it also raises the question of the modification suffered by the abandoned coastline at the height of the last glaciation. In much of south-western Britain, for instance, it appears that periglacial processes rapidly transformed the vertical cliff faces. Freeze–thaw activity reduced the angle of slope and led to accumulation of a wedge of frost-shattered debris on the former beach or shore platform. The results of this interlude are most obvious where the deposit of head is still preserved at the base of the old cliff. However, in many localities the post-glacial marine attack has removed the head and the waves are once more undercutting the bedrock. Nevertheless, the upper part of the cliff retains the sloping form imposed during the periglacial episode, often thinly mantled with frost-shattered debris. The resultant profile is described as 'slope-over-wall', with the height of the wall largely determined by current exposure to storm waves. Although found on many high-latitude coastlines, slope-over-wall profiles are exceptionally well developed in south-western England, possibly because the powerful North Atlantic breakers eroded very high interglacial cliffs which were then subject to unusually intense freeze–thaw activity.

On coasts affected by substantial glacio-isostatic rebound, rather different considerations must apply. The present cycle of marine cliffing can only have commenced after local deglaciation. The period available for periglacial activity is comparatively short. The dominant change in sea-level has generally been an emergence persisting right up to the present day. This means that the idea of

5 000 years of stable sea-level no longer applies. As long as glacio-isostatic recovery remains incomplete, old interglacial cliff lines will be tilted and so cannot be exhumed simultaneously over long distances by current wave action. The evolution of cliffs in such glaciated areas is obviously complex and little detailed work has yet been done. It seems inconceivable that some of the high cliffs in Scotland, for instance, could have been fashioned entirely in the short post-glacial interval, especially when the local sea-level fluctuations are taken into account. Exhumation of buried features seems probable, and this is confirmed by observations in north-eastern Scotland where apparently modern cliffs, complete with sea-stacks, disappear beneath a cover of glacial drift; elsewhere long, structurally guided inlets known as geos contain pockets of till, implying that they too must be older than the last local glaciation.

In the foregoing paragraphs it has been suggested that certain cliff lines appear too large to have developed wholly within the post-glacial period. This implies some appreciation of the rate of cliff evolution, and it is necessary to examine in more detail what is known of the rate of cliff recession. An immediate problem arises from biased sampling, since measurement tends to be confined to coasts undergoing obvious retreat. Even within this category, measured values vary by a factor of 50 or more. The fastest rates are generally found on coasts of weakly coherent material. The Holderness coast is particularly famous in this respect. Many former villages have been lost to inroads of the sea (Fig. 15.8, p. 315) and it has been estimated that since Roman times the coast has retreated some 4 or 5 km. Yet it is only since the advent of accurate surveys that precise measurements have been possible. Valentin examined the cartographic evidence for the period between 1852 and 1952. He estimated the average rate of erosion to be about 1.8 m yr^{-1}, with the figure tending to increase from north to south, and in the vicinity of Easington attaining a local value of 2.75 m yr^{-1}. These mean values hide considerable variations in both space and time. Some sections of the coast remain stable for many decades and are then subject to brief but very severe erosion; around Withernsea, for instance, some 130 m were lost in the relatively brief period between 1852 and 1876. Another area renowned for the rapidity of cliff erosion is the East Anglian coast. At Covehithe, for example, recession exceeded 5 m yr^{-1} between 1925 and 1950, and during the 1953 storm surge, cliffs were driven back a full 12 m in a single day. This pattern was repeated in a storm surge in 1978 and contributed to an annual loss at one site temporarily rising to over 20 m. If continued for several millennia, all the above rates would involve major changes in coastal position. It must be emphasized, therefore, that they are rather exceptional and found only on coasts composed of easily eroded materials. Moreover, they are not sustained and Dunwich, just south of Covehithe, which was famed for erosion in the Middle Ages and where the coast between 1589 and 1753 retreated at an average rate of 1.6 m yr^{-1}, now has cliffs stabilized to the point where they are covered with turf.

On rocky coasts the rates of cliff recession are invariably lower and often so small as to be extremely difficult to measure. Yet some values on bedrock are still far from negligible. For example, cliffs in Miocene sandstones along the Oregon coast are estimated to have been retreating at about 0.6 m yr^{-1} since 1880; cliffs in chalk over the last few decades are believed to have been receding at about 0.3 m yr^{-1} on the Isle of Thanet, 0.5 m yr^{-1} in parts of Sussex and 0.25 m yr^{-1} in northern France. Interesting assessments have been made in North Yorkshire where cliffs composed of till alternate with those composed of bedrock. From cartographic evidence the retreat of the coast since 1892 is estimated to have averaged 0.3 m yr^{-1} on glacial drift and 0.09 m yr^{-1} on the Lias shales. In the 1970s a short segment of the bedrock cliff near Whitby was specially instrumented, and measurements over a period of 12 months demonstrated a recession of the wave-cut notch of up to 0.047 m yr^{-1}. Yet on many coasts of more resistant bedrock, cartographic comparisons fail to reveal any change within the period of a century or more. Ancient structures presumed to have been built near the cliff edge remain untouched. A problem on such coasts is the erratic nature of much cliff destruction. This is well illustrated in North Cornwall. Long sections of that coast show negligible alteration within the last few centuries, but at Tintagel erosion has severely modified the island on which a castle was built about 1145 (Fig. 17.3). Part of the castle originally stood on an isthmus linking the island with the mainland. This collapsed and in the thirteenth century a bridge had to be built in its place. By the sixteenth century the gap had grown too large for bridging and the remains of the castle were isolated from the mainland. Yet the foundations of some of the original castle walls, almost certainly built at the cliff edge, remain intact today. Measurement of cliff recession in such circumstances poses obvious problems, but there seems no doubt that many segments of the coastline are retreating at less than 0.01 m yr^{-1}.

Extrapolating current rates of erosion backwards in time is always perilous, and on some sections of coast Man is now having an obvious effect. In part this is intentional since attempts have often been made to stabilize cliffs undergoing rapid retreat. Yet there are also instances where Man's activities have accidentally exacerbated the erosion. In Devon on the south coast of Britain removal of shingle from the foreshore during the 1890s led to accelerated erosion and destruction of the old fishing village of Hallsands. Groynes intended to diminish longshore drifting have frequently accentuated nearby erosion because a protective beach has been starved of its former shingle supply. Such examples serve as a reminder of the varied factors that can influence cliff recession. It may also be recalled that few workers have regarded sea-level as totally unchanging during the last 5 000 years, and even a slight rise can have a significant effect in maintaining the efficacy of wave erosion. Yet, despite these uncertainties, it is still reasonable to conclude that on resistant coasts the current cycle is often rejuvenating older cliffs and that the modern forms can only be fully understood in the context of this two-stage evolution.

Shore platforms

Many of the factors discussed in the preceding section with respect to cliffs are also apposite in considering the development of shore platforms. For instance, it is clear that certain shore platforms currently swept by the sea are exhumed features and that the present cycle of wave action has done little more than strip off a superficial cover of till or head. Stripping may occur over a relatively wide altitudinal range since waves are capable of washing away weak material high above the level at which they would normally be effective as abrasive agents. This is one reason shore platforms lack any consistent relationship to such a datum as the high-water mark of ordinary spring tides. However, an even more important reason is the complexity of processes involved in the development of platforms.

The classical approach to shore platforms maintained that they are almost exclusively the product of abrasion, firstly by material carried to and fro across the swash zone and later by debris being transported seawards in the offshore zone. Continued cliff recession

Fig. 17.3. The coastal landforms at Tintagel in north Cornwall. The cliffs commonly exhibit a slope-over-wall form, well exemplified in the headland at the top of the photograph. The effectiveness of wave erosion at the cliff base is much influenced by rock structure, producing an intricate pattern of minor bays and headlands.

requires that a wave-cut bench be progressively lowered so that wave energy is not entirely dissipated in a wide area of very shallow water. In the inter-tidal zone there are often indications of abrasion in the form of potholes, smoothed surfaces and well-rounded pebbles, but below low-tide level the evidence is more ambiguous. Abrasion due to wave agitation was once held to be effective at depths up to 180 m, but modern studies suggest the true figure is much closer to 10 m. Below that depth large storm waves may stir fine sediment into movement but its erosive capacity is negligible. Assuming a gradient for the platform of 1 in 100, the greatest width ascribable to abrasion would be 1 000 m in areas of low tidal range, rising to 1 500 m in areas with a 5 m tidal range. Yet, with a stable sea-level, these values would only be attained after very protracted erosion; only with a steady rise would greater values seem feasible.

Several workers have measured the form of the platforms that front the cliffs of southern and eastern Britain. Those at the foot of actively receding cliffs have attracted particular attention since they are clearly being fashioned at the present day and not simply inherited from older forms. In the late 1960s the altitude of the notch at the base of the chalk cliffs on the Isle of Thanet was surveyed by both Wood and Wright who found it to lie some 3 m higher in the embayments than on the headlands, implying a seaward slope in the order of 1 in 15. Since this is due neither to a slope in the water surface nor to recent tectonic tilting, Wood contended that it must represent deeper excavation by the waves breaking on the headlands. On the other hand, So, who made an independent study of the Isle of Thanet, argued that the platforms tend to rise highest where the exposure is greatest, asserting that there are contrasts between different sides of the same headland. This claim has been contested, but Wood and So agreed that the seaward slope of the platforms is much greater in the bays than on promontories. They both dismissed the explanation that rocks in the bays are less resistant than those of the headlands, and maintained that there must be some process by which the platform in the bays is progressively steepened. Wood suggested this may be preferential scouring as a result of the concentration of

sediment in the bays, a view that was later to find echoes in a detailed study in 1977 of shore platforms carved from shale bedrock in North Yorkshire. By means of surveyed traverses between cliff foot and low-water mark, Robinson here distinguished two morphological components of the platforms which he termed the ramp and the plane. The former characteristically slopes seawards at about 6° but the gradient may lie anywhere between 2.5 and 15.5°; the plane, by contrast, slopes on average at about 1°. Robinson contended that the ramp is the product of abrasion at the foot of the cliff, whereas the dominant process on the plane is inter-tidal wetting and drying which breaks up the surface of the shale into small fragments that are readily removed by the sea. In some localities a ramp may pass seawards into a plane, but elsewhere the whole platform may be composed either of a ramp or a plane; a major determinant of form is the availability of suitable debris for the process of abrasion.

A different approach to the study of British shore platforms was adopted by Wright who endeavoured to compare their forms along the whole southern coast of England (Fig. 17.4). He found that eastwards from Torquay the form of the platforms is relatively simple, with gradients ranging from 1 in 14 to 1 in 90, and widths averaging about 100 m but locally attaining 270 m. The junction between platform and cliff is below the uppermost level of marine action, often corresponding with the mean high-water neap position. Occurrences above this datum can generally be ascribed to a structural control in which the upper surface of the platform coincides with a particularly resistant bed; occurrences below it Wright attributed to the effectiveness of locally abundant abrasive materials. Westwards from Torquay platforms are much more complex in form. Two or even three distinct levels can be detected in a single profile, with the upper one often rising above the mean high-water spring level. There is little doubt that many of these multiple platforms arise from rejuvenation of formerly abandoned shorelines.

A yet further factor of potential significance in examining the shore platforms around the British Isles is the Holocene variation in sea-level. The impact of this is difficult to evaluate since there is still dispute about the nature of the variation, but several workers have invoked sea-level changes to account for certain details of platform morphology. For example, the platforms around the headlands on the Isle of Thanet have a convex-upward profile that has been linked either to sea-level fluctuations or to continuing tectonic depression of

Fig. 17.4. The elevation of shore platforms along the south coast of England in relation to various tide levels (after Wright, 1970).

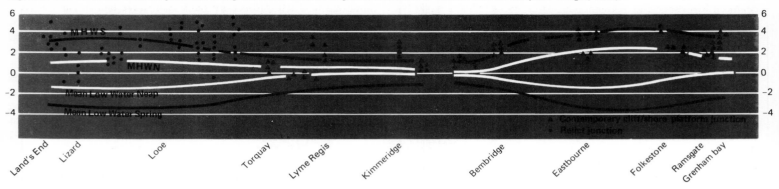

coasts fringing the North Sea basin. In North Yorkshire Robinson explained the grading of the platforms above low-water mark into the submarine floor below that level, without any break of slope, as due to the post-glacial transgression.

The foregoing review of shore platforms in south-eastern Britain serves to emphasize the complex interactions that may be involved in platform development in one small area. Within a global framework the possibility of additional, environmentally controlled processes needs to be acknowledged. Thus on a number of coasts, especially around the Pacific and Indian Oceans, multiple platforms have been described where no question of inheritance from an earlier interglacial period seems to arise. Consisting of nearly horizontal benches separated by short segments of much steeper slope, even miniature cliffs in some cases, they may be divided into two major groups, the high-tide and low-tide platforms. The former occur at or just above mean high-tide level. They may be flooded by the sea at spring tide, but during the rest of the tidal cycle remain dry unless there are large storm waves. They have in fact been ascribed to the action of storm waves, although it is difficult to see why one particular level should be preferentially developed to the exclusion of others. A number of different explanations have been offered. It has been suggested that they are the product of wave action during a slightly earlier and higher Holocene sea-level, possibly kept fresh by modern surf washing over them. An alternative hypothesis ascribes them to water-layer weathering. This process, analogous to

that invoked by Robinson for the plane segments of the North Yorkshire platforms, envisages the gradual enlargement of pools near and just above high-tide level where there is regular replenishment of water by spray. The constant wetting and drying around the edges leads to progressive extension of the pools. For the process to be fully effective occasional washing over by storm waves is necessary to remove the loosened debris. It is unlikely that a platform entirely due to water-layer weathering can exist where rapid cliff recession is taking place, since such recession implies large storm waves actively breaking against the cliff face. This may explain why high-tide platforms are less common around the stormy North Atlantic than on coasts affected mainly by moderate swell.

Low-tide platforms are near-horizontal surfaces exposed only when the sea falls below mid-tide level. Being characteristic of limestone coasts, they are usually attributed to solutional processes. As already pointed out, sea-water is normally saturated with respect to calcium carbonate so that special conditions are required for active solution. Favourable circumstances occur in rock pools temporarily isolated from the sea, and these are most frequent in the inter-tidal zone. The downward limit to corrosion in this way obviously approximates to low-water mark and explains why the platforms should be found at that height. On the other hand, it is equally true that many limestone coasts do not exhibit low-tide platforms. In general they are much better developed in tropical than in stormy temperate areas where it seems that active abrasion often masks the effects of solution.

This is the case with many of the chalk cliffs around the British Isles where the presence of nodular flints arms the storm waves with a powerful abrasive tool; by contrast, in tropical latitudes there are many long reaches of coastline composed exclusively of calcium carbonate and subject mainly to oceanic swell.

Along the coasts of periglacial regions a suite of processes rather different from those so far discussed controls the development of shore platforms. Effective wave abrasion, at least during the cold season, may be precluded by the formation of sea-ice. In many Arctic areas the phenomenon known as 'icefoot' characterizes the water's edge. This accumulates by the progressive freezing-on of ice derived from such varied sources as tidal rise and fall, sea-ice driven shorewards, coastal spray and normal precipitation. There is still much uncertainty regarding the geomorphic effect of the icefoot, with some workers arguing that it may act as a protective blanket while others contend that it will promote a zone of concentrated freeze–thaw activity. Further investigations are required to elucidate the full range of processes associated with the icefoot, but it seems likely that, at least for some part of the year, there will be enhanced surface degradation around high-tide level. It is worth noting that the pace of shoreline development under periglacial conditions has relevance for studies of British coasts, since it has been claimed that one of the prominent raised beaches in western Scotland must have been carved from resistant bedrock during a late-glacial interlude of cold climate that lasted less than a millennium.

It is clear that shore platforms can originate in many different ways and at a variety of levels. Many problems remain to be solved. Compared with measurements of cliff recession there are few assessments of the rate of platform erosion. On one section of the foreshore around the Isle of Thanet it was estimated that between 1904 and 1961 the lowering of the chalk platform averaged about 25 mm yr^{-1}. Such a figure could not be sustained over a long period since it would be inconsistent with local values of cliff recession and platform gradient, but it does indicate the potential efficacy of downwearing by wave action. Some of the most detailed short-term measurements have been made by Robinson on the coast of North Yorkshire. Employing a specially designed 'micro-erosion meter' he was able to demonstrate corrasion of the ramp amounting to 14.6 mm yr^{-1} at one locality but declining to less than 0.3 mm yr^{-1} at another. This wide range of values reflects variations in the thickness and nature of the overlying beach material, the lowest figures being found where large immobile boulders protect the ramp from wave attack. On the plane segment of the platform wetting and drying was found to be lowering the surface at an average rate of 1 mm yr^{-1} but with values varying between zero and 9 mm yr^{-1}; the highest rates were recorded on the highest parts of the plane. Until many more studies of this type have been completed, it is impossible to know how representative the above figures are, especially as it is safe to assume that a very large number of different factors are involved. Another topic requiring further examination is the seaward termination of shore platforms. Many exhibit an abrupt steepening of the gradient, often approaching the dimensions of a low cliff. This has been referred to as the low-tide cliff and has been shown locally to migrate landwards in the same fashion as the high-tide cliff. In such circumstances it is obviously an important factor in determining the width of the shore platform.

Finally, reference may be made to the widest shore platform of all, the strandflat, which still defies satisfactory explanation. Found off the coasts of Norway, Spitsbergen, Iceland and Greenland, it is clearly associated by both distribution and morphology with glaciation. It consists of a platform up to 60 km wide terminating abruptly against coastal mountain ranges. Although lying close to sea-level and undergoing slow alteration by wave action, it still preserves clear signs of glacial moulding. Opinions differ as to whether it is essentially a glacial landform modified by the sea, or a coastal landform modified by the passage of ice. Some workers have suggested that ice and sea may have acted simultaneously, ascribing the strandflat to erosive processes beneath an ice-shelf that rose and fell with the tide.

Forms due to marine deposition

Beaches

Beaches may be regarded as accumulations of marine sediments extending landwards from the low-water mark to some more permanent geomorphological feature. The composition of the materials varies from very coarse cobbles to fine sand, but a useful distinction is that between shingle and sand since intermediate sizes are relatively rare. Davies has mapped the distribution of shingle beaches around the coastlines of the world and shown that they are only common in high latitudes; the popular picture of tropical sandy beaches thus has some justification in fact. Several factors help to explain the distribution pattern, but it is first necessary to identify the four major sources of beach material. The first is the cliff face and adjacent shore platform subject to active wave erosion. The second is the sediment delivered to the coast by agencies involved in the denudation of the land. The third is sediment moving laterally along the coast by longshore transport. The final possible provenance is loose sediment lying in the offshore zone and fed to the beach as a result of wave disturbance. All the primary sources tend to contribute finer materials in low latitudes than in high latitudes. In tropical areas the swell generally lacks the erosive capacity of the steep waves encountered in the temperate storm belt, while the rivers entering the tropical oceans are laden with fine sediment rather than coarse. A further factor in northern latitudes is the widespread cover of glacial, glaciofluvial and periglacial deposits which all constitute a ready source of coarse detritus.

The provenance of shingle can be ascertained by examining the lithology of the pebbles. The majority are generally of local derivation, but many beaches also contain a small proportion of far-travelled constituents that have followed long and intricate routes since first being moved from their point of origin. For example, shingle on the coast of north Cornwall is found to contain material from Wales. This is thought to have been transported first by an ice-sheet from the Welsh upland and deposited on the floor of the Bristol Channel. Later, during the post-glacial marine transgression, it was pushed shorewards by the advancing sea and eventually incorporated as part of the present coastal shingle. On sand beaches the provenance of the sediment may be indicated by the nature of the heavy minerals. Along the Dutch coast most of the coastal sand was formerly assumed to be derived by longshore transport from the Rhine estuary. However, heavy mineral analysis later demonstrated a very significant admixture of material coming from the offshore zone on the floor of the North Sea. It should also be noted that by no means all beaches are predominantly siliceous in composition. This applies most obviously to the coral sands of the tropical beach, but even in temperate latitudes a significant proportion of beaches are highly calcareous. Many sands on the exposed coasts of western Ireland and Scotland consist of comminuted shells derived from abundant organisms living in the offshore zone. In these cases the powerful storm waves are an important factor in driving the shell fragments landwards.

Virtually all sections of a normal coastline have potential sources of beach material in one direction or another. Absence of beaches may in part be a function of time. For instance, rocky coasts subject to continuing glacio-isostatic uplift have often experienced a stable sea-level for too short a period to permit much beach accumulation. Yet the single greatest factor in explaining the absence of a beach is longshore transport. The effect of this transfer from one coastal segment to another is most readily apparent on a 'cape-and-bay' coast where the rocky headlands remain free of sediment while the re-entrants have thick bay-head beaches. Davies has drawn a useful distinction between coasts of free and impeded transport. On the former, material can be distributed long distances by lateral movement from a single source; the beach is virtually continuous. With impeded transport, there is little or no exchange of sediment between adjacent re-entrants; each bay-head beach is an isolated unit fed from a purely local source.

In plan, beaches are usually smoothly curved with their concave side facing seawards. This is most obvious on small bay-head beaches, but may also be seen on a much larger scale along coasts of free transport. There is no doubt that the outline is controlled mainly by wave action. As a general rule it is the powerful waves capable of moving both the greatest quantity and largest size of material that exert most influence; the beach tends to assume an alignment parallel to the crest of these dominant waves. The interaction of factors determining their direction of approach varies slightly from one area to another. Refraction plays the crucial role on coasts affected in the main by oceanic swell. Where there is free transport the beach often exhibits a beautifully regular alignment in accord with the wave pattern as refracted on entering shallow water. On crenulate coasts bay-head beaches may show a more diverse orientation, but again are aligned parallel to the refracted waves. Material is first swept by longshore transport into sharp, rocky re-entrants; initially the beach will be sharply curved, but continued movement towards the centre will eventually reduce the curvature until the trend parallels that of the approaching wave crests. On coasts affected mainly by storm waves refraction is still significant, but needs to be considered in association with the frequency and direction of gale-force winds. Schou advocated calculation of a so-called 'wind resultant' from anemometer data. He suggested that winds of less than Beaufort Scale 4 (21 km hr⁻¹) should be ignored as ineffective. Vectors for those of greater velocity are obtained by multiplying the frequency of winds for each Beaufort class by the cube of the mean velocity for that class, and then summing the products for each wind direction. For any section of coastline attention is often restricted to the three vectors representing onshore winds, permitting calculation of the 'onshore resultant'. After appropriate allowance has been made for refraction, beaches might be expected to lie parallel to the waves associated with the onshore resultant. Although this is often found to be true, where there are marked variations in fetch these tend to be of overriding importance in determining beach orientation; the coasts of the North Sea and Baltic Sea illustrate this situation very clearly.

Although beaches tend to achieve an equilibrium configuration parallel to the crest of approaching dominant waves, regular small-scale forms often develop along their face indicative of a continuing complex interaction between water and sediment movement. The most distinctive are the phenomena known as beach cusps which consist of alternating shallow embayments and transverse ridges aligned at right angles to the shoreline. Spacing between the ridges, which is often remarkably regular, can vary from under 1 m up to about 25 m. Numerous explanations have been offered for the development of beach cusps, but none seems wholly satisfactory and many have been proved entirely erroneous. Cusp development appears to be favoured by waves with long regular crests approaching normal to the coast. Spacing shows some correlation with both wave height and also the distance between wave break-point and the limit reached by the swash, but this throws little light on the actual mechanism of formation. Several workers in the 1970s argued that edge waves, such as those held responsible for the circulation patterns associated with rip currents (see p. 307), may provide the necessary regularity to account for the uniformity of well-developed cusps; more recently, Inman and Guza have hypothesized that edge waves help initiate the pattern but, once established, the form is perpetuated by its obvious capacity to divide the advancing swash into separate cellular units. There is no doubt, however, that much further research is needed. On a number of coasts regular patterns of a rather larger scale have been detected, partly along the beach itself but also within the surf zone; Komar in 1983 termed such patterns 'rhythmic topo-

graphy'. In some instances there is a clear association with rip currents, but in others a succession of long crescentic bars has formed parallel to the coast with no apparent regard to the location of rip currents.

In profile, beach morphology varies according to the nature of the constituent materials. Three common beach types may be distinguished. those composed of sand, those composed of shingle and those in which a shingle ridge forms the landward margin of a broad apron of sand. Shingle beaches are usually much steeper than those formed of sand, while on composite profiles a sharp break of slope often divides shingle ridge from sand apron. Some of these contrasts are attributable not only to the differing wave energies required to set material in motion but also to the dramatic growth in the rate of water percolation through the beach as particle size increases. Beach form is obviously determined by wave action and, to a lesser extent, by tidal influences. In general terms the maximum height of a beach is set by the upper limit of swash action. However, there is a marked difference between shingle and sand in this respect. During major storms large pebbles can be thrown high above the reach of normal waves, building shingle ridges 10 m or more above mean high-tide level; the prominent features thus produced are among the most enduring of beach forms, persisting unaltered while many profile changes occur along their seaward face. Sand, on the other hand, does not normally form prominent storm ridges. Storms tend to be destructive on sand beaches and the higher parts of the profile are usually constructed during periods of moderate swell. The feature built is this way is termed a berm. It typically has a flat upper surface separated by a sharp break of slope from the face regularly washed by the waves.

The seaward face of a beach profile varies according to the nature of recent wave action. As indicated in Chapter 15, repeated surveys can be used to record the changes in form, which may in turn be related to wave and tide observations. In their effect upon the profile, waves have been divided into two major classes, destructive and constructive. The former comb material down the beach and so steepen the gradient near the limit of wave action; the latter return material to the beach and thus build up the surface so as to reduce the profile concavity. In an early influential paper in 1931 Lewis recorded that he had found destructive waves to have a frequency of thirteen to fifteen per minute and constructive waves a frequency of six to eight per minute. By counting the number of waves breaking per unit time Lewis was, in effect, measuring wave length. Later investigations, however, have shown that the primary factor is a further derivative of frequency and length, namely, wave steepness. Studies designed to elucidate the steepness at which wave action changes from constructive to destructive have demonstrated that, for shingle beaches, the critical value appears to lie between 0.016 and 0.020, and for sand beaches between 0.010 and 0.014. The reason for the control by wave steepness seems to be related to the asymmetry of the to-and-fro motion of the water at the sea bed (see p. 306). If only the high onshore velocity exceeds the critical value for sediment entrainment, then the resultant transport should be shorewards; in this way low flat waves with pronounced asymmetry in their sea-floor velocities should tend to drive material landwards. On the other hand, if both onshore and offshore velocities surpass the critical limit, the longer duration of the offshore movement should result in a net seaward transport; high steep waves would thus be associated with sediment being combed down the beach and moved offshore. The variation of threshold velocities with particle size will not only account for the same waves acting differently on sand and shingle beaches, but will also explain why, on the same beach, shingle may be moved landwards at the same time as finer material is being moved seawards.

Repeated surveying of beach profiles reveals many different types of cyclical pattern. Some cycles are annual, with conspicuous differences between winter and summer forms. On many beaches, for instance, winter storms remove a cover of sand to leave exposed a deposit of coarse shingle or even boulders. The summer visitor would often be very surprised if he could see his favourite beach after a severe winter storm. Fortunately, by early summer spells of constructive waves have normally replaced the blanket of finer sediment. Of much shorter duration are cyclical changes that occur between individual storms. For example, if a beach has been steepened by destructive breakers at the time of high spring tide, later constructive action often tends to build a series of minor prograding ridges, each marking the level reached by swash during successive high tides, as the sea regresses from its spring to its neap stage.

Sand beaches can show many deviations from a smooth seaward slope, but probably the most distinctive is that known as ridge-and-runnel. This occurs on gently shelving beaches protected from powerful storm waves. The ridges commonly have a height of about 1 m above the adjacent runnel. They tend to run parallel to the shoreline, although examples are known where they run obliquely across the foreshore. As the tide ebbs, water left in the runnels normally escapes by way of small transverse channels. The ridge-and-runnel pattern is a persistent feature that can survive many tidal cycles. The number of ridges varies. It is not uncommon to find a single large ridge just above low-water mark, but in other cases three separate ridges may be present. They appear to originate as small bars built by the swash in positions where it has the longest opportunity for constructive action, usually the high- and low-water marks of the neap and spring tides. The accumulations formed in this way can survive submergence by the rising tide, and although the pattern they generate may be destroyed by storm waves, it is then re-established during succeeding spells of calm weather.

Spits

Spits are linear depositional forms attached at one end to the coast and most commonly sited where there is an abrupt change in coastal alignment. Material is supplied by longshore transport and shaped into the distinctive form by wave action. General requirements are an abundant supply of sediment and reasonably shallow water into which the spit can grow. Many spits built across river mouths divert the river parallel to the coast. In the case of Orfordness on the East Anglian coast the diversion of the Alde amounts to over 16 km, but when the spit becomes this elongated there is always the risk of storm waves breaching its neck. As mentioned earlier, this is believed on historical evidence to have happened at regular intervals to the spit at Spurn Point.

The alignment of a spit is generally determined by the direction from which the largest storm waves approach. Given an adequate supply of shingle these tend to build high ridges parallel to the crest of the advancing breakers. In many sheltered seas it can be shown that spits have developed normal to the maximum fetch since it is this which controls the size of the breaker. At their distal end spits often develop lateral hooks or recurves. Smoothly rounded hooks may be due to wave refraction as the tip of the spit advances into deeper water. Angular hooks, on the other hand, occur where material swept round the end of a spit is shaped into ridges by waves approaching from a second direction. A well-known example is the fashioning of the distinctive recurves on Hurst Castle spit by waves coming down the Solent (Fig. 17.5). The hooks are here aligned at an angle of 150° to the trend of the main stem.

The pattern of recurves is of great value in deciphering the growth stages of a spit. Development is often spasmodic, with phases of rapid evolution alternating with others of relative stability. At certain periods large recurves may form while there is little lengthwise extension of the feature; at others the dominant change is elongation with few lateral hooks; at yet others there may even be erosional trimming at the tip of the spit. The determinants of such spasmodic evolution are still poorly understood. Many studies have shown elongation near the tip occurring simultaneously with landward recession of the main stem; the point of transition from retreat to continuing accretion is known as the fulcrum. If the site of attachment to the mainland is receding, the spit itself must retreat in order to maintain its orientation. This means that the main storm beach, instead of meeting the old hooks in a smooth curve, will intersect them at an angle. The same landward retreat can also lead to fine sediments originally deposited in the lee of the spit being exposed on the seaward face. It is worth noting that, with very abundant sediment, progradation of a spit can occasionally take place with successive storm beaches being constructed one in front of the other.

Pairs of spits occur at the seaward end of certain estuaries. These pose a problem since they appear to grow towards each other and imply longshore transport from opposing directions. In some instances there may indeed be convergent movement towards the head of an estuary, and where this occurs the ebb and flow of the tide will be an important factor in fashioning the ends of the spits. It should not always be assumed, however, that this is the full explanation of 'double spits'. At Pagham on the English Channel coast Robinson in 1955 described two spits, one trending north-eastwards and the other south-westwards (Fig. 17.5). He presented historical evidence to suggest that the two features derive from a single spit that was breached near its centre by exceptional storm waves. Further east the same author described two shingle accumulations superficially resembling spits that have grown in opposite directions across the mouth of the River Stour. These are the features known as the Stonar Bank and the Sandwich Bay spit. The former extends southwards from the Isle of Thanet, the latter northwards from the cliffs near Deal. The outer feature has all the characteristics of a true spit, but the inner bank is much less typical. Its morphology is not that of an active spit and its composition is not such as can be derived by longshore transport from the adjacent cliffs. It probably originated as an offshore bank that migrated landwards during the post-glacial rise in sea-level and has subsequently been shut in behind a conventional spit growing from the south.

Nesses and cuspate forelands

The terms ness and cuspate foreland have both been applied to a prograding shoreline in which the sedimentary accumulation assumes a triangular shape with a broad base attached to the mainland and the apex facing seawards. It now seems likely that several different processes can be involved in the development of features of this general shape. On the East Anglian coast Robinson related each of five nesses to a distinctive pattern of ebb and flood tidal channels discernible in sandbanks a short distance offshore. He believed that these channels supply sediment to the ness which is thus sustained by onshore rather than longshore movement of material. He was able to show that some of the nesses had shifted their positions within the last few centuries, one apparently migrating 0.8 km between 1827 and 1883 before becoming more stable. He argued that the pattern of movement reflected changes that were taking place in the offshore banks. However, McCave in 1978 challenged some of Robinson's conclusions, arguing that the nesses were points where littoral transport was feeding sediment to the offshore banks, rather than the reverse. Carr later offered some support for McCave's interpretation, but work with tracers has not yet settled the matter. Whichever view is correct, the whole dispute illustrates the necessity of treating the beach and offshore zone as related parts of a single entity with important feedback mechanisms between them. Indeed, Robinson has contended that shifts in the location of the offshore banks may also help to explain why some sections of cliff line suddenly stabilize after having undergone a period of severe erosion.

The other type of triangular-shaped marine accumulation is more closely related to a spit, and will here be distinguished as a cuspate foreland. The best-known example is that of Dungeness on the coast of the English Channel (Fig. 17.5). It has been suggested that this began as a normal shingle spit aligned almost parallel to the trend of the coast. At a later stage it swung round to face the direction from

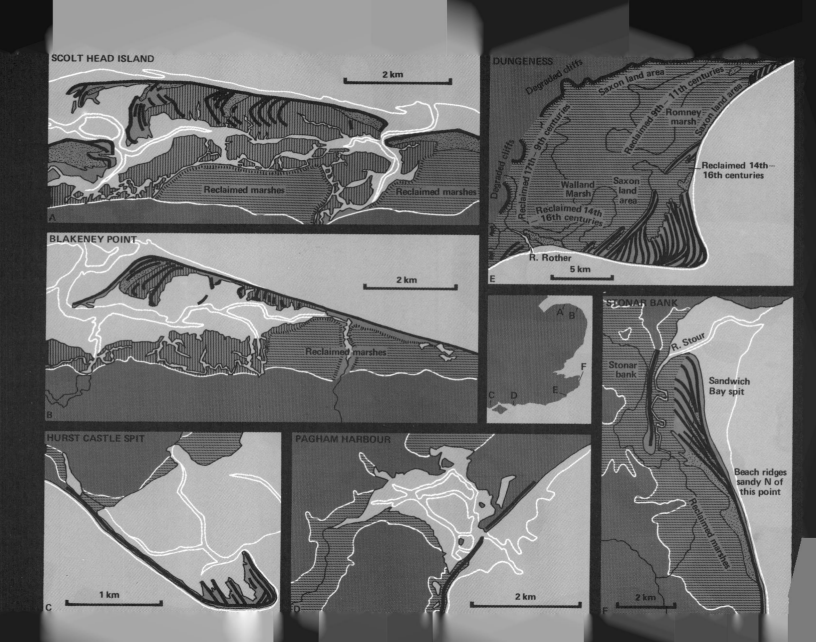

SCOLT HEAD ISLAND

2 km

Reclaimed marshes   Reclaimed marshes

A

BLAKENEY POINT

2 km

Reclaimed marshes

B

HURST CASTLE SPIT

1 km

C

PAGHAM HARBOUR

2 km

D

DUNGENESS

Degraded cliffs

Saxon land area

Reclaimed 9th – 11th centuries

Romney marsh

Saxon land area

Degraded cliffs

Reclaimed 17th – 9th centuries

Reclaimed 14th – 16th centuries

Walland Marsh

Saxon land area

Reclaimed 14th – 16th centuries

R. Rother

5 km

E

A B

F

C   D   E

STONAR BANK

R. Stour

Stonar bank

Sandwich Bay spit

Beach ridges sandy N of this point

Reclaimed marshes

2 km

F

Fig. 17.5. Sketch maps to illustrate the varied morphology of six constructional features located along the coasts of southern and eastern England.

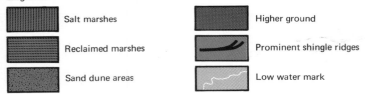

Salt marshes

Reclaimed marshes

Sand dune areas

Higher ground

Prominent shingle ridges

Low water mark

which the dominant storm waves were coming. The early evolution of the foreland may be deduced from historical evidence dating back to Roman times, the more recent evolution from an exceptionally fine suite of shingle ridges. The section facing south-south-west is currently undergoing erosion while accretion continues near and round the point. Recession has taken place under the attack of destructive storm waves, accretion under the influence of constructive waves during periods of high spring tide. The rate of progradation averaged just over 5 m yr^{-1} between 1600 and 1800, but is now much reduced. It appears that a very important factor in maintaining the sharpness of a cuspate foreland like Dungeness is the narrowness of the marine strait in which it is situated. At Dungeness powerful storm waves come either from the south-south-west with a fetch of 200 km to the coast of Normandy, or from the east-north-east through the Straits of Dover with a slightly shorter fetch; the waves approaching from directly across the Channel are relatively impotent and unable to blunt the point of the foreland.

Barrier bars and islands

Offshore accumulations of marine sediment enclosing a lagoon on their landward side are referred to as barriers. They may assume many forms but two of the most common are long narrow bars trending approximately parallel to the coast, and shorter islands that are slightly arcuate in plan with their convex sides pointing seaward. Although barriers of this type are very common, being well exemplified on the Atlantic seaboard of the United States, along the coast of the Gulf of Mexico and on the southern shores of the North and Baltic Seas, their origin is still contentious. They are characteristic of very gently shelving coasts that have a low tidal range. In general, given the same wave energy, the more continuous bars are formed

where the tidal range is least, and the barrier becomes increasingly fragmented as the tidal amplitude rises. As early as the mid-nineteenth century it was suggested that most of the material in a barrier comes from loose sea-floor sediments. This origin has been confirmed by many later investigations, but it leaves unexplained how the features are built above water-level. It is well known that waves can construct a small submarine bar close to their break-point, the requirements being a plentiful debris supply and steeply plunging breakers. However, there seems no simple way in which this process alone can continue to build the bar above water-level, and in any case the position of the break-point will fluctuate with the tide.

Several hypotheses have been proposed to overcome this problem. The first envisages a drop in sea-level, which exposes the bar so that thereafter it can be built up by the swash. One difficulty with this explanation is the frequent lack of any corroborative evidence for a fall in sea-level, together with the siting of many barrier coasts in area of long-term tectonic subsidence. A second hypothesis suggests that very temporary rises in sea-level may form small bars that become the nucleus for later development by swash action; for example, a surge associated with a hurricane might initiate a break-point bar capable of enlargement in this way. There is, however, little direct evidence to support his view, and many workers have sought alternative explanations. When the nature of Holocene sea-level changes is borne in mind, barriers seem more likely to be associated with transgression than regression. Hoyt maintained that barriers off the south-eastern United States are old beach ridge systems, topped by sand dunes, that have been partially submerged and detached from the coast by the continually rising water-level. This concept of beach detachment has been adopted by other workers, some of whom envisage the barriers arising from segmentation of earlier spits. By contrast, other authorities have contended that no change in level is necessary, and Otvos has consistently claimed that many of the barrier islands in the Gulf of Mexico are no more than 5 000 years old and so began to form after the post-glacial rise was virtually complete. He maintains that submerged shoal areas are capable of being raised to a critical inter-tidal level from which the action of swash can take over and construct them into islands. He documents numerous examples of islands that have apparently evolved in this way within historical times.

Although sea-level changes cannot be ignored, it seems unlikely that they are essential for the formation of barriers. The major need is a gently shelving coast with a copious supply of debris. Material may come from the sea floor after inundation of a coastal plain strewn with unconsolidated sediments. Alternatively, material may be furnished by rivers, as off the Mississippi delta, or by glacio-fluvial streams, as along parts of the Icelandic coast. There is no doubt that longshore movement is an important mechanism by which sediment is added to the developing barrier. Many islands move laterally along the coast, being eroded at one end and simultaneously extended at the other. Other barriers may prograde seawards or retreat landwards. Seaward movement of an individual bar is unusual and progradation is normally by addition of an extra ridge on the oceanic margin. More common is a landward migration that takes place by storm waves periodically washing over the barrier, by tidal streams transporting sediment to and fro through inter-island channels, and by onshore winds blowing sand from the beach into the lagoon. These processes have been the subject of intensive study in the eastern United States where their net effect, measured over several decades along 630 km of coastline between New Jersey and North Carolina, has been recession at a mean rate of 1.5 m yr^{-1}; the influence of Man is apparent in some coastal resorts which have actually reversed the natural trend through the construction of groynes and programmes of artificial beach nourishment. The dynamic nature of most barrier islands helps explain the continuing debate over origin since traces of distant history are rapidly erased by the mobility of the features.

The landward migration of offshore barriers can eventually lead to their becoming attached to the mainland. Thereafter they may be very difficult to distinguish from true spits, and it is now clear that features once described as spits could have had a very different origin. The problem is well exemplified by Scolt Head Island and Blakeney Point on the northern coast of East Anglia (Fig. 17.5). Early workers tended to assume these were both spits, but later studies disclosed many details inconsistent with that origin. Steers, for instance, concluded that Scolt Head Island was probably initiated as an offshore feature which migrated landwards until finally lodged in its present position at the coast. This interpretation was suggested in part by a gap which superficially resembles a recent storm breach, but which detailed study shows to be of considerable antiquity. In other words the insular form is not a late modification but an integral part of the structure. That the feature is still not entirely stabilized was shown by a storm surge in 1978 which washed over much of the island and caused the beach to retreat by some 20 m so as to expose fresh root mats on the foreshore. At first sight Blakeney Point, with no break in its continuity, appears a much more characteristic spit than Scolt Head Island. Nevertheless, there are certain puzzling aspects which imply that even this shingle feature is more complex than might be supposed. Observations reveal that the beach material currently moves both eastwards and westwards according to the direction of wave approach; if anything, easterly movement predominates over westerly at the present time. Without replenishment by longshore transport the feature is virtually a fossilized shingle ridge being constantly reshaped by wave action.

Chesil Beach on the coast of the English Channel is another shingle structure that fits uncomfortably into most classificatory schemes. It has been the subject of much research owing to the unusually clear particle-size grading that it exhibits, and is worth fuller description for some of the principles it illustrates. Chesil Beach extends some 28 km westwards from Chesilton on the Isle of Portland to an arbitrary limit near Bridport on the mainland (Fig. 17.6). Although attached at both ends, for about 18 km it is backed by the tidal lagoon known as The Fleet with a maximum depth of 3 m below mean sea-level. The beach is over 150 m wide where it shelters The Fleet, but rather narrower at both extremities. Its crest tends to rise eastwards, reaching a maximum elevation of over 14 m above mean sea-level near Chesilton. At Bridport the constituent pebbles are some 8 mm in diameter, but towards the Chesilton end they reach a mean diameter of about 60 mm. In 1969 the size grading was studied in great detail by Carr who collected samples totalling 89 000 pebbles from 23 sections across the beach. All the samples near high-water mark were well sorted and unimodal, but those at low-water mark tended to show bimodality and a rather greater spread of size values. Skin-divers have demonstrated that size grading scarcely exists immediately offshore, where much of the material is coarser than that found anywhere on the beach, and the mean size of recovered samples is about the same as that occurring only 4 km from Chesilton. From these observations the grading appears to be the product of waves breaking on a bank of poorly sorted debris; in essence, longshore transport in the surf zone carries material to the

Fig. 17.6. A map of, and section through, Chesil Beach. Figures at intervals along the beach indicate the average size of material (in cm) as given by Neate (1967). Section and dotted contour of bedrock surface from Carr and Blackley (1973).

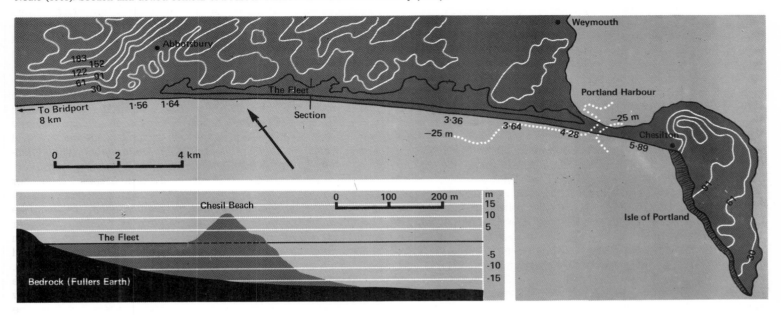

point on the beach appropriate to its size. Many experiments have been undertaken to monitor the movement of specially prepared pebbles. It has been shown that coarse debris tends to move eastwards at a variable rate. Seventeen thousand pebbles of quartz granulite were injected on the beach by Carr in 1971. Taken from a beach in Scotland these consisted of sizes at least as large as those occurring naturally at the point of injection. After 1 day the mean distance travelled was 87.6 m and after 165 days just over 1 km. It appears that sediments move to the point where they are in equilibrium with prevailing wave conditions. Chesil Beach is aligned almost normal to the advance of the largest storm waves, and although many smaller waves approach obliquely it seems improbable that they are the fundamental cause of the size grading. It is more likely that the grading is related to wave-energy variations along the length of the beach. Wave energy is greater at the eastern than at the western end owing to the steeper offshore profile near Chesilton.

In the course of studying the size grading Carr also analysed the composition of the pebbles on Chesil Beach. About 98.5 per cent of the material exposed at the surface consists of flint and chert. A significant proportion of the remainder is composed of quartz, quartzite, sandstone and limestone, together with rarer pebbles of such rocks as granite, serpentine and greywacke. Although it is believed that virtually all the material can be matched somewhere in south-western England, owing to impeded transport it is difficult to conceive of some of the far-travelled fragments being supplied solely by longshore swash transport; moreover, the idea of material currently being added to the western end of the beach is inconsistent with such perfect size grading. The observed degree of sorting is only compatible with a virtually closed system, although this does not of course eliminate the feasibility of longshore transport as a significant factor in the earlier growth of the beach.

Evidence on the development of the beach is provided by several deep boreholes. The marine sediments rest on a gentle bedrock slope composed of a series of separate benches. The lowest lies at about −15 m, and the sequence continues upwards until it encompasses several that lie above current sea-level on the landward side of The Fleet; the whole series must presumably have been cut during one or more warm interludes of the Pleistocene epoch. Beneath the modern shingle, boreholes have proved a layer of sand that is in turn underlain by a pebbly deposit containing up to 45 per cent limestone. This limestone shingle occurs at a height of −13 to −15 m, and appears to be a locally derived shoreline deposit that accumulated either during Pleistocene times or else in Holocene times when the precursor of Chesil Beach still lay some distance out in the English Channel; if the latter, on the basis of eustatic curves this stage was probably reached about 7 500 years ago. Shortly afterwards the shingle accumulation as we know it today first began to affect conditions near the present coast. This is indicated by a bed of peat that underlies much of The Fleet at between −2 and −5 m. Radio-carbon assays show the peat to date from between 6 000 and 4 000 years ago, by which time there must presumably have been a barrier protecting the area from invasion by the sea. Shortly thereafter The Fleet itself was formed, and although the shingle ridge may subsequently have advanced slightly further landwards, since the mid-nineteenth century its location has been relatively constant. A survey of the beach by Sir John Coode in 1852 has been compared with one completed in the 1960s. This reveals that the ridge crest maintained its position almost unchanged for over 100 years, but in that time its height increased along much of its length by about 1.5 m; on the other hand, near Chesilton its height fell by 2.5 m in the same period.

Chesil Beach illustrates the thesis that many modern depositional forms, despite the rapid changes they may exhibit, can only be fully understood in the context of a long and complex history. Two points merit particular emphasis. Firstly, the sea along the Dorset coast has recently reoccupied a position that it held in the past, and relict landforms from that earlier period significantly affect the coastal features observable today; the same may even be true of the more mobile barrier islands off the Atlantic seaboard of the United States. Secondly, during the Holocene transgression constructional forms initiated beyond the limits of the present coastline have been pushed landwards as the sea continued to rise. Chesil Beach is but one of a number of major accumulations that appear to be fossilized in the sense of no longer being actively nourished by longshore transport.

Tidal flats

In localities sheltered from wave action, accretion often assumes the form of mud-flats more or less inundated during each tidal cycle. Such mud-flats are especially common in estuaries and in the lee of spits and barriers. A very significant factor in their development is growth of vegetation adapted to periodic submergence by salt water. The pioneer plant communities vary according to the local climate, but two of the most common are those associated with salt marshes in temperate regions and mangrove swamps in tropical regions.

Salt-marsh plants display a delicate adaptation to the varied environment on a tidal flat. This commonly involves a zonation related to altitude and therefore to the proportion of time during which the flat is submerged. The first colonizer is often *Zostera* spp. or *Spartina* spp. The latter has been a particularly effective pioneer in north-western Europe since the vigorous hybrid *Spartina townsendii* originated in Southampton Water as a cross between native British and American species in the 1870s; within a century this new hybrid had spread across an estimated 12 000 hectares of tidal flat around the coasts of southern Britain. Another early colonizer is *Salicornia* spp. which is unusually well adapted to conditions of high salinity. These pioneers assist in the trapping of fine sediment introduced by the tide, thereby accelerating the rate of surface accretion. This in turn permits other plant communities, less tolerant of frequent submergence, to develop. Ultimately the mature marsh is inundated only at the highest spring tides and is characterized by such species as *Juncus maritimus* and *Artemisia maritima*. A very slow change to freshwater conditions may occur thereafter. This is promoted by invasion of reed-swamp communities dominated by species like *Phragmites communis* that can withstand brackish water.

The plant succession varies considerably from one locality to another, depending on such factors as exposure, climate, tidal range and nature of the substrate. One of the most interesting morphological features of tidal flats is the meandering system of creeks along which discharge on the ebbing tide is concentrated. These are kept clear by scouring while the adjacent flatter areas are raised by accretion. Miniature levees may be constructed by sedimentation around such plants as *Halimione portulacoides* which tends to grow well in that situation. Attempts at measurement have generally shown the rate of accretion to vary significantly over short distances according to the nature of the vegetation cover and other factors. It is usually highest on areas subject to regular tidal inundation and where the plant cover is most dense. On flats covered with *Spartina townsendii* upward growth often attains values between 10 and 50 mm yr^{-1} and it is clear that natural reclamation can take place extremely rapidly; a fully mature salt marsh may evolve in a matter of a century or two. Under estuarine conditions a high proportion of the accumulating sediment may be supplied by the rivers, and in Chesapeake Bay it is estimated that the volume of fluvial material may be up to five times that contributed by coastal erosion.

Mangrove swamps are almost totally confined to the tropics. All mangrove species are readily killed by frost and the swamps only flourish on coasts where the coldest month has an average temperature of at least 20 °C. They cannot tolerate wave action, so growth is normally restricted to low-energy coasts where there is an abundance of fine mud. Two different types of tree are common, those with stilt roots and those that throw up vertical pneumatophores. Both have the effect of reducing water velocity and thereby promoting settlement of any fine suspended load. There has been considerable dispute as to their effectiveness, but most workers agree that they do substantially increase the rate of accretion. As on a salt marsh, drainage on the ebb tide takes place along well-defined channels, in some instances forming a reticulate pattern, in others a dendritic pattern.

Coasts dominated by organic activity: coral reefs

In the preceding section attention was directed to coastal forms in which sedimentation is greatly affected by organic growth. Many further instances might be quoted of interaction between physical and biological processes. However, the outstanding example of a shoreline dominated by organic activity is that constructed by coral polyps.

Reef-building corals have demanding environmental requirements that strictly limit their distribution. Although the main structure of a coral reef is built by the polyps, they live in symbiosis with minute algae, and it appears to be the life requirements of the algae that are often the limiting ecological factor. Global distribution is primarily determined by temperature. The extreme limits are generally taken to be the 18 and 36 °C surface-water isotherms for the coldest and hottest month respectively. However, optimum growth conditions occur within a much narrower band, usually defined as between 25 and 30 °C. Given these temperature requirements, the polyps need salinity values between 27 and 40‰, water that is reasonably free of sediment, and a rapid circulation that constantly replenishes the supply of nutrients and oxygen. Such conditions are met along many tropical coasts with low to moderate swell. Opposite large river mouths, however, dilution with fresh water and increased turbidity may provide locally inimical sites. The final necessity for active reef development is light for algal photosynthesis. This is of crucial importance since it limits the depth of prolific coral growth. Although living corals have been recorded at −100 m, they are very rare at these depths and vigorous reef formation is restricted to within 25 m of the surface. The significance of this last figure lies in the fact that ancient reefs often extend into much deeper water, but before discussing the problem that this poses it is necessary to outline the characteristic reef forms.

In describing coral reefs in the mid-nineteenth century Darwin employed a simple classification which is still in use today. It distinguishes three main types: fringing reefs that directly border the coast; barrier reefs that are separated from the coast by a deep channel; and atolls that form islands encircling a lagoon. Although difficulties occasionally arise in applying this classification, and there are certainly isolated platform reefs that do not fit readily into any of these three categories, it still constitutes a valuable basis for discussing concepts of reef development.

Darwin propounded a hypothesis which neatly linked together the three types of reef he had distinguished and the known fact that corals could not grow in deep water. He argued that the only feasible explanation of atolls, all completely accordant in level yet arising from the deep ocean floors, is subsidence of the foundations on which they rest. Such subsidence would account for fringing reefs round an island first becoming barrier reefs by continued upward growth; eventually, if the process were to persist, the barrier reef would be converted into an atoll by the disappearance of the central peak.

Many of the characteristics which led Darwin to propose the theory of subsidence can be explained equally well by a rising sea-level and it was natural that, as the idea of glacio-eustatic fluctuations gained currency, the implications for the growth of coral reefs should be considered. In 1910 Daly proposed a theory of glacial control. He argued that, during a glacial period, continued reef growth outside a narrow equatorial zone would have been precluded by the drop in water temperature. At the same time any existing reef would have been subject to marine attack during the phase of lowered sea-level; potentially this could truncate oceanic islands and yield platforms from which atolls could later grow upwards as sea-level rose during the next interglacial. In support of this thesis be claimed that atoll lagoons exhibit a general accordance of level, rarely being deeper than about 75 m.

In order to discriminate between the subsidence and glacial control hypotheses it is essential to have information regarding the structure and age of the foundations on which modern coral reefs are built. The earliest deep boring on an atoll was at Funafuti in the Ellice Islands where coral was penetrated to a depth of 339 m without the bottom being reached; later seismic explorations were to suggest that the base of the reef extends to a depth of about 900 m. In 1936 a borehole through the uplifted atoll of Kita Daito Jima, just south of Japan, continued to a depth of 432 m without striking basement rocks. A later drilling on Bikini atoll in the Marshall Islands proved coral to a depth of 780 m, at which level the limestone was believed to be of Oligocene age. On the nearby Eniwetok atoll two boreholes encountered the bedrock foundation at 1 405 m and 1 283 m, in both cases immediately underlying shallow-water Eocene limestones. All these findings leave little doubt regarding the validity of Darwin's original hypothesis, although the sequence of events appears to have been rather more complex than he envisaged. Breaks in the continuity of reef growth indicate that some of the islands were periodically elevated above sea-level and subject to erosion. There is still inadequate evidence to show whether the periods of erosion were contemporaneous over wide areas, but it is generally presumed that they were due to local tectonic movement rather than eustatic changes.

Confirmation of the subsidence hypothesis does not itself refute the glacial control hypothesis. Glacio-eustatic fluctuations are potentially a factor of great significance in the Pleistocene evolution of all reefs. How far an existing reef can be modified during a single glacio-eustatic oscillation obviously depends upon the rates of reef erosion and growth. Daly contended that, during a glacial period of low sea-level, extensive planation might occur. More recent assessments have cast doubt upon this view. Current marine solution along notches at mean sea-level has been found to average about 1 mm yr^{-1}, and if this figure is only approximately correct it still seems to preclude extensive marine planation during a typical glacial stillstand. Recourse cannot be had to subaerial denudation as the agent responsible for reef erosion. If the final glaciation is regarded as representative, a glacio-eustatic cycle lasts some 100 000 years. Adopting this value and an annual runoff of 1 m, it can be calculated that, even if fully saturated, the water would remove a layer only some 10 m thick; a comparable figure is also suggested by measurements of current erosion rates made by Trudgill on Aldabra atoll in the Indian Ocean. Even if only of the right order of magnitude, these assessments demonstrate the inability of erosional processes to destroy an exposed reef within the brief time available. The logical conclusion is that a glacio-eustatic fall in sea-level can modify a reef but not destroy it. As sea-level rises again, renewed growth may induce further change. The rate at which reef growth can take place varies with different species but commonly lies within the range 10–40 mm yr^{-1}. An allowance must be made for the frequency with which coral material is broken from a reef and added to submarine talus slopes, but upwards extension at a rate of 15 mm yr^{-1} appears representative of reasonably favourable conditions. In such circumstances coral growth would have no difficulty in keeping pace with the post-glacial rise in sea-level which occurred at about 10 mm yr^{-1}. The result would presumably be a veneer of recent reef adhering to a much older and partially karstified core; there is increasing evidence from uranium-series and radio-carbon dating to support this hypothetical evolution:

References

Arber, M. A. (1973) 'Landslips near Lyme Regis', *Proc. Geol. Assoc.* **84**, 121–33.

Bloom, A. L. (1965) 'The explanatory description of coasts', *Z. Geomorph.* **9**, 422–36.

Brunsden, D. (1974) 'The degradation of a coastal slope, Dorset, England', in *Progress in Geomorphology*, Inst. Brit. Geogr. Spec. Pub. 7.

Brunsden, D. and **Jones, D. K. C.** (1976) 'The evolution of landslide slopes in Dorset', *Phil. Trans. R. Soc.* **A283**, 605–31.

Byrne, J. V. (1963) 'Coastal erosion, northern Oregon', in *Essays in Marine Geology in Honor of K. O. Emery* (ed. T. Clements), Univ. S. Calif. Press.

Carr, A. P. (1969) 'Size grading along a pebble beach: Chesil Beach, England', *J. Sed. Pet.* **39**, 297–311.

Carr, A. P. (1971) 'Experiments on longshore transport and sorting of pebbles: Chesil Beach, England', *J. Sed. Pet.* **41**, 1084–1104.

Carr, A. P. (1981) 'Evidence for the sediment circulation along the coast of East Anglia', *Mar. Geol.* **40**, M9–22.

Carr, A. P. and **Blackley, M. W. L.** (1973) 'Investigations bearing upon the age and development of Chesil Beach, Dorset, and the associated area', *Trans. Inst. Brit. Geogr.* **58**, 99–111.

Carr, A. P. and **Gleason R.** (1971) 'Chesil Beach, Dorset and the cartographic evidence of Sir John Coode', *Proc. Dorset Nat. Hist. Arch. Soc.* **93**, 125–31.

Davies, J. L. (1980) *Geographical Variation in Coastal Development* (2nd edn), Longman.

Dolan, R. *et al.* (1979) 'Shoreline erosion rates along the middle Atlantic coast of the United States', *Geology* **7**, 602–6.

Gray, J. M. (1978) 'Low-level shore platforms in the south-west Scottish Highlands: altitude, age and correlation', *Trans. Inst. Brit. Geogr.* **3**, 151–64.

Hoyt, J. H. (1967) 'Barrier island formation', *Geol. Soc. Am. Bull.* **78**, 1125–36.

Inman, D. L. and **Guza, R. T.** (1982) 'The origin of swash cusps on beaches', *Mar. Geol.* **49**, 133–48.

Inman, D. L. and **Nordstrom, C. E.** (1971) 'On the tectonic and morphologic classification of coasts', *J. Geol.* **79**, 1–21.

John, B. S. and **Sugden, D. E.** (1975) 'Coastal geomorphology of high latitudes', *Prog. Geogr.* **7**, 53–132.

Komar, P. D. (1975) 'Nearshore currents: generation by obliquely incident waves and longshore variations in breaker height', in *Nearshore Sediment Dynamics and Sedimentation* (ed. J. Hails and A. Carr), Wiley.

Komar, P. D. (1983) 'Rhythmic shoreline features and their origins', in *Mega-geomorphology* (ed. R. Gardner and H. Scoging), OUP.

Leatherman, S. P. (ed.) (1979) *Barrier Islands from the Gulf of St Lawrence to the Gulf of Mexico*, Academic Press.

May, V. J. (1971) 'The retreat of chalk cliffs', *Geogr. J.* **137**, 203–5.

May, V. J. and **Keeps, C.** (1985) 'The nature of change on chalk coastlines' *Z. Geomorph. Supplementband* **57**, 81–94.

McCave, I. N. (1978) 'Grain-size trends and transport along beaches: example from eastern England', *Mar. Geol.* **28**, M43–51.

Neate, D. J. M. (1967) 'Underwater pebble grading of Chesil bank', *Proc. Geol. Assoc.* **78**, 419–26.

Otvos, E. G. (1970) 'Development and migration of barrier islands, northern Gulf of Mexico', *Geol. Soc. Am. Bull.* **81**, 241–6.

Otvos, E. G. (1981) 'Barrier island formation through nearshore aggradation – stratigraphic and field evidence', *Mar. Geol.* **43**, 195–243.

Robinson, A. H. W. (1955) 'The harbour entrances of Poole, Christchurch and Pagham', *Geogr. J.* **121**, 33–50.

Robinson, A. H. W. (1966) 'Residual currents in relation to shoreline evolution of the East Anglian coast', *Mar. Geol.* **4**, 57–84.

Robinson, A. H. W. (1980) 'Erosion and accretion along part of the Suffolk coast of East Anglia, England', *Mar. Geol.* **37**, 133–46.

Robinson, A. H. W. and **Cloet, R. L.** (1953) 'Coastal evolution of Sandwich Bay', *Proc. Geol. Ass.* **64**, 69–82.

Robinson, L. A. (1977) 'Erosive processes on the shore platform of northeast Yorkshire, England', *Mar. Geol.* **23**, 339–61.

Robinson, L. A. (1977) 'Marine erosive processes at the cliff foot', *Mar. Geol.* **23**, 257–71.

Robinson, L. A. (1977) 'The morphology and development of the northeast Yorkshire shore platform', *Mar. Geol.* **23**, 237–55.

Schou, A. (1945) 'Det Marine Forland', *Folia Geogr. Danica* **4**, 1–236.

Shepard, F. P. (1973) *Submarine Geology* (3rd edn), Harper and Row.

Sissons, J. B. (1982) 'The so-called high "interglacial" rock shoreline of western Scotland', *Trans. Inst. Brit. Geogr.* **7**, 205–16.

So, C. L. (1965) 'Coastal platforms of the Isle of Thanet', *Trans. Inst. Brit. Geogr.* **37**, 147–56.

Steers, J. A. (1960) *Scolt Head Island*, Heffer.

Steers, J. A. *et al.* (1979) 'The storm surge of 11 January 1978 on the east coast of England', *Geogr. J.* **145**, 192–205.

Terwindt, J. H. J. (1973) 'Sand movement in the in- and offshore tidal area of the S.W. part of the Netherlands', *Geol. en Mijn.* **52**, 69–77.

Trenhaile, A. S. (1974) 'The geometry of shore platforms in England and Wales', *Trans. Inst. Brit. Geogr.* **62**, 129–42.

Trudgill, S. T. (1976) 'The subaerial and subsoil erosion of limestones on Aldabra Atoll, Indian Ocean'. *Z. Geomorph. Supplementband* **26**, 201–10.

Trudgill, S. T. (1976) 'The marine erosion of limestones on Aldabra Atoll, Indian Ocean', *Z. Geomorph. Supplementband* **26**, 164–200.

Valentin, H. (1952) *Die Kusten der Erde*, Petermanns Geogr. Mitt.

Valentin, H. (1954) 'Der Landverlust in Holderness, Ostengland von 1852 bis 1952', *Die Erde* **3**, 296–315.

Wilson, G. (1952) 'The influence of rock structures on coastline and cliff development around Tintagel, north Cornwall', *Proc. Geol. Ass.* **63**, 20–48.

Wood, A. (1959) 'The erosional history of the cliffs around Aberystwyth', *L'pool Manchr Geol. J.* **2**, 271–9.

Wood, A. (1968) 'Beach platforms in the Chalk of Kent', *Z. Geomorph.* **12**, 107–13.

References

Wright, L. W. (1970) 'Variation in the level of the cliff/shore platform junction along the south coast of Great Britain', *Mar. Geol.* **9**, 347–53.

Selected bibliography

Among valuable texts dealing with the evolution of coastal landforms in general are J. L. Davies, *Geographical Variation in Coastal Development* (2nd edn), Longman, 1980, and J. Pethick, *An Introduction to Coastal Geomorphology*, Arnold, 1984.

Although now rather dated, the standard works on the coastal landforms of the British Isles remain J. A. Steers, *The Coastline of England and Wales*, CUP, 1964, and *The Coastline of Scotland*, CUP, 1973.

Part 5

CONCEPTS OF LANDFORM EVOLUTION

Chapter 18
Rates and patterns of geomorphic change

In the preceding sections of this book, reference has frequently been made to the rates at which individual processes appear to be operating. In most instances the time-scale deployed has intentionally been kept short, even though many of the related landforms might also be viewed as the culmination of a long and involved geomorphic history. The aim of the present chapter is therefore to broaden the time and space perspectives by offering a review of the rate at which continental landscapes as a whole appear to evolve. In addition to being a subject of importance in its own right, the topic is also a necessary precursor to any appreciation of the models of landform development discussed in the final chapter of the book.

Current rates of subaerial denudation

Measurement of fluvial sediment loads

The second half of the twentieth century has seen an immense growth in the systematic monitoring of streams. The most frequently measured variable is water discharge, but an increasing number of stations take regular readings of sediment and solution load. The main reason behind this upsurge is practical rather than scientific; it stems from the need of densely populated and economically advanced nations to assess both the quantity and quality of their water resources. With the expansion in stream monitoring by civil authorities has come an awareness by geomorphologists of the significance of the collected data. The techniques for measuring instantaneous sediment discharge described in Chapter 7 need no elaboration here, but the uncertainties in many of the procedures must be borne in mind when considering the results. Moreover, care needs to be taken that figures from different sources are strictly comparable. Often no attempt is made to measure the bedload, although sometimes a rather arbitrary weighting is applied to the suspension load as an allowance for the unmeasured fraction. Similarly, some estimates of the gross yield from a catchment include the solution load while others do not. These factors can introduce significant differences between apparently compatible sets of data and render regional comparisons very hazardous.

In 1983 Milliman and Meade collated records from the twenty-one rivers of the world believed to be delivering the largest quantity of suspended sediment directly from the continents into the oceans (Table 18.1). Exceptionally high rates of sediment discharge per unit area of catchment are characteristic of the rivers of south-eastern Asia. Moreover, it should be stressed that certain subcatchments are known to exhibit very much higher values than those quoted for the basin as a whole. For instance, large areas of loess plateau draining to the Yellow River in China are estimated to be yielding at least 7 000 tonnes km^{-2} yr^{-1}. It has been calculated that where the Yellow River leaves the loess region it is transporting approximately 1 500 million tonnes of sediment each year, but by the time the river reaches the coast, some 850 km downstream, the figure has fallen to 1 080 million tonnes. The implication is that over 400 million tonnes of material are annually added to the alluvial floodplain between the two gauging points. This example illustrates another problem in the interpretation of sediment transport data, namely that material eroded from the headstream catchments may go into storage lower in the basin and not be delivered immediately to the sea. This helps to explain a noteworthy reduction in annual sediment yield per unit area of catchment as the size of basin under consideration gradually increases.

The rivers listed in Table 18.1 account for just under one-quarter of the total continental land areas draining directly to the oceans. Since the rivers were selected on the basis of their large sediment loads they may be expected to be above average in their rates of sediment yield per unit area. Milliman and Meade have, however, collected data covering a further quarter of the continental land areas, and on this basis they estimate that the total suspended sediment load

discharged each year into the sea reaches just over 13.5 × 10^9 tonnes, or the equivalent of 132 tonnes km^{-2}; when an allowance is made for bedload transport, the authors suggest the aggregate sediment delivery may rise to 16 × 10^9 tonnes. They have prepared a map illustrating what they believe to be the pattern of regional variations (Fig. 18.1). In addition to very high discharges from mountainous regions of low latitudes, another noteworthy feature is the large yield attributable to rivers draining from glaciers. Many early workers commented on the milky appearance of meltwater streams just below the glacier snout, ascribing it to the high content of 'rock flour'. Tarr described how, in Alaska, from a sample collected in a pail, several centimetres of mud and sand settled out within a few minutes. In the same area Reid estimated the amount of silt being removed each year from the Muir glacier to be equivalent to a subglacial rock layer 19 mm thick. Such a figure could only be regarded as a very rough approximation since it was not based upon systematic sampling. A more detailed attempt to obtain representative values was made in 1939 by Thorarinsson who found that the meltwater from the Hoffellsjokul in Iceland was discharging sediment equivalent to a bedrock lowering of 2.8 mm yr^{-1}. Since 1950 many further investigations in various parts of the world have confirmed the exceptional mean rate of surface lowering beneath a glacier. The figures, of course, refer specifically to the subglacial area and the average rates for whole catchments are substantially less.

At the continental scale, the network of stations maintained by government agencies in the United States provides one of the most comprehensive and reliable coverages yet available. It is estimated that, from the conterminous United States, 339 million tonnes of suspension load are currently being delivered into the surrounding oceans each year; this figure discounts the River Colorado which, despite draining an area of exceptionally rapid sediment production (see below p. 368), is now so regulated and depleted by Man that discharge at its mouth has been rendered negligible. The total of 339 million tonnes is equivalent to 49.5 tonnes from each square kilometre of contributory land surface, and assuming a density of 2 700 kg m^{-3} represents a mean lowering of 0.018 mm yr^{-1}. This average, however, conceals marked regional diversity. For example, streams draining to the Alantic seaboard carry suspended sediment equivalent to an annual lowering of their catchments of 0.006 mm. The corresponding figure for the whole Mississippi basin is 0.021 mm,

Table 18.1 Suspended sediment discharge of twenty-one of the world's major rivers (compiled from data in Milliman and Meade, 1983)

River	A Catchment area (10^3 km²)	B Mean water discharge (10^3 m³ s⁻¹)	C Suspended sediment discharge (10^6 tonnes yr⁻¹)	D Sediment concentration (g/l⁻¹) (Col. C/Col. B)	E Sediment yield (tonnes km⁻² yr⁻¹) (Col. C/Col. A)	F Rate of surface lowering (mm yr⁻¹)
Ganges/Brahmaputra	1 048	30.8	1 670	1.72	1 593	0.59
Yellow (Huangho)	770	1.6	1 080	21.38	1 403	0.52
Amazon	6 150	199.8	900	0.14	146	0.05
Yangtze	1 940	28.5	478	0.53	246	0.09
Irrawaddy	430	13.6	285	0.66	662	0.25
Magdalena	240	7.5	220	0.93	917	0.34
Mississippi	3 269	18.4	210	0.36	64	0.02
Orinoco	990	34.9	210	0.19	212	0.08
Hungho (Red)	119	3.9	160	1.31	1 344	0.50
Mekong	795	14.9	160	0.34	201	0.07
Indus	969	7.5	100	0.43	103	0.04
Mackenzie	1 810	9.7	100	0.33	55	0.02
Godavari	310	2.7	96	1.11	310	0.11
La Plata	2 830	14.9	92	0.19	33	0.01
Haiho	50	0.1	81	26.00	1 620	0.60
Purari	310	2.4	80	1.04	258	0.10
Copper	60	1.2	70	1.83	1 167	0.43
Zhu Jiang (Pearl)	440	9.6	69	0.23	157	0.06
Danube	810	6.5	67	0.32	83	0.03
Choshui	30	0.2	66	10.50	2 200	0.81
Yukon	840	6.2	60	0.31	71	0.03
Total	24 210	414.9	6 254	0.48	258	0.10

Fig. 18.1. Sediment delivery to the oceans based upon the work of Milliman and Meade (1983). Great caution should be used in interpreting this map, since it refers specifically to current conditions that are much affected by the activities of Man. Thus the Nile used to carry an estimated 10^8 tonnes to the sea each year, but since the construction of the Aswan dam that figure has declined virtually to zero. Moreover, as Milliman and Meade emphasize, the figures can be misleading when translated into denudational lowering since they take no account of temporary storage of sediments in floodplains along the lower reaches of river valleys. Nevertheless, despite these limitations, the broad regional pattern does accord with many modern assessments of the spatial distribution of denudational lowering by sediment removal.

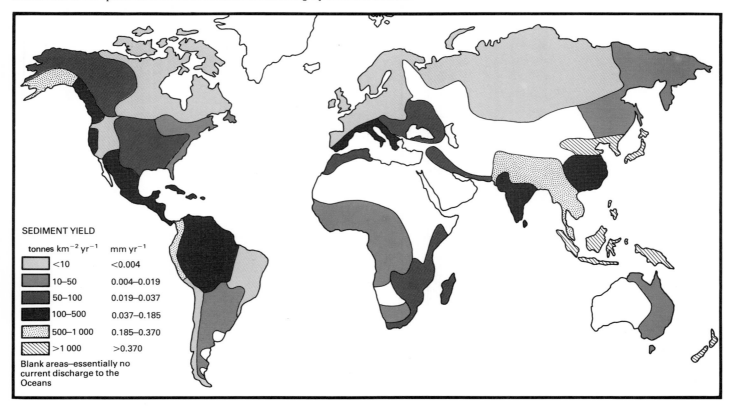

SEDIMENT YIELD

tonnes km^{-2} yr^{-1}	mm yr^{-1}
<10	<0.004
10–50	0.004–0.019
50–100	0.019–0.037
100–500	0.037–0.185
500–1 000	0.185–0.370
>1 000	>0.370

Blank areas–essentially no current discharge to the Oceans

while for a group of catchments on the north Californian coast it is over 0.4 mm. These values must be regarded as conservative for total denudation rates, since they include no allowance for either bedload or solution load.

Measurement of fluvial solution loads

Assessment of denudation rates from the chemistry of stream waters is a field that has also made substantial advances in the last decade or two. In 1979 Meybeck published an analysis of data for the solution load of the world's major rivers (Table 18.2). His data cover sixty large rivers which are estimated to carry approximately 63 per cent of the total global discharge to the oceans; for the remaining areas, Meybeck made estimates based upon anticipated loads computed from values of runoff, temperature and relief. After corrections were

River	Mean concentrations in mg l⁻¹		Total solution load	Rate of chemical denudation
	SiO$_2$	Σ ions*	(10^6 tonnes yr^{-1})	(mm yr^{-1})†
Ganges	12.8	154.6	75.33	0.020
Brahmaputra	(11.0)	127.2	84.16	0.039
Amazon	11.2	41.0	288.67	0.012
Yangtze	(10.0)	(101.2)	100.00	0.013
Magdalena	12.6	105.6	27.78	0.031
Mississippi	7.5	208.5	125.28	0.009
Orinoco	11.5	23.8	33.39	0.009
Mekong	8.85	90.0	57.04	0.019
Indus	5.1	166.1	16.26	0.016
Mackenzie	3.0	207.5	63.99	0.009
Parana	14.3	54.9	39.24	0.004
Uruguay	15.0	60.2	11.66	0.009
Danube	5.0	301.5	62.22	0.020
Yukon	6.4	167.3	33.87	0.012

Table 18.2 Estimates of the current dissolved loads of fourteen of the world's major rivers (raw data mainly from Meybeck, 1979). These have been converted into values for total annual solution load and, with much more uncertainty, into chemical denudation rates

* Ions included in monitoring: Ca^{2+}, Mg^{2+}, Na^+, K^+, Cl^-, SO^{2-}, HCO_3^- (estimated values in brackets).
† After a rather arbitrary allowance of 30 per cent for dissolved material derived from the atmosphere and from Man-made pollution.

made for such factors as atmospheric CO_2 and Man-made pollution, mean chemical denudation of the continents outside the desert areas was calculated to vary from 3 to 80 tonnes km^{-2} yr^{-1}. Highest values are found in mountainous areas of both humid temperate and equatorial regions (Fig. 18.2). Such wet uplands cover only 12.5 per cent of the earth's surface, but nevertheless furnish about 45 per cent of the silica and 41 per cent of the major ions (Ca, Mg, Na, K, Cl, SO$_4$, H$_2$CO$_3$) transported to the oceans. Meybeck estimated that the mean chemical erosion for the continents as a whole is 19 tonnes km^{-2} yr^{-1}, but that if areas of internal drainage are excluded the value rises to 24 tonnes km^{-2} yr^{-1}; these figures correspond to a lowering of the land surface amounting respectively to 0.007 and 0.009 mm yr^{-1}. The same author suggested that mechanical erosion products might overall be six to eight times as important as chemical, but that the

figure almost certainly varies greatly, and over large plainlands there may be little difference in the effectiveness of mechanical and chemical denudation.

As pointed out in Chapter 7, the concentration of different solutes fluctuates as a function of many contrasting variables, so that for a detailed local catchment study monitoring of a wide range of ions is desirable. Only a limited number of comprehensive surveys have yet been completed, and instead attention has often been concentrated on either monitoring water conductivity as a surrogate for the total dissolved solids, or measuring just one or two ions that can be confidently related to the local bedrock. Attempts at full geochemical budgets have generally been restricted to small catchments. An example is provided by the investigation of Waylen into the chemical weathering of an Old Red Sandstone catchment, some

Fig. 18.2. Current chemical denudation rates as computed by Meybeck (1979). Comparison with Fig. 18.1 reveals a regional pattern that shows significant similarities with that of physical denudation and, over extensive areas, the values for chemical and physical denudation are of the same order of magnitude. However, note the very different incremental scales employed on the two maps, and nowhere do the chemical rates appear to approach those characteristic of mechanical erosion in low-latitude mountainous regions.

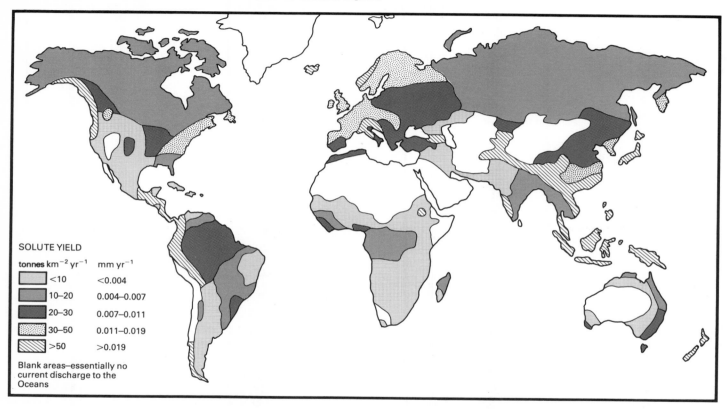

SOLUTE YIELD

tonnes km⁻² yr⁻¹	mm yr⁻¹
<10	<0.004
10–20	0.004–0.007
20–30	0.007–0.011
30–50	0.011–0.019
>50	>0.019

Blank areas—essentially no current discharge to the Oceans

0.18 km² in extent, forming part of the Mendip Hills in south-western England. He first constructed solute rating curves for calcium, magnesium, sodium, potassium, silicon, bicarbonate, sulphate, chloride and pH, which together account for over 95 per cent of the material leaving the basin in solution. By combining these curves with continuous flow records, he was able to calculate the total of all the major solutes leaving the catchment in stream flow during a single year. He further estimated the input of solutes to the basin through the natural precipitation, and the likely uptake of plant nutrients by a vegetation cover that had not yet established an equilibrium with its environment. From these three values he was able to compute a balance that presumably reflected the quantity of solutes taken up or released by weathering reactions. He interpreted his analysis as indicating that 3.6 tonnes of primary minerals had suffered alteration in the course of the year, but only 0.77 tonnes of true denudational solutes had been released from the catchment. The latter figure converts into a mean lowering of the basin of about 0.0016 mm yr⁻¹. Such full geochemical budgets are obviously a very powerful tool in understanding the way denudation is operating. Yet the topic becomes one of immense complexity when a large catchment underlain by a diversity of rocks is considered; moreover, considerable annual variations must be anticipated, so that observations should preferably be sustained over a minimum of several years.

An example of a study in a much larger, more varied basin is provided by the work of Walling and Webb in Devonshire. They had twelve gauging stations within the Exe catchment that already provided relatively full data on solute transport. However, owing to the geological diversity of the region they wished to determine more precisely the areas from which the solutes were coming. During a period of low flow they therefore collected water samples not only from the gauging stations but also from 500 additional sites scattered across the 1 462 km² of the basin. The specific conductance of each sample was measured and the reading converted into total dissolved solids by means of a specially prepared calibration curve. By manipulating the data thus obtained the authors were able to derive estimates of the annual solute load at each of the extra 500 sites in the spatial survey. However, as with all estimates of chemical erosion, care had to be taken to distinguish between the denudational and non-denudational components of the solution load. The latter in this case included not only atmospheric inputs but also Man-made pollution through such activities as the application of fertilizer. After allowance had been made for these factors, it was argued that the current rates of chemical denudation range from under 0.01 mm yr⁻¹ on some of the upland areas of weakly soluble bedrock to over 0.05 mm yr⁻¹ on certain lowland areas of Permian marl.

Limestone solution is the single denudational process most widely investigated by monitoring dissolved river loads since, in theory, it can be calculated from records of the concentration of calcium and magnesium ions. Complications arise where other rocks occur within the catchment, but assessments have generally clustered within a comparatively narrow band from 0.015 to 0.085 mm yr⁻¹ (Table 18.3). If, as is often the case, limestone is regarded as a relatively resistant material, many other rocks can presumably be lowered at a faster rate, although not necessarily by purely chemical means.

Another specific lithology receiving increasing attention from those concerned with chemical denudation is siliceous bedrock; in 1983 Saunders and Young collated the data then available which showed rates of surface lowering, estimated from river loads, in the range from 0.002 to 0.05 mm yr⁻¹.

Historical rates of subaerial denudation

Sediment traps

All particulate material in transit along a stream course must eventually be deposited. While it may temporarily come to rest within the stream channel, more permanent deposition normally takes place in either a lake or the sea. It is these sites of longer-term accumulation that are described as sediment traps. They vary enormously in both dimensions and the length of time they have acted as receptacles for detrital material. At one end of the scale are small artificial reservoirs that have been acting as traps for no more than a decade or so. At the other end are the ocean basins which have been receiving the products of continental erosion for millions of years. The basic principles of study are the same for all. The amount of sediment must be determined, the potential source area defined and the period of accumulation delimited. Thereafter it is a relatively easy matter to compute the mean rate of surface lowering represented by the accumulated material. However, each of the preliminary calculations may introduce some degree of uncertainty into the final result. The subaqueous position and high water content of the sediments makes their volume difficult to assess accurately; the period for which a sediment trap has functioned may be subject to doubt; while a final possible source of error is leakage either from or into a sediment trap. The last is important, for instance, in the oceanic environment where marine processes, besides providing a potential extra source of debris, can redistribute material supplied by the rivers. It can also be significant in a small reservoir, where sometimes so much of the finest suspension load fails to settle that a correction factor, known as the 'trap efficiency', needs to be applied. Despite these limitations, sediment traps still afford one of the best means of assessing denudation rates.

In a few areas small reservoirs have been installed for the specific purpose of trapping and measuring sediment. These are particularly

Area		Rate of solutional lowering (mm yr⁻¹)
British Isles	*Yorkshire	0.049–0.050
	*Derbyshire	0.075–0.083
	*East Anglia	0.025
	*South Wales	0.018
	*Co. Clare	0.055
Continental Europe	†Poland	0.032
	*Swiss Alps	0.015
	‡Hungary	0.020
	*Yugoslavia	0.077–0.080
North America	*Jamaica	0.072
	§California	0.017–0.021
	‡Arctic Canada	0.002
Other regions	‡Indonesia	0.083
	‖New Zealand	0.061–0.088
	‡Algeria (Sahara)	0.003

Table 18.3 Estimates of current rates of solution in limestone areas under a variety of climates

* From compilation by Sweeting, *Karst Landforms* (Macmillan), 1972.
† Pulina, *Geogr. Polonica* (1972).
‡ Compilation by Smith and Atkinson, in *Geomorphology and Climate* (ed. Derbyshire), Wiley, 1976. This reference includes a very full list of earlier work.
§ Marchand, *Am. J. Sci.* (1971).
‖ Gunn, *Earth Surf. Processes Landf.* (1981).

appropriate in regions of intermittent stream flow where the total volume of debris deposited during the wet season, or even during an individual storm, can be readily assessed once the water in the reservoir has evaporated. However, reliance normally has to be placed on reservoirs constructed for other purposes. A good example is afforded by Lake Mead on the Colorado River in Arizona. Impounded in 1935 behind the 180 m Hoover Dam, this was the subject of a classic investigation by the United States Geological Survey in 1948. Sediment volumes were computed by comparing bathymetric charts accurate to ±0.6 m with earlier contoured maps. A further check on sediment thickness was provided by low-frequency echo-sounders capable of penetrating the superficial cover but subject to reflection from the original bedrock surface. It was found that, in 14 years, debris had accumulated near the lake head to a thickness of 83 m, while even at the foot of the dam it had reached about 30 m through the action of turbidity currents. To determine the density and composition of the highly fluid sedimentary mantle, 46 cores and over 300 surface samples were recovered for laboratory analysis. From the results it was computed that the Colorado had delivered 1 775 million tonnes of particulate material into the lake at an average annual rate of 127 million tonnes. This figure agrees remarkably well with the 129 million tonnes of suspension load recorded at an old-established gauging station a short distance higher upstream in the Grand Canyon, and suggests that bedload transportation by the Colorado may be comparatively small. For the 14-year period investigated the overall denudation rate of the Colorado catchment upstream from Lake Mead was estimated at 0.16 mm yr^{-1}. It is worth noting that the Lake Mead study could never be repeated in its original form since a major new dam, constructed just upstream from the Grand Canyon in the 1960s, has entirely altered the water and sediment regime of the Colorado, with the suspension load being reduced by no less than 87 per cent.

The techniques adopted on Lake Mead are only appropriate where massive volumes of sediment are involved. Under normal circumstances the cover on a reservoir floor is too thin for indirect measurement, and advantage has to be taken of periods when the water is drained for remedial engineering work. Even then, measuring the volume of new sediment poses substantial problems. Sampling procedures must take account of the many variations in the character of the material. Among factors needing careful evaluation are the water content, the percentage of organic matter and the volume of desiccation cracks. In each of these respects fine muds differ from coarse silts and sands. Given this complexity, estimation of the sediment yield to an accuracy of ±10 per cent still demands a detailed and painstaking survey. In England reservoir studies have yielded estimates of surface lowering ranging from 0.001 2 mm yr^{-1} on the subdued relief of the Midlands, to 0.013 mm yr^{-1} in the more hilly southern Pennines. In central Scotland the same technique has provided figures of 0.010 mm yr^{-1} for a sandstone and mudstone catchment near Edinburgh, and of 0.018 mm yr^{-1} for a sandstone catchment west of Glasgow. These Scottish values seem rather low by comparison with measured suspension loads on nearby rivers, and this may reflect the inefficiency of the traps in catchments that cover only 7 and 3.4 km^2 respectively. All the above British investigations refer to periods of sedimentation lasting between 83 and 121 years.

Assessment of the volume of sediment contained in natural lakes usually depends either upon coring techniques or upon echo-sounding at frequencies capable of penetrating to the underlying rock. The length of time during which accumulation has been taking place is often very problematical, and is one reason that the potential of lacustrine sedimentation has not yet been fully exploited. In glaciated areas it is frequently assumed that lacustrine accumulation began immediately the basin was divested of its ice cover. In such circumstances outwash from the retreating ice would probably lead to rapid deposition at first, with a gradual reduction in rate as the climate ameliorated. To eliminate the effect of glaciation, extensive coring would be necessary so as to identify the separate sedimentary units. Few reliable surveys of such a nature have yet been completed, and most studies have relied instead upon dating horizons within a small number of cores and from this deducing relative rather than absolute rates of accumulation. Dating has usually been achieved by pollen analysis and radio-carbon assays, but two methods of increasing importance are based upon the magnetic properties of the sediment. The first employs the fact that the shifts in the position of the magnetic poles during Holocene time are clearly imprinted on many lake samples, and because the pattern of changes is well known and dated, it can be used as an indication of age. The second method depends upon the property known as the magnetic susceptibility of the sediments. In effect this is the 'magnetizability' of the material, which is found to vary according to the proportion of ferri-magnetic minerals that it contains. The property is easily measured on a core sample, and has proved to offer a sensitive and very rapid means of correlation between cores from different parts of a lake; by using the method in combination with radio-carbon dating, Oldfield and his associates were able to trace changing patterns of sedimentation in lakes in Papua New Guinea for the last 10 000 years. An additional advantage of the technique is that it may eventually prove feasible to identify the exact provenance of lacustrine mineral detritus from the magnetic susceptibility that it displays.

The most important form of sediment trap within the ocean basins is the deep-sea fan. In many instances these are built of detritus emanating in the main from a single continental catchment. The primary mechanism of transfer from the continental shelf to the deep ocean floor is the turbidity current capable of carrying debris 100 km or more in a single movement. Admittedly there are many possible causes of loss or accession of material between the river mouth and the fan surface, but in general the volume of a fan appears to offer a reasonable guide to the rate of mechanical denudation, provided that the date of initiation can be firmly established. The latter often proves a difficult task, although circumstances occasionally lead to a fan being fed with debris for only a short period. This is believed to have happened in the case of the Navy fan off the coast of southern California. Owing to accidents of the local submarine topography, the supply of debris was restricted to a brief interlude of very low sea-level during the last glaciation. The total volume of sediment is about 56 km³ derived from a catchment of 6 500 km²; if accumulation took place for the estimated 15 000 years, the rate of mechanical denudation must have been about 0.6 mm yr⁻¹.

With increased oceanographic research, and particularly with more deep-sea drilling, additional light is being shed on the age of large submarine fans and thereby on continental erosion rates. Geophysical surveys of the huge Ganges–Brahmaputra fan, which covers an area of 3 000 × 1 000 km in the Bay of Bengal, have shown that the sediments often exceed 3 km in thickness. The age of the rocks at the base of the fan suggests it was initiated in Eocene times. Making the necessary allowances for density variations, it was calculated in the early 1970s that some 8 million km³ of rock have been eroded from the Ganges–Brahmaputra basin in the last 20 million years. Of the total area of the catchment about one-quarter is composed of alluvial floodplain and other depositional forms. From the remainder a layer of rock averaging 5.3 km thick has apparently been removed, yielding a long-term rate of mechanical denudation of around 0.26 mm yr⁻¹. However, within the fan a major unconformity points to intermittent deposition that may well be related to separate phases of Himalayan orogenesis. On the basis of known geology a period of rapid uplift has been assigned to early Pleistocene times, and if that part of the fan above the unconformity is entirely Pleistocene in age it implies a recent denudation rate of about 0.7 mm yr⁻¹.

Another huge fan that may eventually prove similarly informative is that of the Amazon. The maximum dimensions of this feature are 700 × 600 km, with a proven thickness of up to 4.2 km. Accumulation started in early Miocene times, possibly in response to major tectonic developments in the Andes. The average long-term accumulation rate has been estimated at between 0.25 and 1.00 mm yr⁻¹, but a significant aspect of the Amazon fan is that, at the present time, huge quantities of clay are carried north-westwards from the river mouth by marine currents and so make no contribution to fan growth. It has been suggested that, during Pleistocene times, sediment accumulation was largely confined to the glacial periods when sea-level was much lower than today.

It is evident that many uncertainties persist regarding large deep-sea fans, but potentialities are still exciting. Moreover, with the

accelerating exploration of the continental shelves it may well become feasible to identify other forms of trap, such as tectonic basins, into which the products of continental erosion have been poured. One possible illustration is the North Sea basin where the search for oil has revealed a major structural depression filled with Cenozoic sediments up to 3.5 km thick and with a total volume approaching 400 × 10³ km³. Many of these are clastic sediments that are undoubtedly the product of erosion from the lands around the North Sea, although without much further research it is still impossible to quantify the relationships; nevertheless, the prospects of ultimately integrating the marine sedimentary record with the erosional history of, say, the Scottish Highlands remain good.

Calculations based upon geological relationships

Many assessments of long-term denudation rates have been founded either on the destruction of some datable topographic feature or on the dissection of some parent material of known age. In each case the time constraints are provided by geological evidence. For example, where horizontal marine strata overlie rocks severely deformed in an orogenesis, the time interval represented by the unconformity indicates the maximum period taken to destroy any mountains that were formed; using this approach it has been argued that on a continental margin a major mountain chain can be completely erased in a period of 20–40 m.y. Likewise, where plateaux are formed by dissection of uplifted strata, incision of the drainage cannot have begun prior to deposition of the youngest continuous bed. In both these examples basic principles define the maximum period available for fluvial erosion; in many instances supplementary evidence permits a closer definition of the varying rate at which denudation has operated.

In the foregoing pages recent erosion rates have been described from the Colorado valley in the United States and the southern Pennines in Great Britain, and it is instructive to review the light shed by geological considerations on the long-term rates in these two areas. On the simple principle enunciated in the preceding paragraph, excavation of the Grand Canyon (Fig. 18.3) cannot have

Fig. 18.3. A rim-to-rim view of the Grand Canyon showing, in the distance, the remarkably smooth plateau surface formed by the Permian Kaibab limestone (compare with Fig. 18.4).

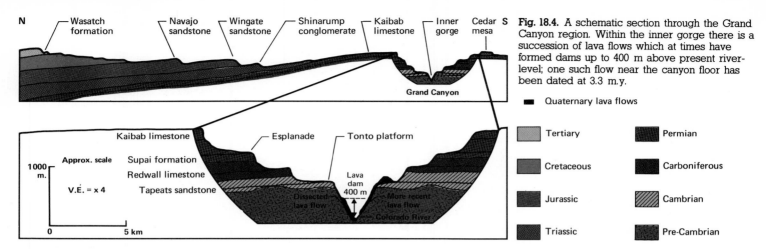

Fig. 18.4. A schematic section through the Grand Canyon region. Within the inner gorge there is a succession of lava flows which at times have formed dams up to 400 m above present river-level; one such flow near the canyon floor has been dated at 3.3 m.y.

- ■ Quaternary lava flows

Tertiary	Permian
Cretaceous	Carboniferous
Jurassic	Cambrian
Triassic	Pre-Cambrian

begun before Mesozoic times since its rim is formed by the Permian Kaibab limestone (Fig. 18.4). Beyond the rim are dissected remnants of formerly continuous marine strata of which the youngest are Eocene in age; nearly all workers have assented to the proposition that this constitutes the earliest possible date for initiation of the canyon. Even so, much uncertainty still surrounds the exact period at which the Colorado first assumed its present course and began incising its valley. Where the river now emerges from the Grand Canyon near Lake Mead, mid-Cenozoic limestones dated by interbedded volcanics with a K/Ar age of 8.7 m.y. bear no sign of accumulation at the debouchment of a major sediment-carrying river; on this evidence it seems improbable that the major incision began more than 8.7 m.y. ago. Further important time constraints are provided by lava flows within the canyon itself, one of which has been dated as 3.3 m.y. old and lies only 100 m above present river-level. In other words, of a total incision of 1 500 m, over 90 per cent seems to have been achieved in an interval of less than 5.4 m.y. at an average rate of about 0.26 mm yr^{-1}. Although it is difficult to compare this figure with a denudation rate for the whole catchment, all assessments point to the Colorado basin as an area of rapid geomorphic change.

In the southern Pennines the rocks at present undergoing erosion are of Carboniferous age, but it has been suggested that a chalk cover formerly extended across the whole region (Fig. 18.5). The minimum feasible elevation for the Cretaceous rocks is provided by the 600 m plateau of Kinderscout, and if the general argument is sound it implies that during the last 70 m.y. denudation has removed the Mesozoic cover and penetrated at least 450 m into the underlying strata. Although such a sequence may be correct, it provides a very inadequate basis for attempts to assess local denudation rates. For that purpose a much more closely controlled chronological framework is required. This is not easily obtained in the southern Pennines, even though many workers have identified one or more erosion surfaces truncating the dome of Carboniferous rocks (Fig. 18.6). The most prominent is the undulating Upland Surface which bevels the limestone in the core of the dome at an elevation of between 300 and 360 m. Within deep solution hollows on this plateau lie detrital sediments containing distinctive floral remains of middle or upper Cenozoic age. On the assumption that the Upland Surface was originally fashioned by subaerial processes some 5 m.y. ago, the subsequent incision by the streams has been achieved at a mean rate of no more than 0.03 mm yr^{-1}. This cannot, of course, be more than a very crude estimate, but further refinement may be possible if sufficient uranium-series dating of speleothems in some of the limestone caverns can be achieved. In the central Pennines a short distance to the north over eighty spelaeothems have been dated and

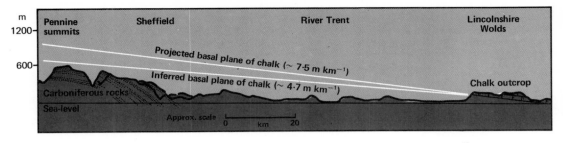

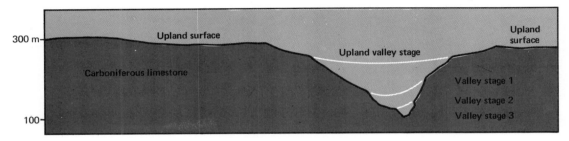

Fig. 18.5. The geomorphic evolution of the Pennines as visualized by Linton (1956) who suggested that the present drainage was initiated on a former cover of chalk. On the top diagram, the upper line represents a simple extrapolation of the present dip of the Chalk in Lincolnshire, the lower line a position for the Chalk base which Linton regarded as more likely. On the bottom diagram, detailed stages in the denudational history subsequent to the unroofing of the Derbyshire dome are depicted. From the valley cross-profiles it is inferred that the region has been subject to intermittent uplift and rejuvenation of the drainage.

these begin to provide a more detailed chronology with assessments of stream incision varying between 0.05 mm yr^{-1} and a maximum of 0.2 mm yr^{-1}. These figures for the last 350 000 years must obviously be regarded as averages of rates that could have fluctuated dramatically under glacial and interglacial climatic regimes.

Spatial variability in denudation rates

Global patterns

As soon as the number of sediment-gauging stations throughout the world began to grow, it became feasible to examine the factors likely to influence the rate of sediment yield. As already seen, both Milliman and Meade in the case of fluviatile sediments and Meybeck in the case of solution loads have attempted to extrapolate from the evidence of reliable gauged catchments to other areas so as to predict aggregate global values. Yet it is worth noting that the topic of environmental controls of current denudation rates has occasioned substantial conflict in the past. In the 1960s several workers claimed to be able to discern patterns of global erosion related to either climate or topography or some combination of the two. Corbel, for instance, published an analysis of eighty catchments in which he identified annual temperature and precipitation as the dominant climatic factors, and relief amplitude as the dominant topographic factor (Fig. 18.7). In each precipitation class the denudation rate was believed to vary inversely with temperature; at the same time in each temperature class it varies directly with precipitation. This meant that, for the same topographic form, the highest rates would be in cold, wet regions and the lowest in hot, arid regions, with as much as a hundredfold difference between the extremes. Mountainous regions were thought to undergo erosion at a rate between two and five times that of plainlands in the same climatic class, but the very highest values were reserved for areas currently experiencing glaciation; this view he supported by detailed work in the French Alps where, after several years of monitoring, he concluded that the floor of the St Sorlin Glacier is being lowered at a rate of 2.2 mm yr^{-1}, a

Fig. 18.6. The undulating plateau of the southern Pennines truncating the limestone outcrop in the core of the Derbyshire dome (cf. Fig. 18.5). Steeply incised valleys like Dovedale here, have been held to indicate the polycyclic evolution of the landscape, although Pitty (1968) has offered an alternative interpretation for the origin of the upland plateau (p. 401).

figure many times that recorded from adjacent valleys. Corbel calculated the mean annual denudation rate for all non-glacial lands as about 0.028 mm.

Another French worker, Fournier, plotted the sediment yield from seventy-eight individual catchments against selected precipitation indices representing annual total, seasonality and intensity. He found particular significance attaching to a weighted measure of seasonality defined as the square of the precipitation in the rainiest month divided by the mean annual total. When this measure was plotted against sediment yield, the catchments appeared to fall into four distinct categories: (a) gently sloping temperate regions; (b) gently sloping tropical, subtropical and semi-arid regions; (c) steeply sloping humid regions; and (d) steeply sloping semi-arid regions. By assigning all major drainage basins to one of these categories, and calculating for each the seasonality measure referred to above, Fournier estimated the mean rate of lowering for all continental areas as 0.4 mm yr^{-1}, a figure more than a full order of magnitude greater than that given by Corbel and implying a strikingly different distribution pattern. The highest sediment yields according to Fournier occurred in tropical monsoon and savanna regions, although the latter, according to Corbel, were among the areas of least erosion. Conversely, Fournier ranked cold regions as zones of minor sediment yield, whereas Corbel argued that they suffer above-average denudation.

Russian workers were also active in the analysis of gauging records. The best known was Strakhov who, in 1967 emphasized the wide range of values for surface lowering implied by the data then available for sixty of the world's largest rivers. He maintained that two fundamentally different zones were identifiable. The first comprises the temperate moist belt of the Northern Hemisphere generally receiving 150–600 mm annual precipitation and defined on its southern margin by the 10°C mean annual isotherm; within this zone denudation rates rarely exceed 0.006 mm yr^{-1}. The second zone embraces the moist tropical and subtropical regions with annual rainfall in excess of 1 200 mm and mean annual temperature above

10°C; here the denudation rate normally lies between 0.02 and 0.04 mm yr^{-1}, but in large parts of south-eastern Asia surpasses 0.35 mm yr^{-1}. Strakhov regarded relief as a further major factor, holding that erosion rates vary directly with the amount of tectonic uplift. He also emphasized the importance of seasonality in the rainfall regime, arguing that a prolonged dry spell thoroughly loosens the surface soil which, in consequence, is more easily removed during later downpours. From the available data Strakhov formulated the additional rule that chemical denudation, despite regional variations, tends to increase with rising sediment yield. He thought lowland rivers normally carry more material in solution than in suspension, while the reverse is true of mountain streams. Yet most upland rivers, he believed, carry more solutes per unit area of catchment than do plainland rivers. His calculation of gross annual yield from all non-glacial continental areas implied a mean denudation rate of approximately 0.05 mm yr^{-1}. Strakhov thus agreed reasonably closely with Corbel on the general magnitude of global erosion but almost exactly reversed his climatic weighting; on the other hand, Strakhov's regional distribution pattern resembled that of Fournier but with the absolute values much reduced.

At the same time as the workers mentioned above were employing global statistics, the American geomorphologist Schumm was analysing the more consistent data available from the varied climatic regions of the United States and concluding that relief and runoff were the major determinants of sediment yield. He contended that the yield is an exponential function of relief when the latter is defined as basin relief divided by basin length. He also argued that the highest denudation rates occur in semi-arid areas with adequate rainfall to promote active runoff, but too little to sustain a dense cover of protective vegetation. The actual precipitation values necessary to achieve maximum erosion will depend upon temperature (Fig. 18.8); with a mean annual temperature of 10°C the required rainfall is 300 mm, but with the temperature raised to 20°C precipitation must exceed 600 mm. Schumm concluded that the maximum feasible denudation rate approximates 1.0 mm yr^{-1} and that the mean rate probably lies between 0.03 and 0.09 mm yr^{-1}, both figures relating to combined suspension and solution loads.

The possibly overriding importance of relief was stressed in a number of analyses of published data undertaken by Young around 1970. Grouping the figures into two classes, he showed that they are

Fig. 18.7. An interpretation of the views of Corbel (1964), Fournier (1960) and Strakhov (1967) on current rates of erosion (for purposes of comparability a density of $2\,700\ \mathrm{kg\,m^{-3}}$ has been assumed for the material removed).

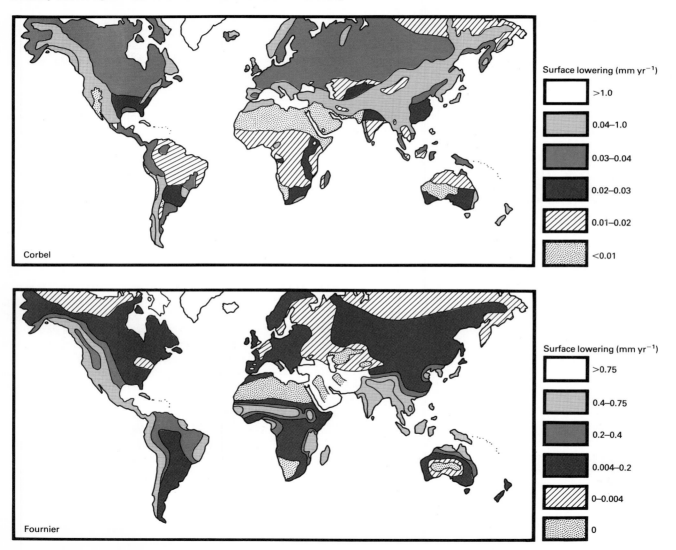

separated by an order of magnitude in terms of denudation rate (Fig. 18.9): for areas of normal relief he quoted a median rate of 0.046 mm yr^{-1}, and for areas of steep relief a median rate of 0.5 mm yr^{-1}. In a further study in collaboration with Saunders, Young in 1983 suggested that, owing to the large ranges being reported from different areas, the figures might better be quoted as 0.01–0.1 mm yr^{-1} for normal relief and 0.1–1.0 mm yr^{-1} for steep relief.

This summary of work by a wide variety of authors has revealed some glaring discrepancies in both postulated controls and overall rates of denudation. It is of interest, therefore, to compare the findings with those of Milliman and Meade in their collation of the available sediment-transport data for the period up the early 1980s. Their estimate for the gross annual sediment load being delivered to the oceans, 16×10^9 tonnes, is a slight reduction on the figure computed by Strakhov, but incorporates no allowance for the solution load. The regional distribution pattern (Fig. 18.1) bears reasonably close resemblance to that deduced by Fournier and Strakhov. It would be premature to assume that future research will not disclose unexpected patterns in current denudation rates, but a consensus does appear to be emerging that many areas of moderate relief are at present being lowered by about 0.05 mm yr^{-1} and mountainous regions by about 0.5 mm yr^{-1}. What remains much less clear is the nature and degree of climatic control.

Local patterns

In the collection of data relating to current denudation rates, it is common practice to convert the measured load into an average lowering of the whole catchment. This procedure often has to be adopted merely for want of any alternative, since the precise sources of the material are not known. However, it must not be inferred from the quotation of mean catchment figures that there may not be great disparities within individual river basins. It has already been mentioned, for instance, that parts of the Yellow River basin are aggrading while others are being denuded, and that, even on the basis of solution load, there is at least a fivefold difference between various parts of the Exe basin. Occasionally, geological evidence can afford insights into long-established contrasts between different segments of a single catchment. One example is provided by the extrusion of a fluid lava that may flow along existing drainage lines and, after consolidation, act as such a protective cover that the old valley floors are preserved almost intact. This has happened in the

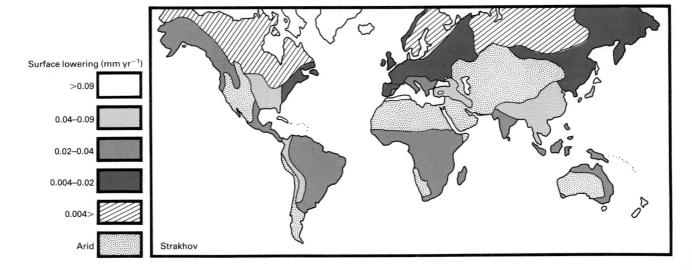

Surface lowering (mm yr^{-1})

>0.09

0.04–0.09

0.02–0.04

0.004–0.02

0.004>

Arid

Strakhov

Fig. 18.8. The relationship between denudation rates and precipitation as computed (above) by Schumm in 1965 from US data, and (below) by Walling and Kleo in 1979 from more extensive global data. Schumm suggested that denudation reaches its maximum in semi-arid lands, with the critical figure varying according to the mean annual temperature. Walling and Kleo identified a comparable semi-arid peak, but also detected evidence for further peaks associated with highly seasonal rainfall in areas of Mediterranean and monsoonal climate.

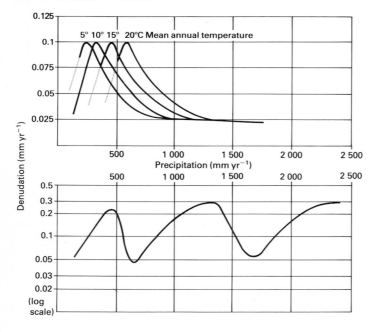

Fig. 18.9. Rates of surface lowering as measured by various workers and collated by Saunders and Young (1983). On the right are shown the median values and inter-quartile ranges for steep and normal relief as calculated by Young in 1972 from a more limited data set.

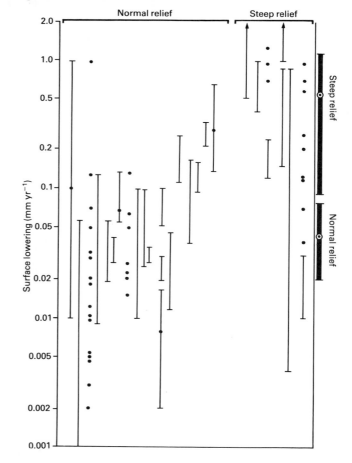

case of the Little Colorado valley which joins the main Colorado just upstream from the Grand Canyon. Here the Little Colorado has managed to incise its course almost 200 m during the last 2 m.y., while some of its tributaries in the same period have achieved negligible downcutting. The same basic technique of comparing a fossilized sublava topography with the modern relief can also demonstrate startling contrasts in the long-term denudation rates on different lithologies. To quote but one example from the Little Colorado basin, a lava flow with a K/Ar date of 500 000 years now winds as a prominent ridge across an area of denuded mudstones.

When first extruded the lava must have flowed down a steep-walled valley at least 20 m deep, but the whole topography has since been inverted with the lava remaining largely unmodified but erosion of the mudstone averaging in excess of 0.04 mm yr^{-1}.

Temporal variability in denudation rates

The impact of Man

As more refined assessments of current denudation rates become available, it is extremely tempting to extrapolate the values backwards in time. Yet such a procedure is fraught with difficulty. This is primarily because, in the last few centuries, the impact of Man has been so profound and so pervasive. There is scarcely a river basin in the world where he has not had a marked influence on present-day geomorphic processes. Through all stages of the hydrological cycle his influence, although often difficult to evaluate, is undoubtedly present. Activities like reservoir construction and urbanization have had unplanned but measurable effects upon local climates, to say nothing of deliberate interference as in attempts to alleviate aridity by cloud-seeding. Even the chemical composition of the rainfall has been altered so that downwind from many industrial areas the water is demonstrably more acidic. Yet the greatest changes wrought by Man arise from his modifications of land-use. These affect not only the nature of the sediment supply to the streams but also the stream regimes themselves.

Man first had a significant impact on the natural erosional processes at the time of the Neolithic revolution. Land under crop production is laid bare for at least a short period each year during which it is subject to increased rainsplash and surface wash. This is a particularly severe problem on steep slopes, well exemplified by the experience of farmers in the Tennessee valley during the first half of this century. Even under mature crops, areas of bare soil may still lie open to attack. A further consideration is the disturbance by ploughing which tends to accelerate the rate of creep on cultivated hillslopes. Although the conversion of land from forest to permanent pasture may have fewer obvious and immediate consequences, in theory it will tend to modify hillslope processes by its influence on such factors as rainfall interception, soil organic activity and leaching. It has even been claimed that in upland Britain deforestation promoted extension of the blanket peat; this would clearly alter the hillslope processes quite drastically. In less humid regions over-grazing can produce dramatic changes when the protective grass sod is broken and areas of soil are laid bare to erosion by both wind and rain.

The magnitude of the changes arising from these agricultural practices is still not firmly established. Ideally the impact of human activity might be regarded as a third variable to be considered alongside relief and climate as determinants of current erosion rates; ultimately some form of multivariate analysis may be possible, but at present the data are inadequate to sustain such a sophisticated approach. Strakhov commented on what he saw as anomalously high denudation rates on the European plainlands and in the Mississippi valley, contending that these show agricultural activities to multiply natural erosion rates two, three or four times. This approach presupposes that the writer can accurately forecast the natural rate, but the continuing doubt over the significance of various climatic controls indicates the need for caution. In the long term a more reliable approach is likely to be that of comparing sediment yields from two small catchments similar in all respects save land-use; for example, one drainage basin may be subject to arable farming while the other remains under woodland. In pursuit of this idea, suspension load data have been compiled for a series of small catchments along the middle Atlantic seaboard of the United States. These suggest that conversion from forest to arable cultivation in this area increases the sediment yield tenfold. Yet the severity of the enhanced erosion will obviously depend on local farming practices. Sediment yield from cultivated land in parts of Mississippi has been estimated at up to 100 times that from forested land, whereas from mixed farmland in Oregon the corresponding figure is nearer three times. The impact of grazing has been similarly assessed by controlled experiments in an arid region of Colorado. Of eight minor catchments, four were left open to grazing animals and four were fenced off. By means of small reservoirs specially constructed as sediment traps, those protected from grazing were found to produce much less material than those which remained unfenced.

Urbanization and industrialization also transform the natural processes. When the ground is first disturbed during construction work the rate of sediment yield is often enormously increased. Mean lowering of some small catchments in the Washington metropolitan area of the United States has temporarily attained a rate as high as 5 mm yr^{-1}. Once building is complete the extensive cover of concrete and other resistant materials causes a rapid fall in sediment production. However, an equally important consequence of the impermeable cover is the change in river regime. The proportion of rainfall entering the stream channels as surface runoff is greatly increased and the result is a much more flashy flow. This effect has long been recognized by engineers concerned with the design of artificial conduits, but its geomorphic implications have yet to be fully investigated. More frequent flooding has been reported downstream from many urban areas, and in theory the character and velocity of the sediment load should be altered. It must also be remembered that, in many agricultural regions, tile drains and other artificial means of removing water contribute their share to the change in river regime.

The use of natural stream channels for the disposal of urban and industrial waste can lead to a radical alteration in the composition of river water. One manifestation is the dramatic change in the quality of lake water in many parts of the world. The deterioration of Lakes Erie and Ontario, for instance, has necessitated international legislative action. In cases of bad pollution, evaluation of chemical denudation may be virtually impossible. In agricultural regions, lime added to the fields to combat leaching is more readily mobilized than bedrock calcium carbonate, a potentially significant factor when attempting to assess solution rates in limestone areas. It must also be remembered that chemical and biochemical activity can modify processes within the stream channels themselves; for example, waters enriched with plant nutrients may promote algal and other aquatic growth that would not otherwise be present. Lest this be regarded as a catalogue of deteriorating conditions, it should be recorded that the quality of the waters in a number of rivers has been improved during the last few years. However, this has scarcely rendered them suitable for assessing long-term denudation rates; for that particular purpose, clean urban effluent is little better than dirty urban effluent.

A final illustrative example of the impact of Man upon current denudational processes is provided by the pressure of recreational activities. It is a matter of common observation that heavily used pathways, particularly on steeper slopes, are now suffering severe degradation. The width of pathways, regarded as strips of bare ground, has grown rapidly as usage by walkers has increased, and in many instances the destruction of the protective vegetation cover has led to severe gullying by storm runoff. In the worst instances whole areas have had to be closed to public access, while in many others remedial measures have had to be put in hand to prevent further deterioration of the environment. All such cases, however, highlight the inadvertent effect that Man's activities can have.

The significance of climatic change

If the consideration of denudation rates over the last few centuries must pay due regard to the impact of Man, their consideration over longer periods must equally take cognizance of the potential effects of climatic change. Although, as already seen, it has proved difficult to quantify the precise nature of the climatic controls on denudation rates, almost all investigators have accepted that they do exert a profound influence. This means that once consideration is being given to periods in excess of a few millennia, the likely effects of climatic change must receive due attention.

As previously mentioned in Chapter 2, the last few decades have seen a remarkable advance in our appreciation of the complexity of global climatic history during the Pleistocene epoch. These advances have been stimulated in particular by work on deep-sea cores with their relatively complete stratigraphic record. It is now apparent, moreover, that almost no area of the world has escaped significant environmental fluctuations. It was at one time believed that the equatorial regions might have suffered no change, but there is now sufficient evidence from a variety of sources to demonstrate important fluctuations in these latitudes as well. In Papua New Guinea during the last 20 000 years major vegetation zones have risen in elevation by a much as 1 700 m, while in the same period equatorial forest in the Amazon basin has spread across vast tracts of what were formerly savanna grasslands. In the circumstances it would obviously be impossible to review all the evidence on with current ideas about climatic change are based, and it seems prudent to restrict discussion to two related topics, namely the time-scales of climatic fluctuations, and attemps to reconstruct climatic distribution patterns for specific intervals in the past.

One of the most striking consequences of the relatively full climatic record preserved in deep-sea cores has been the revival of an old hypothesis associated with the name of the Yugoslavian mathematician Milankovitch. He had argued, during the first few decades of the present century, that astronomical variations in the movement of the earth round the sun would generate cyclical changes in the insolation received at different latitudes, and that these would be capable of inducing climatic fluctuations. The astronomical varia-

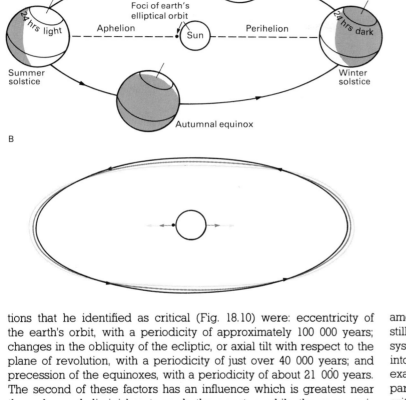

A

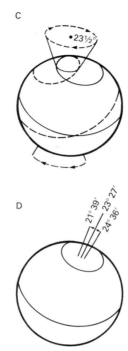

Fig. 18.10. Schematic portrayal of the major variables invoked in the astronomical theory of climatic change. (A) Basic relationships as they are at the present day, i.e. low eccentricity, axis tilted at 23° 27' relative to the plane of the ecliptic, winter solstice close to perihelion. (B) Eccentricity of the earth's orbit, periodicity *c* 100 000 years. The sun lies at one focus of the elliptical orbit, the other focus is unoccupied. (C) Axial precession, periodicity *c*. 21 000 years, leading to the astronomical phenomenon known as precession of the equinoxes. In about 11 000 years the winter solstice will be close to aphelion. (D) Axial tilt, periodicity *c*. 41 000 years.

tions that he identified as critical (Fig. 18.10) were: eccentricity of the earth's orbit, with a periodicity of approximately 100 000 years; changes in the obliquity of the ecliptic, or axial tilt with respect to the plane of revolution, with a periodicity of just over 40 000 years; and precession of the equinoxes, with a periodicity of about 21 000 years. The second of these factors has an influence which is greatest near the poles and diminishes towards the equator, while the reverse is true of the precessional changes. Moreover, at any point on the globe, the combined effect will be different at different seasons of the year so that these three variables by themselves are capable of generating very complex patterns of incident solar radiation.

To comprehend how any such pattern may be converted into a climatic change must ultimately involve consideration of the atmospheric and oceanic circulations of the globe which transfer massive amounts of heat energy from one location to another. However, we still have limited understanding of the dynamics of these circulatory systems, and the reason that the deep-sea cores breathed new life into the Milankovitch hypothesis was that they afforded a means of examining the time-scales of environmental change. The upper parts of a deep-sea core could be dated by radio-carbon, and a critical transition at greater depth could be identified in the Brunhes–Matuyama polarity reversal; interpolation within the rest of the Brunhes magnetic epoch thus permitted a time-scale to be applied to the saw-tooth record of O^{18}/O^{16} changes over the last 700 000 years. A number of cores with fast accumulation rates were then carefully selected so that the timing of oxygen isotope fluctuations could be subjected to rigorous statistical examination. Using the technique known as spectral analysis, it was shown that the

Fig. 18.11. A reconstruction by the CLIMAP team (1976) of global conditions during a Northern hemisphere summer 18 000 years ago. The primary aim in preparing this map was to estimate the surface albedo at the height of the last glaciation. The categories in the key reflect this objective, and are not always readily translated into vegetation zones corresponding in detail to those of the present day. Nevertheless, future research will undoubtedly permit a progressive refinement of this map, with further subdivision of some of the broad categories currently depicted.

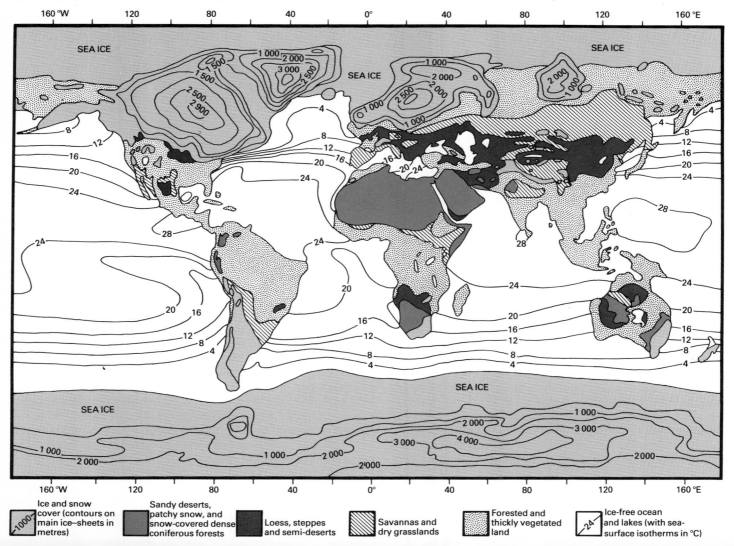

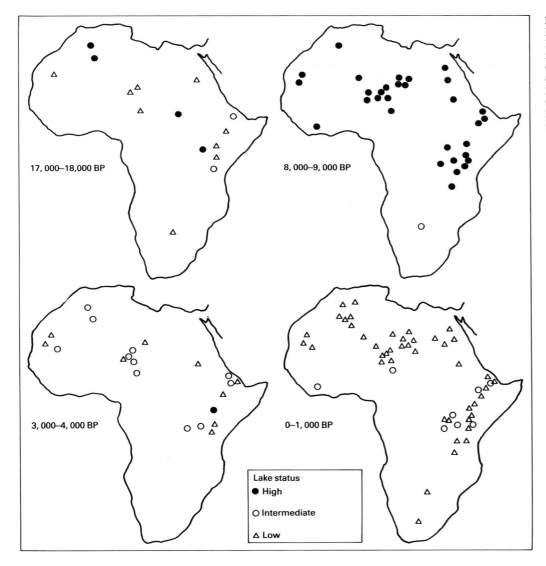

17,000–18,000 BP

8,000–9,000 BP

3,000–4,000 BP

0–1,000 BP

Lake status
● High
○ Intermediate
△ Low

Fig. 18.12. The changing pattern of lake-levels in Africa during the last 18 millennia. Each lake is characterized by its status relative to its long-term minimum water-level; dates are determined by radio-carbon assays; the changing distribution of data points reflects availability of evidence. The outstanding contrast is that between the present day and 8 000–9 000 years ago (after Street and Grove, 1976).

O^{18}/O^{16} record contains peaks corresponding to cycles with period-icities of 100 000, 43 000 and around 21 000 years, almost exactly matching the timing predicted by the astronomical theory. It was after an investigation along these lines that Hays, Imbrie and Shackle-ton in 1976 confidently described the variations in the earth's orbit as the 'pacemaker of the ice ages'. Of course, this does not preclude the possibility of other factors influencing the climatic record, nor does it afford any real insight into the detailed mechanisms of climatic change. However, it does imply rapid and dramatic fluctuations with a frequency in the range 10^4–10^5 years.

Some workers have hypothesized that there may be two or three stable circulatory systems between which the atmosphere 'flips' when triggered by changes in the insolation pattern. As a guide to conditions during a glaciation, the CLIMAP team (p. 31) endeav-oured to reconstruct the climatic environment of the earth as it existed 18 000 years ago (Fig. 18.11). This date was selected because it appears to coincide with the culmination of the Scandinavian and North American ice-sheets during the last glaciation and, equally important, to lie well within the range of radio-carbon dating which can therefore be used to correlate the record from many different regions of the world. In the Northern Hemisphere of 18 000 years ago, one of the most striking differences from today, in addition to the huge land-based ice-sheets, was a pronounced increase in the extent of permanent pack ice and even the development of marine-based ice-sheets. There was a dramatic steepening of the thermal gradient across the surface waters of the North Atlantic in latitudes 40–50°N, with an almost complete suppression of the North Atlantic Drift. The waters were locally over 10 °C colder than at the present day, although the average for the surface waters of the globe as a whole was probably less than 2.5 °C; this is partly because the main oceanic gyres in mid to low latitudes experienced deviations from modern temperatures of less than 1 °C. On the continents, grass-lands, steppes and deserts all expanded at the expense of the forests. It is difficult to convert the vegetational changes to precise temperature values, although, as with the oceans, the cooling within the tropics appears to have been substantially less than that at high latitudes; near the equator estimates of the depression in the mean annual temperature often lie in the range 3–6 °C, while close to the ice-sheets the corresponding figures are 10–15 °C or even more.

Attempts to reconstruct, for a specific interval of time, former environmental patterns at both the global and regional scale are growing in number. Some of the most striking results have come in modern desert areas from efforts to plot the changing distribution of water-levels in enclosed basins that have periodically been the sites of 'pluvial' lakes. Using radio-carbon dates, Street and Grove in 1976 plotted, for each millennium since 18 000 BP, the conditions of the lake basins of Africa (Fig. 18.12). They were able to demonstrate patterns which completely undermined an earlier view that wet periods here coincided with glacial periods in Europe. Eighteen thousand years ago most lakes in central Africa were low, and only those near the northern and southern margins of the continent were relatively high. This phase of Saharan aridity has been confirmed by Atlantic deep-sea cores which have yielded large quantities of contemporaneous aeolian silt, and by groundwater investigations which have revealed little aquifer recharge in the period between 20 000 and 14 000 years BP. The major phase of high lake-levels in Africa appears to have occurred around 8 000–9 000 years ago, at a time when the climate in northern Europe had already begun to assume an interglacial rather than a glacial aspect. It would be erroneous, however, to treat the African deserts as representative of all arid regions, since later work by Street and Grove in 1979 amply demonstrated contrasting sequences in other parts of the world. One clear conclusion that can be drawn is that not all desert regions have experienced the same chronology of climatic change.

This brief review of climatic change, a complex subject with numerous ramifications, serves to underline both the frequency and the amplitude of the fluctuations. The thrust of much recent research has been to emphasize the speed with which changes occur, and it now seems clear that palynological studies often smooth the abrupt-ness of the climatic record since there is a lag while vegetation adapts to the altered conditions. It is obvious that climatic change must enter as a major factor both in long-term denudation rates, and in the concepts of land form evolution to be considered in the final chapter of this book.

References

Al-Ansari, N. A. and McManus, J. (1979) 'Fluvial sediments entering the Tay estuary: sediment discharge from the River Earn', *Scot. J. Geol.* **15**, 203–16.

Blair, W. N. and Armstrong, A. K. (1979) 'Hualapai Limestone member of the Muddy Creek formation: the youngest deposits pre-dating the Grand Canyon', *U.S. Geol. Surv. Prof. Pap.* 1111.

CLIMAP Project Members (1976) 'The surface of the ice-age earth', *Science N. Y.* **191**, 1131–44.

Corbel, J. (1964) 'L'érosion terrestre, étude quantitative', *Ann. Geogr.* **68**, 97–120.

Cummins, W. A. and Pottter H. R. (1967) 'Rate of sedimentation in Cropston reservoir, Charnwood Forest, Leicestershire', *Mercian Geol.* **2**, 31–9.

Curray, J. R. and Moore, D. G. (1971) 'Growth of the Bengal deep-sea fan and denudation in the Himalayas', *Geol. Soc. Am. Bull.* **82**, 563–72.

Damuth, J. E. *et al.* (1983) 'Age relationships of distributary channels on the Amazon deep-sea fan: implications for fan growth patterns', *Geology* **11**, 470–3.

Flenley, J. R. (1979) 'The late Quaternary vegetational history of the equatorial mountains', *Prog. Phys. Geogr.* **3**, 488–509.

Flenley, J. R. (1984) 'Andean guide to Pliocene–Quaternary climate', *Nature, London* **311**, 702–3.

Fournier, F. (1960) *Climat et érosion: la relation entre l'érosion du sol par l'eau et les précipitations atmosphériques*, Presses Univ. de France.

Gascoyne, M., Ford, D. C. and Schwarcz, H. P, (1983) 'Rates of cave and landform development in the Yorkshire dales from speleothem age data', *Earth Surf. Processes Landf.* **8**, 557–68.

Gibbs, R. J. (1977) 'Distribution and transport of suspended particulate material of the Amazon River in the ocean', in *Estuarine Processes II* (ed M. Wiley), Academic Press.

Hammen T. van der (1974) 'The Pleistocene changes of vegetation and climate in tropical South America', *J. Biogeogr.* **1**, 3–26.

Hays, J. D., Imbrie J. and Shackleton N. J. (1976) 'Variations in the earth's orbit: pacemaker of the ice ages', *Science N. Y.* **194**, 1121–32.

Kerr, R. A. 1984 'Climate since the ice began to melt', *Science N. Y.* **226**, 326–7.

Ledger, D. C., Lovell, J. P. B. and Cutttle, S. P. (1980) 'Rate of sedimentation in Kelly reservoir, Strathclyde'. *Scot. J. Geol.* **16**, 281–5.

Lewin, R. (1984) 'Fragile forests implied by Pleistocene pollen', *Science N. Y.* **226**, 36–7.

Linton, D. L. (1956) 'Geomorphology', in *Sheffield and its Region*, Brit. Ass. Adv. Sci.

Lovell, J. P. B. *et al.* (1973) 'Rate of sedimentation in the North Esk reservoir, Midlothian', *Scot. J. Geol.* **9**, 57–61.

Lubsy, G. C. *et al.* (1963) 'Hydrologic and biotic characteristics of grazed and ungrazed watersheds of the Badger Wash basin in western Colorado', *U.S. Geol. Surv. Water Supply Pap.* 1532–B.

Meade, R. H. (1969) 'Errors in using modern stream-load data to estimate natural rates of denudation', *Geol. Soc. Am. Bull.* **80**, 1265–74.

Meybeck, M. (1979) 'Concentrations des eaux fluviales en elements majeurs et apports en solution aux oceans', *Rev. Geol. Dyn. Geogr. Phys.* **21**, 215–46.

Milliman, J. D. and Meade, R. H. (1983) 'World-wide delivery of river sediment to the oceans', *J. Geol.* **91**, 1–21.

Moore, P. D. (1973) 'The influence of prehistoric cultures upon the initiation and spread of blanket bog in upland Wales', *Nature, London* **241**, 350–3.

Moore, D. G. *et al.* (1974) 'Stratigraphic-seismic section correlations and implications to Bengal fan history', in *Initial Rep. Deep Sea Drilling Project*, Vol. 22, pp. 403–12.

Normark, W. R. and Piper, D. J. W. (1972) 'Sediments and growth pattern of Navy deep-sea fan, San Clemente basin, California borderland', *J. Geol.* **80**, 198–223.

Oldfield, E., Appleby, P. G. and Thompson, R. (1980) 'Palaeoecological studies of lakes in the highlands of Papua New Guinea', *J. Ecol.* **68**, 457–77.

Pitty, A. F. (1968) 'The scale and significance of solutional loss from the limestone tract of the southern Pennines', *Proc. Geol. Ass.* **79**, 153–77.

Rice, R. J. (1980) 'Rates of erosion in the Little Colorado valley, Arizona', in *Timescales in Geomorphology* (ed. R. A. Cullingford, D. A. Davidson and J. Lewin), Wiley.

Saunders, I. and Young, A. (1983) 'Rates of surface processes on slopes, slope retreat and denudation', *Earth Surf. Processes Landf.* **8**, 473–501.

Schumm, S. A. (1965) 'Quaternary paleohydrology', in *The Quaternary of the United States*, (ed. H. E. Wright and D. G. Frey) Princeton Univ. Press.

Smith, W. O. *et al.* (1960) 'Comprehensive survey of sedimentation in Lake Mead, 1948–9', *U.S. Geol. Surv. Prof. Pap.* 295.

Strakhov, N. M. (1967) *Principles of Lithogenesis* (trans. J. P. Fitzsimmons), Oliver and Boyd.

Street, F. A. (1981) 'Tropical palaeoenvironments', *Prog. Phys. Geogr.* **5**, 157–85.

Street, F. A. and Grove, A. T. (1976) 'Environmental and climatic implications of Late Quaternary lake level fluctuations in Africa', *Nature London* **261**, 385–90.

Street, F. A. and Grove, A. T. (1979) 'Global maps of lake-level fluctuations since 30,000 years B.P.', *Quat. Res.* **12**, 83–118.

Thompson, R. *et al.* (1980) 'Environmental applications of magnetic measurements', *Science N.Y.* **207**, 481–6.

Walling, D. E. and Webb, R. W. (1978) 'Mapping solute loadings in an area of Devon, England', *Earth Surf. Processes* **3**, 85–99.

Walling, D. E. and Kleo A. H. A. (1979) 'Sediment yields of rivers in areas of low precipitation: a global view' in *The Hydrology of Areas of Low Precipitation*, IASH pub. 128, 479–93.

Walsh P. T *et al.* (1972) 'The preservation of the Neogene Brassington formation of the southern Pennines and its bearing on the evolution of upland Britain', *J. Geol. Soc. Lond.* **128**, 519–59.

Waylen, M. J. (1979) 'Chemical weathering in a drainage basin underlain by Old Red Sandstone', *Earth Surface processes* **4**, 167–78.

Williams, G. P. and Wolman, M.G. (1984) 'Downstream effects of dams on alluvial rivers', *U.S. Geol. Surv. Prof. Pap.* 1286.

Woodland, A. W. (ed.) (1975) *Petroleum and Continental Shelf of Northwest Europe*, Appl. Sci. Pub.

References

Young A. (1958) 'A record of the rate of erosion on Millstone Grit', *Proc. Yorks. Geol. Soc.* **31**, 149–56.

Young, A. (1969) 'Present rate of land erosion', *Nature, London* **224**, 851–2.

Young, A. (1974) 'The rate of slope retreat', in *Progress in Geomorphology,* Inst. Brit. Geogr. Spec. Pub. 7.

Selected bibliography

The literature on denudation rates remains very scattered so that supplementary reading is difficult to specify. However, a collection of relevant published papers has been gathered together in J. B. Laronne and M. P. Mosley, *Erosion and Sediment Yield*, Benchmark Papers in Geology, Hutchinson Ross, 1982.

A wide-ranging review of Quaternary environmental fluctuations is provided by A. Goudie, *Environmental Change* (2nd edn), OUP, 1983. A popular history of the changing scientific assessment of the evidence for climatic variation is offered by J. and K. P. Imbrie, *Ice Ages: solving the mystery*, Macmillan, 1979.

Chapter 19
Models of landform development

Much of the history of geomorphology in the first half of the present century was dominated by a small group of theories concerned with long-term landform evolution. Propounded by workers of different nationality who had pursued their investigations in contrasting parts of the globe, each attracted its influential adherents in a fashion that gave geomorphology, as a scientific discipline, a curiously divided appearance. The primary objective of these early workers was to offer an explanation of present-day landform assemblages in terms of their supposed evolutionary history. In the 1960s, however, geomorphological investigations broadened in scope and were addressed to a wider range of tasks than had previously been the case. In part this was probably a response to the unduly constraining influence of the prevailing models of landform history, although many other influences were doubtless at work. Yet the expanding scope of the subject did not by itself generate a strong conceptual framework, and it has been argued that geomorphology at that time was long on fact and short on theory. In the 1970s a resurgence of interest in possible theoretical frameworks for landform studies manifested itself and has continued to the present day.

In the following discussion, consideration will first be given to the early classic models of landform development. Certain variants of these ideas will then be reviewed, before attention is finally directed to some more recent concepts that have been suggested as appropriate for geomorphological investigations.

Classic models of landform evolution

The cycle concept of W. M. Davis

It was the American worker, Davis, who had the most profound effect upon the early development of geomorphology in the English-speaking world. In a series of extremely influential publications between 1889 and 1934 he argued that, although changes within a lifetime may be too slight for measurement, landscapes can be arranged in evolutionary sequences illustrative of cyclical changes that take place over a much longer period (Fig. 19.1). Ordering landscapes in this manner rests upon fundamental premises which are difficult to substantiate by direct observation but which Davis sought to justify by a simple deductive approach. Perhaps the single most crucial concept was that of base-level or a level below which streams cannot erode their valley floors. Construed by many as merely an extension of sea-level beneath the continents this has little practical value, but several contemporary workers suggested an alternative formulation which envisaged a critical minimum gradient below which a streamcourse cannot be reduced. Conceived in this way, base-level would slope gently upwards away from the coast. No satisfactory definition of this limiting gradient has ever been proposed, and Davis was compelled to supplement the idea of base-level with a further fundamental concept, that of 'grade'. He argued that, once uplift of a land mass is complete, the subsequent erosional history may be divided into two distinct phases. At first, rapid stream incision produces narrow, steep-sided valleys. However, a point is reached at which the pace of downcutting abruptly diminishes, to be replaced in large measure by lateral erosion with widening valley floors and ever more extensive floodplains. Treated as the transition from a youthful to a mature landscape stage, the change was equated with the attainment by the streams of a notional state termed grade.

For purposes of defining grade, Davis generally had recourse to the idea of a balance between erosion and deposition, or between capacity to do work and the work needing to be done. Implicit in his writings was the view that, during early downcutting, more energy is available than is being employed in transportation, and the surplus is expended in corrasion of the valley floor. Following incision of the valley, the load supplied to the stream increases while the available energy diminishes, and eventually a balance is struck between these

Fig. 19.1. The cycle concept of W. M. Davis. He normally described the sequence as depicted on the left although the available relief, i.e. the altitudinal difference between the initial surface and the level at which floodplains begin to form, will clearly have an important effect on landform evolution, as illustrated on the right.

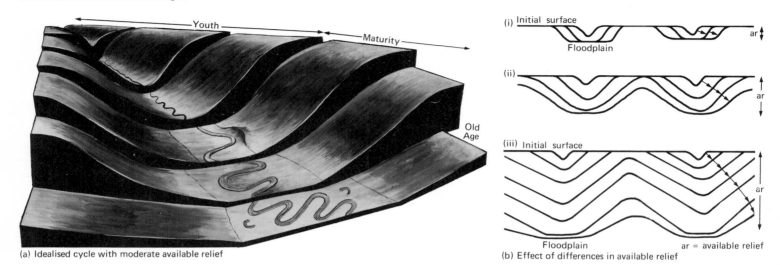

(a) Idealised cycle with moderate available relief

(b) Effect of differences in available relief

two quantities. Thereafter any changes are relatively slow and depend on a gradual reduction in load as the valley-side slopes decline. Unfortunately, in this context the idea of a balance between available energy and work to be done is both vague and confused. Work to be done is commonly assumed to refer to sediment transportation, while energy expenditure is regarded as divisible between that function and erosion of the stream bed. Such a division is obviously unreal because, on the one hand, downcutting of a bedrock channel is itself dependent on sediment movement and, on the other, it is impossible to envisage conditions in which all energy is devoted to transportation and there is no possibility of channel corrasion. Owing to such inconsistencies, several workers have argued that the concept of grade is best abandoned. While this may be so, there is none the less an indisputable decrease in potential energy as streams incise their valleys, accompanied presumably by a reduced rate of downcutting. If the independent speed of valley-side retreat remains constant, this deceleration will produce much broader transverse profiles. In this limited respect one of the basic postulates of the

Davisian model may be accepted, even though much of the original explanation is rejected.

The change from mature to senile landscapes was regarded by Davis as being characterized by floodplains sufficiently broad to accommodate fully-developed meander trains. In his analysis of landform evolution Davis made frequent reference to stream patterns, but later work has shown many of his inferences to be totally unjustified. For instance, free-swinging meanders are not necessarily indicative of old age, as brief consideration of the lower Mississippi meanders immediately demonstrates (see p. 187). On the other hand, it is difficult to contest his general thesis that, with prolonged tectonic stability, relief amplitude will continue to diminish until eventually an extensive surface of low relief or peneplain is formed. Once again the fundamental idea seems sound, but many of the embellishments are highly questionable.

For purposes of exposition Davis made many simplifying assumptions that were later the target of critical comment. Among the most important was a division of uplift and subsequent denudation into entirely separate episodes. Yet he recognized the unreality of this assumption, and in some of his writings in German specifically discussed the likely consequences of simultaneous uplift and erosion. From modern work on the relationship between relief and denudation rates, it might be anticipated that the greater the total uplift, the greater the impact on contemporaneous erosion. However, current uplift rates so surpass those for denudation that erosion does not normally appear to constitute a limit for continued surface elevation. Indeed, in this respect the model proposed by Davis seems more realistic than some of his sternest critics have been prepared to admit. Greater weight appears to attach to the argument that stability is unlikely to endure long enough to permit reduction of an upland area to a peneplain. Accumulating data on crustal mobility strengthen this belief, and underline the need to consider persistent tectonic activity as a factor in landform evolution. The only stage in the Davisian cycle not reputedly represented on the earth's surface today is the peneplain, despite numerous supposed uplifted examples. This has led many geomorphologists to conclude that the erosion cycle envisaged by Davis rarely reaches culmination in a peneplain, and that alternative explanations should be sought for the presumed elevated examples.

One factor to be borne in mind when considering the complete sequence from uplifted mountain mass to peneplain is isostatic adjustment. The contrast of large Bouguer gravity anomalies over modern mountain chains with small anomalies over ancient denuded remnants implies deep-seated changes within the crust and upper mantle. On the assumption that a crustal 'root' of density $2\ 700$ kg m^{-3} is replaced by mantle material of density $3\ 400$ kg m^{-3}, the normal volume of rock to be removed in the erasure of a mountain chain is increased almost fourfold. This is obviously important in estimating the speed with which erosion can destroy such a feature. Equally significant is the nature of the movement associated with isostatic adjustment. Two possibilities present themselves. The first is that uplift takes place continuously *pari passu* with denudation. The second is that it occurs spasmodically and represents periodic adjustment to slowly accumulating stresses as the surface layers are removed by erosion. The evidence is far from conclusive, but many geomorphological investigations have revealed marginal benches suggestive of episodic uplift long after the main orogenic phase was complete; these could well be due to isostatic adjustments triggered by denudation. A further guide is the incomplete isostatic compensation of certain cordilleran remnants, suggesting that erosion has as yet failed to activate the final uplift required for equilibrium to be restored.

In expounding his ideas of cyclical development, Davis devoted considerable attention to the evolution of valley-side slopes. He contended that during youthful stream incision the valley sides will be steep, the controlling factor being the relative rates of fluvial downcutting and backwearing by slope processes. However, once rapid valley deepening has ceased slope processes will become almost the sole influence on form, leading to a gradual decline in angle (Fig. 19.2). On youthful slopes crags often mark the outcrop of resistant rocks, and debris descending towards the stream is coarse. With the passage of time moderately angled and more regular hillsides become mantled with finer detritus until, in old age, the very gentle peneplain slopes are covered with deep and almost immobile waste. Unfortunately, Davis restricted his discussion of hillslopes to generalities and provided little guidance to his thinking on the complex relationship between process and form. Moreover, he attempted to apply the notion of grade to valley sides and, as already seen, this concept defies satisfactory definition. In consequence most attention has been directed at his efforts to integrate slope development with other aspects of landscape evolution. He suggested that, in many instances, the rectilinear valley sides of the youthful stage will just be consuming the last remnants of the original surface when the streams cease their active downcutting. He acknowledged, however, that these two critical stages might not coincide, since their timing depends on such factors as the amount of uplift and drainage density. One estimate is that, with normal stream spacing, simultaneity is likely when uplift is between 60 and 90 m. Particular interest attaches to the landform development that follows uplift in excess of this. Regular valley spacing and constant slope angles might theoretically sustain an accordance of interfluve heights long after the original surface has been consumed (Fig. 19.1b). This implies that evenness of divides is not by itself an adequate criterion for reconstructing an uplifted surface; identifiable fragments of the old surface must be preserved.

Davis examined the effect of repeated minor uplift in considerable detail, arguing that the majority of landscapes are polycyclic in origin. However, his discussion concentrated on changes along the

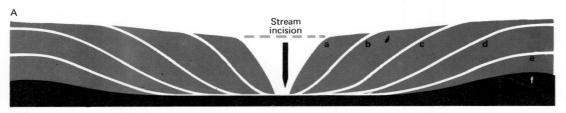

A

Stream
incision

a   b   c   d   e   f

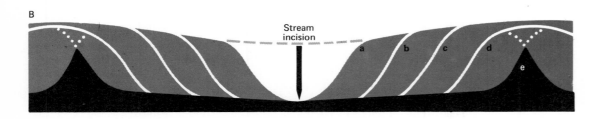

B

Stream
incision

a   b   c   d   e

C

a   b   c   d   e

Fig. 19.2. Three models of valley-side evolution: (A) W. M. Davis; (B) W. Penck as interpreted by Davis; (C) W. Penck from an original diagram.

valley floor and paid little attention to the effects on valley-side slopes. He visualized a relative fall in base-level initiating a knick-point that slowly migrates upstream. During this migration the valley floor is dissected to leave the former floodplain deposits as elevated terraces. This model of rejuvenation, affording one possible interpretation of the long profiles and associated terraces, has been widely employed in the study of valley-floor evolution (see, e.g. Fig. 9.12). However, extension of these ideas to encompass evolution of the valley as a whole presents formidable problems. Recent minor uplift, suitably attested by river terraces, has presumably had little direct effect upon valley-side development; the very presence of the terraces signifies isolation of the higher hillslopes from events on the

valley floor. Nevertheless, it is difficult to conceive that in the longer term major changes in the rate of river downcutting will not be reflected in the form of valley cross-profiles. If a moderately long phase of stability has resulted in gentle gradients near the river, rejuvenation will produce a new generation of steeper slopes with an initial angle determined by the relative rates of stream incision and slope backwearing. Once the downcutting slackens the character-istic angle will decline, and the segment of valley cross-profile related to the new phase of stability will progressively broaden. The resultant 'valley-in-valley' form is often regarded as diagnostic of long-term intermittent uplift, but one crucial question not really addressed by Davis was the subsequent evolution of such profiles.

Implicit in some later work has been the idea that the older slopes above each hillslope convexity are virtually fossilized. One common practice, for instance, has been to interpolate smooth curves between paired breaks of slope (as in Fig. 18.5). It is easy to be deceived into assuming that these 'reconstructions' accurately depict the valley as it existed at some earlier period, but such an assumption can only be founded on the belief that slopes do not continue to evolve once the valley floor has been rejuvenated. There appears to be no logical reason for a cessation in slope development, and any satisfactory model must incorporate the possibility of continued evolution.

Finally, it is worth recalling that modern estimates for the periodicity of Pleistocene climatic oscillations lie in the range 10^4–10^5 years. Davis essayed no precise figures for the time required for an erosion cycle to run its course, but on the basis of known denudation rates it can hardly be less than 10^6 years. This means that any area will almost certainly have experienced numerous climatic fluctuations during a single cycle, with consequent changes in the erosional processes. It follows that landscapes, besides often being polycyclic, must also be viewed as 'polygenetic'.

This summary review has endeavoured to highlight both strengths and weaknesses in the Davisian model of landform evolution. It is appropriate to end it by reaffirming the fundamental importance of the historical approach to landform description and explanation, and the outstanding contribution to this topic made by the American author.

The geomorphological concepts of W. Penck

The objectives of Penck, a German worker who published his most important book, *Die Morphologische Analyse*, in 1924, differed fundamentally from those of Davis. He regarded surface morphology as the outcome of competition between crustal movement and denudational processes, and as a geologist he saw in landform analysis a potential tool for deciphering recent tectonic history. His ideas aroused much controversy, partly because he used such obscure language that it is often difficult to tell exactly what he intended, and partly because the book was incomplete on his death so that certain critical chapters are missing.

Penck viewed the primary control of landform evolution as the relationship between the rates of river downcutting and tectonic uplift. This relationship in turn influences the form of valley-side slopes. He contended that three crustal states may be distinguished. The first is that of stability in which no active displacement is occurring; this he regarded as characteristic of extensive continental regions at the present time. The second is domed uplift which starts by affecting a relatively small area but later expands to encompass much wider tracts. The third is intense crustal upheaval, characteristic of mountain areas where tangential compression is combined with regional arching. An important aspect of Penck's thesis was that the rate of crustal movement varies greatly at different times. If it should remain constant for a prolonged period, the rate of river downcutting will adjust until an equilibrium is established. For example, if the speed of uplift initially exceeds that of stream incision, river gradients will steepen until the downcutting can keep pace with the crustal movement; conversely, if the rivers can cut down more rapidly than the land is being elevated, their gradients will be reduced until the two rates are equalized. Such adjustments take some time to achieve, and if the speed of tectonic movement undergoes frequent changes disequilibrium will be common.

Penck maintained that conditions of equilibrium and disequilibrium will produce contrasting slope forms. So long as the rates of uplift and downcutting remain equal, slope profiles will tend to be rectilinear; if downcutting accelerates, increasing steepness of the newly formed segments will generate a convex profile; if downcutting decelerates, declining steepness of the newly formed segments will generate a concave profile. The final possibility is that of tectonic stability, and it is here that much misrepresentation of Penck's ideas has occurred (e.g. Fig. 19.2B). Often quoted as believing in the parallel retreat of slopes, he actually appears to have held quite different views. He argued that the steep portions of a slope profile retreat more rapidly than the gentle. He considered the simplest case of a steep rectilinear face rising directly from the valley floor. Weathering attacks all parts of the face equally so that it tends to retreat without change of angle. At its foot there emerges a sloping surface, designated a *Haldenhang*, across which debris is transported to the stream. This less steep unit is itself subject to gradual replacement from below by an even more gentle slope. The eventual effect is a concave profile in which the constituent elements appear to migrate upwards and away from the valley floor. As Fig. 19.2 shows, Penck even visualized a convex slope evolving towards a concave form, with the steep basal segment overrunning the more gentle upper portion and a

Haldenhang simultaneously developing at its foot. There seems no doubt that, irrespective of the original form, Penck envisaged stable conditions leading to concave profiles. Although the intersection of such profiles might be expected to produce sharp-crested interfluves, he argued that in practice they would be rounded by weathering.

The most fundamental difference from the model proposed by Davis lies in the emphasis on concurrent uplift and erosion. A surface of low relief may evolve in two contrasting circumstances. That which develops during prolonged tectonic stability is characterized by widespread gentle slopes of concave form; to such a feature Penck gave the name *Endrumpf*. That which forms during slow but accelerating uplift has broad convex interfluves and is termed a *Primarrumpf*. Implicit in this concept is the notion of accordant crestlines arising through a mechanism other than dissection of an uplifted erosion surface. Penck elaborated on the idea of a *Primarrumpf* in considering landform development on an upheaved dome. He visualized uplift gradually accelerating and spreading outwards, so that any point temporarily on the periphery would first experience slow elevation and then progressively faster movement. During the slow phase an annular *Primarrumpf* would tend to form, only to be dissected as uplift accelerated and the dome expanded. Penck maintained that ultimately a succession of concentric benches might be formed, constituting an assemblage of features he labelled a *Piedmonttreppen*. Most workers have found Penck's explanation of *Piedmonttreppen* inadequate, although the elevated massif fringed by a series of benches has certainly been recognized as a common relief form.

For many years the ideas of Penck received scant support from English-speaking geomorphologists. The obscure language in which they were expressed contributed to this neglect, but a secondary factor was reliance on an interpretation of Penck's views offered by Davis. The American undoubtedly misconstrued some of these views, until finally this led to a mistaken account of Penck's ideas on slope development being castigated as 'one of the most fantastic errors ever introduced into geomorphology'. More recently workers have returned to the original German exposition and have found it thought-provoking, even though at times muddled and contradictory. Moreover, some of its underlying ideas have recently been given fresh impetus by Scheidegger who has argued that one of the most funda-mental geomorphic principles must be what he terms 'antagonism' between endogenetic and exogenetic forces. The former are characterized by their non-randomness, the latter by their randomness, and this contrast may provide the basis for distinguishing between their effects on relief patterns at the surface of the globe.

The cycle of erosion according to L. C. King

The ideas of the South African geomorphologist King, expounded in his book *South African Scenery*, were more closely allied to those of Davis than of Penck. Acknowledging his debt to Davis, he formulated a model of landform development differing from that of the American mainly in its ideas about slope evolution. He argued for the pre-eminence of cyclic erosion in the development of continental landscapes, and maintained that the chief defect of the Davisian concept was the absence of any idea of parallel slope retreat.

Like Davis, King contended that once an area has been uplifted, knickpoints generated on the rivers near the coast begin to migrate far inland. During this youthful stage rapid incision produces gorges, but as downcutting slackens the valley-side slopes decline to a stable angle determined by lithology and the physical agencies acting on them. At the onset of maturity downcutting virtually ceases and is replaced by lateral corrasion. The chief means of landscape modification becomes the migration of the valley-side slopes away from the rivers without significant alteration in angle. At their base develop relatively smooth flat areas known as pediments. These have gentle concave profiles which may locally attain an angle of over $10°$ but generally do not surpass $5°$. Gradual extension of the pediments eventually produces a pediplain. This surface of low relief comprises a series of subdued intersecting concavities above which rise occasional steep-sided residuals. Protracted weathering may slowly transform the divides into broad convexities so that the senile landform ultimately comes to resemble the peneplain envisaged by Davis.

King's major thesis was that the form of a migrating slope is determined primarily by the processes operating on it. So long as these remain constant, a slope profile should keep its shape unchanged as it retreats across the landscape. He divided hillslopes into four elements – crest, scarp, debris slope and pediment – each associated with a distinctive process (Fig. 19.3). The crest or summit area is usually convex and shaped by soil creep. The scarp, characterized

Fig. 19.3. Hillslope development according to L. C. King: the morphological elements (A) of a hillslope formed under the action of running water (B) and gravitational mass movement (C) (after King, 1962).

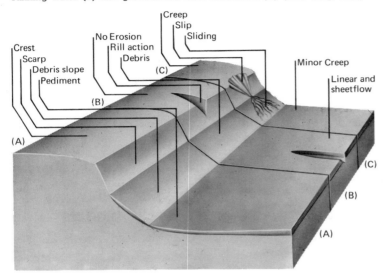

concavo-convex profile subject to very slow denudation. A similar profile may characterize areas of weak bedrock. King contended that there is a 'basic homology of landforms' and that climate is not a fundamental control but influences only the degree to which the various slope elements are developed. The most efficient erosional system operates under conditions of heavy intermittent rainfall. With increased humidity rapid weathering tends to blanket the landscape with a thick immobile regolith; with increased aridity the water is unable to move all the debris to the stream channels, and large fans fringing the uplands severely restrict pedimentation. Of course, whether to stress the similarities or dissimilarities of slopes in different environments is largely a matter of choice. Moreover, attribution of a common slope form to a single process can lead to semantic problems; the term pediment, for example, has been extended by some writers to basal concavities in humid temperate regions without any proof of fashioning by sheet-floods in such areas.

A vital consequence of the King model of slope evolution is that scarps continue to migrate unaltered in form even when isolated from the modern valley floor by renewed uplift. In essence the landscape consists of a massive mobile staircase eating back into the upland regions. By contrast with the Davisian model, a prolonged phase of stability does not necessarily destroy earlier surfaces but simply adds another tread to the staircase. Only when a particular scarp has migrated across the whole land surface is the evidence for earlier evolutionary phases erased. Because the stairway of scarps is regarded as mobile rather than static, each pediment surface may be viewed as having both a local age, that is, the time when it was first formed at the particular point under consideration, and a regional age referring to its date of initiation either at the coast or on the flanks of a crustal fold. King normally inferred regional ages from either superficial deposits or coastal sedimentary sequences, and by this means identified three major surfaces on the African continent: the Gondwana surface of Jurassic age, the African surface of Cenozoic age, and the post-African surface of late Cenozoic age. His contention that landform evolution is fundamentally similar in different environments encouraged him to seek analogous histories for other continents. The great age assigned to the Gondwana surface in Africa is particularly intriguing because it implies that, at the time of formation, the southern continents may still have been joined together as Gondwanaland. King therefore paid special attention to the lands of the

by rocky outcrops, retreats by means of rockfall, landslips and gullying, and is the most active element of the whole profile. The debris slope, comprising detritus coming from above, has its gradient controlled by the appropriate angle of repose. The pediment is a smooth feature cut in bedrock and fashioned mainly by turbulent sheet-flooding; the abrupt change of gradient at the base of the debris slope denotes a sudden transformation in the processes shaping the landscape.

King formulated his ideas in a predominantly semi-arid region that has experienced little recent tectonic deformation, and controversy surrounded his attempts to extend them to other parts of the world. He argued in 1962 that the 'four-element hillslope is the basic landform that develops in all regions of sufficient relief and under all climates wherein water-flow is a prominent agent of denudation' (*Morphology of the Earth*, p. 139). Sufficient relief is important because its absence will lead to elision of the scarp and debris slope: the crest will then grade directly into the pediment to yield a

Southern Hemisphere where he noted striking similarities with the evolutionary sequence in Africa. Later claims of analogous sequences in both Europe and North America led him to draw the obvious inference of rhythmic global tectonics controlling topographic development on all the major continents.

The ages assigned by King to the earlier of his cyclic surfaces are greater than many geomorphologists believe likely. Measurements suggest erosional processes operate too fast to permit preservation of such ancient features. However, if King's model of landscape evolution is correct, computed denudation rates could be misleading. With downwearing as the dominant process, current rates of sediment yield would render preservation of elevated landforms for 100 m.y. or so most improbable. On the other hand, scarp recession as envisaged by King would permit removal of immense quantities of material without effacing remnants of the earliest surfaces. Yet it is difficult to deny that a serious conflict still exists. This may be resolved in the future by better dating methods, but for now it is informative to enquire how far King's ideas receive support from detailed local studies. One region that might in theory be expected to preserve ancient landscapes is the arid and tectonically stable heart of Australia. Here mapping and dating is greatly assisted by the presence of duricrusts. The most distinctive is a silcrete caprock, up to 9 m thick, that required for its formation a more humid climate than prevails today and a stable plainland over which the crust could develop. Adequate conditions seem to have occurred only once and can be dated with some confidence, since the crust is formed in part from Cretaceous rocks and is locally overlain by Miocene lacustrine deposits. Rising above the silcrete caprock are bevelled summits, believed by certain workers to have been fashioned in a period of erosion extending back continuously as far as Palaeozoic times. The general time-scale of King appears to be corroborated by these Australian investigations. By comparison, the whole interior of North America seems vastly more changeable, with some workers contending that nothing in the landscape can date from earlier than Pliocene times. The contrast with Australia might be explained by greater crustal instability and more humid climate. Yet difficulties arise when these same arguments are applied to Africa. King identified remnants of the Gondwana surface within 200 km of the coast in Lesotho, Angola and Cameroon. Elevations range from 1 500 to 3 000 m and local precipitation from 500 to 2 500 mm yr^{-1}. In Lesotho the mean temperature of the coldest month falls below 5 °C, while in Angola there are pronounced seasonal variations in rainfall. These are all factors regarded by one author or another as favouring rapid erosion, and most estimates of current denudation seem inconsistent with the preservation of really ancient erosion surfaces.

By emphasizing the homology of all water-eroded landscapes King tended to minimize the significance of climatic change. On the other hand, the great age he assigned to cyclic landforms would imply that they evolved not only under the climatic fluctuations of the Pleistocene epoch but also under the more enduring changes occasioned by the slow drift of the continents across the surface of the globe. If validated, King's thesis would clearly have immense and exciting consequences, but it must be admitted that there are formidable objections to its general acceptance. At best the evidence seems to point to huge contrasts in the length of time for which uplifted erosion surfaces may be preserved.

Climatic variants of the cyclic models

Both Davis and King argued that their models of landscape evolution could be adapted, with only minor modifications, to a wide variety of climatic environments. Nevertheless, as early as 1909 Davis was convinced of the need to formulate a separate cycle for arid regions, and since that date many writers have proposed variants of the original Davisian cycle which they believe appropriate to specific climatic zones. Peltier, for instance, argued that at least nine 'morphogenetic regions' should be distinguishable, each in theory characterized by a unique combination of frost action, chemical weathering, mass movement and pluvial erosion. Yet it is worth noting that Peltier's scheme started from an assumption, that different processes will produce different forms, which has so far received little rigorous testing. Peltier himself showed by morphometric studies that the relationship between mean hillside slope and channel spacing varies from one environment to another, while Chorley and Morgan have demonstrated that a number of morphological distinctions between England and the south-eastern United States are ascribable to climatic rather than lithological contrasts. However, much more research is required to establish the full relationship between climate and morphology.

In more recent times the most influential exponent of climatically differentiated geomorphic regions has been the German worker Büdel. He recognized seven non-glacial 'morphoclimatic zones': equatorial zone of partial planation; peritropical zone of excessive planation; subtropical zone of mixed relief development; warm arid zone of preserved plains; winter-cold arid zone of surface transformation; extra-tropical zone of retarded valley cutting; and polar zone of excessive valley cutting. As the rather unwieldy descriptive terms imply, Büdel placed great emphasis on contrasts between processes of regional planation that characterize humid tropical climates and those of vertical downcutting that characterize cold climates. Under hot, seasonally wet conditions intense weathering produces a thick regolith whose surface is subject to concurrent rainwash that transports the fine debris towards the main stream channels. Nearly all these channels carry only fine sediment and show almost no tendency towards vertical incision, certainly not at a speed exceeding that at which chemical erosion can lower the surrounding terrain. In the equatorial zone where there is no dry season, and also in the subtropical regions, similar although rather less extreme conditions are thought to obtain. By contrast, in polar environments the dominant erosional process is stream downcutting which is promoted by, among other factors, permafrost disrupting the bedrock beneath the valley floors which are then swept clear by regular springtime flooding; similar conditions periodically affected the extra-tropical zone as permafrost extended equatorwards during the Pleistocene glacial intervals. Büdel divided deserts into warm and cold types: in the former, current erosional processes are held to have achieved little more than a stripping away of the regolith from old planation surfaces fashioned under earlier humid conditions; in the latter, frost action generates prolific coarse debris and the resultant processes have overprinted the original humid landscape with extensive pediments and veneers of loose sediment.

Büdel elevated his scheme of morphoclimatic zones into a global view of landform development, since he argued that in early Cenozoic times virtually all continental areas were subject to tropical denudational processes, and it is on to the planation surfaces produced at that time that, in different areas, forms due to arid and cold climatic conditions have been superimposed. It is impossible in such a brief space to do full justice to Büdel's arguments, but it is pertinent to note that he interprets virtually all landscapes in terms of tropical, arid and periglacial process assemblages, and it is these three environments that have most commonly been identified as offering variants on the 'normal cycle' expounded by Davis. With the possible exception of tropical regions, no great utility can be claimed for these climatic variants of the cycle concept; yet they do merit brief review because, historically, they drew attention to landform groupings regarded as peculiar to specific climatic regimes.

Desert areas

Davis was originally persuaded to recognize a distinctive desert cycle of erosion by three considerations: the absence of the normal base-level control in areas without perennial streams draining to the coast; the increasing importance of wind action in intensely arid regions; and finally the belief that a unique combination of processes may lead to slope retreat without angular decline. In formulating details of the cycle Davis was much influenced by his familiarity with the block-faulted landforms of the south-western United States. He maintained that the lack of drainage to the coast will result in much internal redistribution of sediment. The floor of any structural basin will act as a local base-level, but instead of being stable its elevation will slowly rise as more and more detritus accumulates. Integration of adjacent basins may take place by either capture or overflow of the sedimentary fill. The margins of uplifted areas will retreat in the form of steep scarps fringed by progressively widening pediments. Relief amplitude will steadily decline, and as the power of running water diminishes the relative importance of wind will increase, primarily as a transporting medium but also as a local abrasive agent at the foot of bedrock slopes. This represents a further noteworthy weakening of the normal base-level control. Nevertheless, the original accidented topography will ultimately be reduced to a surface of low relief diversified by occasional steep-sided residuals.

The value of recognizing such a cycle of desert erosion is certainly questionable since aridity is a state that may befall any region regardless of structure or relief. In practice deserts are extremely diverse, topographically, geologically, biologically and even climatically. Moreover, as already seen (p. 384), most if not all have been subject to pronounced climatic fluctuations. Whereas extreme dryness may retard denudational processes and thereby fossilize an inherited landscape, semi-aridity often leads to rapid transformation of such a landscape. Many workers have therefore queried whether a desert cycle can ever run its full course without the intervention of more humid interludes, and some have contended that it would be better to treat aridity as a special condition associated with an unusual combination of erosional and depositional processes. Normal base-level control is at least temporarily suspended, and the usual assumption of downstream increase in fluvial discharge is no longer justified. Pedimentation appears to become especially prevalent, and although no single explanation for desert pediments has yet been advanced, their development does seem to be favoured by the sparseness of the vegetation cover and the high proportion of bare ground. Yet the most distinctive characteristic is the increasing significance of wind as a geomorphic agent. Perhaps as a reaction against a widespread belief around the turn of the century that most desert landforms are attributable to wind, for many decades its effects in arid regions were treated as negligible. However, since about 1970 there has been growing awareness of the potential importance of wind; this view is reflected in the discussion of aeolian processes in Chapter 10.

Humid tropical regions

These constitute a second climatic zone in which the cycle of erosion has been deemed to vary significantly from the schemes outlined by Davis and King. Although King's ideas attracted wide support in the drier tropics, they have been held to pay too little heed to rapid chemical weathering to be applied unaltered to hot, moist conditions. Most models of landscape evolution assume an approximate equality between the rates of weathering and surface lowering. Yet in many humid tropical areas the mantle of rotted bedrock locally surpasses

50 m in thickness, so that it appears that the downward advance of the weathering front must have outpaced removal of the regolith. This has encouraged a number of geomorphologists to suggest that two relatively independent surfaces, the ground surface and the weathering front, may interact to produce a distinctive set of landforms. For instance, at the end of a long phase of tectonic stability surface processes will be altering the ground elevation very little, but chemical action can still be lowering the weathering front. Subsequent uplift may then lead to stripping of the regolith at a much faster rate than it is being generated by further rotting. If the weathering front is approximately level, complete stripping will eventually produce an 'etchplain' of low relief. Yet the junction between sound and rotted bedrock is often so irregular that partial removal of the regolith tends to expose cores of unaltered rock between more extensive areas of decomposed material. The appropriate model of landform evolution depends to a certain extent upon the nature of the denudational agencies involved in the stripping.

Pugh argued that in West Africa pedimentation, primarily by surface runoff, is the normal process and he envisaged a fall in base-level starting a new cyclic surface which rapidly dissects the old regolith (Fig. 19.4). Where the new surface encounters buried masses of sound rock it may first sweep past to leave prominent residuals. Eventually, however, the margins of the residuals will themselves be subject to pedimentation, finally producing a pedi-plain that transects both sound and decomposed rock and leaves upstanding only the largest and most resistant of the original unaltered masses. Pugh distinguished two types of residual. Born-

Fig. 19.4. Potential evolution of humid tropical regions by deep chemical weathering followed by pedimentation.

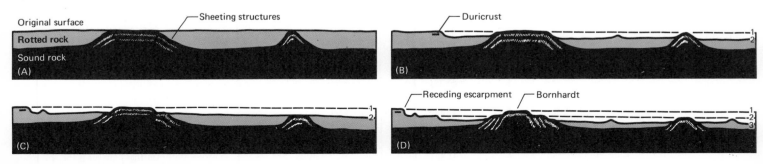

hardts are rounded hills characterized by curved sheeting structures, whereas kopjes are irregular in outline and possess a dominantly rectangular jointing system. The difference between the two is usually attributable to geological contrasts, bornhardts forming in massive rock and kopjes in more closely jointed rock. Detailed studies along the edges of residuals occasionally reveal what appears to be a second phase of chemical weathering; narrow zones of decomposed rock are being removed by intermittent surface flow to produce annular hollows, or even overhangs in extreme cases. Many minor variants of this basically cyclic concept have been proposed, but nearly all have as a fundamental component the possibility of deep chemical weathering proceeding independently of surface lowering.

One further way in which chemical processes can affect landform development in the humid tropics is by the generation of lateritic crusts. In areas of deep rotting the upper layers of the regolith might be expected to succumb particularly rapidly to stripping since they would presumably be the most thoroughly decomposed. However, in many regions it has been noted that the surface horizons of a thick regolith are converted into a hard lateritic crust *pari passu* with undercutting by a retreating scarp. This induration appears to be due to an enrichment with iron associated with improved drainage and aeration as the local amplitude of relief increases; it has the effect of making the decomposed rock of tropical regions much more resistant than might otherwise be supposed. Receding scarps steepened by their cap of hardened crust can even produce landforms reminiscent of mesas fashioned from horizontally stratified bedrock. Recemented lateritic detritus may also accumulate as a secondary deposit on the lower parts of hillslopes and there develop into conspicuous smaller scarps.

Periglacial areas

Little is known in detail about the Pleistocene climatic history of Arctic Canada and Siberia, but it is reasonable to infer that unglaciated enclaves in these two regions may have experienced exceptionally lengthy periods during which conditions were at least as severe as those today. The belief that such prolonged operation of periglacial processes might produce a unique assemblage of land-forms encouraged several workers to envisage a distinctive peri-glacial cycle of erosion. One of the foremost exponents of this idea

was Peltier, who visualized an initial landscape composed of rounded hills of moderate relief with an almost continuous cover of regolith and soil. During the youthful stage, solifluction results in bedrock being exposed on the upper slopes where the debris mantle is removed more rapidly than it is replaced by weathering. The exposed rock faces are subject to frost-sapping so they recede rapidly and soon consume the whole of each hilltop. In a mature periglacial landscape the bedrock faces have all been eliminated and the morphology is dominated by concave slopes mantled with solifluction sheets. Further stages in the cycle see a gradual reduction of relief amplitude and mean slope angle, but without major changes in geometrical form or dominant process. However, if the periglacial regime is sufficiently persistent, the debris cover may be so comminuted that aeolian deflation assumes an important role.

Although the Peltier cycle may depict a realistic sequence of land-form changes its practical application is fraught with difficulty. It assumes a moderately hilly relief to start with, but clearly many other initial conditions could be envisaged which would follow rather different patterns of geomorphic evolution. It pays little attention to geological structure although the relative susceptibility of rocks to frost-riving, and of the weathered mantle to frost-heaving, is of fundamental concern in modern periglacial regions. In an area of shale or clay, for instance, solifluction is likely to dominate all other processes; it is scarcely conceivable that free faces will develop because the accelerated movement of the debris cover will not exceed the rate at which the bedrock is mobilized to form part of the solifluction mantle. Moreover, no distinction is drawn between regions with and without permafrost, although it can be argued that this factor will be of great significance in the long-term evolution of the landforms. In general, therefore, the cycle concept has proved of only very limited value in the study of periglacial landscapes.

More recent conceptual frameworks for landform studies

By about 1960 the influence of the cyclic concept in geomorphology had gone into sharp decline. As mentioned earlier, many reasons may be adduced for this change, but perhaps the most fundamental was a feeling that adherence to the traditional cycle of erosion was acting as an undue strait-jacket to the full development of the subject. In effect it had circumscribed both the temporal and spatial scales of landform investigations, since it had resulted in research being confined very largely to regional landscapes evolving over protracted periods of geological time. Linked with this was a profound neglect of process studies, for many geomorphologists had tended to follow the dictum of King and 'to sit passively upon hills just letting the scenery "soak in" and teach the beholder.

The explosion in the nature and scope of geomorphology that followed relegation of the cyclic paradigm to a minor role left a conceptual vacuum that has never been entirely filled. This is scarcely surprising when one considers that the field of study widened dramatically so as to span spatial scales ranging from regional landscapes covering millions of square kilometres down to individual features only a few square metres in extent, and temporal scales ranging from millions of years down to a matter of days or even less. A conceptual framework that will encompass such diversity needs to be much more flexible than the rather rigid cyclic models that had previously sufficed. Yet it also runs the risks of being so diffuse that any unifying influence is almost totally lost. None the less, it is possible to identify one central idea that has underpinned much geomorphological thinking during the last few decades, and that is General Systems theory. In the following account, the basic precepts of systems theory will first be outlined and then some of the ways in which they have been applied to landform studies will be more closely examined.

Systems analysis

Systems analysis is based upon the premiss that the real world, such as the environment at the surface of the globe, can be disassembled into a series of interlocking systems, each composed of a structured set of objects and/or attributes. The latter phrase is often employed as a definition of a system, and merits closer examination. Use of the word 'set' implies boundaries, although whether these are better viewed as 'natural' or mere artefacts of the observer can be a fine philosophical point. For practical purposes, and because of all the inter-linkages, boundaries have to be defined so that the system under investigation can be kept within manageable proportions. The sets of objects and attributes must be structured in the sense that they operate together as a functional whole. A homely example sometimes quoted is that a bicycle is a system, but that the components of a

bicycle laid out on the ground do not constitute a system because they lack the necessary structure.

Two major ways of classifying individual systems are envisaged in systems theory. The first is a structural classification, and geomorphic systems are commonly regarded as falling into one of four structural categories:
1. Morphological systems, comprising the network of internal linkages between constituent parts of the system;
2. Cascading systems, in which the output of one subsystem becomes the input to the next;
3. Process-response systems, in which at least one morphological system is linked to at least one cascading system by means of feedback loops; and
4. Control systems, in which there is significant intervention of an intelligence into the workings of the system.

The second means of classification is a functional one concerned with the nature of the boundaries of the system. Three categories may be envisaged:
1. An isolated system, in which neither mass nor energy crosses the boundary;
2. A closed system, in which there may be import and export of energy, but not of mass; and
3. An open system, in which there may be import and export of both mass and energy.

It should be stressed that these functional categories are rather idealized since, for instance, it is difficult to conceive of a totally isolated system unless it be the entire universe itself. It is somewhat easier to visualize a closed system, but in most geomorphic studies the appropriate model appears to be an open system of one of the four structural classes listed above.

Many open systems are self-regulatory in the sense that, when change is introduced through one of the inputs, adjustment occurs via a negative feedback so as to absorb the effects of the change. By this means a state of equilibrium is often maintained. However, several different types of equilibrium have been recognized (Fig. 19.5). Many geomorphic systems appear to display a steady-state equilibrium, in which conditions show short-term fluctuations about a long-term constancy; one obvious reason for the pervasiveness of the steady-state equilibrium lies in the vagaries of the input from the hydrological cascade. Other systems show comparable

Fig. 19.5. Diagrammatic portrayal of equilibrium conditions that are commonly encountered in geomorphic studies. The time-scale being considered in any investigation may well influence the most appropriate 'model' to be employed (cf. Fig. 19.7.).

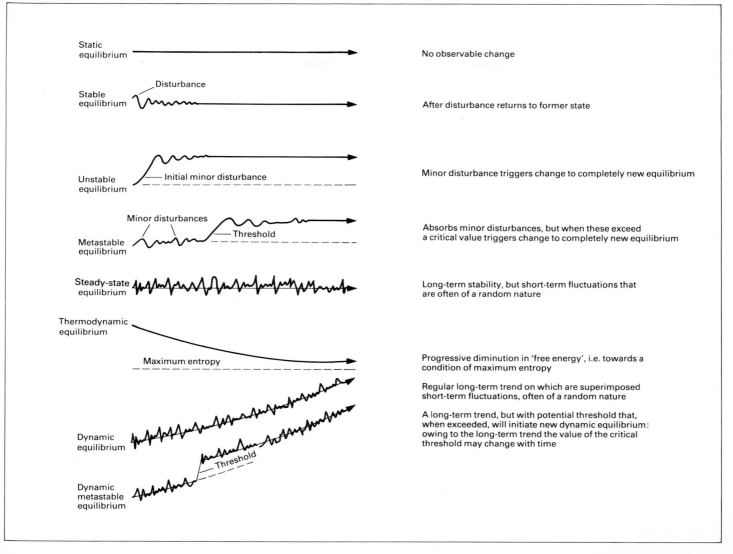

short-term oscillations about a long-term trend, a condition termed dynamic equilibrium. In practice the distinction between steady-state and dynamic equilibria will often be very problematic, since the transient oscillations may effectively mask any long-term trend; much will depend upon the rate of change and the time-scale under consideration. Another critical system condition is that of metastable equilibrium. This occurs when a system is stable unless there is some disturbing trigger sufficient to carry it over a threshold into a new equilibrium regime. By extension, this leads to further recognition of a dynamic metastable equilibrium. The period required for the transition from one equilibrium condition to another is known as the relaxation time. An especially potent destroyer of the equilibrium state is a positive feedback by which an initial perturbation is progressively magnified until the existing relationships are totally disorganized.

Before proceeding to examine, from within the systems framework, selected ideas that have proved of particular interest to geomorphologists, it may be helpful to illustrate the abstract concepts of the foregoing paragraphs by a few simple examples. Assuming a small area of fluvially dissected terrain, the drainage basins obviously constitute functionally organized segments of the land surface. They may be viewed as open morphological systems, with inputs of both mass and energy, in which the linkages lend themselves to morphometric analysis as already shown in Chapter 7. Within an individual basin, the output of water and sediment from the valley sides becomes the input for the stream system, so that fluvial activity is conditioned by what happens on the hillslopes. There is thus a cascading system of great complexity, in which are enmeshed numerous smaller subsystems whose internal structure may either be analysed separately or treated as a 'black box'; in the latter case the internal workings are ignored and attention confined to the inputs and outputs of the subsystem. Hillslopes and stream channels also provide many examples of process-response systems. Surface runoff, part of the hydrological cascade, is a major factor influencing slope morphology. The proportion of the incident rainfall discharged as overland flow varies as a function of infiltration capacity, which in turn is affected by the runoff processes. This can afford a good illustration of positive feedback. If, for any reason, overland flow temporarily increases, it is liable to strip away the more absorbent surface soil horizons, thereby increasing the tendency for runoff generation, and

thereby promoting yet further erosion. Eventually this may lead to rilling and gullying so that the whole hillside slope morphological system is transformed. After a phase of rapid change, the duration of which is the relaxation time, a new steady state based upon a relatively constant pattern of rills and gullies may be established. This example also affords an obvious case where Man may intervene in a process-response system to prevent occurrences that he sees as detrimental to his interests; he may thus establish a control system in which the operation of the natural processes is so moderated that a Man-controlled equilibrium eventuates.

Dynamic equilibrium

The term dynamic equilibrium has been used by various authors with different shades of meaning and not necessarily in strict conformity with the definition used in systems theory. It was employed by the American geomorphologist Hack in 1960 as the antithesis of the cyclic evolutionary models. It encapsulated the view that landforms may be interpreted as adapted to present-day conditions rather than inherited from the past. As already seen, geomorphologists were increasingly concerning themselves with the relationship between form and process and adopting an agnostic attitude towards landform history. This was not to deny the existence of a prolonged period of landform evolution, but to contend that its details are now lost beyond recall. Two instances of the contrast with Davisian thought may be given. Davis and his disciples maintained that certain parts of the landscape are so fossilized after rejuvenation that, by climbing to a broad upland interfluve, one might literally stand upon a preserved segment of a landscape dating back 1 m.y. or more. The geomorphologist who views landforms as constantly evolving argues that such complete inactivity is impossible. Precipitation over such a long period of time must inevitably have had some effect, at the very least in the form of chemical rotting. If such rotted rock is not present it means that material has been removed and therefore the landscape is not fossilized. In the field of channel patterns a similar sharp conflict may be discerned. Davis tended to explain modern forms by reference to presumed conditions in the past, interpreting meanders as developments arising from chance perturbations in a youthful streamcourse. Yet such explanations are demonstrably inadequate for the beautifully regular pattern of meanders in which geometry is closely related to current discharge. In the foregoing examples, the weaknesses of the Davisian argument are readily apparent, yet the question remains whether they are sufficient to undermine the whole cyclical concept.

Hack was the most extreme advocate of a non-cyclic approach to landform studies. Whereas Penck may be said to have viewed landscape as the result of competition between uplift and erosion, Hack viewed it as the product of competition between the resistance of crustal materials and the forces of denudation. The evenness of Appalachian ridge crests which Davis explained as due to rejuvenation of an old peneplain, Hack saw as the manifestation of equal resistance to the forces of erosion. He argued that the orderliness of stream organization first discerned by Horton will naturally lead to a regularity in the relief patterns. Within a single climatic region where stream and slope profiles are both controlled by the nature of the bedrock, similar geological conditions should produce similar topography. By extension of this argument, Hack explained the even-crested Appalachian ridges as a simple reflection of uniform bedrock resistance along the strike of major sandstone beds. A corollary of this view is that very little can be inferred about their evolution.

The point is well illustrated by considering the simple case of closely spaced streams cutting down through a succession of thick clays and sandstones. While being excavated in the clays, the valleys will be broad and gently sloping, but in the sandstones they will be narrow and steep-sided. Examining the valleys at any stage in their history, a geomorphologist will be able to infer nothing about their previous evolution from their current form. If one imagines the same rock succession uniclinally tilted, the effect of downcutting will be to produce asymmetric valleys which conserve their shape while migrating down-dip, and so again reveal nothing about their history. These two illustrations assume that downcutting can proceed without hindrance, but in practice base-level must constitute a limit. This will inevitably lead to progressive modification of valley cross-profiles, but does not alter the basic thesis that they still represent a balance between erosional forces and bedrock resistance. As the land surface is lowered, the available energy diminishes and the new forms are merely a response to the changing conditions; once again no inferences can be drawn about landform history. In systems theory, this uncertainty is sometimes referred to as the problem of equifinality, by which it is acknowledged that the same end state may conceivably have arisen from convergence along totally different

pathways. Admittedly, in an area uplifted after protracted stability the former slopes may be briefly preserved before succumbing to reinvigorated erosion and so provide some guidance to geomorphic history; yet Hack maintained that such conditions are too restricted to constitute a general framework for landform studies.

As mentioned earlier, current denudation rates under humid climates have reinforced doubts about the antiquity of many landforms. Characteristic sediment yields equivalent to surface lowering at $0.05–0.5$ mm yr^{-1} represent long-term changes at a rate of $50–500$ m per m.y. Of course these figures are means for catchments and do not necessarily apply, for instance, to interfluve crests; yet it is difficult to conceive of rapid erosion continuing to affect valley floors and slopes but leaving watershed areas totally intact. This specific question was raised by Pitty in 1968 in a discussion of the limestone plateau of the southern Pennines (see p. 372–4). He endeavoured to show that there is no adequate source for the calcium carbonate currently carried by the streams other than the interfluve regions, and concluded that the plateau surface must be subject to lowering at a rate of approximately 0.08 mm yr^{-1}. If his arguments are valid, it is difficult to see how the Upland Surface can be 5 m.y. old and still preserve details of its original form. Pitty, in fact, proposed a model whereby steady solution acting on a progressively unroofed limestone dome might produce a central plateau (Fig. 19.6). If this model is applicable to the southern Pennines the dated organic remains in solution hollows on the Upland Surface are presumably not in their original position; that they have been subject to some unknown amount of solutional lowering is indicated by their generally disturbed condition. It is worth recording that the concept of dynamic equilibrium has also been applied to landforms in two other British areas mentioned in this book, Wales and the Weald, in each case casting legitimate doubt on a traditional cyclic denudation chronology.

It is intrinsic to Hack's argument that landforms adapt rapidly to changing environmental controls. While that may be a tenable position, if the thesis is pushed to extremes it verges on the absurd since it would preclude, for instance, the identification through morphological criteria of formerly glaciated areas. Furthermore, some of the basic precepts of dynamic equilibrium are very difficult to test and there is the risk that in individual cases they may be assumed rather than demonstrated; the mere fact that a particular

Fig. 19.6. A sketch of Pitty's (1968) concept of solutional lowering producing a bevelled surface across a domed limestone outcrop.

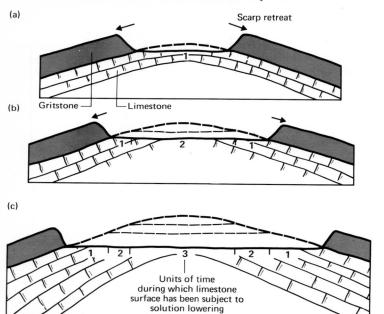

(a)

Scarp retreat

Gritstone —   — Limestone

(b)

(c)

Units of time during which limestone surface has been subject to solution lowering

process can be shown to be operating is no guarantee that it is responsible for fashioning the landforms on which it is found. It seems that some compromise may be necessary between the over-simple cycle concept as outlined by Davis and the agnostic attitude championed by Hack.

Cyclic, graded and steady time

In an influential paper in 1965 Schumm and Lichty sought to reconcile the increasingly polarized attitudes being adopted by the historical and process-orientated schools of geomorphology. One of the first conceptual difficulties that arose when emphasis shifted from long-term landform evolution to short-term process investigations was the inordinate difference in the time-scales employed. Moreover, in some respects the various cyclic models did not fit happily into the fashionable systems framework, and there was disagreement whether they could or could not be aptly classified as closed systems. In a review of these problems Schumm and Lichty argued that a whole spectrum of time-scales is inherent in geomorphological investigations, but that different relationships, and therefore different modes of explanation, must be expected to apply according to the time-scale being employed. They recognized what they termed cyclic, graded and steady time (Fig. 19.7).

Cyclic time encompasses the geological spans required for the denudational evolution of the landscape. Ignoring the possible intervention of tectonic movements, Davis must surely have been right to

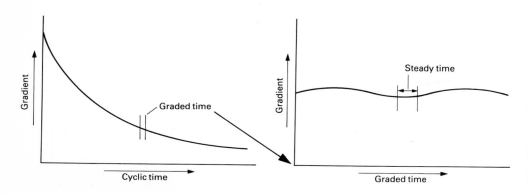

Gradient

Graded time

Cyclic time

Gradient

Steady time

Graded time

Fig. 19.7. The concepts of cyclic, graded and steady time as applied by Schumm and Lichty (1965) to changes in stream gradient. During cyclic time there is a progressive reduction in gradient. During graded time, by contrast, the gradient remains relatively constant, although there may be fluctuations above and below a mean value. Gradient is constant in the brief duration of steady time. As explained in the text, the nature of the dependent and independent variables alters according to the time-span under consideration.

envisage, for instance, a progressive decline in stream gradients although this would only become apparent at the scale of cyclic time; the appropriate model is an open system undergoing consistent one-way change. At this scale the only independent variables are time, climate, geology and initial relief. Graded time refers to a much briefer span, during which a state of dynamic equilibrium may be applicable to individual landscape components, but not to the landscape as a whole. At first sight this runs counter to the ideas of Hack, but the problem is mainly a semantic one since Schumm and Lichty were using the term dynamic equilibrium in a more restricted sense than that author. They argued that the landscape, viewed as a total system, can only change in one direction owing to the continual loss of mass by denudation; in terms of the states illustrated in Fig. 19.5 the appropriate model would be thermodynamic, rather than dynamic, equilibrium. In investigations at the graded time-scale such variables as run-off, sediment yield and relief become independent, although during the span of cyclic time they are subject to progressive change and therefore regarded as dependent. At the scale of steady time, a true steady state may be visualized for selected components of a river basin but not for the basin in its entirety. Network and hillslope morphologies, regarded as dependent variables at the graded time-scale, now become independent, and the only dependent variable is the water and sediment discharge.

Both temporal and spatial scales must therefore influence the choice of the proper conceptual model. Landscapes can be treated either as entities or as separate components, and may be viewed either as the product of past events or as the response to modern erosional agents. These views are not mutually exclusive, but when dealing with cyclic time it is necessary to accept a high level of generalization, and as investigations endeavour to become more and more specific a reduction in both temporal and spatial scales becomes inevitable. It must also be recognized that some components of a landscape have a longer relaxation time than others, and therefore bear a longer historical record. In studying an individual component, an appropriate time-scale and therefore mode of explanation must be adopted. For instance, many rivers of North America and Europe changed from a braided to a meandering habit less than 10 000 years ago; it is totally inappropriate to seek to explain their present style by reference to a cycle of erosion initiated 1 m.y.

or more ago. On the other hand, flat-topped interfluves may evolve so slowly that, although modified in detail, they still preserve in essentials the form attained as part of an ancient landscape prior to rejuvenation. This propensity of forms to change in response to a variation in geomorphic controls constitutes what is known as their sensitivity. The examples quoted lie close to opposite extremes in the range of geomorphic sensitivity. The valley floors are mobile, fast-responding subsystems that relax quickly to new states; they are morphologically complex areas that may be capable of acting as energy filters so as to absorb part of the impetus for change and pass on only minor modifications to contiguous subsystems. The interfluves, on the other hand, are characterized by low concentrations of flows of energy, water and sediment, and in consequence are relatively insensitive to changing system controls. Brunsden and Thornes in 1979 suggested that landform sensitivity, conceived as the ratio of mean relaxation time to the mean recurrence interval of significant formative events, may ultimately provide a more rational basis for scaling geomorphological time than that proposed by Schumm and Lichty.

Geomorphic thresholds

Most of the ideas reviewed up to this point have been concerned with states of equilibrium and the maintenance of form. However, another group of ideas has concentrated instead on the initiation of disequilibrium and the resulting periods of change. These conditions of transformation from one state to another have been divided into two types, termed extrinsic and intrinsic thresholds. The former depend upon a change in an external variable, whereas the latter do not since they are triggered when the geomorphic system itself evolves to a critical condition. Obvious extrinsic thresholds are those arising from climatic, tectonic or land-use alterations which affect the inputs of either energy or mass, or both, into the system. On the other hand, as pointed out on p. 189, controlled experiments by Schumm and his associates have demonstrated that alluvial fans can go through alternate phases of trenching and aggradation without any changes in water or sediment discharge. Under natural conditions this internal instability is married to random rainfall fluctuations, a combination capable of producing complex patterns of cut-and-fill since there is no guarantee that adjacent fans will be at the same critical phase in the trenching and aggradation cycle. This idea that the same event, possibly of high magnitude and low frequency, may have very different consequences on superficially similar landforms helps explain local diversity that can be such a characteristic but puzzling feature of geomorphic systems.

Not only are the forcing events that may generate a threshold very intricate, but the pathway subsequently followed by the geomorphic system may be extremely complicated. This has been termed the principle of complex response. Again the underlying idea was well illustrated by a scale model constructed by Schumm and described in 1975. He caused a minor rejuvenation of about 10 cm in a small artificial catchment and noted the changes that ensued. An

immediate incision worked up-valley along the trunk stream, rejuvenating successive tributaries and scouring earlier alluvium (Fig. 19.8). However, as erosion progressed upstream the main channel became the conveyor of increasing sediment loads until eventually aggradation occurred in the newly cut channel. Yet as the tributaries gradually adjusted to their new local base-level, sediment loads diminished and a fresh phase of incision was initiated. Thus channel incision and terrace formation was followed in succession by deposition of an alluvial fill, channel braiding and lateral erosion, renewed incision and finally formation of another low terrace. Although this sequence was recorded on a scale model, there is

Fig. 19.8. Sketches illustrating the concepts of complex response and episodic erosion, as outlined by Schumm (1979). (A) Results of experiments in laboratory channel under conditions of constant discharge: (i) channel flows on alluvium; (ii) base-level lowered 0.1 m, inducing channel incision; (iii) bank erosion widens channel, increase in sediment from higher up-valley leads to progressive infill and culminates in channel braiding; (iv) reduced sediment from higher up-valley promotes single-thread channel which then incises itself into the earlier infill. (B) Cross-section through Douglas Creek valley, Colorado, USA. A staircase of twentieth-century terraces implies alternating incision and aggradation without corresponding changes in external variables (figures above each terrace indicate its height relative to the present creek, and also the age of the aggradation). (C) Application of the idea of complex response to the Davisian cycle of erosion: (i) changing elevation of interfluves and valley floors – dashed lines reflect the ideas of W. M. Davis, continuous lines the potential effect of periodic isostatic adjustment; (ii) episodic changes on the valley floors resultant upon occasional isostatic uplift; (iii) successive phases of erosion and deposition separated by longer periods of dynamic equilibrium. All these complex events can theoretically be triggered by infrequent isostatic adjustment.

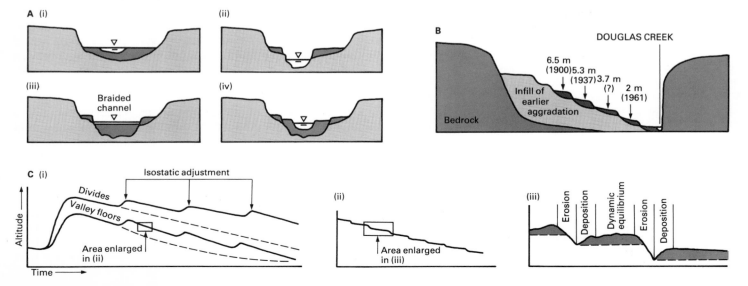

abundant evidence that conditions on real river systems are more, rather than less, complicated. When a major rejuvenation occurs, it appears that the fluvial system may be overwhelmed by the amount of sediment generated and the result of a single disturbance is a multiplicity of erosional and depositional episodes; these will, of course, be out of phase in different parts of the catchment.

Concluding summary

Many of the foregoing ideas were encapsulated by Brunsden and Thornes in 1979 in four basic propositions which together epitomize much current geomorphological thinking:

1. For any given set of environmental conditions, through the operation of a constant set of processes, there will be a tendency over time to produce a set of characteristic landforms.
2. Geomorphological systems are continually subject to perturbations which may arise from changes in the environmental conditions of the system or from structural instabilities within. These may or may not lead to a marked unsteadiness or transient behaviour of the system over a period of 10^2–10^5 years.
3. The response to perturbing displacement away from equilibrium is likely to be temporally and spatially complex and may lead to a considerable diversity of landform.
4. Landscape stability is a function of the temporal and spatial distributions of the resisting and disturbing forces and may be described by the landscape change safety factor, interpreted as the ratio of the magnitude of barriers to change to the magnitude of the disturbing forces.

As the brief and inevitably partial review of the preceding few pages serves to demonstrate, the vacuum left by the decline of the cycle concept has now been filled by a reasonably robust framework for landform studies. Admittedly emphasis here has been placed upon processes of subaerial denudation, but one of the advantages of the systems framework is that it is readily adaptable to the study of coastal and glacial landforms. Of course, systems analysis does not provide a model of landform evolution in the sense that the Davisian cycle of erosion did, and therefore one cannot be viewed as a straight replacement for the other. Indeed the systems framework must be capable of embracing the cycle concept if the latter is seen as describing one possible sequence of landform changes.

Schumm in 1979 undertook a re-examination of the Davisian erosion cycle in the light of modern geomorphic systems theory. He argued that the reason most long-term evolutionary models are unsatisfactory for short-term landform interpretation is that they are grossly over-simplified. He suggests that periodic isostatic adjustment should be incorporated in any realistic model. This will inject pulses of uplift, and if the concepts of geomorphic thresholds and complex reponse are then applied, an intricate sequence of incision and aggradation may be expected to ensue (Fig. 19.8). This is approximated by a system in dynamic metastable equilibrium. It implies, moreover, that many landscape details such as low terraces and recent alluvial fills do not necessarily signify changing external variables since they can develop as an integral part of system evolution. However, as relief and sediment yield diminish in the later part of the cycle, conditions will become less changeable and fluvial activity will conform more closely to that of simple dynamic equilibrium. Obviously, the internal complexity of this cyclic model is likely to be complemented by changes in external variables, most notably climate, and one of the problems for the geomorphologist must be to distinguish between internally and externally generated fluctuations.

Finally it is worth commenting that systems theory provides a powerful analytical tool, yet still has limited predictive value. This situation arises in part because the underlying principles governing the operation of complex geomorphic systems are as yet poorly understood. The concept of entropy, borrowed from the field of thermodynamics, is widely employed in systems studies and is reasonably assumed to be applicable to geomorphic systems as well. The idea of entropy was originally formulated in relation to the second law of thermodynamics which states that an isolated system always evolves spontaneously in the direction of the most probable distribution of energy. Expressed another way, this means that there is always an increase in entropy which is a measure of the unavailability of energy to do work. This is an irreversible process that continues until the system is characterized by a uniform energy distribution and entropy is at a maximum. Geomorphology is normally concerned with open systems, with a constant throughput of mass and energy; maximum entropy then denotes an equal probability of encountering a given energy level throughout the system, and represents a diminution of free energy capable of performing work. The system cannot 'run down' in the sense that an isolated system ultimately must, and the constant input of energy has been equated with the introduction of negative entropy. In a steady state this means that the rate of increase in entropy is reduced to zero, an idea that has been used, *inter alia*, in attempts to explain the form of stream long profiles. Nevertheless, there is still a gap between our theoretical understanding of how systems work and the practical application of that understanding to the operation of complex geomorphic systems, especially over longer time-spans. Only when that challenge has been met will there be a fully integrated theory of landform development.

References

Ahnert, F. (1970) 'Functional relationships between denudation, relief and uplift in large mid-latitude drainage basins', *Am. J. Sci.* **268**, 243–63.

Brunsden, D. and **Thornes, J. B.** (1979) 'Landscape sensitivity and change', *Trans. Inst. Brit. Geogr.* **4**, 463–84.

Büdel, J. (1982) *Climatic Geomorphology* (trans. L. Fischer and D. Busche), Princeton Univ. Press.

Chorley, R. J. (1962) 'Geomorphology and general systems theory', *U.S. Geol. Surv. Prof. Pap.* 500-B.

Chorley, R. J. and **Kennedy, B. A.** (1971) *Physical Geography, a systems approach*, Prentice Hall.

Chorley, R. J. and **Morgan, M. A.** (1962) 'Comparison of the morphometric features, Unaka Mountains Tennessee and N. Carolina, and Dartmoor', *Geol. Soc. Am. Bull.* **73**, 17–34.

Hack, J. T. (1960) 'Interpretation of erosional topography in humid temperate regions', *Am. J. Sci.* **558-A**, 80–97.

Hack, J. T. (1975) 'Dynamic equilibrium and landscape evolution' in *Theories of Landform Development* (ed. W. N. Melhorn and R. C. Flemal), State Univ. N.Y.

Kiewietdejonge, C. J. (1984) 'Büdel's geomorphology', *Prog. Phys. Geogr.* **8**, 218–48.

King, L. C. (1962) *Morphology of the Earth*, Oliver and Boyd.

Leopold, L. B. and **Langbein, W. B.** (1962) 'The concept of entropy in landscape evolution', *U.S. Geol. Surv. Prof. Pap.* 500-A.

Ollier, C. D. 1979 'Evolutionary geomorphology of Australia and Papua New Guinea', *Trans. Inst. Brit. Geogr.* **4**, 516–39.

Peltier, L. C. (1950) 'The geographic cycle in periglacial regions as it is related to climatic geomorphology', *Ann. Ass. Am. Geogr.* **40**, 214–36.

Pitty, A. F. (1968) 'The scale and significance of solutional loss from the limestone tract of the southern Pennines', *Proc. Geol. Ass.* **79**, 153–77.

Pugh, J. C. (1966) 'The landforms of low latitudes', in *Essays in Geomorphology* (ed. G. H. Dury), Heinemann.

Scheidegger, A. E. (1979) 'The principle of antagonism in the Earth's evolution', *Tectonophysics* **55**, T7–10.

Scheidegger, A. E. (1983) 'Instability principle in geomorphic equilibrium', *Z. Geomorph.* **27**, 1–19.

Schumm, S. A. (1963) 'The disparity between present rates of denudation and orogeny', *U.S. Geol. Surv. Prov. Pap.* 454- H.

Schumm, S. A. (1975) 'Episodic erosion: a modification of the geomorphic cycle', in *Theories of Landform Development* (ed. W. N. Melhorn and R. C. Flemal), State Univ. N.Y.

Schumm, S. A. (1979) 'Geomorphic thresholds: the concept and its applications', *Trans. Inst. Brit. Geogr.* **4**, 485–515.

Schumm, S. A. and **Lichty, R. W.** (1965) 'Time, space and causality in geomorphology', *Am. J. Sci.* **263**, 110–19.

Strahler, A. N. (1952) 'Dynamic basis of geomorphology', *Geol. Soc. Am. Bull.* **63**, 923–38.

Twidale, C. R. and **Bourne, J. A.** (1975) 'Episodic exposure of inselbergs', *Geol. Soc. Am. Bull.* **86**, 1473–81.

Wolman, M. G. and **Miller, J. P.** (1960) 'Magnitude and frequency of forces in geomorphic processes', *J. Geol.* **68**, 54–74.

Worssam, B. C. (1973) 'A new look at river capture and at the denudation history of the Weald', *Inst. Geol. Sci. Rep.* 73/17.

Selected bibliography

The classic models of landform evolution outlined in the foregoing pages are elaborated in: W. M. Davis, *Geographical Essays*, Dover Publ., 1954 (reprint of 1909 edition); W. Penck, *Morphological Analysis of Landforms: a contribution to physical geology* (trans. H. Czech and K. C. Boswell), Macmillan, 1953; and L. C. King *Morphology of the Earth*, Oliver and Boyd, 1962.

In addition to the books on periglacial geomorphology listed on p. 298, other texts devoted to the specific climatic zones discussed in this chapter are M. F. Thomas, *Tropical Geomorphology*, Macmillan, 1974, and R. U. Cooke and A. Warren, *Geomorphology in Deserts*, Batsford, 1973.

The resurgence during the 1970s of interest in the conceptual framework for landform studies is reflected in the publication, among other volumes, of W. N. Melhorn and R. C. Flemal (eds), *Theories of Landform Development*, State Univ. N.Y., 1975, and D. R. Coates and J. D. Vitek (eds), *Thresholds in Geomorphology*, Allen and Unwin, 1980.

Appendix A

Throughout the text measurements are quoted in SI units, that is, in conformity with the Système International d'Unités which is both an extension and a refinement of the traditional metric system and is now the recognized standard for scientific work. However, older reference works often employ earlier styles of the metric system, such as the centimetre-gram-second version, and the following table is therefore given to facilitate conversion between SI and other units.

Physical quantity	SI unit	Traditional metric unit	Conversion formula
Length	metre (m)	metre (m)	—
Mass	kilogram (kg)	kilogram (kg)	—
Time	second (s)	second (s)	—
Density	kilogram per cubic metre ($kg\ m^{-3}$)	gram per cubic centimetre ($g\ cm^{-3}$)	$1\ g\ cm^{-3}$ $= 10^3\ kg\ m^{-3}$
Energy	joule (J) ($J = kg\ m^2\ s^{-2}$)	heat energy: gram-calorie (cal) potential energy: kilogram force elevation (kgf m)	$1\ cal$ $= 4.186\ 8\ J$ $1\ kgf\ m$ $= 9.81\ J$
Force	newton (N) ($N = J\ m^{-1}$)	kilogram force (kgf)	$1\ kgf$ $= 9.81\ N$
Power	watt (W) ($W = J\ s^{-1}$)	calories per second ($cal\ s^{-1}$)	$1\ cal\ s^{-1}$ $= 4.186\ 8\ W$
Pressure	pascal (Pa) ($Pa = N\ m^{-2}$)	either kilogram force per square centimetre ($kgf\ cm^{-2}$) or bar*	$1\ kgf\ cm^{-2}$ $= 9.81 \times 10^4\ Pa$ $1\ bar$ $= 10^5\ Pa$
Dynamic viscosity	pascal second ($Pa\ s$ or $N\ s\ m^{-2}$)	poise (P)*	$1\ P = 10^{-1}\ N\ s\ m^{-2}$ $= 10^{-1}\ Pa\ s$
Kinematic viscosity	square metres per second ($m^2\ s^{-1}$)	stokes (St)*	$1\ St$ $= 10^{-4}\ m^2\ s^{-1}$

* Although not technically part of SI usage, bar, poise and stokes are regarded as permissible in conjunction with SI units.

Prefixes commonly used to indicate either decimal fractions or multiples of SI units include:

μ (micro-) 10^{-6}
m (milli-) 10^{-3}
k (kilo-) 10^3
M (mega-) 10^6

Index